UCLA Symposia on Molecular and Cellular Biology, New Series

Series Editor
C. Fred Fox

Volume 1
Differentiation and Function of Hematopoietic Cell Surfaces, Vincent T. Marchesi and Robert C. Gallo, *Editors*

Volume 2
Mechanisms of Chemical Carcinogenesis, Curtis C. Harris and Peter A. Cerutti, *Editors*

Volume 3
Cellular Recognition, William A. Frazier, Luis Glaser, and David I. Gottlieb, *Editors*

Volume 4
Rational Basis for Chemotherapy, Bruce A. Chabner, *Editor*

Volume 5
Tumor Viruses and Differentiation, Edward M. Scolnick and Arnold J. Levine, *Editors*

Volume 6
Evolution of Hormone-Receptor Systems, Ralph A. Bradshaw and Gordon N. Gill, *Editors*

Volume 7
Recent Advances in Bone Marrow Transplantation, Robert Peter Gale, *Editor*

Volume 8
Gene Expression, Dean H. Hamer and Martin J. Rosenberg, *Editors*

Volume 9
Normal and Neoplastic Hematopoiesis, David W. Golde and Paul A. Marks, *Editors*

Volume 10
Mechanisms of DNA Replication and Recombination, Nicholas R. Cozzarelli, *Editor*

Volume 11
Cellular Responses to DNA Damage, Errol C. Friedberg and Bryn A. Bridges, *Editors*

Volume 12
Plant Molecular Biology, Robert B. Goldberg, *Editor*

Volume 13
Molecular Biology of Host–Parasite Interactions, Nina Agabian and Harvey Eisen, *Editors*

Volume 14
Biosynthesis of the Photosynthetic Apparatus: Molecular Biology, Development and Regulation, J. Philip Thornber, L. Andrew Staehelin, and Richard B. Hallick, *Editors*

Volume 15
Protein Transport and Secretion, Dale L. Oxender, *Editor*

Volume 16
Acquired Immune Deficiency Syndrome, Michael S. Gottlieb and Jerome E. Groopman, *Editors*

Volume 17
Genes and Cancer, J. Michael Bishop, Janet D. Rowley, and Mel Greaves, *Editors*

Volume 18
Regulation of the Immune System, Harvey Cantor, Leonard Chess, and Eli Sercarz, *Editors*

Volume 19
Molecular Biology of Development, Eric H. Davidson and Richard A. Firtel, *Editors*

Volume 20
Genome Rearrangement, Ira Herskowitz and Melvin Simon, *Editors*

Volume 21
Herpesvirus, Fred Rapp, *Editor*

UCLA Symposia Published Previously

(Numbers refer to the publishers listed below.)

1972
Membrane Research **(2)**

1973
Membranes **(1)**
Virus Research **(2)**

1974
Molecular Mechanisms for the Repair of DNA **(4)**
Membranes **(1)**
Assembly Mechanisms **(1)**
The Immune System: Genes, Receptors, Signals **(2)**
Mechanisms of Virus Disease **(3)**

1975
Energy Transducing Mechanisms **(1)**
Cell Surface Receptors **(1)**
Developmental Biology **(3)**
DNA Synthesis and Its Regulation **(3)**

1976
Cellular Neurobiology **(1)**
Cell Shape and Surface Architecture **(1)**
Animal Virology **(2)**
Molecular Mechanisms in the Control of Gene Expression **(2)**

1977
Cell Surface Carbohydrates and Biological Recognition **(1)**
Molecular Approaches to Eucaryotic Genetic Systems **(2)**
Molecular Human Cytogenetics **(2)**
Molecular Aspects of Membrane Transport **(1)**
Immune System: Genetics and Regulation **(2)**

1978
DNA Repair Mechanisms **(2)**
Transmembrane Signaling **(1)**
Hematopoietic Cell Differentiation **(2)**
Normal and Abnormal Red Cell Membranes **(1)**
Persistent Viruses **(2)**
Cell Reproduction: Daniel Mazia Dedicatory Volume **(2)**

1979
Covalent and Non-Covalent Modulation of Protein Function ***(2)***
Eucaryotic Gene Regulation **(2)**
Biological Recognition and Assembly **(1)**
Extrachromosomal DNA **(2)**
Tumor Cell Surfaces and Malignancy **(1)**
T and B Lymphocytes: Recognition and Function **(2)**

1980
Biology of Bone Marrow Transplantation **(2)**
Membrane Transport and Neuroreceptors **(1)**
Control of Cellular Division and Development **(1)**
Animal Virus Genetics **(2)**
Mechanistic Studies of DNA Replication and Genetic Recombination **(2)**

1981
Immunoglobulin Idiotypes **(2)**
Initiation of DNA Replication **(2)**
Genetic Variation Among Influenza Viruses **(2)**
Developmental Biology Using Purified Genes **(2)**
Differentiation and Function of Hematopoietic Cell Surfaces **(1)**
Mechanisms of Chemical Carcinogenesis **(1)**
Cellular Recognition **(1)**

1982
B and T Cell Tumors **(2)**
Interferon **(2)**
Rational Basis for Chemotherapy **(1)**
Gene Regulation **(2)**
Tumor Viruses and Differentiation **(1)**
Evolution of Hormone-Receptor Systems **(1)**

Publishers

(1) Alan R. Liss, Inc.
150 Fifth Avenue
New York, NY 10011

(2) Academic Press, Inc.
111 Fifth Avenue
New York, NY 10003

(3) W.A. Benjamin, Inc.
2725 Sand Hill Road
Menlo Park, CA 94025

(4) Plenum Publishing Corp.
227 W. 17th Street
New York, NY 10011

Symposia Board

The members of the board advise the director in identification of topics for future symposia.

MOLECULAR BIOLOGY OF DEVELOPMENT

MOLECULAR BIOLOGY OF DEVELOPMENT

Proceedings of a CETUS-UCLA Symposium held April 1–7, 1984, in Steamboat Springs, Colorado

Editors

ERIC H. DAVIDSON
Division of Biology
California Institute of Technology
Pasadena, California

RICHARD A. FIRTEL
Department of Biology
University of California, San Diego
La Jolla, California

Alan R. Liss, Inc. • New York

Address all Inquiries to the Publisher
Alan R. Liss, Inc., 150 Fifth Avenue, New York, NY 10011

Printed in the United States of America.

Library of Congress Cataloging in Publication Data

Main entry under title:

Molecular biology of development.

Includes index.
1. Developmental biology—Congresses. 2. Molecular biology—Congresses. I. Davidson, Eric H., 1937-
II. Firtel, Richard A. III. Cetus Corporation.
IV. University of California, Los Angeles.
QH491.M64 1984 574.3 84-21792
ISBN 0-8451-2618-0

Contents

Contributors

Bruce M. Alberts, Department of Biochemistry and Biophysics, University of California, San Francisco, San Francisco, CA **[263]**

Hamish Allen, Biogen Research Corporation, Cambridge, MA **[527]**

Batya Aloni, Department of Cell Biology, The Weizmann Institute of Science, Rehovot, Israel **[647]**

Serge Alonso, Department of Molecular Biology, Pasteur Institute, Paris, France **[275]**

Frederick W. Alt, Department of Biochemistry and Institute for Cancer Research, Columbia University College of Physicians and Surgeons, New York, NY **[537]**

Frederick M. Ausubel, Department of Molecular Biology, Massachusetts General Hospital, Boston, MA **[611]**

Jean C. Baker, Department of Biochemistry, University of Georgia, Athens, GA **[591]**

Aimee H. Bakken, Division of Basic Sciences, Fred Hutchinson Cancer Research Center, Seattle, WA; present address: Zoology Department, University of Washington, Seattle, WA **[199]**

M. Barinaga, Molecular Biology and Virology Laboratory, The Salk Institute, San Diego, CA and The University of California at San Diego, La Jolla, CA; present address: Department of Biological Sciences, Stanford University, Stanford, CA **[561]**

Paul Barton, Department of Molecular Biology, Pasteur Institute, Paris, France **[275]**

Guriqbal S. Basi, Department of Biological Chemistry, University of Illinois Health Sciences Center, Chicago, IL **[407]**

D.C. Beebe, Department of Anatomy, Uniformed Services University for the Health Sciences, Bethesda, MD **[331]**

L.M. Bilezikjian, Peptide Biology Laboratory, The Salk Institute, San Diego, CA **[561]**

T. Keith Blackwell, Department of Biochemistry, Columbia University College of Physicians and Surgeons, New York, NY **[537]**

Mark Boardman, Department of Biological Chemistry, University of Illinois Health Sciences Center, Chicago, IL; present address: Department of Biological Sciences, University of Warwick, Coventry, England **[407]**

The number in brackets is the opening page number of the Contributor's article.

T. Borrás, Laboratory of Molecular and Developmental Biology, National Eye Institute, National Institutes of Health, Bethesda, MD **[331]**

Anne M. Boulet, Department of Biochemistry and Biophysics and Hormone Research Laboratory, University of California at San Francisco, San Francisco, CA **[481]**

Sean Brennan, CRC Molecular Embryology Unit, Department of Zoology, Cambridge University, Cambridge, England **[109]**

Roy J. Britten, Division of Biology, California Institute of Technology, Pasadena, CA **[119,621]**

R.D. Brodeur, Center for Developmental Biology, Department of Zoology, The University of Texas at Austin, Austin, TX **[145]**

P. Brûlet, Unité de Génétique cellulaire du Collège de France et de l'Institut Pasteur, Paris, France **[253]**

Margaret E. Buckingham, Department of Molecular Biology, Pasteur Institute, Paris, France **[275]**

Gabriele Bugaisky, Department of Molecular Biology, Pasteur Institute, Paris, France; present address: Department of Medicine, University of Chicago, Chicago, IL **[275]**

Sharon Busby, Division of Basic Sciences, Fred Hutchinson Cancer Research Center, Seattle, WA; present address: Zymogenetics, Seattle, WA **[199]**

Frank J. Calzone, Division of Biology, California Institute of Technology, Pasadena, CA **[119]**

Barry Caplan, Division of Biology, California Institute of Technology, Pasadena, CA **[511]**

Joe Cappello, Department of Biology, Massachusetts Institute of Technology, Cambridge, MA **[491]**

James A. Cardelli, Department of Bacteriology, University of Wisconsin, Madison WI **[427,447]**

Sunil Chada, Department of Biology, University of California at Los Angeles, Los Angeles, CA **[459]**

A.B. Chepelinsky, Laboratory of Molecular and Developmental Biology, National Eye Institute, National Institutes of Health, Besthesda, MD **[331]**

Caryl A. Chlan, Department of Biochemistry, University of Georgia, Athens, GA **[591]**

Chuan-Chu Chou, Department of Biology, University of California at Los Angeles, Los Angeles, CA **[459]**

Arlette Cohen, Department of Molecular Biology, Pasteur Institute, Paris, France **[275]**

Stephen M. Cohen, Department of Biology, Massachusetts Institute of Technology, Cambridge, MA **[491]**

G.C. Das, Laboratory of Molecular and Developmental Biology, National Eye Institute, National Institutes of Health, Bethesda, MD **[331]**

Nina Dathan, CRC Molecular Embryology Unit, Department of Zoology, Cambridge University, Cambridge, England **[109]**

Philippe Daubas, Department of Molecular Biology, Pasteur Institute, Paris, France **[275]**

Eric H. Davidson, Division of Biology, California Institute of Technology, Pasadena, CA **[xxv, 119, 621]**

Norman Davidson, Department of Chemistry, California Institute of Technology, Pasadena, CA **[657]**

Richard C. Davis, Molecular Biology Institute, University of California at Los Angeles, Los Angeles, CA **[459]**

Igor B. Dawid, Laboratory of Molecular Genetics, National Institute of Child Health and Human Development, National Institutes of Health, Bethesda, MD **[93]**

Antonia T. Destree, Department of Cardiology, Children's Hospital, Boston, MA **[437]**

James J. Devlin, Biogen Research Corporation, Cambridge, MA **[527]**

Randall L. Dimond, Department of Bacteriology, University of Wisconsin, Madison, WI **[427,447]**

James Douglass, Department of Chemistry, University of Oregon, Eugene, OR **[573]**

Marietta Dunaway, Division of Basic Sciences, Fred Hutchinson Cancer Research Center, Seattle, WA **[199]**

P. Duprey, Unité de Génétique cellulaire du Collège de France et de l'Institut Pasteur, Paris, France **[253]**

Leon Dure III, Department of Biochemistry, University of Georgia, Athens, GA **[591]**

Thomas Edlund, Department of Biochemistry and Biophysics and Hormone Research Laboratory, University of California at San Francisco, San Francisco, CA; present address: Department of Microbiology, University of Umea, Umea, Sweden **[481]**

Cynthia A. Edwards, Department of Biology, University of California at San Diego, La Jolla, CA; present address: The Laboratory of Cellular and Developmental Biology, National Institute of Arthritis, Diabetes, Digestive, and Kidney Diseases, National Institutes of Health, Bethesda, MD **[361]**

Juanita D. Eldridge, Laboratory of Biochemistry, National Cancer Institute, National Institutes of Health, Bethesda, MD **[383]**

L. Engstrom, Department of Developmental Genetics and Anatomy, School of Medicine, Case Western Reserve University, Cleveland, OH **[185]**

R.M. Evans, Molecular Biology and Virology Laboratory, The Salk Institute, San Diego, CA **[561]**

Karen Fahrner, Biogen Research Corporation, Cambridge, MA **[527]**

Sharon Fairman, CRC Molecular Embryology Unit, Department of Zoology, Cambridge University, Cambridge, England **[109]**

Richard A. Firtel, Department of Biology, University of California at San Diego, La Jolla, CA **[xxv, 293,361,633]**

Richard A. Flavell, Biogen Research Corporation, Cambridge, MA **[527]**

Constantin N. Flytzanis, Division of Biology, California Institute of Technology, Pasadena, CA **[621]**

Nevis Fregien, Department of Chemistry, California Institute of Technology, Pasadena, CA **[657]**

Xin-Yuan Fu, Department of Biological Sciences, Columbia University, New York, NY **[351]**

Michael L. Fuller, Department of Biology, University of California at Los Angeles, Los Angeles, CA **[459]**

Glenn A. Galau, Department of Biochemistry, University of Georgia, Athens, GA; present address: Department of Botany, University of Georgia, Athens, GA **[591]**

Joseph G. Gall, Department of Biology, Yale University, New Haven, CT; present address: Department of Embryology, Carnegie Institution of Washington, Baltimore, MD **[227]**

Ian Garner, Department of Molecular Biology, Pasteur Institute, Paris, France **[275]**

Walter J. Gehring, Department of Cell Biology, Biozentrum, University of Basel, Basel, Switzerland **[3]**

John C. Gerhart, Department of Molecular Biology, University of California, Berkeley, CA **[51]**

Frank J. Gonzalez, Laboratory of Developmental Pharmacology, National Institute of Child Health and Human Development, National Institutes of Health, Bethesda, MD **[309]**

Robert M. Grainger, Department of Biology, University of Virginia, Charlottesville, VA **[241]**

Warwick N. Grant, Max Planck Institut für Biochemie, Martinsried bei München, Federal Republic of Germany; present address: Imperial Cancer Research Fund, Lincolns Inn Fields, London, England **[75]**

David Greenberg, Department of Cell Biology, The Weizmann Institute of Science, Rehovot, Israel **[647]**

Jeffrey M. Greve, Department of Biological Chemistry, Harvard Medical School, Boston, MA **[213]**

J.B. Gurdon, CRC Molecular Embryology Unit, Department of Zoology, Cambridge University, Cambridge, England **[109]**

J.W. Hawkins, Laboratory of Molecular and Developmental Biology, National Eye Institute, National Institutes of Health, Bethesda, MD **[331]**

J.F. Hejtmancik, Howard Hughes Medical Institute, Houston, TX **[331]**

Edward Herbert, Department of Chemistry, University of Oregon, Eugene, OR **[573]**

Barbara R. Hough-Evans, Division of Biology, California Institute of Technology, Pasadena, CA **[621]**

Richard O. Hynes, Center for Cancer Research and Department of Biology, Massachusetts Institute of Technology, Cambridge, MA **[417]**

F. Jacob, Unité de Génétique cellulaire du Collège de France et de l'Institut Pasteur, Paris, France **[253]**

Milan Jamrich, Laboratory of Molecular Genetics, National Institute of Child Health and Human Development, National Institutes of Health, Bethesda, MD **[93]**

W.R. Jeffery, Center for Developmental Biology, Department of Zoology, The University of Texas at Austin, Austin, TX **[145]**

Erzsebet Jonas, Laboratory of Molecular Genetics, National Institute of Child Health and Human Development, National Institutes of Health, Bethesda, MD **[93]**

Karen S. Katula, Division of Biology, California Institute of Technology, Pasadena, CA **[621]**

Brian K. Kay, Laboratory of Molecular Genetics, National Institute of Child Health and Human Development, National Institutes of Health, Bethesda, MD **[93]**

Alan R. Kimmel, Laboratory of Cellular and Developmental Biology, NIADDKD, National Institutes of Health, Bethesda, MD **[373]**

Shioko Kimura, Laboratory of Developmental Pharmacology, National Institute of Child Health and Human Development, National Institutes of Health, Bethesda, MD **[309]**

C.R. King, Laboratory of Cellular and Molecular Biology, National Cancer Institute, National Institutes of Health, Bethesda, MD **[331]**

David A. Knecht, Department of Bacteriology, University of Wisconsin, Madison, WI; present address: Department of Biology, University of California at San Diego, La Jolla, CA **[427]**

K. Konrad, Department of Developmental Genetics and Anatomy, School of Medicine, Case Western Reserve University, Cleveland, OH **[185]**

Marianne Krefft, Max Planck Institut für Biochemie, Martinsried bei München, Federal Republic of Germany; present address: Biochemistry Department, University of Wuppertal, Wuppertal, Federal Republic of Germany **[75]**

Paul Labhart, Division of Basic Sciences, Fred Hutchinson Cancer Research Center, Seattle, WA **[199]**

E. Diann Lewis, Department of Biological Sciences, Columbia University, New York, NY **[351]**

Garrick Little, Department of Chemistry, University of Oregon, Eugene, OR **[573]**

George P. Livi, Department of Bacteriology, University of Wisconsin, Madison, WI; present address: Cold Spring Harbor Laboratory, Cold Spring Harbor, NY **[427,447]**

Harvey F. Lodish, Department of Biology, Massachusetts Institute of Technology, Cambridge, MA **[491]**

Kathleen A. Mahon, Department of Biology, Yale University, New Haven, CT; present address: Laboratory of Molecular Genetics, National Institutes of Health, Bethesda, MD **[227]**

A.P. Mahowald, Department of Developmental Genetics and Anatomy, School of Medicine, Case Western Reserve University, Cleveland, OH **[185]**

James L. Manley, Department of Biological Sciences, Columbia University, New York, NY **[351]**

C. Marle, Unité de Génétique cellulaire du Collège de France et de l'Institut Pasteur, Paris, France **[253]**

Belinda Martineau, Department of Genetics, University of California, Berkeley, CA **[601]**

Stephen P. Mayfield, Department of Genetics, University of California, Berkeley, CA **[601]**

David R. McClay, Department of Zoology, Duke University, Durham, NC **[165]**

Andrew P. McMahon, Division of Biology, California Institute of Technology, Pasadena, CA; present address: National Institute for Medical Research, London, England **[621]**

Russell M. Medford, Department of Medicine, Beth Israel Hospital, Harvard Medical School, Boston, MA **[437]**

Mona C. Mehdy, Department of Biology, University of California at San Diego, La Jolla, CA **[293]**

Andrew L. Mellor, Biogen Research Corporation, Cambridge, MA **[527]**

Danielle Melloul, Department of Cell Biology, The Weizmann Institute of Science, Rehovot, Israel **[647]**

Robert C. Mierendorf, Department of Bacteriology, University of Wisconsin, Madison, WI **[427,447]**

Adrian Minty, Department of Molecular Biology, Pasteur Institute, Paris, France; present address: Veterans Administration Medical Center, Palo Alto, CA **[275]**

Seiji Miyatani, Laboratory of Molecular Genetics, National Institute of Child Health and Human Development, National Institutes of Health, Bethesda, MD **[93]**

T.J. Mohun, CRC Molecular Embryology Unit, Department of Zoology, Cambridge University, Cambridge, England **[109]**

Garry Morgan, Division of Basic Sciences, Fred Hutchinson Cancer Research Center, Seattle, WA; present address: Department of Genetics, University of Nottingham, Nottingham, England **[199]**

Bernardo Nadal-Ginard, Department of Cardiology, Children's Hospital and Department of Pediatrics, Harvard Medical School, Boston, MA **[395,437]**

Daniel W. Nebert, Laboratory of Developmental Pharmacology, National Institute of Child Health and Human Development, National Institutes of Health, Bethesda, MD **[309]**

Wolfgang Nellen, Department of Biology, University of California at San Diego, La Jolla, CA **[633]**

Robert Nelson, Molecular Biology Institute, University of California at Los Angeles, Los Angeles, CA **[459]**

Hanh T. Nguyen, Department of Cellular and Developmental Biology, Harvard University, Cambridge, MA **[437]**

J.M. Nickerson, Laboratory of Molecular and Developmental Biology, National Eye Institute, National Institutes of Health, Bethesda, MD **[331]**

Uri Nudel, Department of Cell Biology, The Weizmann Institute of Science, Rehovot, Israel **[647]**

Bruce M. Paterson, Laboratory of Biochemistry, National Cancer Institute, National Institutes of Health, Bethesda, MD **[383]**

Jeremy I. Paul, Center for Cancer Research and Department of Biology, Massachusetts Institute of Technology, Cambridge, MA **[417]**

Muthu Periasamy, Department of Cardiology, Children's Hospital and Department of Pediatrics, Harvard Medical School, Boston, MA **[395]**

Rosario M. Perona, Department of Biological Chemistry, Harvard Medical School, Boston, MA **[213]**

N. Perrimon, Department of Developmental Genetics and Anatomy, School of Medicine, Case Western Reserve University, Cleveland, OH **[185]**

Carey R. Phillips, Department of Zoology, University of California, Berkeley, CA **[23]**

William H. Phillips, Department of Biology, University of Virginia, Charlottesville, VA **[241]**

J. Piatigorsky, Laboratory of Molecular and Developmental Biology, National Eye Institute, National Institutes of Health, Bethesda, MD **[331]**

Marina R. Picciotto, Department of Biological Sciences, Stanford University, Stanford, CA **[551]**

Steven C. Pruitt, Division of Basic Sciences, Fred Hutchinson Cancer Research Center, Seattle, WA **[199]**

Kathleen Ranney, Department of Chemistry, University of Oregon, Eugene, OR **[573]**

Ronald H. Reeder, Division of Basic Sciences, Fred Hutchinson Cancer Research Center, Seattle, WA **[199]**

J.D. Richter, Department of Biochemistry, University of Tennessee, Knoxville, TN **[129]**

C. Rivier, Peptide Biology Laboratory, The Salk Institute, San Diego, CA **[561]**

Benoît Robert, Department of Molecular Biology, Pasteur Institute, Paris, France **[275]**

Richard J. Roller, Department of Biological Chemistry, Harvard Medical School, Boston, MA **[213]**

M.G. Rosenfeld, Eukaryotic Regulatory Biology Program, University of California at San Diego, La Jolla, CA **[561]**

Ellen Rothenberg, Division of Biology, California Institute of Technology, Pasadena, CA **[511]**

William J. Rutter, Department of Biochemistry and Biophysics and Hormone Research Laboratory, University of California at San Francisco, San Francisco, CA **[481]**

Rochelle D. Sailor, Division of Biology, California Institute of Technology, Pasadena, CA **[511]**

Winston Salser, Department of Biology and Molecular Biology Institute, University of California at Los Angeles, Los Angeles, CA **[459]**

George S. Salzmann, Department of Biological Chemistry, Harvard Medical School, Boston, MA **[213]**

Thomas D. Sargent, Laboratory of Molecular Genetics, National Institute of Child Health and Human Development, National Institutes of Health, Bethesda, MD **[93]**

Charles L. Saxe III, Department of Biology, University of California at San Diego, La Jolla, CA **[293]**

Stanley R. Scharf, Department of Molecular Biology, University of California, Berkeley, CA; present address: Botany Department, University of Calirfornia, Berkeley, CA **[51]**

Richard H. Scheller, Department of Biological Sciences, Stanford University, Stanford, CA **[551]**

Einhard Schierenberg, Department of Molecular, Cellular, and Developmental Biology, University of Colorado, Boulder, CO **[37]**

Jean E. Schwarzbauer, Center for Cancer Research and Department of Biology, Massachusetts Institute of Technology, Cambridge, MA **[417]**

Anne Seiler-Tuyns, Laboratory of Biochemistry, National Cancer Institute, National Institutes of Health, Bethesda, MD **[383]**

Rosemary J. Shott-Akhurst, Division of Biology, Calfornia Institute of Technology, Pasadena, CA; present address: Biochemistry Department, St. Mary's Hospital Medical School, London, England **[119]**

Colleen Silan, Department of Biology, University of California at San Diego, La Jolla, CA **[633]**

Janet P. Slovin, University of California at Los Angeles, Los Angeles, CA; present address: Plant Hormone Laboratory, USDA Agriculture Research Center, Beltsville, MD **[459]**

L. Dennis Smith, Department of Biological Sciences, Purdue University, W. Lafayette, IN **[129]**

Barbara Sollner-Webb, Division of Basic Sciences, Fred Hutchinson Cancer Research Center, Seattle, WA; present address: Department of Physiological Chemistry, Johns Hopkins Medical School, Baltimore, MD **[199]**

Robert V. Storti, Department of Biological Chemistry, University of Illinois Health Sciences Center, Chicago, IL **[407]**

Emanuel E. Strehler, Department of Cardiology, Children's Hospital and Department of Pediatrics, Harvard Medical School, Boston, MA **[395]**

Susan Strome, Department of Molecular, Cellular, and Developmental Biology, University of Colorado, Boulder, CO **[37]**

Eric Summers, Harvard Medical School, Boston, MA; present address: Albert Einstein College of Medicine, Bronx, NY **[437]**

Venkatesan Sundaresan, Department of Molecular Biology, Massachusetts General Hospital, Boston, MA; present address: Department of Genetics, University of California at Berkeley, Berkeley, CA **[611]**

Wynne W. Szeto, Department of Molecular Biology, Massachusetts General Hospital, Boston, MA **[611]**

John W. Tamkun, Center for Cancer Research and Department of Biology, Massachusetts Institute of Technology, Cambridge, MA **[417]**

Ronald Taussig, Department of Biological Sciences, Stanford University, Stanford, CA **[551]**

M.A. Taylor, Department of Biological Sciences, Purdue University, W. Lafayette, IN **[129]**

William C. Taylor, Department of Genetics, University of California, Berkeley, CA **[601]**

Arlen Thomason, University of California at Los Angeles, Los Angeles, CA; present address: AMGen Inc., Newbury Park, CA **[459]**

C.R. Tomlinson, Center for Developmental Biology, Department of Zoology, The University of Texas at Austin, Austin, TX **[145]**

Michael Uhler, Department of Chemistry, University of Oregon, Eugene, OR; present address: Department of Biochemistry, University of Washington, Seattle, WA **[573]**

W.W. Vale, Peptide Biology Laboratory, The Salk Institute, San Diego, CA **[561]**

M. Vasseur, Unité de Génétique cellulaire du Collège de France et de l'Institut Pasteur, Paris, France **[253]**

Jean-Paul Vincent, Department of Molecular Biology, University of California, Berkeley, CA **[51]**

Ludwig Voet, Max Planck Institut für Biochemie, Martinsried bei München, Federal Republic of Germany **[75]**

Claire T. Wake, Biogen Research Corporation, Cambridge, MA **[527]**

Michael D. Walker, Department of Biochemistry and Biophysics and Hormone Research Laboratory, University of California at San Francisco, San Francisco, CA **[481]**

Marika F. Walter, Department of Biochemistry and Biophysics, University of California at San Francisco, San Francisco, CA **[263]**

Paul M. Wassarman, Department of Biological Chemistry, Harvard Medical School, Boston, MA **[213]**

Elizabeth H. Weiss, Biogen Research Corporation, Cambridge, MA; present address: Institut für Immunologie der Universität München, München, Federal Republic of Germany **[527]**

Dennis L. Welker, Max Planck Institut für Biochemie, Martinsried bei München, Federal Republic of Germany **[75]**

Gary M. Wessel, Department of Zoology, Duke University, Durham, NC **[165]**

André Weydert, Department of Molecular Biology, Pasteur Institute, Paris, France **[275]**

Georg Widera, Biogen Research Corporation, Cambridge, MA **[527]**

Keith L. Williams, Max Planck Institut für Biochemie, Martinsried bei München, Federal Republic of Germany; present address: School of Biological Sciences, Macquarie University, North Ryde, Sydney, NSW, Australia **[75]**

Fred H. Wilt, Department of Zoology, University of California, Berkeley, CA **[23]**

Jeffrey A. Winkles, Laboratory of Molecular Genetics, National Institute of Child Health and Human Development, National Institutes of Health, Bethesda, MD **[93, 241]**

William B. Wood, Department of Molecular, Cellular, and Developmental Biology, University of Colorado, Boulder, CO **[37]**

Jocyndra Wright, Molecular Biology Institute, University of California at Los Angeles, Los Angeles, CA **[459]**

David Yaffe, Department of Cell Biology, The Weizmann Institute of Science, Rehovot, Israel **[647]**

G. Yamomoto, Peptide Biology Laboratory, The Salk Institute, San Diego, CA **[561]**

George D. Yancopoulos, Department of Biochemistry, Columbia University College of Physicians and Surgeons, New York, NY **[537]**

Glenn Yasuda, University of California at Los Angeles, Los Angeles, CA; present address: Department of Genetics, University of Washington, Seattle, WA **[459]**

P.S. Zelenka, Laboratory of Molecular and Developmental Biology, National Eye Institute, National Institutes of Health, Bethesda, MD **[331]**

J. Lynn Zimmerman, Department of Molecular Biology, Massachusetts General Hospital, Boston, MA; present address: Department of Biological Sciences, University of Maryland Baltimore County, Catonsville, MD **[611]**

Charles Zuker, Department of Biology, Massachusetts Institute of Technology, Cambridge, MA **[491]**

Preface

The Cetus-UCLA Symposium "Molecular Biology of Development" was held at Steamboat Springs, Colorado, April 1–7, 1984. The emphasis of the program was on major conceptual problems of development. Several recent similarly titled conferences had either narrow topical coverage or more limited perspective. We tried to avoid these pitfalls. We also sought to avoid multiple presentations of similar data in order to provide for maximal diversity, though we were not invariably successful in this. Many excellent laboratories could not be represented on the program simply because others in the same field were. The result of these precepts was a conference of great diversity. Thus, the sessions ranged from cytoplasmic localization and oogenesis to whole organism transformation, and from differentiation in nerve cells, plants, slime molds, and adult cell types to the early development of embryos of many forms. The various avenues of approach described in the presentations and reviewed in this volume display the current elegance of the science of developmental biology, its full breadth, and its multiple foci of interest. One of the most interesting aspects for us was the melding of knowledge joined by the disparate disciplines of genetics, cytology, and molecular biology.

We are beholden to the more than 60 speakers and 200 additional conferees who contributed posters for a most interesting and stimulating intellectual experience. It is hoped that the papers in this volume will communicate to a wider audience some of the novel insights, new discoveries and achievements, and the general scientific excitement that were enjoyed by those of us that were fortunate enough to participate.

We gratefully acknowledge sponsorship of this conference by Cetus Corporation. Also we thank the National Institute of Child Health and Human Development and Biogen Research Corporation for additional funding which was used to defray the expense of speakers. It is our pleasure to acknowledge the superb efficiency with which the UCLA Symposia staff handled the many practical demands of arranging a conference of the dimensions of this one. In particular, we would like to thank Robin Yeaton, Hank Harwood, and Sandy Malone for their seemingly tireless efforts both before and during the event itself.

Eric H. Davidson
California Institute of Technology

Richard A. Firtel
University of California, San Diego

I. CYTOPLASMIC LOCALIZATIONS AND PATTERN FORMATIONS

Molecular Biology of Development, pages 3–22

HOMEOTIC GENES AND THE CONTROL OF CELL DETERMINATION

Walter J. Gehring

Department of Cell Biology
Biozentrum, University of Basel
CH-4056 Basel, Switzerland

The spatial organization of the embryo is under genetic control. In order to understand the molecular mechanisms underlying regional specification, we have cloned the homeotic *Antennapedia* gene of *Drosophila*. On the basis of cross homology several other homeotic genes and the segmentation gene *fushi tarazu* have been isolated. The homology is confined a precisely defined DNA segment called the *homeo box* which has also been found in other metazoans including vertebrates and man. Sequencing data suggest that the homeo box codes for a highly conserved protein domain, which may be involved in the control of other genes. The localization of transcripts of homeotic and segmentation genes in the blastoderm corresponds to the fate map. This indicates that the state of determination of the blastoderm cells is reflected in the pattern of activity of these control genes. The mechanisms of cell determination are discussed.

INTRODUCTION

The question of how a single cell, the fertilized egg, develops into a complex multicellular organism, is one of the most fundamental problems of developmental biology. Since the work of E.B. Wilson (1) and T.H. Morgan (2) it has become increasingly clear that the developmental program resides in the genome, and that in most cases the environment provides only general stimuli and relatively little specific information. Development is a highly ordered process which requires a precise temporal and spatial control of gene expression. In order to understand the principles of development, we have to identify the genes involved in the control of development and to elucidate the mechanisms and circuits by which they are regulated. The genetic regulatory circuits have to ensure that cell division and cell differentiation are precisely timed and localized in the developing organism, so that ordered structures are formed which allow the organism to function properly. The control circuits are such that changes in the environment can be compensated to some extent, and in some cases their flexibility is high enough to counteract direct interference by the environment, as for example in the regeneration of defective or missing parts.

The first step in the analysis of the genetic control of development is the identification of the controlling genes. Only few organisms are suitable for this purpose, Drosophila being one of them. Gene cloning techniques can be used to isolate such genes and to study their mechanism of action. In this article the current status of the work carried out in our laboratory on homeotic genes and their involvement in cell determination will be discussed, and a working hypothesis concerning their mechanism of action will be presented.

THE SPATIAL ORGANIZATION OF THE DROSOPHILA EMBRYO: DETERMINATIVE EVENTS AT THE BLASTODERM STAGE

The secrets of development lie in the egg. In Drosophila it takes approximately 10 days to produce a mature egg, whereas embryogenesis only lasts for 22 hours. How much developmental information is preformed in the egg and how much develops "epigenetically" is a question which is as old as the science of embryology. In the mature Drosophila egg the future body axes, anterior-posterior, dorsal-ventral, left and right, can clearly be distinguished by morphological criteria. However, relatively little is known about cytoplasmic localization in precise molecular terms although this is now amenable to molecular analysis.

Following fertilization the Drosophila embryo develops as a syncytium of synchronously dividing nuclei, which are localized in the central yolk-filled area of the egg. Following the eighth nuclear division, the nuclei migrate to the periphery of the egg cytoplasm, and at the 512 nuclear stage a small group of nuclei at the posterior pole becomes enveloped by plasma membranes to form the pole cells, which are the future germ cells. Most of the remaining nuclei line up in the egg cortex and after three additional synchronous divisions a perfect monolayer of some 6000 cells, the cellular blastoderm, is formed. These cells will give rise to the somatic tissues of the fly.

On the basis of gynandromorph studies, Sturtevant (3) showed that the cleavage nuclei are equivalent and totipotent. A given nucleus can colonize any region of the egg cortex, and differentiates according to its location rather than its origin. This has been confirmed by nuclear transplantation studies in several laboratories (4-6). However, when the cleavage nuclei reach the cortex and cellularization occurs, the blastoderm cells are determined, as was first shown by culturing dissociated and reaggregated blastoderm cells in vivo (7). Subsequent cell lineage studies indicated that the blastoderm cells are segmentally determined, i.e. clones of genetically marked cells induced at the blastoderm stage are confined to one body segment, and do not cross the boundary to the next segment, even if they comprise a large number of cells (8). The observation that the cleavage nuclei are equivalent and not committed to any specific developmental pathway, whereas the blastoderm cells are determined to contribute to a specific body segment, leads to the conclusion that there is an interaction between the cortical cytoplasm and the immigrating nuclei which leads to the determination of the nuclei and the cells which they form. Since the blastoderm cells differentiate according to their location in the egg, the cortical cytoplasm is thought to contain positional information or determinants which specify the spatial coordinates within the egg. This conclusion is supported by the analysis of maternal effect mutants (see below). At present, it is not known how precisely the spatial coordinates of the egg are defined and what the nature of the determinants of spatial organization is. Nuclear transplantation experiments have failed to conclusively demonstrate nuclear determination. However, the transplantation of syncytial blastoderm nuclei with adhering cortical cytoplasm to ectopic sites in another embryo of the same stage suggests that these nuclei have acquired at least an anterior versus posterior committment (9). The nature of this nucleo-cytoplasmic interaction is crucial to our understanding of cell determination.

GENETIC CONTROL OF EARLY EMBRYOGENESIS

Basically two classes of mutants controlling early embryogenesis can be distinguished, mutants showing a maternal effect and zygotic mutants (10). Maternal effect genes are involved in the "construction" of the egg during oogenesis in the mother. Some of these genes seem to be active exclusively during oogenesis, others may also be active later in the developing zygote. Several maternal effect mutants that alter the spatial coordinates of the embryo have been described (11). They affect the general polarity of the egg, as is the case for bicaudal (12, 13), dicephalic (14), and dorsal (11) which affect the anterior-posterior or the dorso-ventral polarity. Other maternal effect mutants are defective in pole cell formation like grandchildless (15) or show relatively gross morphological defects "gap mutants" (16). This may suggest that the positional information localized in the egg cytoplasm is rather general, perhaps in the form of polarity gradients, and does not specify the spatial organization of the embryo in all details. However, in the bicaudal embryos, which lack head and thoracic segments and give rise to an embryo with two abdomens in mirror-image symmetry, pole cells are formed at the authentic posterior pole only. This indicates that the germ cell determinants are positioned independently of the spatial determinants for the somatic cells.

Among the early zygotic mutants, two classes of mutants are known to affect the spatial organization of the embryo into segments, the segmentation mutants affecting segment number and polarity (17) and the homeotic mutants, which transform one segment into another, or parts of a segment into the corresponding part of another segment, e.g. an antenna into a leg.

The term homoeosis goes back to Bateson's "Materials for the study of variation" (18) although at that time the genetic basis of this phenomenon was not understood. The pioneering work on the genetics of homeotic mutants, in particular the *bithorax* complex, was carried out by E.B. Lewis (19), who proposed a model for the action of the *bithorax* genes which is based on evolutionary considerations. In primitive arthropods, from which insects have evolved, the numerous body segments are uniform, each one carrying e.g. a pair of legs. In insects, the segments have become extensively diversified, especially in the head region and in the abdomen, and in the adults only the three thoracic segments have retained their legs. Therefore, Lewis postulates that in the course of evolution, mutations have accumulated which remove the legs from the abdominal segments. Similarly, fossil insects are known which have wings on all three thoracic segments, the prothoracic ones being rather small. Living insects generally have two pairs of wings, but the highly evolved diptera, including *Drosophila*, have only one pair on the mesothorax, the second pair having been transformed to small halteres. Lewis argues that in the course of evolution, mutations transforming the metathoracic pair of wings into halteres have accumulated. Using a combination of mutants which are located in the *bithorax* complex, Lewis has in fact been able to "construct" four-winged flies, or flies with eight legs. Therefore, the evidence that the genes of the *bithorax* complex are involved in controlling the development of the posterior thoracic and the abdominal segments is very strong. T. Kaufman (20) has proposed that the *Antennapedia* complex represents the extension of the *bithorax* complex into the prothoracic and head region, and there is evidence that the two complexes interact in the determination of segments (21-23).

The phenotypic effects of mutations at the *Antennapedia* (*Antp*) locus can be subdivided into dominant effects which, according to the terminology of Lewis, are considered as a gain of function. They involve the partial or complete transformation of the antennae into mesothoracic legs. The deletion of the locus does not show this dominant phenotype, which indicates that at least a part of the gene is required for generating the dominant phenotype, which probably results from an alteration in gene expression and/or an altered gene product. The loss of function of the *Antp* gene leads to a recessive lethal phenotype at late embryonic or larval stages associated with the transformation of the second thoracic segment towards the first (24), and into head structures (25). This suggests that *Antp* is required for the formation of the mesothoracic segment, and that it specifies this segmental identity.

CLONING OF HOMEOTIC GENES

The importance of homeotic genes for our understanding of development has been obvious, but until recently it was not possible to elucidate the molecular basis of such mutations. With the development of gene cloning techniques, this situation has changed markedly. In *Drosophila* it is now possible to clone essentially any gene, even in the absence of biochemical information about its products, as long as mutants are available which allow the physical mapping of the mutational lesion on the DNA. Since D. Hogness and his group did the pioneering work on the *bithorax* genes (26), we decided to clone the *Antennapedia* locus by using a combination of "chromosomal walking and jumping" (27). The results are summarized in Figure 1. More than 220 kb of overlapping DNA segments were cloned from a genomic *Drosophila* DNA library in Lambda from the 84B1-2 section of the third chromosome, including the *Antp* locus. The locus was first defined physically by means of chromosomal rearrangements. Subsequently, chromosomal DNA segments were used for the isolation of cDNA clones from embryonic and pupal cDNA libraries. The transcribed region spans over 100 kb and consists of at least five exons separated by introns, some of which are extremely long. By means of Northern blotting at least two transcripts of 3.4 kb and 5.8 kb respectively can be detected (A. Kuroiwa et al. unpubl.) which indicates that there is differential transcription and/or splicing of the *Antp* locus. It is interesting to note that chromosomal inversions like $Antp^{PW}$ which split the locus approximately in the middle, give both the recessive loss of function and the "dominant gain of

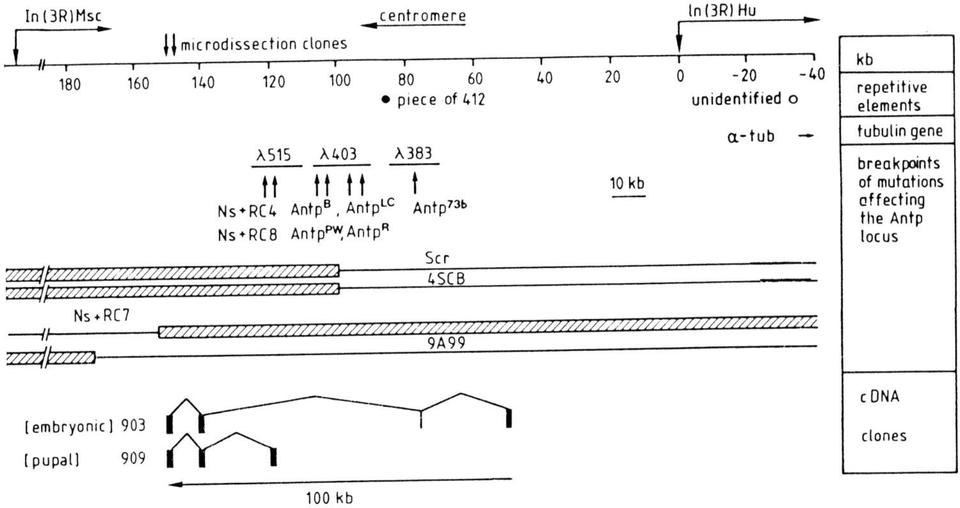

Figure 1. Molecular cloning of the *Antennapedia* locus. The 0 position is defined by the breakpoint of the inversion In(3R)Hu which was used to jump into the region. Over 180 kb of overlapping DNA segments were cloned to the left and 40 kb to the right of this breakpoint. The positions of two repetitive elements and an α-tubulin gene are indicated. The breakpoints of six inversions and one translocation disrupting the *Antp* locus are indicated by vertical arrows, and the extent of four deletions is indicated by the cross hatched bars. The *Antp* locus is delimited by the breakpoint of the 9A99 deletion which leaves the locus intact on the left, and by the *Humeral* breakpoint and the α-tubulin gene on the right. Two cDNA clones derived from embryonic (903) and pupal (909) poly$(A)^+$ RNA are indicated at the bottom. The black bars indicate exons, the connecting lines introns. The horizontal arrow indicates the direction of transcription of the *Antp* locus which spans at least 100 kb (after Ref. 27).

function" phenotypes. The Antp locus has been cloned independently by Scott et al. (28). On the basis of the cross homology to be described below several other homeotic genes and the segmentation gene fushi tarazu (ftz) have also been cloned.

CROSS HOMOLOGY BETWEEN HOMEOTIC GENES: THE HOMEO BOX

In the course of mapping the exons of Antp by hybridization of the cDNA clone 903 to the chromosomal DNA, weak cross homology was detected between the Antp cDNA and a sequence mapping 30 kb to the left of Antp (Fig. 2). A 1.9 kb poly(A)$^+$ RNA transcript originating from this locus, different from the Antp transcripts, was discovered by Northern blotting (29). This transcript originates from the locus of the segmentation gene fushi tarazu (29, 28), which will be discussed below. Using the 903 cDNA clone as a hybridization probe on whole genome Southern blots at reduced stringency conditions, a repetitive DNA sequence was detected which occurs at 5 to 10 locations in the Drosophila genome (30). The region of cross homology was mapped to a small DNA segment, 180 bp long, which is present not only in Antp and ftz but also in the Ultrabithorax (Ubx) transcription unit of the bithorax complex (Fig. 2). Therefore, it was called the homeo box and used as probe for the cloning of additional cross hybridizing sequences. Two of these, p93 and p99, have been characterized, and on the basis of their chromosomal location, their time of expression and the localization of their transcripts (see below), they were shown to be homeotic genes. One of them, p93, originates from the right half of the bithorax complex and was tentatively assigned to the infra-abdominal-2 locus (F. Karch and W. Bender, pers. commun.), whereas p99 maps to the left of Antp in the Antennapedia complex and may correspond to the Deformed locus. So far, we have not found the homeo box in the bithoraxoid/postbithorax unit of the bithorax complex, which indicates that it is not present in all homeotic genes. However, another subset of hoemotic genes may share another repeat.

The homeo box was also used to probe the genomes of various other metazoans and cross hybridization was detec-

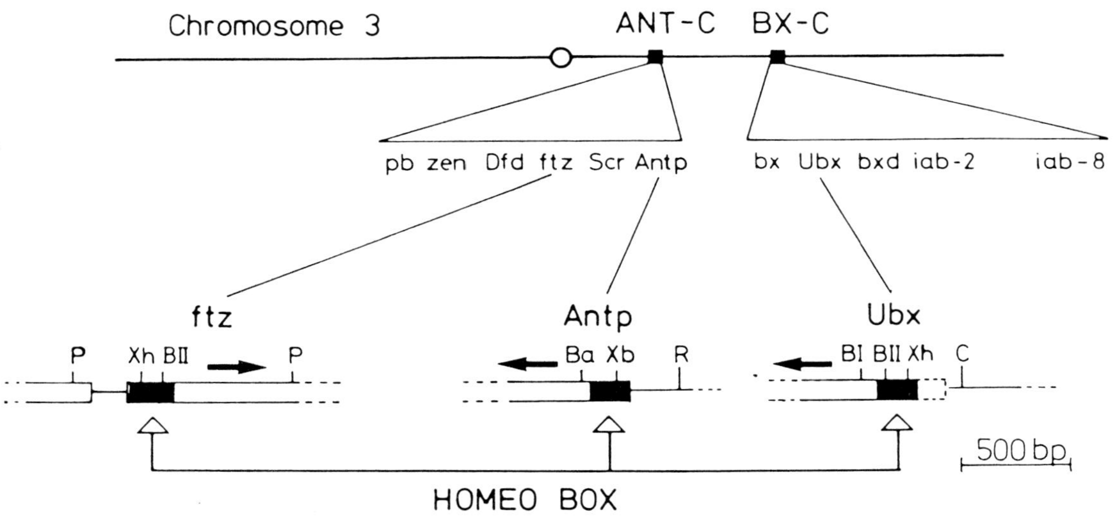

Figure 2. Chromosomal map of the Antennapedia- (ANT-C) bithorax complexes and the location of the homeo boxes in fushi tarazu, Antennapedia and Ultrabithorax (BX-C) (after Ref. 31). Not all of the known loci are shown in the chromosomal map (upper part). The lower part of the figure dipicts the location of the homeo boxes (black bars). Exons are bars, introns are lines, and uncertainties are marked with dotted lines. The direction transcription at each locus is indicated by a horizontal arrow.

ted not only in arthropods and their putative ancestors, the annelids, but also in vertebrates, including man (31, 32). There may be a correlation between the presence of segmentation and the presence of the homeo box in the genome of a given taxonomic group of animals, since we have not detected homeo box sequences in the genomes of nematodes and of sea urchins, which lack overt segmentation. In collaboration with E. De Robertis' group a putative "homeotic" gene, AC1, was cloned from the frog Xenopus laevis (32) and in collaboration with F. Ruddle several chromosomal DNA clones from the mouse were isolated which cross hybridize with the homeo box.

The homeo box region has been sequenced for the Drosophila genes, Antp, ftz, and Ubx (31). In all three cases the homeo box is located in the most 3' exon and confined to 180 bp. Outside this 180 bp box, the sequence homology drops off precipitously to random levels. The degree of homology is striking and varies between 73 and 79% at the nucleotide level. There is one common open reading frame and the conceptual translation of the nucleotide sequence into the amino acid sequence shows that the amino acid sequence is even more conserved than the nucleotide sequence, since approximately half of the observed nucleotide changes do not alter the amino acid sequence. Therefore, we conclude that the homeo box codes for a precisely defined protein domain, the homeo domain, in which 44 out of 60 amino acids are identical in all three genes.

The conservation of the homeo box sequence during evolution is even more remarkable. Sequencing the corresponding region of the AC1 gene of Xenopus laevis (32) indicates that the homeo box is also confined to 180 bp and the degree of homology between AC1 and Antp, for example, is even greater than the homology between the two Drosophila genes Antp and ftz. As the homeo domain sequence is highly conserved even between vertebrates and invertebrates, it is likely to serve an important role. Since one third of the amino acids in the homeo domain are basic, we speculate that it might have a DNA or chromatin binding function. We have as yet found no extensive homology with previously reported protein sequences. However, a weak homology to the a1 and $\alpha 2$ proteins from the yeast mating-type locus (MAT) has been found (33). These proteins have been implicated in the control of expression of other genes in yeast. The suggested DNA or chromatin binding property and the homology with the *MAT* genes are compatible with the idea that homeotic genes control batteries of other genes. It also raises the possibility that the homeo box serves as a "casette". At the present time, we have no indication of DNA rearrangements during ontogeny, but the homeo box perhaps might serve as a casette during evolution, and be subject to gene conversion. Alternatively, the homeotic genes may have arisen from a common ancestral gene by duplications and subsequent divergence, with the homeo box being the most conserved part. In any case, the discovery of this novel class of genes, which share sequences that are highly conserved, may provide the key to the understanding of the genetic regulatory circuits which control development, not only in *Drosophila*, but also in higher vertebrates.

LOCALIZATION OF TRANSCRIPTS

One of the most important problems in developmental genetics is the question of when and where a particular gene is expressed in development. Since no satisfactory methods were available to detect rare mRNAs in tissue sections, we developped an improved procedure for *in situ* hybridization of cloned tritiated DNA probes to transcripts in frozen sections and subsequent detection of the hybridization signals by autoradiography (34). Using this method, the transcripts of the *Antp*$^{+}$ locus were analyzed during development in wildtype *Drosophila* (34, 35). For the first time, the developmental profile of a homeotic gene could be analyzed, and a few points will be mentionned here (for a detailed description see Ref. 35). As early as the blastoderm stage, a slight accumulation of *Antp*$^{+}$ transcripts is observed in the future thoracic region, mostly over the cell nuclei. At later embryonic stages, the ventral nervous system becomes labelled in a spatially restricted manner. Initially, transcripts accumulate in all thoracic and abdominal ganglia. However, at later stages the highest concentration of transcripts is clearly detected in the ganglion of the mesothorax. This correlates well with the postulated role of *Antp*$^{+}$ in establishing segment identity of the mesothoracic segment (see above). A strong accumulation of *Antp*$^{+}$ transcripts is also observed in the cells surrounding the anterior spiracles of the larva. In homozygous *Antp*^{PW} larvae which develop to the late third instar, the anterior spiracles are degenerated, indicating that the *Antp*$^{+}$ gene product is required for proper development of the anterior spiracles (25). In general, a good correlation between the sites of mutant defects due to loss of gene function, and the accumulation of transcripts in the wildtype has been found. The phenotypic effects due to a "dominant gain of function" remain to be explained.

Further evidence that *Antp*$^{+}$ transcripts are associated with the formation of the mesothorax has been obtained by analyzing the distribution of *Antp*$^{+}$ transcripts in *bithorax* mutant embryos that lack the entire *bithorax* gene complex (23). Mutant embryos homozygous for this deletion show a transformation of the metathorax and the first seven abdominal segments into mesothoracic segments. In this case *Antp*$^{+}$ transcripts are accumulated at late embryonic stages in all the mesothoracic segments to a level characteristic of the normal mesothorax. This observation strengthens the correlation between *Antp*$^{+}$ gene products and the mesothoracic developmental pathway, and it suggests that the genetic control circuits are such that in the absence of *bithorax* gene products, the *Antp*$^{+}$ gene is expressed. This is a first step towards the elucidation of the genetic control circuits operating during cell determination.

The segmentation gene *fushi tarazu* (*ftz*) was isolated on the basis of its cross homology to *Antp* (30). Homozygous *ftz* embryos die prior to hatching from the egg and lack alternate body segments (Fig. 3; see also 24, 16). The posterior part of one segment and the anterior part of the next segment are missing, and the remaining parts are fused. As shown by Northern blotting experiments the *ftz*$^{+}$ gene is expressed exclusively during the early blastoderm to gastrula stages of embryogenesis, at the time when segments are determined, but before segmentation becomes morphologically visible (29). Hybridization of a cloned genomic DNA segment containing most of the RNA coding region of the *ftz*$^{+}$ gene to wildtype

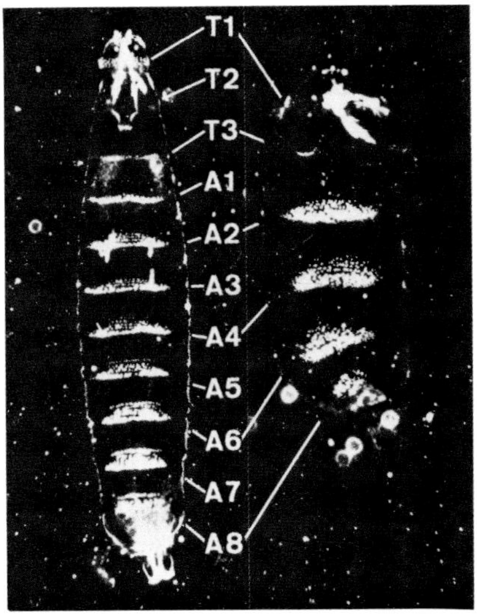

Figure 3. Comparison of the cuticular segmentation pattern of a wildtype first instar larva (on the left) and a lethal fushi tarazu (ftz 9093) mutant embryo (on the right). The ventral denticle belts are shown in darkfield illumination. T1-3 designates the three thoracic segments, A1-A8 the abdominal segments. The ftz embryo lacks alternate body segments (after Ref. 36).

blastoderm embryos gave a remarkable result (Fig. 4; 36). At the cellular blastoderm stage ftz^+ transcripts are localized precisely in seven evenly spaced bands of cells between approximately 15 and 65% egg length (0% corresponds to the posterior end of the egg). The size of each band corresponds closely to the size of the segment primordia of the blastoderm, as determined by fate mapping (37). The labeled cells are most likely the progenitors of the segmental portions that are missing in ftz mutant embryos. A careful examination of earlier developmental stages indicates that the localized accumulation of transcripts begins prior to the formation of cell membranes during syncytial blastoderm stages. Sections through embryos that have completed the 11th synchronous

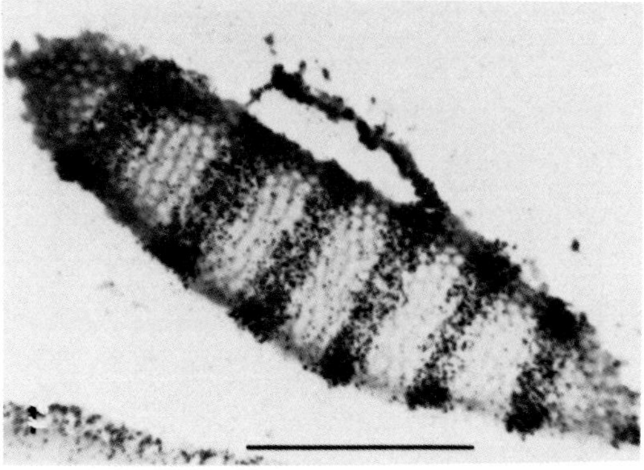

Figure 4. Localization of fushi tarazu (ftz^+) transcripts in wildtype blastoderm embryos. Frozen tissue sections were hybridized to a tritiated cloned ftz^+ chromosomal DNA probe (p523B) and autoradiographed for 21 days. A superficial section through the blastoderm epithelium shows seven belts of hybridization (after Ref. 36).

nuclear division exhibit a significantly higher accumulation of silver grains between 15 and 65% egg length which is presumably due to newly transcribed ftz^+ transcripts, since clusters of silver grains are localized over the nuclei in this region. After one further nuclear division, regional differences in signal intensity are observed within the labelled domain and labelling is also found over the cytoplasm. After completion of the 13th nuclear division the segmental pattern of hybridization becomes more obvious. At this stage the nuclei begin to elongate and cell membranes grow inwards from the egg surface enclosing the nuclei. At this stage, before completion of the cell membranes, the regular segmental pattern of hybridization is established. This provides direct evidence for nuclear determination during blastoderm formation. At the cellular blastoderm stage there is an excellent agreement between the fate map and the localization of transcripts (36) indicating that the transcriptional activity of the nuclei reflects the fate map.

Hybridization of the p99 probe, which was isolated on the basis of its cross homology (see above) and tentatively identified as the Deformed$^+$ (Dfd) locus, gave equally strong signals at the blastoderm and early gastrula stages. In this case only one belt of hybridization, approximately two segments wide, is observed in the posterior head region. This suggests that the p99 locus may have a role in establishing segmental identity in these posterior head segments.

NUCLEO-CYTOPLASMIC INTERACTIONS DURING CELL DETERMINATION

The earliest determinative events in the egg concern the spatial organization of the embryo. The Drosophila embryo is first subdivided into a series of repeating segments whose fate is determined at the blastoderm stage. This regional specification preceeds the determination of the different cell types which are usually defined on the basis of their ultimate function, like neurons, muscle cells, epidermal cells etc. Many such cell types are found in all segments. However, the various segments are not exactly repeated, but rather diversified. On the basis of their phenotypic effects and the localization of their transcripts, two classes of genes can be identified, those which are involved in segmentation in general and those which are thought to specify the identity of the various segments.

Based upon earlier experiments, I have proposed a working hypothesis according to which a given state of determination reflects a certain pattern of gene activity (38). The localization of the transcripts of ftz or Dfd in the blastoderm strongly support this hypothesis. These genes are clearly expressed in some of the future segments and not in others. The possibility that these genes are transcribed in all blastoderm cells and that the transcripts are rapidly degraded in some regions and not in others, is highly unlikely, since we detect no labelling over the nuclei of those cells which do not accumulate the transcripts. On the other hand, the first ftz transcripts can be detected over the nuclei at early syncytial stages, which indicates that our method is sensitive enough to detect the nascent transcripts in the nuclei. Therefore, the pattern of in situ hybridization in the blastoderm cells reflects differential gene activity at least at the level of accumulation of transcripts, and most probably at the level of transcription. Since the localization of the transcripts corresponds rather precisely to the fate map of the blastoderm, we may conclude that the state of determination is reflected by the pattern of activity of the developmental control genes like for example ftz and Dfd. Since ftz is expressed prior to cellularization, the localized accumulation of transcripts provides direct evidence for nuclear determination.

The mechanism of determination is largely a matter of speculation, but the observations on ftz raise some new possibilities. Since the cleavage nuclei prior to their migration into the cortical cytoplasm are totipotent and the blastoderm cells are segmentally determined, we have concluded above that there is a nucleo-cytoplasmic interaction which leads to the determination of the nuclei and the cells which they form. The cortical cytoplasm apparently contains positional information or determinants which specify the spatial coordinates within the egg and interact with the immigrating nuclei which leads to their determination. On the basis of our observations on ftz and Dfd, I would like to propose the idea that such genes may serve as sensors which allow the nucleus to read its position within the egg. In the simplest case, the determinants specifying positional information may contain DNA-binding proteins (or their respective mRNAs) that are differentially distributed in the cortical cytoplasm and released when the nuclei reach the cortex. Such proteins might then enter the nucleus and bind to the regulatory DNA sequences in the ftz locus and other controlling genes, thereby regulating their activity depending on the position of the nucleus in the cortex. The combinational pattern of activity of the various controlling genes would then specify the state of determination. This model can be tested experimentally by identifying the factors which control ftz or by cloning the genes which specify the positional information. Also, it should be possible to find those genes which are controlled by ftz and presumably represent the next lower level in the hierarchy of controlling genes. The cloning of homeotic genes and the localization of their transcripts certainly opens up completely new possibilities to analyze the molecular control mechanisms of development.

ACKNOWLEDGMENTS

I would like to thank all the members of my research group for their contribution to this work, the stimulating discussions, and the continued efforts to solve these exciting problems which can only be solved by teamwork. My thanks also go to Iain Mattaj for critical reading of the manuscript and to Erika Wenger-Marquardt for typing it. This work has been supported mostly by the Swiss National Science Foundation and the Kanton Basel-Stadt for which I am very grateful.

REFERENCES

(1) Wilson EB (1928). "The Cell in Development and Heredity." New York: Macmillan.
(2) Morgan TH (1934). "Embryology and Genetics." New York: Columbia University Press.
(3) Sturtevant AH (1929). The claret mutant type of Drosophila simulans: a study of chromosome elimination and cell-lineage. Z Wiss Zool 135:323-356.
(4) Zalokar M (1971). Transplantation of nuclei in Drosophila melanogaster. Proc Natl Acad Sci USA 68:1539-1541.
(5) Okada M, Kleinman IA, Schneiderman HA (1974). Chimeric Drosophila adults produced by transplantation of nuclei into specific regions of fertilized eggs. Dev Biol 39:286-294.
(6) Illmensee K (1978). Drosophila chimeras and the problem of determination. In Gehring WJ (ed): "Genetic Mosaics and Cell Differentiation," Berlin-Heidelberg: Springer, p 51-69.
(7) Chan L-N, Gehring W (1971). Determination of blastoderm cells in Drosophila melanogaster. Proc Natl Acad Sci USA 68:2217-2221.
(8) Wieschaus E and Gehring W (1976). Clonal analysis of primordial disc cells in the early embryo of

Drosophila melanogaster. Dev. Biol. 50:249-263.

(9) Kauffman SA (1980). Heterotopic transplantation in the syncytial blastoderm of Drosophila: Evidence for anterior and posterior nuclear commitments. Wilh Roux's Arch 189:135-145.

(10) Gehring WJ (1973). Genetic control of determination in the Drosophila embryo. In Ruddle FH (ed): "Genetic Mechanisms of Development," New York: Academic Press, p 103-128.

(11) Nüsslein-Volhard C (1979). Maternal effect mutations that alter the spatial coordinates of the embryo of Drosophila melanogaster. In Subtelny S, Konigsberg IR (eds): "Determinants of Spatial Organization," New York: Academic Press p 185-211.

(12) Bull AL (1966). Bicaudal, a genetic factor which affects the polarity of the embryo in Drosophila melanogaster. J Exp Zool 161:221-242.

(13) Nüsslein-Volhard C (1977). Genetic analysis of pattern-formation in the embryo of Drosophila melanogaster: Characterization of the maternal-effect mutant bicaudal. Wilh Roux's Arch 183:249-268.

(14) Lohs-Schardin M (1982). Dicephalic - A Drosophila mutant affecting polarity in follicle organization and embryonic patterning. Wilh Roux's Arch 191:28-36.

(15) Spurway H (1948). Genetics and cytology of Drosophila subobscura. IV. An extreme example of delay in gene action causing sterility. J Genet 49:126-140.

(16) Nüsslein-Volhard C, Wieschaus E, Jürgens G (1982). Segmentierung bei Drosophila - Eine genetische Analyse. Verh Dtsch Zool Ges 1982: 91-104.

(17) Nüsslein-Volhard C and Wieschaus E (1980). Mutations affecting segment number and polarity in Drosophila. Nature 287:795-801.

(18) Bateson W (1894). "Materials for the Study of Variation treated with Especial Regard to Discontinuity in the Origin of Species." London: MacMillan, p 598.

(19) Lewis EB (1978). A gene complex controlling segmentation in Drosophila. Nature 276:565-570.

(20) Kaufman TC and Wakimoto BT (1982). Genes that control high level developmental switches. In Bonner JT (ed): "Evolution and Development," Berlin, Heidelberg, New York: Springer, p 189-205.

(21) Struhl G (1982). Genes controlling segmental specification in the Drosophila thorax. Proc Natl Acad Sci USA 79:7380-7384.
(22) Struhl G (1983). Role of the esc^+ gene product in ensuring the selective expression of segment-specific homeotic genes in Drosophila. J Embryol exp Morph 76:297-331.
(23) Hafen E, Levine M, Gehring WJ (1984). Regulation of Antennapedia transcript distribution by the bithorax complex in Drosophila. Nature 307:287-289.
(24) Wakimoto BT and Kaufman TC (1981). Analysis of larval segmentation in lethal genotypes associated with the Antennapedia gene complex in Drosophila melanogaster. Dev. Biol. 81:51-64.
(25) Schneuwly S and Gehring WJ (1984) in preparation.
(26) Bender W, Akam M, Karch F, Beachy PA, Pfeifer M, Spierer P, Lewis EB, Hogness DS (1983). Molecular genetics of the bithorax complex in Drosophila melanogaster. Science 221:23-29.
(27) Garber RL, Kuroiwa A, Gehring WJ (1983). Genomic and cDNA clones of the homeotic locus Antennapedia in Drosophila. The EMBO J 2:2027-2036.
(28) Scott MP, Weiner AJ, Hazelrigg TI, Polisky BA, Pirrotta V, Scalenghe F, Kaufman TC (1983). The molecular organization of the Antennapedia locus of Drosophila. Cell 35:763-776.
(29) Kuroiwa A, Hafen E, Gehring WJ (1984). Cloning and transcriptional analysis of the segmentation gene fushi tarazu of Drosophila. submitted to CELL.
(30) McGinnis W, Levine M, Hafen E, Kuroiwa A, Gehring WJ (1984). A conserved DNA sequence found in homeotic genes of the Drosophila Antennapedia and bithorax complexes. Nature, in press.
(31) McGinnis W, Garber RL, Wirz J, Kuroiwa A, Gehring WJ (1984). A homologous protein-coding sequence in Drosophila homeotic genes and its conservation in other metazoans. submitted to CELL.
(32) Carrasco AE, McGinnis W, Gehring WJ, De Robertis EM (1984). Cloning of a Xenopus laevis gene expressed during early embryogenesis that codes for a peptide region homologous to Drosophila homeotic genes: implications for vertebrate development. submitted to CELL.

(33) Shepherd J (1984) unpublished results.

(34) Hafen E, Levine M, Garber RL, Gehring WJ (1983). An improved in situ hybridization method for the detection of cellular RNAs in Drosophila tissue sections and its application for localizing transcripts of the homeotic Antennapedia gene complex. The EMBO J 2:617-623.

(35) Levine M, Hafen E, Garber RL, Gehring WJ (1983). Spatial distribution of Antennapedia transcripts during Drosophila development. The EMBO J 2:2037-2046.

(36) Hafen E, Kuroiwa A, Gehring WJ (1984). Spatial distribution of transcripts from the segmentation gene fushi tarazu during Drosophila embryonic development. submitted to CELL.

(37) Lohs-Schardin M, Cremer C, Nüsslein-Volhard C (1979). A fate map for the larval epidermis of Drosophila melanogaster: Localized cuticle defects following irradiation of the blastoderm with an ultraviolet laser microbeam. Dev. Biol. 73:239-255.

(38) Gehring WJ (1976). Determination. In Allfrey VG, Bautz EKF, McCarthy BJ, Schimke RT, Tissières A (eds): "Organization and Expression of Chromosomes," Dahlem Konferenzen, Berlin: Springer p 97-113.

Molecular Biology of Development, pages 23–36

LOCALIZATION OF POLY (A) IN XENOPUS EMBRYOS[1]

Fred H. Wilt and Carey R. Phillips[2]

Department of Zoology, University of California
Berkeley, California 94720

ABSTRACT The distribution of yolk platelets in eggs and early embryos of *Xenopus laevis* was followed by microscopy. There are extensive rearrangements of coherent masses of yolk platelets. There is an especially prominent movement of large yolk platelets of the vegetal hemisphere toward the animal pole, which produces a columnar projection of yolk. This column is always on the prospective dorsal side of the embryo. The distribution of poly (A) was studied by hybridization with ^{3}H poly (U) followed by autoradiography. There are also extensive changes in the distribution of poly (A) during early development. Among the changes in the distribution of poly (A) is the appearance of an area rich in poly (A) associated with the column of large yolk platelets; this poly (A) appears at stage 8. Changes in poly (A) distribution are probably due to changes in the number and/or length of the poly (A) tails on pre-existent RNA molecules. Feasible approaches to evaluate the biological role of localized RNAs are also discussed. Clones may be isolated from a cDNA library that appear highly enriched in the RNA of the prospective ventral or dorsal side of the embryo. Manipulation of the internal contents of the egg by centrifugation show a correlation between the position of certain yolk platelet areas and the reoriented embryo axis.

[1]This work was supported by grants from NIH (HD 15043) and NSF (PCM 8022836).
[2]Supported in part by NIH Fellowship HD 095945.

INTRODUCTION

In E.B. Wilson's summary of the state of cell and developmental biology in 1925 (1), a concise statement of a central problem of embryology was already available, as well as a general solution to the problem. Wilson said (p. 1037), "The fundamental problem, therefore, which includes all the others, is that of determination". He goes on (p. 1059), "Specification of the blastomeres cannot, therefore, be due to specific nuclear differences produced by a fixed order of qualitative nuclear divisions, but must be sought in the condition of the ooplasm....The specification of the blastomeres is due to the nature of the specific cytoplasmic materials which they receive during cleavage."

Not only are the ideas venerable, they are still the only general and credible theory for the origin of cell type determination in early embryogenesis. It is the cytoplasmic endowment of the egg, elaborated in oogenesis, and its subsequent distribution and action upon the nucleus and other cell parts, that entrains the differences in time and space that arise in the developing embryo.

Subsequent to Wilson's critique, 60 years of evidence show that there are, indeed, substances in the egg that are distributed differentially to the different blastomeres. There is also ample evidence for the assymetric distribution of developmental potential to different blastomeres. But without exception the crucial link between a given organelle, structure, or chemical and a given developmental potential is lacking. Can the methods of modern cell and molecular biology be brought to bear productively on the issues? Certainly we can and are learning more about what is localized, and how such substances come to be localized and subsequently distributed to blastomeres. Less obvious is how to firmly establish that a given localized material entrains a sequence of events leading to a specific cell type. Even on this score there are now beginnings, as is testified by work reported in this symposium.

We have undertaken to study the localization of RNA in developing *Xenopus laevis*; it is our hope that by learning what RNA is localized, if any, and how it is distributed, that the description may serve as a spring board to critically examine if such localizations play a causative role in development. The observations reported here are a beginning. We have examined the cytoplasm of the early embryo by carefully following the areas occupied by yolk platelets of different sizes. The distribution of poly (A) has been followed by autoradiography following in situ hybridization with ^{3}H poly (U). The feasibility of isolating specific sequences localized in various parts of the Xenopus embryo by screening cDNA libraries has been examined. And manipulations of the embryo which alter the pattern of cell type determination and cause rearrangements of the internal contents of the egg have also been examined.

METHODS

Embryos were processed by soaking them in saline containing 7% glycerol and plunging them into liquid nitrogen. The frozen embryos were then brought to -20°C and fixed for 20 minutes at this temperature in 17% glacial acetic acid, 43% ethanol and single strength DeBoer's saline, and then dehydrated in the alcohol series at -20°C. Subsequent paraffin infiltration and histological procedures were carried out by conventional methods. Saturating levels of ^{3}H poly (U) were hybridized to tissue sections in buffered 0.2 M NaCl at 50°C for four hours, followed by treatment with RNAse, and washing in saline and then dilute TCA. Yolk platelet distributions were examined using Nomarksi optics and outlines of areas in photographs were traced on a computerized graphics tablet. Grain distributions of autoradiograms were examined with darkfield optics, and hence, the grains appear as bright spots on a darker background. Estimates of silver grain density were made by scanning sections with a computer controlled (Apple 2) Cohu video monitor. The size of each area being scanned is arbitrary and may be adjusted. A grid was drawn on the photograph after it was scanned and shows the position and size of each individual area scanned in the example. A sample photo of an autoradiogram and the grid is shown in figure 1. Potentiometer settings were adjusted against a linear grey scale so that the minimum brightness detected was equivalent to the brightness of a single silver grain. The brightness readings were adjusted to linearity, and then the relative brightness of each square was measured and stored as a 2 dimensional matrix. The software was written and provided by Dr. John Russ (Scientific Microprograms). The data may be displayed as an isomorphic map, as shown in figure 2, or any row or column of grain concentrations may be displayed individually as a histogram, as shown in figure 3. The relative grain densities across a section, or any specific cytoplasmic area of interest, can also be measured by scanning an area, such as the one marked by the diagonal bar (Figure 1). In this example, the area within the bar was divided into a grid of 30 x 5. The lower portion (first 11 portions) of the bar is enlarged and shown in figure 4. Each column is scanned and the average brightness and standard deviations shown in figure 5. Brightness measurements may be converted to grain densities and these are marked on the left ordinate. This computerized treatment of grain density data facilitated quantitative comparisons of large numbers of autoradiograms.

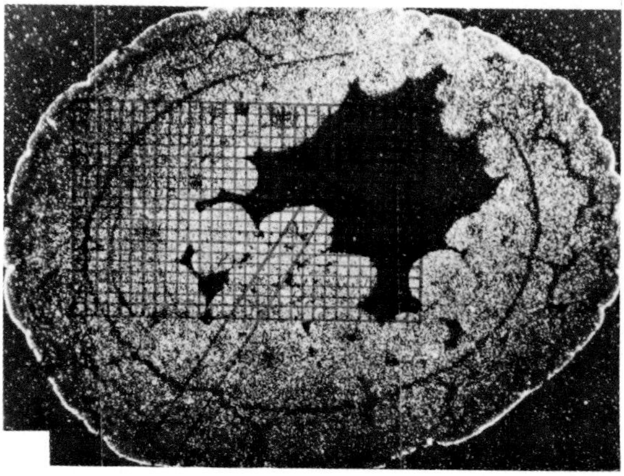

FIGURE 1. Autoradiogram of a section cut nearly perpendicular to the animal-vegetal axis, approximately 800 μm (80% of distance) from the vegetal pole of a stage 8 mid-blastula is shown. The bright grains represent hybrids between ^{3}H poly (U) and poly (A) in the section. A grid and diagonal bar is drawn on the photo to aid in orientation in subsequent figures. Areas in the grid were scanned with a video camera to determine grain density as explained in the text.

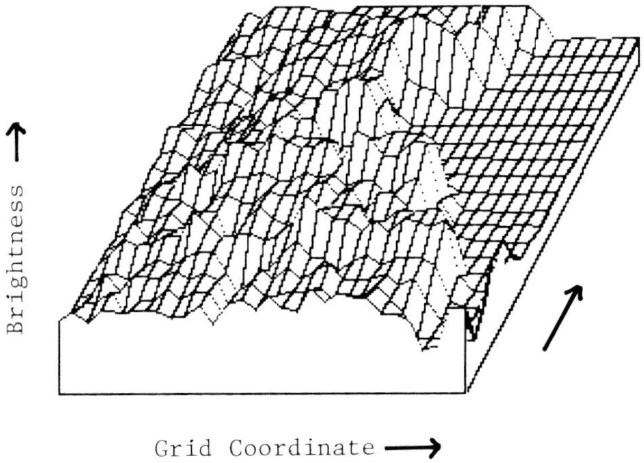

FIGURE 2. An isomorphic map generated by computer of the brightness of the sections of the grid shown in figure 1.

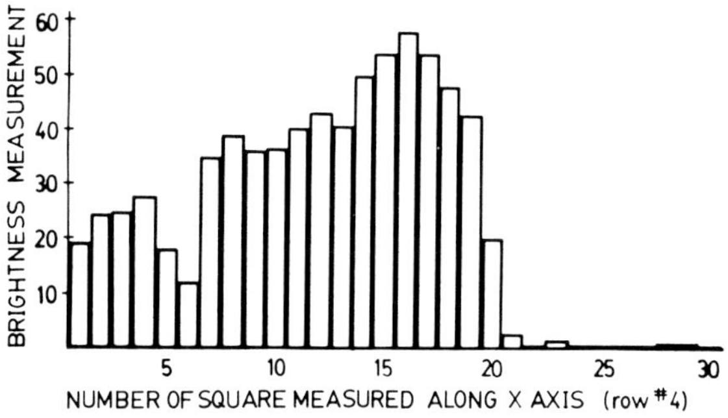

FIGURE 3. The data from row #4 in figure 2 is displayed in the format of the histogram.

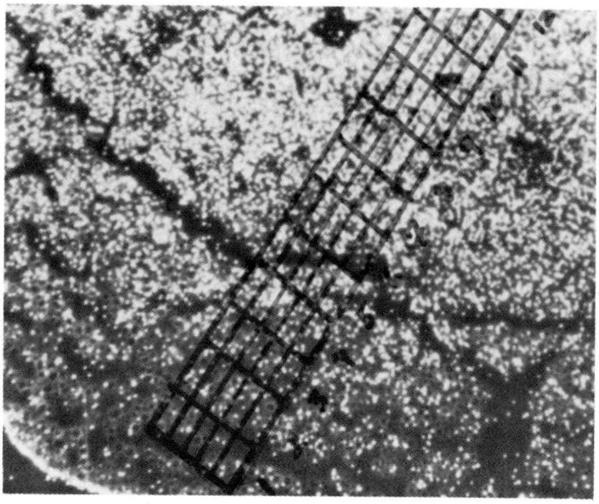

FIGURE 4. An enlargement of the diagonal bar shown in figure 1 is displayed. Individual silver grains may be visualized. The area within the diagonal bar has been divided into a grid of 30 sections long and 5 wide. Only the first 11 lengthwise columns of the diagonal are shown here.

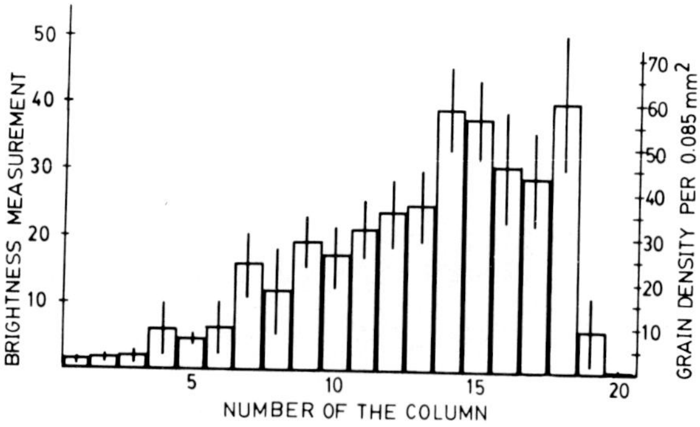

Figure 5.

FIGURE 5. The average brightness of each of the first 20 segments of the diagonal shown in figure 1 and 4, together with the standard deviations of the brightness within the 5 subsections of each lengthwise segment, are shown here. The cells in the center of the grid (top of the diagonal) contain the column of large yolk platelets and have 6-8 times more poly (A)/unit area than the cells in the vegetal end of the embryo.

RESULTS

Well fixed eggs and embryos show clearly demarcated areas that contain yolk platelets of a characteristic average size and closeness of packing. These areas remain coherent during pregastrular development. The shape of the areas changes somewhat, and the different areas of yolk platelets move relative to one another during the first few cleavage divisions, which probably indicates movement of the cytoplasm in which the platelets are embedded. Of course, as the embryo becomes cellularized, cytoplasmic rearrangements occur by cell movement rather than by intracellular cytoplasmic rearrangements.

There are a number of yolk platelet rearrangements, of which we shall emphasize only one. The class of yolk platelets of largest size is found in the vegetal hemisphere. Prior to first cleavage, the central and upper aspects of this yolk mass begins to extend in the animal direction, along the first cleavage furrow. This extension comes to produce a column of very large yolk platelets, extending animalward, and continuous vegetally with the large platelet yolk mass. As the 8 cell stage is attained, the forming blastocoel becomes apparent, and the yolk column is found along the side of the blastocoel that will become the dorsal side of the embryo. Other characteristic and well known movements of yolky cytoplasm, such as the subcortical movement of yolk platelets along the prospective dorsal side prior to first cleavage (forming the so-called vitelline wall) may also be observed.

The yolk platelet positions provide relative internal landmarks in the early embryos. Sections through embryos at various stages have been examined for their content of poly (A) by in situ hybridization in the belief that this serves as an indication of the distribution of one type of maternal mRNA molecules in the embryo. Sagata (2) et al.

have shown virtually all the poly (A) in the early embryo is attached to heterodisperse cytoplasmic RNA. Poly (A) is detected in the tissue sections by utilizing the ability of ^{3}H poly (U) to efficiently hybridize to the poly (A). Careful studies of this reaction in solution have been carried out (3), and the factors that influence the reaction in tissue sections of sea urchin embryos (4) and Xenopus oocytes (7) are also known. It is not known, however, what is the minimum length of poly (A) in tissues that will nucleate a poly (U):poly (A) hybrid which produces a signal under the conditions employed. Hence, changes in grain density can be due to changes in the average length of the poly (A) isostich, or to the number of poly (A) tails, or both. However, it is known (2) that the average size distribution of poly (A) isostichs from whole eggs and embryos does not change much during cleavage. In some instances there may also be a considerable overlap in the RNA complexity between the so-called poly (A) - and poly (A) + fractions. Consequently, grain distributions give us general information about the density of poly (A) isostichs, and by inference, of polyadenylated RNA. This method is a useful way to see if there are marked changes in the concentration of poly (A) in different regions of the egg and early embryo.

We do find that there are extensive, reproducible and easily detectable changes in poly (A) concentration. For example, the newly fertilized egg has a high concentration of poly (A) in a region adjacent to the animal cortical area. By the time of first cleavage, however, this high local concentration of poly (A) is no longer apparent, and the vegetal area containing large yolk platelets now shows a striking enrichment of poly (A). This change from a predominant sub-cortical localization in the animal hemisphere to a predominant internal, vegetal localization occurs at the one and two cell stages, and hence, could be due to physical movement of molecules or to changes in extent of adenylation of relatively immobile RNAs. This change in measurable local poly (A) concentrations is consistent with the previous experiments of Phillips (5) in which RNA isolated from larger sections of the egg were subjected to solution hybridization procedures.

During cleavage the prominent vegetal enrichment of poly (A) gradually disappears. A striking new localization appears in the blastula. At stage 8, within the columnar area of the animal-dorsal extension of the large yolk platelets to which we referred earlier, there is a marked increase in poly (A). This is shown in figure 6. Here the columnar area of large yolk platelets is cut orthogonal to its long axis and the high grain density may be observed along one side of the blastocoel. Thus, there is an association of poly (A) with the cytoplasm containing the column of large yolk platelets that is correlated with the forming dorsal side of the embryo.

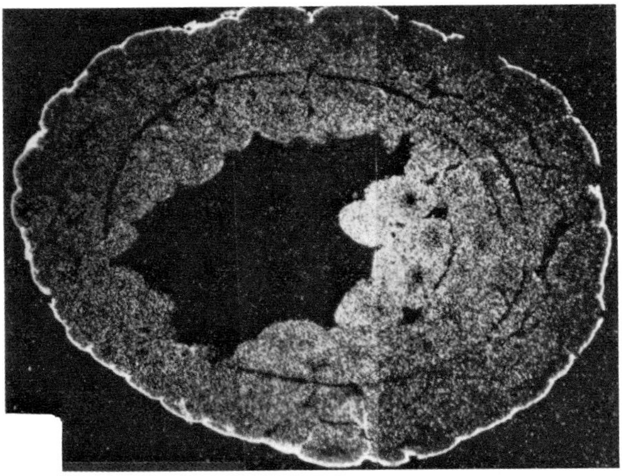

FIGURE 6. Autoradiogram of a section cut perpendicular to the animal-vegetal approximately 75% of the distance from the vegetal pole of a stage 8 embryo. The area containing the column of large yolk platelets shows a high concentration of poly (A).

Since the embryo is now highly cellularized, changes in poly (A) can only be due to new transcription or adenylation of previously existent RNA. We observe the high concentration of poly (A) in the cytoplasm of the column of large yolk platelets at stage 8. This is when new transcription just begins (6). We injected ^{3}H uridine into fertilized eggs and examined RNA synthesis by autoradiography of sections at late stage 8. Cells in the area of the column of large yolk were no more active in accumulation of new RNA than cells in other regions. Hence, different areas of the blastula are probably carrying out adenylation and/or deadenylation of maternal mRNA molecules to different extents.

As might be expected, ^{3}H poly (U) is detecting a molecule that acts like authentic poly (A). The signal is still obtained when samples are treated with RNAse at 200 mM salt, but poor hybridization is obtained if sections are digested with RNAse at 10 mM salt. Hybridization with poly (C) shows no assymetric grain distributions and the signal is close to background. If the sea urchin H-3 histone gene, which cross hybridizes extensively with Xenopus histone mRNA, is used as a probe, assymetric distributions of grains are not observed.

It would also be useful to study particular sequences that might be localized. To this end, early embryos have been sectioned along either the animal-vegetal or the forming dorsal-ventral axis; the sections from many embryos were pooled, RNA extracted from them, and the RNA used to make reverse transcripts that may be used as probes in an analysis of a cDNA library employing differential colony hybridization. This approach could conceivably allow one to find sequences that are highly enriched in one portion or another of the egg or early embryo. We have not yet found any sequences that seem to be assymetrically distributed along the animal-vegetal axis (however, see 7). There are indications, however, that some sequences present in the cDNA library are much enriched in areas of the embryo along the dorsal-ventral axis. The data is shown in Table 1. Though we have no further information on these sequences at present, we believe this preliminary result indicates the feasibility of such an approach.

Finally, it is also useful to impose conditions on the embryo that might alter the pattern of cell type determination and relocate any localized RNAs. An analysis has begun in collaboration with Dr. Steve Black using centrifugation to accomplish this purpose. Zygotes were oriented and immobilized in gelatin and centrifuged perpendicular to the animal-vegetal axis (30 xg, 4 min.) at a time corresponding to 40% of the interval between fertilization and first cleavage. The zygotes were recentrifuged with the original vegetal pole centrifugal (10 xg, 4 min.) at a time corresponding to 60% of the interval from fertilization to first cleavage. Eggs subjected to this regiment produce a very high percentage of bifurcated axes resulting in Siamese twins (figure 7). We have examined the internal structure of such eggs and it is observed that the column of large yolk platelets that appears during cleavage is found on both sides of the blastocoel. This is in distinct contrast to the situation in controls where the large yolk platelet column is found only on the prospective dorsal side. Hence, this manipulation may provide an experimental tool to learn about the localization of RNA sequences and their association with the forming embryonic axes.

Table 1

REGION SPECIFIC cDNA CLONES

Library	Colonies Screened	Dorsal Specific 1st Screen	Dorsal Specific Recheck	Ventral Specific 1st Screen	Ventral Specific Recheck
Gastrula	12000	19	12	14	10
	7200	17	11	12	9
Cleavage	6500	16	9	10	8

A gastrula cDNA library (5000 transformants) or cleavage library (3000 transformants) in the vector pBR 322 were subjected to colony hybridization procedures using reverse transcripts copied from RNA extracted from the prospective dorsal or ventral third of a 4 cell embryo. The number of colonies that showed a positive signal with dorsal cDNA but not with ventral cDNA on a replicate filter is shown in the column termed "First Screen" under "Dorsal Specific". The filters were then washed at melting temperatures to remove the probe and rehybridzed with the alternative probe, i.e., the filter originally hybridized to dorsal cDNA was nexthybridized to ventral cDNA. The number of original colonies which proved to be positive by this assay are shown in the column designated "Recheck". Corresponding data are shown for the "Ventral Specific".

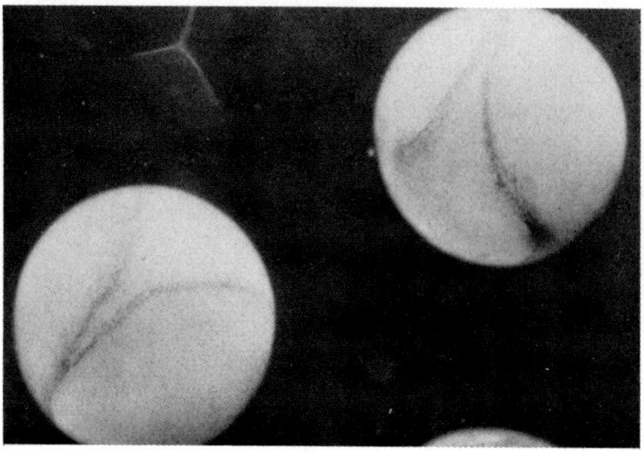

FIGURE 7. Zygotes were subjected to centrifugation prior to first cleavage as described in the text. The resultant anterior bifurcation of the neural plate may be seen in the figure. These embryos go on to form Siamese twins with a twinned axis anterior to the mid-body level.

DISCUSSION

We were surprised to find the marked local changes in poly (A) concentration in very early Xenopus development. Since there is no transcription at these stages, we favor the idea that there is local control of the extent of adenylation of RNA molecules synthesized during oogenesis. In very early stages there may also be movement of maternal RNA. It is not clear whether such changes are causes, or results, or both, of the observed events of early development. These results are concordant with the earlier work of Phillips (5). In these experiments eggs and embryos were sectioned into 6 regions along either the dorsal-ventral or animal-vegetal axis, and poly (A) concentrations measured by conventional solution hybridization.

Though much more detail is revealed in the present work using in situ hybridization, the results of the two methods are completely consistent. In earlier work of Capco and Jeffery (8) localization of poly (A) was not observed in early embryos. It is possible that differences in fixation and other procedures might account for the difference in results. We believe the fact that differences in poly (A) are observed both with solution hybridization and the hybridization of tissue sections supports the idea of local variations in poly (A) concentrations. Sagata et. al (9) examined poly (A) concentrations in different blastomeres of the 8 cell embryo, and the animal, vegetal, dorsal and ventral regions of the blastula. The methods do not have the resolution of the autoradiographic procedures, but insofar as the results may be compared with the present work, the results are probably consistent. The possibility that there are specific sequences that are localized, and the presence of local changes in extent of polyadenylation of maternal mRNA molecules, lead us to believe that Xenopus affords a good opportunity to study the relationship between specific RNA molecules and cell type determination.

ACKNOWLDEGEMENTS

We appreciate the techincal assistance of Ms. Dionne Darracq and Ms. Britta Kreps.

REFERENCES

1. Wilson EB (1925) The Cell in Development and Heredity New York: The Macmillan Company.
2. Sagata N, Shiokawa K, and Yamana K (1980) A study on the steady-state population of poly (A)+ RNA during development Xenopus laevis Dev Biol 77:431.
3. Wilt FH (1977) The dynamics of maternal poly (A) containing mRNA in fertilized sea urchin eggs. Cell 11:673.
4. Angerer LM, and Angerer RC (1981) Detection of poly A+ RNA in sea urchin eggs and embryos by quantitative in situ hybridization. Nuc Acid Res 9:2819.
5. Phillips CR (1982) The regional distribution of poly (A) and total RNA concentrations during early Xenopus development. J Exp Zool 223:265.

6. Newport J and Kirschner M (1982) A major developmental transition in early Xenopus embryos: II control of the onset of transcription. Cell 30:687. Bacharova R & Davidson EH 1966 Nuclear activation at the onset of amphibian gastrulation. J Exp Zool 163:285.
7. Carpenter CD and Klein WH (1982) A gradient of poly (A)+ RNA sequences in Xenopus laevis eggs and embryos. Develop Biol 91:43.
8. Capco, DG and Jeffery WR (1982) Transient localizations of messenger RNA in Xenopus laevis oocytes. Dev. Biol 89:1.
9. Sagata N, Okuyama K, and Yamana K (1981) Localization and segregation of maternal RNAs during early cleavage of Xenopus laevis embryos. Develop Growth, and Differ 23:23.

Molecular Biology of Development, pages 37–49

LOCALIZATION AND DETERMINATION IN EARLY EMBRYOS OF *CAENORHABDITIS ELEGANS*[1]

William B. Wood, Einhard Schierenberg, and Susan Strome

Department of Molecular, Cellular and Develomental Biology, University of Colorado, Boulder, CO 80309

ABSTRACT Developmental fates of blastomeres in early *C. elegans* embryos appear to be governed by internally segregating, cell-autonomous determinants. To ascertain whether previously described gut-lineage determinants are nuclear or cytoplasmic, laser microsurgery was used to show that exposing the nucleus of a non-gut-precursor cell to gut-precursor cytoplasm can cause the progeny of the resulting hybrid cell to express gut-specific differentiation markers, supporting the view that the determinants are cytoplasmic.

In attempts to obtain molecular probes for such determinants, a library of monoclonal antibodies to early embryonic antigens was generated and screened by immunofluorescence microscopy for antibodies reacting with lineage-specific components. Three of the antibodies react with cytoplasmic granules (P granules) that segregate specifically with the germ line in early cleavages and are found uniquely in germ-line cells throughout the life cycle. Experiments on unfertilized eggs, on mutant embryos with defects in early cleavage, and on normal embryos treated with various cytoskeletal inhibitors indicate that P-granule segregation depends upon

[1]Supported by grants from the National Institutes of Health (HD-11762, HD-14958) and the American Cancer Society (CD-96). S.S. also received postdoctoral fellowship support from the Anna Fuller Foundation and the National Institutes of Health, and E.S. from the Deutsche Forschungsgemeinschaft.

fertilization and requires the function of actin microfilaments, but is independent of spindle and microtubule functions. Work on the biochemical nature and function of the P granules is in progress.

INTRODUCTION

A fundamental unsolved problem of animal development is how the fates of individual cells are determined, at the appropriate time and the appropriate position in the early embryo. We have tried to approach this problem using the nematode Caenorhabditis elegans as a model. C. elegans has several advantages for this purpose, in

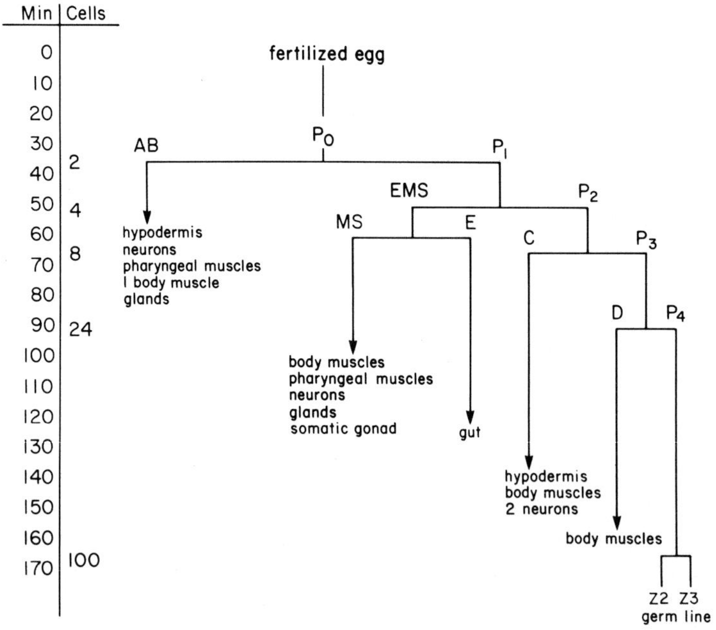

FIGURE 1. Cell names, lineage relationships, division times, and progeny cell fates in C. elegans early cleavage divisions. Fates are based on data from (2). Cleavages shown all occur approximately in an anterior-posterior direction; in this and all subsequent figures, anterior is to the left and posterior to the right. Modified from (3).

particular its short life cycle, relative simplicity, and suitability for light microscopy and genetic analysis (1). The lineal relationships of all the cells that arise during its development have been completely determined by observation of living animals with the light (Nomarski) microscope (2).

The early embryonic lineage is shown in Figure 1. Micrographs of an early embryo at several stages between fertilization and the two-cell stage are shown in Figure 2. In the early embryo, the pattern and timing of

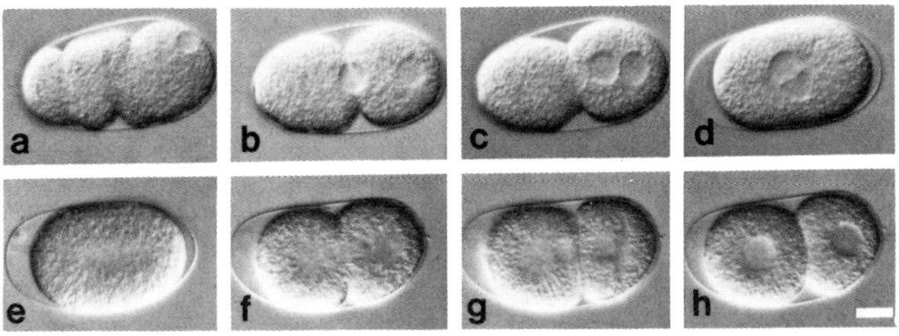

FIGURE 2. Nomarski photomicrographs of living embryos at several stages between fertilization and first cleavage. (a) Contractions of the anterior membrane (approx. 27 min after fertilization at 25°C). The egg pronucleus at the anterior pole is not visible in this focal plane. (b) Migration of the egg pronucleus through the pseudocleavage constriction. (c) Meeting of the two pronuclei. (d) Movement to the center and rotation of the pronuclei. (e) Pronuclear fusion and formation of the first mitotic spindle. (f) Anaphase and beginning of cleavage. Asters are visible as round, granule-free regions; their positions show the asymmetric location of the spindle along the anterior-posterior axis. (g) Telophase, showing smaller disk-shaped aster in the posterior (P1) cell. (h) Two-cell embryo, showing larger anterior AB cell and smaller P1 cell. Bar: 10um. Reproduced from (4) with permission. © M.I.T. Press.

cleavages and the developmental fates of individual blastomeres are invariant, (5), so that by observation of normal embryos alone it is impossible to ascertain whether fates are determined according to position or lineage. However, several lines of evidence support the view that many cell fates throughout _C. elegans_ development, and perhaps all fates in the early embryo are determined lineally, that is, according to internal determinants acquired by a cell from its ancestors (summarized in ref. 2).

In an earlier study from this laboratory, Laufer et al. (6) used cleavage-blocked embryos to show that a determinant for subsequent expression of a convenient differentiation marker, autofluorescent granules found only in gut cells, is segregated in early cleavages first to the P1 cell, then to the EMS cells, and then to the E cell and its progeny, which give rise to the gut (see Figure 1). Expression of the marker is observed only in cells on the gut line of descent, and is not dependent on the normal embryonic environment at the time when the marker normally appears (after about 4 hr of development), indicating that determination occurs in response to internally segregating, cell-autonomous factors. In this paper we present results from two experimental approaches to elucidating the nature and function of these determinants.

RESULTS

Microsurgical Experiments on the Location of Gut Cell Determinants.

Conventional wisdom would suggest that developmental determinants in the early embryo might be cytoplasmic, in view of the many reported cases of cytoplasmic localization and a few clear cut cases of determination by cytoplasmic factors in both somatic (7) and germ-line cells (8, 9). Alternatively, however, determinants could be chromosomal factors (10) which are somehow asymmetrically segregated in early cleavages. To distinguish between these alternative locations for gut-cell determinants, we took advantage of techniques developed previously for microsurgery on *C. elegans* embryos using a dye laser. Irradiation of dyed eggs at an appropriate wavelength punctures the egg shell, allowing extrusion of embryonic contents (11). Shorter wavelength irradiation of the cell membranes separating two adjacent blastomeres can result in destruction of the membranes and fusion of the two cells (12).

In the experiment diagrammed in Figure 3, the shell of a 2-cell embryo is punctured at the posterior end with

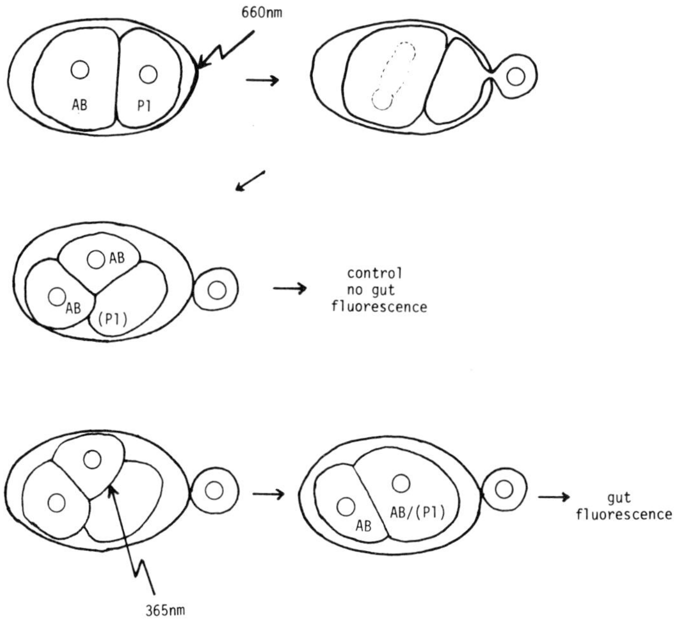

FIGURE 3. Diagram of laser microsurgery experiments. For explanation, see text.

the laser, and the P1 nucleus is extruded by pressure on the microscope coverslip; the cell membrane then often fuses at the puncture, separating the nucleated P1 fragment (which degenerates after a few cleavages) from the embryo, and leaving an intact, larger, enucleated P1 fragment (cytoplast) inside the shell. The P1 cytoplast divides no further, but the AB cells continue division for several hours. Under these conditions, gut autofluorescence does not develop (Fig. 4, panels a-c), indicating that the P1 cytoplasm alone does not have the capacity to express the differentiation marker. However, if after one or two AB cleavages one of the AB descendants is fused to the P1 cytoplast by means of a second laser shot, then localized autofluorescence typical of the developing gut does appear in over half the experimental embryos (Fig. 4, panels d-f). This result shows that exposure of a non-gut-precursor nucleus to gut-precursor cytoplasm allows expression of the gut-specific marker, arguing strongly that at least some gut-cell determinants are cytoplasmic.

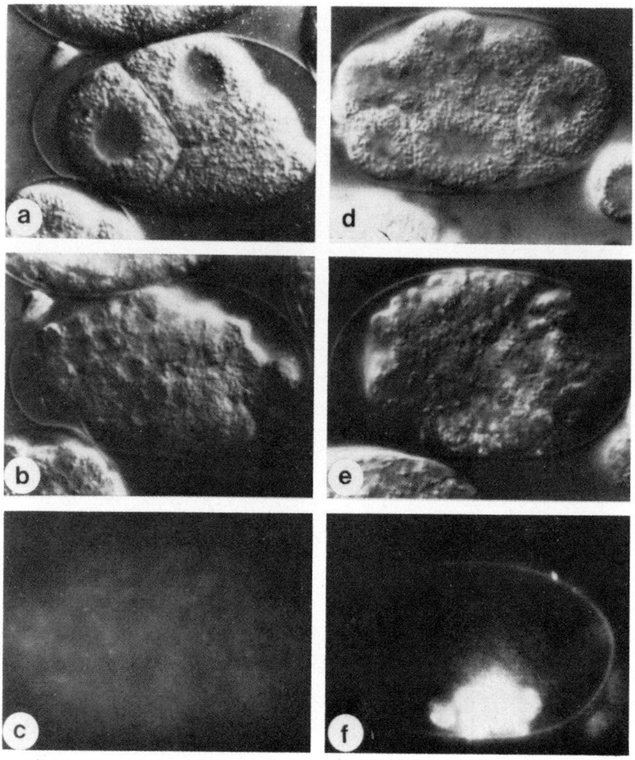

FIGURE 4. Results of laser microsurgery experiments. (a) Embryo after extrusion of the P1 nucleus at the 2-cell stage and subsequent division of the AB cell, as diagrammed in Figure 3. (b) Several hours later, the AB cell has given rise to many progeny cells, while the P1 cytoplast is still visible in the lower right region of the embryo. (c) An embryo similar to that shown in panel (b), photographed with epi-illumination to show lack of localized autofluorescence. (d) Embryo after extrusion of the P1 nucleus at the 2-cell stage and subsequent fusion of the P1 cytoplast with one of the four AB granddaughter cells. The resulting hybrid cell has divided once to give the two large cells in the lower right region of the embryo. (e) The same embryo as in panel (d), photographed several hours later. (f) The embryo of panel (e), photographed with epi-illumination to show the localized autofluorescence typical of the developing gut.

Study of Localization Using Antibodies.

In the hope of obtaining molecular probes for such determinants, we have generated a library of monoclonal antibodies (MAb's) directed against antigens in the early embryo, and screened them by immunofluorescence microscopy for recognition of components that are segregated or synthesized in a lineage-specific manner (4). Screening of about 1200 hybridomas produced from spleen cells of mice injected with whole embryonic homogenates (S. Strome, unpublished results) has yielded three clones producing MAb's with the desired properties. All three MAb's appear to recognize the same component: cytoplasmic granules, also stained by certain rabbit sera (3, 4), that segregate asymmetrically in early cleavages and are found specifically in germ-line cells throughout the life cycle. Because the granules are restricted to the P lineage in the early embryo (see Fig. 1), we have called them P granules. Two apparently similar MAb's were obtained in an independent screen (13).

At fertilization the P granules are uniformly distributed throughout the egg cytoplasm (Fig. 5, a panels). By the time the two pronuclei meet, the granules appear to have moved to the posterior periphery of the egg (Fig. 5, b panels). After first cleavage, all the granules are found in the P1 cell (Fig. 5, c panels). At prophase of P1 mitosis, the granules are already prelocalized so that they will be segregated to P2 when P1 divides. In a similar manner, they are segregated to P3, P4, and then to Z2 and Z3 (see Figure 1), which divide no further until gonadogenesis begins in the first stage larva, when they proliferate to give rise to the entire germ line.

P granules appear to be present in all germ-line cells throughout the life cycle, with the possible exception of mature sperm. Sperm are not stained by our MAb's, but are stained by an independently isolated MAb (13) and some rabbit antisera (3) that react with P granules. The granules begin to coalesce around the nuclear envelope in P3, and remain perinuclear for the rest of the life cycle. As oocytes mature, the granules dissociate from the nucleus and become dispersed throughout the oocyte cytoplasm (3).

In electron micrographs of early embryos, granular cytoplasmic inclusions are seen whose sizes and

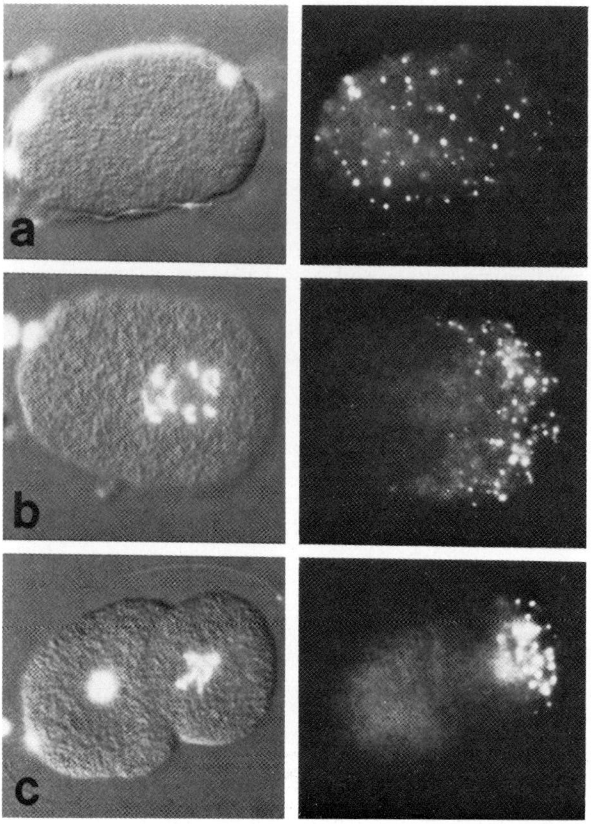

FIGURE 5. Localization of P granules after fertilization. Embryos were cut out of hermaphrodites, fixed in cold methanol and acetone, and stained with DAPI, MAb, and fluorescein-labeled secondary antibody as described in (4). Left panels show Nomarski images with DAPI-stained chromosomes; right panels show immunofluorescence images. (a) Fertilized egg after completion of meiosis (see DAPI-stained polar bodies next to egg pronucleus at the anterior pole) and prior to pronuclear migration. P granules (right panel) are dispersed throughout the cytoplasm. (b) Zygote at time of pronuclear meeting, showing localization of P granules around posterior periphery. (c) Two-cell embryo in which the P1 cell is in prophase, and P granules are prelocalized in the region of cytoplasm destined for the next P-cell daughter, P2. Reproduced from (4) with permission. © M.I.T. Press.

distributions at different stages closely resemble those of the P granules seen by immunofluorescence (14, 15). These inclusions appear not to be bounded by membranes, and are similar in appearance to bodies previously described in the germ-line cytoplasm of several other species and referred to variously as germinal plasm, pole plasm, polar granules, and nuage (reviewed in ref. 16).

The convenient immunofluorescence assay for following P granules was used in experiments to investigate the mechanism of their asymmetric segregation in early cleavages (4). Several directed movements occur in the egg during the period between fertilization and first cleavage (see Figure 2). In order of occurrence, these include extrusion of the polar bodies at the anterior pole as the oocyte nucleus completes meiosis, contractions of the anterior membrane and formation of a pseudocleavage furrow, migration of the maternal pronucleus from the anterior pole to a point about two-thirds of the distance from anterior to posterior, apparent migration of the P granules to the posterior region, 90-degree rotation of the pronuclei after their meeting, formation of an asymmetric first spindle in which the posterior aster is disk-shaped in contrast to the spherical anterior aster, and then first cleavage, giving rise to the larger AB cell and smaller P1 cell.

Experiments with spermless or fertilization-defective mutants show that fertilization is required for asymmetric P-granule segregation. However, this segregation is independent of spindle formation and movement, as shown by immunofluorescence staining of both components and by experiments with a mutant in which the first-cleavage spindle remains perpendicular to its normal orientation, but the P granules nevertheless migrate normally to the posterior. Moreover, P-granule migration appears to be independent of any microtubule function. In embryos treated with any of three microtubule inhibitors at concentrations sufficient to block mitosis and up to five times higher, pronuclear migration and spindle formation are blocked, but other early events including P-granule migration to the posterior occur essentially as usual. In embryos treated with cytochalasins B or D, however, the P granules fail to become asymmetrically localized and instead coalesce near the center. The pronuclei migrate, but meet near the center of the embryo rather than toward the posterior, and several of the other usual manifestations of asymmetry are lacking, such as extrusion of the second polar body, membrane contractions, and the asymmetry of the first-cleavage spindle.

These results are consistent with the view that fertilization sets in motion two fairly independent kinds of cytoskeletal machinery: one, microtubule-mediated, which is responsible for movement of the pronuclei and chromosomes, and the other, microfilament-mediated, which is responsible for generating several asymmetries in the early zygote, including movement of the P granules. Drug experiments can be misleading; in these experiments, however, one set of movements is blocked while the other proceeds, indicating that the drug is acting, but that its effect is not caused simply by nonspecific damage to the embryo.

Although the segregation behavior of the P granules is as expected for a lineage-specific determinant, there is so far no evidence that the granules are determinative for the germ line. As an approach to elucidating P-granule functions, a screen for ethyl-methane-sulfonate-induced mutations that affect the presence, morphology, or behavior of P granules is being carried out in a manner that will allow recovery of lethal or sterile mutants. So far one such mutant has been isolated as a homozygous-viable, fertile animal whose germ-line cells fail to stain with one of the three MAb's (K76), but do stain with the other two. The molecular basis for this result is still unclear. However, the mutant has been useful for determining how long maternal P-granule components (specifically the K76 cognate antigen) persist in embryos, as well as the time at which newly synthesized K76 cognate antigen can first be detected by immunofluorescence during development. In homozygous mutant embryos produced by hermaphrodites heterozygous for the mutation, the maternal K76 cognate antigen persists throughout embryogenesis. Conversely, when homozygous mutant hermaphrodites are mated to wild-type males, staining of P granules in the resulting outcross progeny is first seen in first-stage larvae shortly after somatic gonad development begins and the germ-line cells Z2 and Z3 begin to proliferate. Attempts are in progress to identify components of wild-type and mutant P granules by immunoprecipitation with MAb's. Details of these experiments will be published elsewhere.

DISCUSSION

The nature of developmental determinants in early animal embryos remains mysterious. There is evidence for the cytoplasmic or cortical location of both somatic (7; these experiments) and germ-line (8, 9) determinants of specific fates, as well as more global determinants of embryonic polarity (e.g., 17, 18). But although these determinants appear to be properties of specific regions of early embryonic cytoplasm, they have not been identified with any specific cytoplasmic components. There is no clear experimental basis for the popular notion that lineage-specific regulators of gene expression are present in oocyte cytoplasm as maternally synthesized gene products that become appropriately localized and segregated to different lineages during the early cleavages. In fact, there is suggestive negative evidence against this view. First, screening of substantial numbers of maternal-effect embryonic lethal mutants in both Drosophila (19, 20) and Caenorhabditis (21, 22) has so far failed to identify mutations that specifically eliminate a particular lineage or class of differentiated cell, with the possible exception of the grandchildless (23) and tudor (24) mutations in Drosophila. These appear specifically to eliminate the germ line (which could represent a special case), but they show other effects as well, suggesting that their effects could result from defects in ooplasmic organization rather than absence of specific determinants. Second, our screen for lineage-specific embryonic antigens failed to identify any components that segregate uniquely to somatic lineages. While this screen was not exhaustive, and was not designed to detect low level antigens, our results again are consistent with the possibility that the determinative properties of early embryonic cytoplasm may depend not on the presence or absence of specific regulatory macromolecules, but rather on more complex attributes such as potential for cytoskeletal organization or metabolic activity, which could be difficult to identify by either mutational or serological analyses. Monoclonal antibodies against P granules provide new approaches to studying one striking example of localization, and will allow further characterization of components that show at least the behavioral properties expected of cytoplasmic determinants.

REFERENCES

1. Brenner S (1974). The genetics of Caenorhabditis elegans. Genetics 77:71.
2. Sulston JE, Schierenberg E, White JG, Thomson JM (1983). The embryonic cell lineage of the nematode Caenorhabditis elegans. Devel Biol 100:64.
3. Strome S, Wood WB (1982). Immunofluorescence visualization of germ-line-specific cytoplasmic granules in embryos, larvae, and adults of Caenorhabditis elegans. Proc Natl Acad Sci USA 79:1558.
4. Strome S, Wood WB (1983). Generation of asymmetry and segregation of germ-line granules in early C. elegans embryos. Cell 35:15.
5. Deppe U, Schierenberg E, Cole T, Crieg C, Schmitt D, Yoder B, vonEhrenstein G (1978). Cell lineages of the embryo of the nematode Caenorhabditis elegans. Proc Natl Acad Sci USA 75:376.
6. Laufer JS, Bazzicalupo P, Wood WB (1980). Segregation of developmental potential in early embryos of Caenorhabditis elegans. Cell 19:569.
7. Whittaker JR (1982). Muscle lineage cytoplasm can change the developmental expression in epidermal lineage cells of ascidian embryos. Devel Biol 93:463.
8. Illmensee K, Mahowald AP (1974). Transplantation of posterior polar plasm in Drosophila. Induction of germ cells at the anterior pole of the egg. Proc Natl Acad Sci USA 71:1016.
9. Wakahara M (1978). Induction of supernumerary primordial germ cells by injecting vegetal pole cytoplasm into Xenopus eggs. J Exp Zool 203:159.
10. Cairns J (1975). Mutation selection and the natural history of cancer. Nature 255:197.
11. Laufer JS, vonEhrenstein G (1981). Nematode development after removal of egg cytoplasm: Absence of localized unbound determinants. Science 211:402.
12. Schierenberg E (1984). Altered cell-division rates after laser-induced cell fusion in nematode embryos. Devel Biol 101:240.
13. Yamaguchi Y, Murakami K, Furusawa M, Miwa J (1983). Germ line-specific antigens identified by monoclonal antibodies in the nematode Caenorhabditis elegans. Develop, Growth and Differ 25:121.

14. Krieg C, Cole T, Deppe U, Schierenberg E, Schmitt D, Yoder B, vonEhrenstein G (1978). The cellular anatomy of embryos of the nematode Caenorhabditis elegans. Devel Biol 65:193.
15. Wolf N, Priess J, Hirsh D (1983). Segregation of germline granules in early embryos of Caenorhabditis elegans: an electron microscopic analysis. J Embryol Exp Morphol 73:297.
16. Eddy EM (1975). Germ plasm and the differentiation of the germ cell line. Int Rev Cytol 43:229.
17. Kalthoff K (1983). Cytoplasmic determinants in Dipteran eggs. In Jeffery WR, Raff RA (eds): "Time, Space and Pattern in Embryonic Development," New York: Alan R Liss, p 313.
18. Scharf SR, Gerhart JC (1983). Axis determination in eggs of Xenopus laevis: A critical period before first cleavage, identified by the common effects of cold, pressure and ultraviolet irradiation. Devel Biol 99:75.
19. Gans M, Audit C, Masson M (1975). Isolation and characterization of sex-linked female-sterile mutants in Drosophila melanogaster. Genetics 81:683.
20. Zalokar M, Audit C, Erk I (1975). Developmental defects of female-sterile mutants of Drosophila melanogaster. Devel Biol 47:419.
21. Wood WB, Hecht R, Carr S, Vanderslice R, Wolf N, Hirsh D (1980). Parental effects and phenotypic characterization of mutations that affect early development in Caenorhabditis elegans. Devel Biol 74:446.
22. Schierenberg E, Miwa J, vonEhrenstein G (1980). Cell lineages and developmental defects of temperature-sensitive embryonic arrest mutants in C. elegans. Devel Biol 76:141.
23. Mahowald AP, Caulton JH, Gehring WJ (1979). Ultrastructural studies of oocytes and embryos derived from female flies carrying the grandchildless mutation in Drosophila subobscura. Devel Biol 69:118.
24. Boswell RE, Mahowald AP (1983). Cytoplasmic determinants in embryogenesis. In Kerkut GA, Gilbert LI (eds): "Comprehensive Insect Physiology, Biochemistry, and Pharmacology," New York, Pergamon Press, Vol 1 (in press).

Molecular Biology of Development, pages 51–73

AXIS DETERMINATION IN THE XENOPUS EGG[1]

Stanley R. Scharf,[2] Jean-Paul Vincent, and John C. Gerhart

Department of Molecular Biology, University of California
Berkeley, California 94720

ABSTRACT Between fertilization and first cleavage the radially symmetric Xenopus egg undergoes a global polarization which predicts the location of the future dorsal structures of the body axis. We have found that impairment of one aspect of this process results in embryos which are deficient in axial structures, and perturbation of another aspect results in embryos which show supernumerary axes. In both cases the treatments used would be expected to affect the egg cytoskeleton. We have also made observations on the reorganization of the egg cortex and endoplasm in normal and treated eggs. On the basis of these data we propose a model for a coordinated cortico-endoplasmic rearrangement by which potential axis determining agents might be localized in the amphibian egg prior to first cleavage.

INTRODUCTION

One proposed mechanism for regional differentiation in early development is the localization of specific cytoplasmic materials [1,2]. This may occur in the frog egg where an early cytoplasmic reorganization process is thought to confer what we shall call dorsal potential on a particular egg region [3,4]. In later development cellularized regions with dorsal potential can carry out, and organize neighboring regions to carry out, the complex series of morphogenetic movements that result in the formation of the embryonic body

[1] This work was supported by NIH grant GM 19363 to JCG.
[2] Present address: Botany Department, University of California, Berkeley, CA. 94720. To whom reprint requests should be addressed.

axis [5,6]. This early reorganization can be considered a process of axis determination, since, as we shall see, failure of the egg to create a unique differentiated region in its very early development results in the complete failure of the embryo to develop axial structures.

Before fertilization the Xenopus egg has only a single axis, the animal-vegetal (A-V) axis which is set up in oogenesis, probably on the basis of the nucleus-centriole-Golgi arrangement [7]. This A-V axis is defined by the line connecting the poles of the darkly pigmented, less yolky animal hemisphere and the lighter, more yolk filled, vegetal hemisphere. At fertilization, the entry of a single sperm at a random point in the animal hemisphere breaks the radial symmetry and defines a plane of bilateral symmetry (Fig. 1).

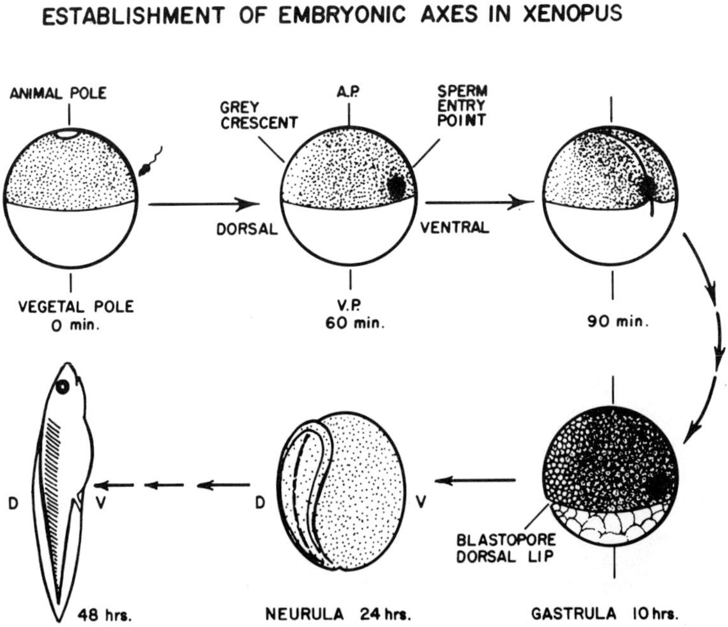

FIGURE 1. General schematic of Xenopus development. Note that the first cleavage plane passes through the point of sperm entry and that dorsal axial structures are formed from material initially opposite the sperm entry point.

In the 90 min that follow (at 20°C), prior to first cleavage, the egg becomes polarized along this plane such that the future dorsal structures of the embryo arise mainly from material on the side of the egg opposite the point of sperm entry (SEP), and ventral structures from the SEP side.

It has been known for many years that egg polarization occurs by a process of cytoplasmic reorganization in which the internal contents of the egg are shifted relative to the surface [4,8]. In some species (e.g., Rana pipiens [8]), although not in Xenopus, this cortico-endoplasmic shift is quite obvious upon inspection. It has long been referred to as "grey crescent formation" since, as it progresses, a crescent-shaped region of intermediate pigment density forms on the future dorsal side in the sub-equatorial region [9]. On the basis of its topographic relatedness to the point of origin of future dorsal structures, the grey crescent region has long been considered to be a locus of dorsal or axial determinants [4,9,10]. What has been lacking however, is a direct causal link between this morphological entity and the later formation of the embryonic body axis. Although some years ago Curtis [11] claimed that secondary dorsal structures were induced in embryos from eggs receiving a transplanted grey crescent cortex, the validity of these experiments has been cast is some doubt [4].

Aside from this early redistribution of pigment, no further overt manifestation of dorsal-ventral polarity is seen until the formation of the dorsal lip of the blastopore (Fig. 1) some 10 hours later (at 20°C). This initial differentiation of specialized cell types marks the onset of a series of morphogenetic movements which culminates in axis formation. These movements are led by the dorsal mesodermal cells, which migrate up the blastocoel roof, followed only later by lateral and ventral mesodermal cells. The most dorsal aspect of the mesodermal mantle thus formed then interacts with the overlying ectoderm to induce neural structures.

We have investigated the early process of egg reorganization and attempted to relate it causally to the later events of axis formation. We have been able to generate two types of axis abnormality which we suggest are caused by impairment or enhancement of the initial egg polarization process. Our results indicate that egg polarization consists of two separable processes, one during an earlier orientation phase and a second during a later rearrangement phase.

RESULTS

Axis Deficient Embryos

Many years ago Spemann [12] examined the developmental capacity of isolated single blastomeres obtained from newt embryos at the 2-cell stage. Most of these halves would make complete larvae, but some (20%) made only "belly pieces", that is, ventralized embryos lacking dorsal axial structures and differentiating only mesenchyme, blood islands, gut tissue and a ciliated epidermis. These embryos were proposed to represent the developmental fate of those half-eggs which lacked the axial determinants of the grey crescent region. Such halves would occur in that small percentage of cases in which the first furrow formed frontally. These results were among the first to suggest that dorsal developmental potential had become localized as early as the first cell cycle.

Embryos deficient in dorsal axial structures were generated in a quite different way by Grant and Wacaster [10] and Malacinski and colleagues [13]. They found that eggs exposed on the vegetal hemisphere to 250-280 nm light (UV-irradiation) prior to first cleavage gave embryos with reduced axial structures, ranging from slightly microcephalic to totally "aneural". These latter most extreme cases superficially resembled the "belly pieces" of Spemann. The more intermediate grades resembled the incomplete secondary axes often seen following organizer grafts [12]. It was initially suggested that this effect was due to the destruction of a component of the neural induction system, i.e., a specific determinant, initially present in the egg and later directly utilized in the induction of neural structures [13]. The action spectrum of the UV effect suggested a protein was the primary target [14].

The UV effect was cast in a different light by our work showing that embryos from irradiated eggs could be completely rescued to normal tadpoles by being obliquely oriented, that is, held with the animal-vegetal axis perpendicular to the gravitational vector [15]. Normally the A-V axis is parallel to the gravity vector, since the egg is free to rotate in its fertilization envelope. Such a procedure therefore serves to redistribute the dense vegetal yolk as it falls in response to gravity. We took this approach because of earlier work which showed that oblique orientation could define the position of the future embryonic axis [4,8]. These rescue results suggest that UV acts to impair a reorganization process of the egg critical for later axis formation, rather than destroy a determinant utilized only some time later by the embryo. This assertion was supported by our subsequent work showing that an identical syndrome is caused by brief treatment of the pre-cleavage egg with low temperature or high hydrostatic pressure [3]. These treatments are known to depolymerize microtubules and, in the case of pressure, microfilaments as well [16,17,18]. It is easy to imagine that the function of these cytoskeletal elements might be crucial to any cytoplasmic reorganization, given their roles in maintaining cell structure and in cell motility [19]. Eggs subjected to cold or pressure could also be rescued by oblique orientation [3], again reinforcing the notion that the damage done was only to an immediate process, not a lasting product.

The morphology of the embryos produced by UV, cold or pressure, as pictured in Fig. 2, suggested they be called axis deficient. The stereotyped series common to all three treatments can be quantitatively graded on the basis of the reduction and loss of dorsal axial structures (see legend, Fig. 2 and [3]). As dose is increased, bilaterally placed structures such as eyes, ear vesicles and somites first fuse medially, then become simplified, and finally disappear. This reduction occurs in an obvious anterior-posterior progression which is also seen in ventral structures, such as heart and gut, although these are not used in scoring.

The morphology of these embryos can be understood by assuming that the normally dorsal mesoderm behaves in a progressively more ventral manner. During gastrulation in normal embryos ventral mesoderm does not migrate as far anteriorly along the blastocoel roof as does dorsal mesoderm [6]. In axis deficient embryos mesodermal migration is reduced [13]. A lack of more anterior mesoderm at the time of neural induction would result in the absence of anterior neural structures. This supposition is supported by examination of embryos from irradiated eggs at the time of blastopore lip formation. As dose is increased, the appearance of the dorsal lip is progressively delayed. In the case of radially ventral embryos (Fig. 2G), the blastopore appears as a circle at the time of control ventral lip formation.

The effects of UV-irradiation, cold and pressure are limited to the pre-first cleavage egg, as shown in Fig. 3. All treatments are effective until approximatly 0.75 of the time to first cleavage, on a normalized scale in which fertilization equals 0.0 and first cleavage equals 1.0. The width of the cold and pressure sensitive period is somewhat dose dependent, but generally begins about 0.45. UV, on the

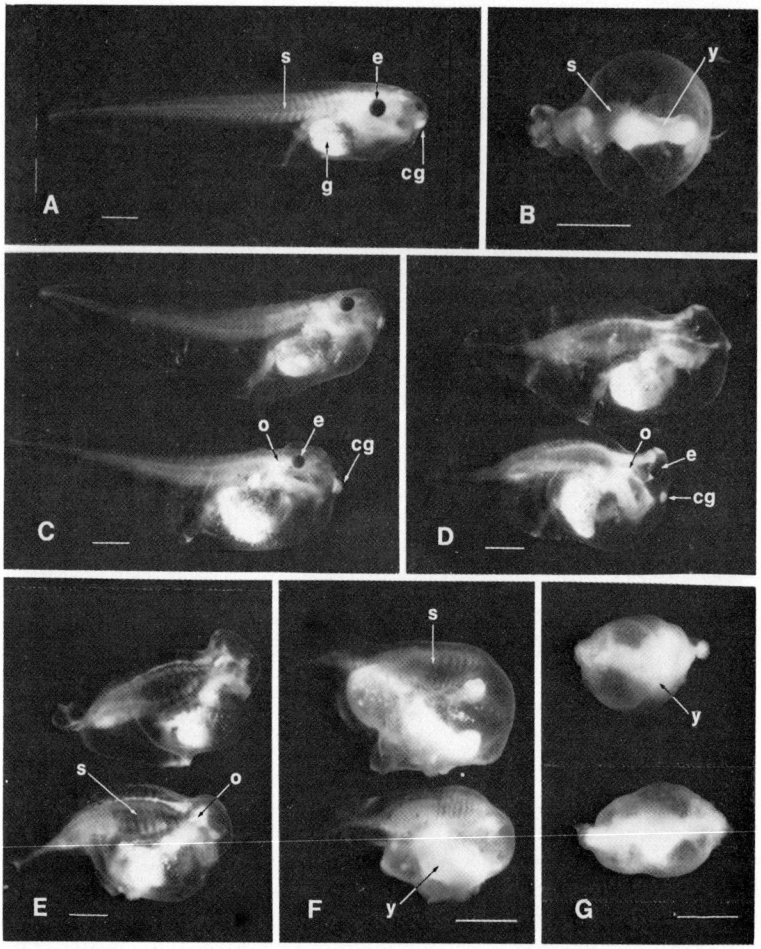

FIGURE 2. Axis deficient embryos. A: normal tadpole; B: Index of axis deficiency (IAD) grade 4 embryo produced by pressure treatment (8000 psi, 5 min); C through G: IAD grade 1 through grade 5 embryos, respectively, produced by cold (lower embryo) and UV-irradiation (upper embryo). UV dose was 1.3-2.7 x 10^4 ergs/mm^2. Cold treatment was 0.5° for 4 min. Grade 5 embryos (G) are oriented with the old blastopore leftward. (cg) cement gland; (e) eye(s); (g) gut; (o) otic vesicle(s); (s) somites; (y) yolk mass. Bar length: 1mm.

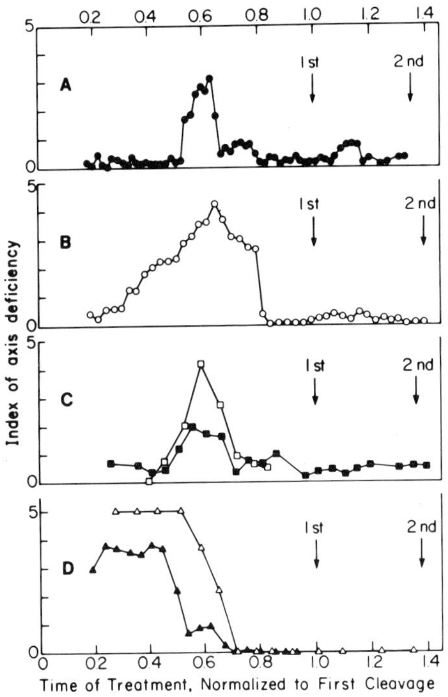

FIGURE 3. Stage dependence of cold, pressure, and UV effects on axis determination. Eggs were treated at the times shown with A: cold, low dose (1.0°C, 4 min); B: cold, high dose (1.0°C, 4 min followed by slow warming); C: pressure, low dose (solid squares, 8000 psi, 4 min); high dose (open squares, 9000 psi, 5 min); D: UV irradiation, low dose (solid triangles, 1.3 x 10^4 ergs/mm); high dose (open triangles, 2.7 x 10^4 ergs/mm^2). The index of axis deficiency is defined in Fig. 2 and in [3], where details of treatments may be found.

other hand is effective at all times prior to 0.75. Other experiments [3], in which eggs are treated with UV, cold or pressure at various times from second cleavage to gastrulation demonstrate that axis deficiency only results following treatment before 0.75. These data can be explained by a model in which an early egg reorganization that is critical to axis determination requires cold, pressure and UV-sensitive elements, and takes place from 0.45-0.75. Damage to these elements would be rapidly reversible in the case of cold and pressure, but irreversible or only slowly reversible in the case of UV-irradiation. This would provide for a difference in the onset of a sensitive period, but a similar end point.

With this explanation in mind it is easy to understand how the simple procedure of off-axis orientation could rescue impaired eggs. It would effect a gravitationally-induced egg reorganization mimicking the impaired normal reorganization.

This is clearly a process which has a simple quantitative aspect, thus providing for continuous variation in the extent of the eventual body axis. We present data below directly correlating the degree of reorganization with the extent of the body axis.

Axis Enhanced Embryos

The organizer grafts of Spemann and Mangold [12] demonstrated that the total possible committment of embryonic material to dorsal axial structures is not limited to a single set of such structures. Rather cells of the early gastrula normally fated to form lateral or ventral structures could be redirected to the dorsal state upon receiving proper "instruction", in this case from an implanted organizer (see also [5]). We now ask at what earlier stages of development it is possible to re-program the embryo so that the production of axial structures is enhanced, that is supernumerary dorsal axial structures are formed.

Respecification can be accomplished at the 32-64 cell stage, as shown by Gimlich and Gerhart [20], by transplanting vegetal blastomeres from the future dorsal side of a donor embryo into the ventrally fated side of a recipient from which the corresponding blastomeres have been removed. Secondary axes of varying degrees of completeness result. This same implantation procedure serves to rescue embryos from UV-irradiated eggs [20]. These experiments differ from the organizer graft in that the vegetal blastomeres will not contribute material to dorsal axial structures, but only to the future gut. Thus these vegetal cells apparently act to induce the formation of an organizer by some intercellular interaction.

Embryos with duplicated dorsal structures have also been produced by perturbing the uncleaved egg. In general, the techniques have involved artificially redistributing the egg's contents by gravity [8] or double centrifugation [4], the same procedures that are effective in rescuing cold, pressure and UV-treated eggs. These procedures are presumably effective because they serve to create a second region of axis determining agents, either by physically redistributing material to another region of the egg or by locally activating factors initially uniformly distributed.

To complement these studies we wanted to generate embryos with exaggerated axis forming capacity in a manner more easily interpretable in terms of a target molecule or specific cellular structure. Our previous cold results [3] had suggested that depolymerization of egg microtubules during a certain critical stage causes axis deficiency. We thus examined the effects of heavy water (D_2O), an agent which acts to stabilize microtubules and to enhance polymerization of tubulin monomer [21]. It was initially found that D_2O cotreatment protects eggs from cold and UV-induced axis deficiency [3], a result which reinforced the suggestion that microtubules were important in the egg reorganization process. Control experiments using D_2O alone, however, proved especially interesting. Treatment of eggs for brief intervals during a well defined pre-cleavage period caused the formation of secondary and even occasional tertiary axes in embryos. In some cases embryos with near radial symmetry were produced. Examples of these novel embryos are shown in Fig. 4.

It is apparent from the morphology of these embryos that this syndrome has associated with it two distinct aspects. One is axial duplication: minimally, the most anterior structures are duplicated (e.g. Fig. 4A-C), and in more severe cases, the point of bifurcation is progressively more caudal (not shown). Usually one and sometimes both axes show some anterior axis deficiency. The second aspect of the syndrome is posterior reduction: In the "Janus" form (Fig. 4E), for example, axial structures rearward of the brain are not seen, but anterior head structures are fully duplicated. In a less extreme case (Fig. 4D), somites are reduced and tail structures lacking. In almost radially symmetric embryos (thus far only seen at a frequency of 5% under our current conditions) the most anterior structure, the cement gland, occurs as a band which completely surrounds the embryo (Fig. 4G).

How are we to understand this bizarre morphology? Observation of these embryos during gastrulation provides an answer. At the time of control dorsal lip formation embryos from D_2O-treated eggs destined to form simple twins are often seen to initiate two separate dorsal lips. Embryos destined for more extreme forms initiate progressively wider and eventually radial blastopore lips (Fig. 5B) at the time of control dorsal lip formation. The subsequent morphogenetic movements of gastrulation and neurulation in embryos showing this radial symmetry are quite remarkable. Following closure of the circular blastopore to about 1/2 of its initial diameter, the blastopore region begins to pucker and extend. As gastrulation and neurulation procede, this extension elongates to form a "proboscis" roughly as long as the body of the

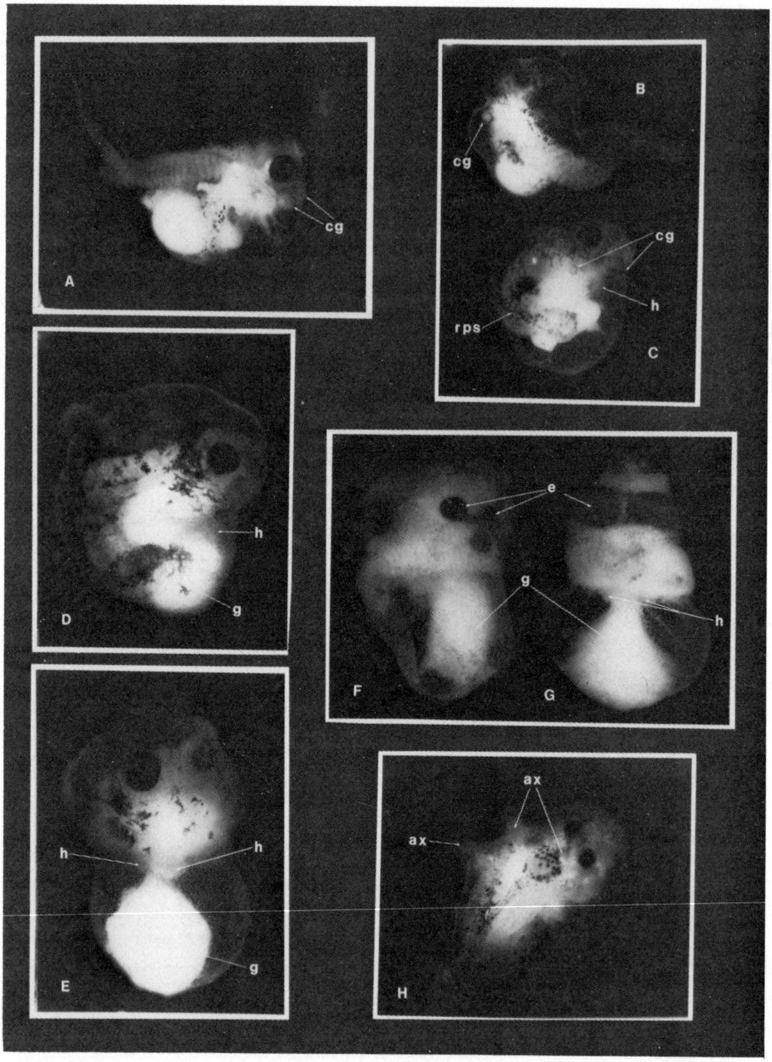

FIGURE 4. Hyperdorsal embryos from D_2O-treated eggs. Embryos were produced by exposure of eggs to 50% D_2O for 5 min at time 0.3. A-C: mild anterior duplication showing double cement gland; D: no duplication, but reduced posterior structures; E, F: "Janus" twins showing head duplication and extreme posterior reduction; G: near radial embryo; H: triplet. ax (axis); cg (cement gland); e (eye); h (heart); rps (reduced posterior structures).

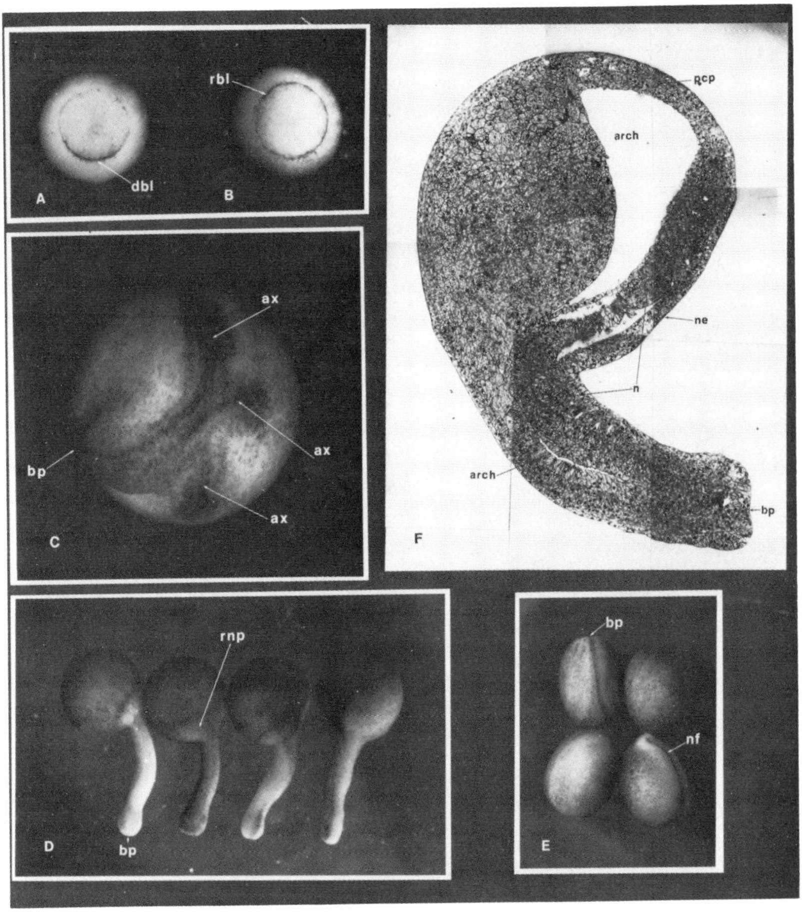

FIGURE 5. Gastrulation and neurulation in extremely hyperdorsal embryos from D_2O-treated eggs. A and B: blastopore lip formation showing control with normal dorsal blastopore lip, and future hyperdorsal embryo with radial blastopore lip forming simultaneously. C: triplet at late neurula stage. D: "Proboscis" embryos. E: Control neurulae from the same egg batch as D. F: Median section through proboscis form. Eggs were exposed to 50% D_2O for 5 min at time 0.3. ax (axis); arch (archenteron); bp (blastopore); dbl (dorsal blastopore lip); n (notochord); ne (neurectoderm); nf (neural folds); pcp (pre-cordal plate); rbl (radial blastopore lip); rnp (radial neural plate).

embryo (Fig. 5D). Histologically the proboscis includes notochord-like material (Fig. 5F). Following its formation, the proboscis, consisting of what would normally be material destined for posterior structures, is lost to the embryo. The proboscis literally tears itself apart as its entire surface attempts to participate in the invagination events of neurulation. Thus the embryo is left with only the most anterior structures, and these occur in a duplicated or radial configuration.

In sum, these embryos appear to develop as if they had two or more organizer regions at the time of blastopore lip formation. Since the organizer or dorsal marginal zone is considered to be a region of high dorsal potential, we have chosen to call these embryos "hyperdorsal".

Hyperdorsal embryos result from D_2O treatment only when eggs are exposed between 0.25 and 0.45 of the first cell cycle (Fig. 6). Treatment during the period of cold and pressure sensitivity (0.45-0.75) is completely ineffective. The basis for the remarkable difference in stage specificity of these treatments, all of which would be expected to affect the egg cytoskeleton, is considered in the following section.

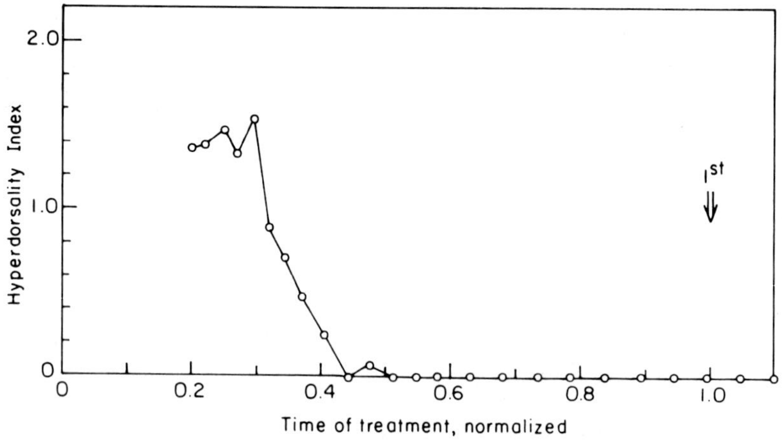

FIGURE 6. Stage dependence of D_2O-induced hyperdorsality. Eggs were treated with 50% D_2O for 5 min at the times shown. Scoring of hyperdorsality index was as follows: grade 0, normal; grade 1, wide cement gland; grade 2, duplicated head structures or trunk structures; grade 3, multiple or radial anterior structures, posterior structures lost. The time point at maximal sensitivity included 8% grade 0, 33% grade 1, 44% grade 2 and 15% grade 3.

The Process of Egg Reorganization

We now present data showing that the discrete periods of egg sensitivity to cold, pressure and D_2O are also distinguishable by totally independent criteria, namely cell cycle-dependent changes in the egg cortex and endoplasm.

In the period of D_2O sensitivity the sperm aster enlarges and moves centrad, its most obvious function being to bring the male and female pronuclei together [22]. Aster microtubules may interact with the animal hemisphere cortex at this time [23]. Coincident with these events is a change in the configuration of cortical pigment granules from that of random granular aggregates into a radiate array centered on the SEP (Fig. 7). The presence of the large aster at this

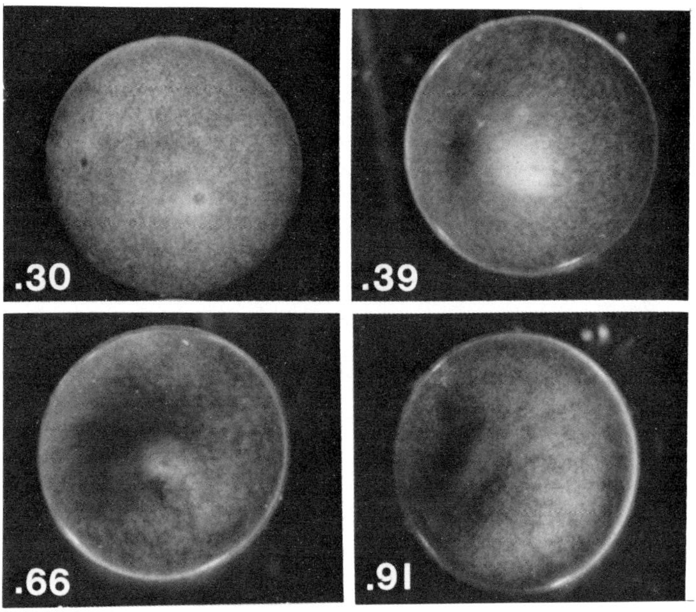

FIGURE 7. Changes in animal hemisphere pigment configuration in the first cell cycle. Normalized time is shown. Each photograph is of a different egg and in all cases the SEP is left. .30, SEP first visible, remainder of egg granular; .39, pigment becoming aligned around SEP; .66, translocation of pigment in progress; .91, pigment aggregation seen.

stage of the first cell cycle causes the redistribution of animal hemisphere yolk platelets (Fig. 8A). In D_2O-treated eggs the sperm aster is less extensive, but numerous cytasters are found in the animal hemisphere cytoplasm [24,25]. The pigment configuration in the cortex of these treated eggs reflects this abnormality -- the radiate array surrounding the SEP is unusually small and dense [24].

This sperm aster phase is characterized by a unique state of high endoplasmic fluidity. Vegetal hemisphere cytoplasm is easily displaced with respect to the surface [24], and animal hemisphere cytoplasm flows more easily than at any other time in the cell cycle [26]. In D_2O-treated eggs, the cytoplasm is far more rigid [24,26]. Thus D_2O can affect two aspects of egg structure particular to this phase, the aster configuration and cytoplasmic fluidity. We discuss below the possible role of these effects in causing the hyperdorsal syndrome.

A number of amphibian species form an obvious grey crescent during the second half of the first cell cycle; e.g., in Rana the crescent forms from 0.5 to 0.8 [27]. Since this interval coincides precisely with egg sensitivity to cold and pressure and to the decline in UV sensitivity, we searched

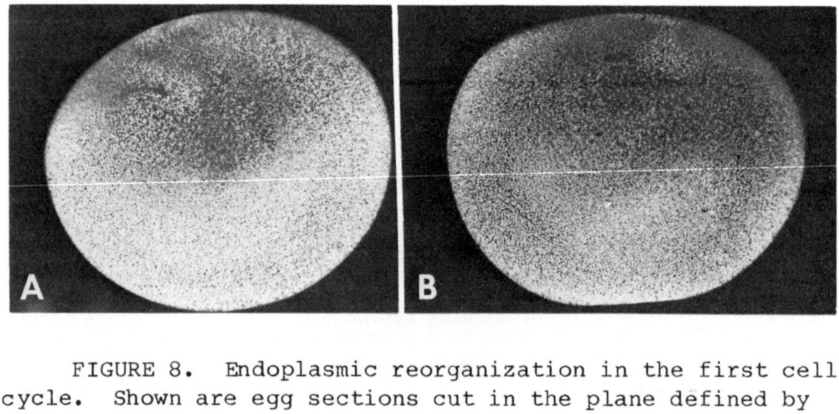

FIGURE 8. Endoplasmic reorganization in the first cell cycle. Shown are egg sections cut in the plane defined by the sperm entry point and the animal pole. Autofluorescent yolk platelets show cytoplasmic organization. SEP is leftward in both cases. A: Egg at time 0.38. Note the sperm trail and suggestion of large aster. B: Egg from same fertilization as A, at time 0.74. Note the asymmetric distribution of large platelets and yolk-poor cytoplasm.

for a comparable event in Xenopus, where a grey crescent is generally not detectable.

Time-lapse video microscopy of animal and vegetal cortical pigment showed that beginning at time 0.45 animal hemisphere cortical granules begin to move away from the SEP and vegetal hemisphere granules to move toward the SEP [24]. These movements continue in the animal hemisphere until 0.8 and in the vegetal hemisphere until 0.9. The typical appearance of the animal hemisphere during this interval is shown in Fig. 7. Importantly, granules are seen to move relative to the plasma membrane, not in tandem with it. This was shown by videotaping pigment movement relative to fine chalk or carime particles placed on the egg surface [24].

In order to assess directly the relative movement of the deeper endoplasm with respect to the egg surface, one of us (J.-P. V.) developed a means to impose a geometrical pattern of vital dye on the vegetal hemisphere cytoplasm, staining material to a depth of about 50 μm. Dye movements are then visualized by time-lapse video microscopy of eggs, the surface of which is immobilized in transparent gelatin. A very clear pattern is seen. Marks are translocated relative to the plasma membrane in a coherent array and in a unidirectional manner through approximately 30° of arc (Fig. 9). An unstained crescent-shaped area results on one side of the egg -- presumably the Xenopus equivalent of the grey crescent, which in Rana extends over roughly 30° of arc [8]. The extent of this endoplasmic shift is the same as that observed for cortical pigment granules, indicating that the layer of cortex which contains granules is moving in tandem with the endoplasm. These data thus imply that both this pigmented cortex and the deeper endoplasm are moving with respect to a more superficial unpigmented cortical layer that is tightly connected to the plasma membrane, held immobile in gelatin.

When the directionality of dye spot movement is compared to the future location of the dorsal midline, an extremely good correlation is seen. The average angle between the point away from which the spots move and the midline of the neural folds is only 5°, an even better correlation than that normally seen between the SEP and the dorsal midline (an average of 26°, [28]).

As seen in Fig. 10 these vegetal hemisphere endoplasmic movements take place in a well defined interval between 0.4 and 0.9, essentially coincident with cortical pigment granule movements. Median sections of eggs fixed between 0.8 and 0.9 show that indeed the yolk mass has shifted toward the SEP, when compared to its placement at time 0.4 (Fig. 8). On the

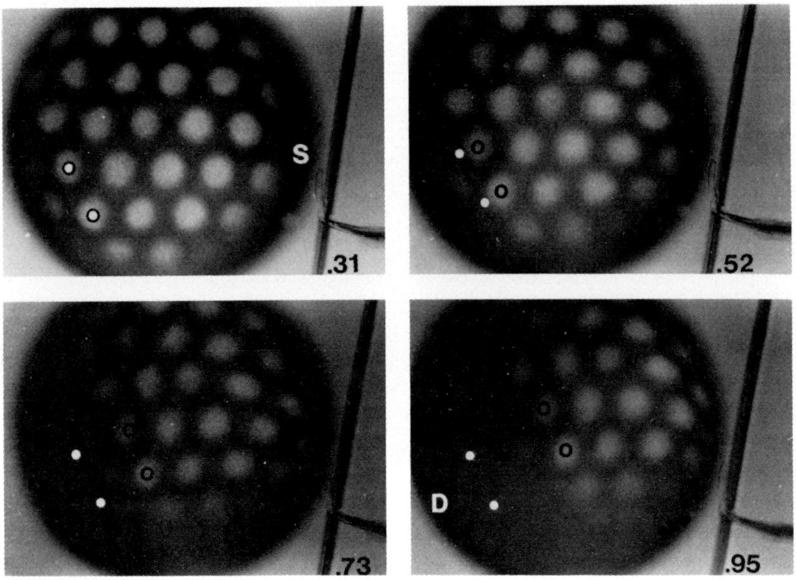

FIGURE 9. Successive frames showing the movement of subcortical dye marks. Vegetal hemisphere view. A solution of Nile Blue was applied for 5 min at time 0.3 through a hemispherical cup made in a perforated sheet of metal. Normalized time is shown. S denotes the meridian of the egg which, in the animal hemisphere, bore the sperm entrance point. D refers to the meridian where the neural groove was later identified.

basis of the behavior of both the animal and vegetal cortex and endoplasm in the 0.4 to approximately 0.8 interval, we have termed this interval the translocation phase.

Our presumption of a phase separable from the earlier sperm aster phase and the following attachment phase is bolstered by the observation of Elinson [26] that the egg cytoplasm in both *Rana* and to a lesser extent in *Xenopus* has unique characteristics during this period, which he called "phase 3". The cytoplasm is firm, in contrast to the sperm aster phase, when it is fluid. Furthermore, the cytoplasm is liquified by the microtubule poison colchicine, which is not the case in the later attachment phase.

We suggested above that vegetal hemisphere UV irradiation might directly impair cytoplasmic reorganization in the

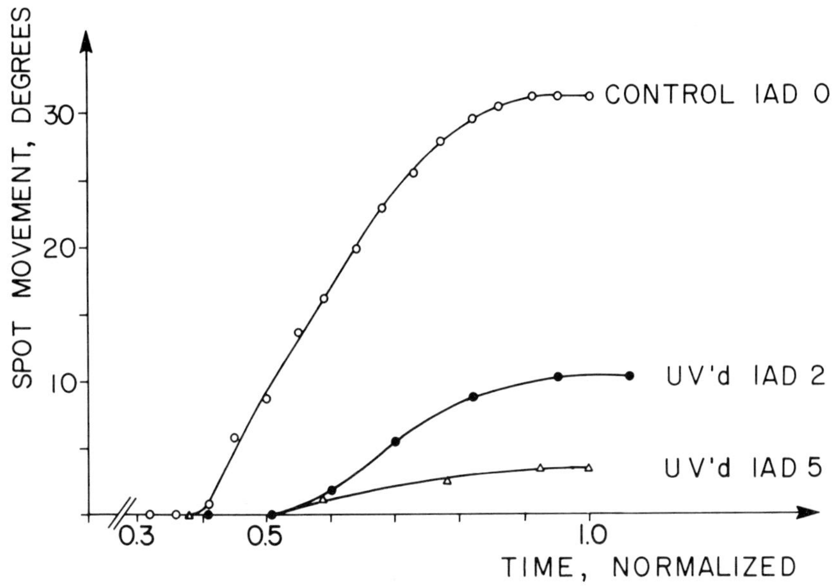

FIGURE 10. Time course of the vegetal dye spot movement for one control (not UV-irradiated) egg and two eggs that received an intermediate dose of UV irradiation at 0.2. Axis deficiency is correlated to a reduction in movement.

Xenopus egg. Manes and Elinson had in fact previously shown that UV prevents grey crescent formation in *Rana* [27]. We thus measured endoplasmic translocation in eggs receiving a range of UV doses and compared the extent of movement with the degree of axis deficiency. It was found that eggs destined for radially ventral embryos, showed very little movement while intermediate grades showed subnormal movement (Fig. 10). This result directly correlates the future capacity of the embryo to develop axial structures with the degree of egg reorganization.

After time 0.75, the egg is no longer sensitive to any of the treatments which perturb or impair axis determination. This transition is characterized by dramatic changes in the relationship of cortex and endoplasm. From time 0.81 to 0.9 pigment granules in animal hemisphere form aggregates (Fig. 7). This change in configuration, first observed by Rzehak [29], propagates as a wave beginning at the animal pole [24]. In experiments in which the relative movement of the plasma membrane and the animal hemisphere pigment granules was mapped (by marking the membrane with attached particles) it was found that just when granules aggregate, they no longer move independently of the egg surface, as they had done during the translocation phase [24]. As mentioned above, translocation of the vegetal hemisphere cytoplasm and pigment also stop at this time (0.9). Thus this transition marks a new association between the cortical layer adjacent to the surface and a deeper pigment layer which seems to move in tandem with the endoplasm.

Again, the data of Elinson [26] support the presumption of a transition at this time. He found that the previously colchicine sensitive animal hemisphere cytoplasm now becomes sensitive to cytochalasin B, a microfilament inhibitor. Thus the interval from approximately 0.8 to 1.0, which we call the *attachment phase*, and which was referred to by Ilinson as "phase 4", is distinguishable on the basis of both cortical and endoplasmic changes. Its developmental significance is discussed below.

DISCUSSION

The data presented in this report suggest that in the first cell cycle the *Xenopus* egg carries out a coordinated process of cytoplasmic reorganization which is critical for the future development of embryonic axial structures. In some way this process causes the localization of factors, or the creation of some specific cytoarchitecture, which "dorsalizes" the side of the egg opposite to the point of sperm entry. Used in this sense, "dorsalization" does not involve the immediate expression of any overtly dorsal characteristics, but rather the attainment of *dorsal potential*. This unique developmental potential is later expressed in stereotypic cell behavior, e.g., early blastopore lip formation and the extensive mesodermal migration and convergence that typify the behavior of the dorsal mesoderm at gastrulation [6]. We cannot at this time identify the factors responsible, nor even assert that the localization created in the first cell cycle is more than indirectly responsible for the later character of the dorsal mesoderm. We can, however, characterize the nature of the early cytoplasmic localization in the following general terms: 1) it behaves in a continuously variable rather than incremental manner in terms of its influence; and 2) the localization is likely to result from the physical shifting of deeper egg materials relative to more superficial substances or structures.

The reorganization process is comprised of three distinct phases which we have identified as the sperm aster, translocation, and attachment phases. These correspond to three phases previously identified by Elinson in activated and fertilized eggs on the basis of cytoplasmic response to cortical peeling [26]. The first phase, from 0.25-0.45, is defined in terms of the presence of the sperm aster, the occurrence of the sperm aster wave (in which pigment granules are aligned), and a fluid cytoplasm. The second phase from 0.45-0.8 is characterized by the translocation of pigment granules and vegetal hemisphere endoplasm with respect to the plasma membrane, and by a firm cytoplasm which is colchicine sensitive [26]. The final phase from 0.8-1.0 is characterized by aggregated pigment granules tightly associated with the egg surface, the end of translocation, and by firm cytoplasm which is cytochalasin B sensitive [26].

These observations allow us to suggest mechanisms for the production of hyperdorsal and axis deficient embryos. Recall that hyperdorsal embryos, which show secondary or multiple axes, are generated by exposure of eggs to the microtubule stabilizing agent D_2O during the sperm aster phase, but at no other time. Axis deficient embryos, which show the progressive reduction and loss of axial structures, are produced by exposure of eggs to cold or pressure, which depolymerize microtubules, only during the translocation phase. They are also produced by UV irradiation both during that phase and at earlier times.

We suggest that the critical event in localizing axial determinants is the movement of the endoplasm, or perhaps the pigmented cortex, relative to the surface during the translocation phase. This process would be impaired by brief exposure of the egg to cold or pressure since temporary destruction of the egg cytoskeleton would limit the ability of the egg to generate force and/or maintain a cytoplasm with sufficient integrity to respond to force generation. A role for microtubules in such a process was previously suggested by the finding that colchicine inhibits grey crescent formation in *Rana* [30]. Cold and pressure treatment would have no effect prior to 0.45, since the cytoskeletal elements would have sufficiently recovered by the time they were required. We would suggest that UV-irradiation impairs this translocation process in a different way. Vegetal hemisphere irradiation would result in an abnormally tight association between vegetal cortical layers which normally shear past one another during the translocation phase. Thus even though force generation could occur, the normal cortico-endoplasmic shift would be prevented. This supposition is supported by two findings: 1) the irradiated vegetal cortex is unusually difficult to peel away from the egg [31] and, 2) in irradiated eggs, the vegetal cytoplasm is less able to shear with respect to the surface than control eggs, when forces are imposed which tend to displace it [24]. Assuming damage to the UV target to be only very slowly reversible, UV would be effective at all times prior to translocation, but would become progressively less effective during the 0.45-0.9 period as rearrangement is accomplished.

The extent of endoplasmic translocation is a parameter which is obviously continuously variable. This quality and the fact that it is directly correlated with the extent of axis deficiency, also continuously variable, suggest that the two are intimately related, perhaps via the extent of dorsalization of the mesoderm. In support of this suggestion is our result that cold-, pressure-, or UV-treated eggs can be completely rescued by an artifically imposed cortico-endoplasmic shift, as is possible with oblique orientation. It is worthwhile noting that the endoplasmic arrangement which results from oblique orientation is identical to that which follows from the translocation event we are proposing here [9]. (Oddly enough, both rather resemble the ancient yin-yang motif for polarity!). Furthermore, the location of dorsal axial structures in obliquely oriented eggs (at the point uppermost with respect to gravity) also follows from this model.

How may we now explain hyperdorsality? The morphology of hyperdorsal embryos suggests that dorsal localizations sufficient for normal axis formation are created, but that the circumferential extent of these localizations is abnormally wide. D_2O treatment is effective during the sperm aster period but not during the translocation period or any other time. Since in the unperturbed egg the location of the aster and the direction of translocation are correlated, it would not be surprising that a perturbance of aster activity or endoplasmic response to aster activity might modify the subsequent egg response. It has in fact been shown that D_2O impairs pronuclear movement, by affecting the sperm aster [32].

Finally, we may speculate on the structural basis of the reorganization process. Recent work on the Xenopus cortex has shown that it is composed of two layers, a thin outer microfilament rich layer 0.1-1.0 µm thick, and a thicker inner layer, 5-7 µm thick, composed of intermediate filaments extending into the deeper endoplasm [33]. At least one of these layers is thought to have contractile capacity [34]. It may be that during translocation these two layers "walk over" one another. Such a model would be consistent with our observations of the relative movements of the egg surface and the deeper pigmented layer during the translocation phase. During the attachment phases, these two layers would form a tight association. We are currently examining this possibility.

In summary, we have described an early cytoplasmic reorganization process in the Xenopus egg which seems to be critical for the establishment of the future embryonic body axis of the embryo. This process is tightly coupled to the cell cycle and is composed of three distinct phases. Perturbation of egg cytoskeletal structures during one of these phases causes axis enhancement and during a second phase, axis deficiency. The development of these abnormal forms can be attributed to the failure of the pre-first cleavage egg to properly localize materials which confer dorsal potential to a unique region of the egg and embryo.

ACKNOWLEDGEMENT

We would very much like to thank Richard Elinson, Robert Gimlich, Michael Danilchik, and Zac Cande for useful and stimulating discussions and careful review of the manuscript.

REFERENCES

1. Davidson EH (1976). "Gene activity in early development," New York: Academic Press, p 249.
2. Slack JMW (1983). "From egg to embryo," Cambridge: University Press, p 31.
3. Scharf SR, Gerhart JC (1983). Axis determination in eggs of Xenopus laevis: A critical period before first cleavage, identified by the common effects of cold, pressure and ultraviolet irradiation. Dev Biol 99:75.
4. Gerhart J, Ubbels G, Black S, Hara K, Kirschner M (1981). A reinvestigation of the role of the grey crescent in axis formation in Xenopus laevis. Nature 292:511.

5. Gimlich RL, Cooke J (1983). Cell lineage and the induction of second nervous systems in amphibian development. Nature 306:471.
6. Keller RE (1976). Vital dye mapping of the gastrula and neurula of Xenopus laevis. II. Prospective areas and morphogenetic movements of the deep layer. Dev Biol 51:118.
7. Gerhart JC (1980). Mechanisms regulating pattern formation in the amphibian egg and early embryo. In Goldberger RF (ed): "Biological regulation and development," New York: Plenum, p 133.
8. Ancel P, Vintemberger P (1948). Cited in [4].
9. Pasteels J (1964). The morphogenetic role of the cortex of the amphibian egg. Advan Morphol 3:363.
10. Grant P, Wacaster JF (1972). The amphibian grey crescent - A site of developmental information? Dev Biol 28:454.
11. Curtis ASG (1962). Cited in [4].
12. Spemann H (1938). "Embryonic development and induction," New Haven: Yale University.
13. Malacinski GM, Brothers AJ, Chung HM (1977). Destruction of components of the neural induction system of the amphibian egg with ultraviolet irradiation. Dev Biol 56:24.
14. Youn BW, Malacinski GM (1980). Action spectrum for ultraviolet irradiation inactivation of a cytoplasmic component(s) required for neural induction in the amphibian egg. J Exp Zool 211:369.
15. Scharf SR, Gerhart JC (1980). Determination of the dorsal-ventral axis in eggs of Xenopus laevis: Complete rescue of uv-impaired eggs by oblique orientation before first cleavage. Dev Biol 79:181.
16. Messier PE, Seguin C (1978). Cited in [3].
17. Tilney LG, Cardell, Jr. RR (1970). Cited in [3].
18. Tilney LG, Porter KR (1967). Cited in [3].
19. Brinkley, BR (1982). Organization of the cytoplasm. Cold Spring Harbor Symp Quant Biol 46:1029.
20. Gimlich RL, Gerhart JC (1984). Early cellular interactions promote embryonic axis formation in Xenopus laevis. Dev Biol in press.
21. Houston LL, Odell J, Lee YC, Himes RH (1974). Cited in [3].
22. Stewart-Savage J, Grey RD (1982). The temporal and spatial relationships between cortical contraction, sperm trail formation, and pronuclear migration in fertilized Xenopus eggs. Wilhelm Roux's Arch 191:241.

23. Ubbels GA, Hara K, Koster CH, Kirschner MW (1983). Evidence for a functional role of the cytoskeleton in determination of the dorsoventral axis in Xenopus laevis eggs. J Embryol Exp Morphol 77:15.
24. Scharf SR (1983). Ph.D. Thesis, Department of Molecular Biology, University of California, Berkeley.
25. Heidemann SR, Kirschner MW (1975). Aster formation in eggs of Xenopus laevis. Induction by isolated basal bodies. J Cell Biol 67:105.
26. Elinson RP (1983). Cytoplasmic phases in the first cell cycle of the activated frog egg. Dev Biol 100:440.
27. Manes ME, Elinson RP (1980). Ultraviolet light inhibits grey crescent formation on the frog egg. Wilhelm Roux's Arch 189:73.
28. Black SD (1982). Ph.D. Thesis, Department of Genetics, University of California, Berkeley.
29. Rzehak K (1972). Cited in [23].
30. Manes ME, Elinson RP, Barbieri FD (1978). Formation of the amphibian grey crescent: Effects of colchicine and cytochalasin B. Wilhelm Roux's Arch 185:99.
31. Elinson RP, Manes ME (1979). Grey crescent formation and its inhibition by ultraviolet light. J Cell Biol 83:213a.
32. Briedis A, Elinson RP (1982). Suppression of male pronuclear movement in frog eggs by hydrostatic pressure and deuterium oxide yields androgenetic haploids. J Exp Zool 222:45.
33. Gall L, Picheral B, Gounon P (1983). Cytochemical evidence for the presence of intermediate filaments and microfilaments in the egg of Xenopus laevis. Biol Cell 47:331.
34. Meriam RW, Sauterer RA, Christensen K (1983). A subcortical, pigment-containing structure in Xenopus eggs with contractile properties. Dev Biol 95:439.

Molecular Biology of Development, pages 75–90

ANALYSIS OF CELL SURFACE AND EXTRACELLULAR MATRIX ANTIGENS IN D.DISCOIDEUM PATTERN FORMATION

Keith L. Williams,[1] Warwick N. Grant,[2] Marianne Krefft,[3] Ludwig Voet, and Dennis L. Welker

Max Planck Institut für Biochemie, D-8033 Martinsried bei München, Fed. Rep. Germany

ABSTRACT The migratory "slug" stage of *Dictyostelium discoideum* is a cylinder of ~10^5 cells enclosed in an extracellular matrix, the slime sheath. The slug exhibits a pattern of prestalk cells at the front and prespore cells in the rear. Components of the patterning system probably include morphogenetic signalling, cell surface interactions and cell-matrix (sheath) interactions. To investigate the roles of the cell surface and matrix, monoclonal antibodies MUD1, MUD50, MUD51, MUD52, recognizing four different antigenic determinants on slime sheath, have been characterized. Three of these determinants, recognized by MUD1, MUD50 and MUD52, are also found on the surface of prespore (but not prestalk or vegetative) cells. MUD1 recognizes a single 32kd glycoprotein on prespores and in the slime sheath. Two size polymorphisms of this protein have been identified and the gene mapped to linkage group 1. The other monoclonal antibodies recognize antigenic determinants that are present on more than one protein. A further monoclonal antibody, MUD9, which recognizes vegetative cells, can be used to identify prestalk cells in the slug since these cells carry more MUD9 determinants than prespore cells. These results emphasize that differentiating cells acquire characteristic surfaces and some relationship between slime sheath and prespore cells is apparent.

[1]Present address: Sch.Biol.Sci., Macquarie University, North Ryde, 2113. N.S.W. Australia.
[2]Present address: Imp.Cancer Res.Fund, London, U.K.
[3]Present address: Biochem.Dept., University of Wuppertal, 5600 Wuppertal, F.R.G.

INTRODUCTION

The slug of Dictyostelium discoideum is one of the simplest multicellular eukaryotes. Unlike more complex organisms the slug is formed largely by aggregation of solitary cells using self-generated gradients of 3'5'cyclic AMP (1). At the end of the aggregation phase, the group of cells establish a definite tip region and cover themselves with sheath material which is continuously synthesised. Therefore the slug cells migrate through a tube of slime sheath until they proceed through a series of morphogenetic movements leading to the formation of a well proportioned fruiting body consisting of spores, stalk and basal disk cells (2,3,4). The cells are already predifferentiated in the slug stage with a pattern of prestalk cells at the front and prespore cells (plus some anterior-like and basal disk cells) at the rear. This is a suitable system for studies on the establishment of polarity, proportion regulation and pattern formation. This cycle can be carried through as a haploid or a diploid, and all aspects of the developmental cycle are accessible for genetic study using a well developed parasexual genetic system (5,6) and transformation (7,8).

Meinhardt (9) has proposed that two morphogenetic gradients are involved in pattern formation, one which controls the proportioning mechanism, the other which specifies polarity (i.e. tip) formation. There is controversy concerning the timing and nature of proportion regulation. Some workers (10,11) believe that proportions of prestalk and prespore cells are established during aggregation, before tip formation. In this model proportion regulation occurs locally in small groups of cells and at tip formation the already differentiated cells sort out from each other.

Figure 1. Detailed time course of the period during which the MUD1 antigen first appears on prespore cells. This sequence is illustrated by (a) 5 µm frozen sections of aggregates stained with MUD1 and goat anti-mouse IgG $F(ab')_2$-FITC. The white dotted line shows the outer contour of the section which is not easily visible when cells are unlabelled. Individual cells are 10 µm in diameter. (b) Parallel contour plots of data from the flow cytometer. The small pictures at the left of (a) show the morphology of the aggregates taken for flow cytometry. Note that in this experiment tip formation occurred at 6.25 h, psp = prespore cells, pst = prestalk and other unlabelled cells (from Ref.12).

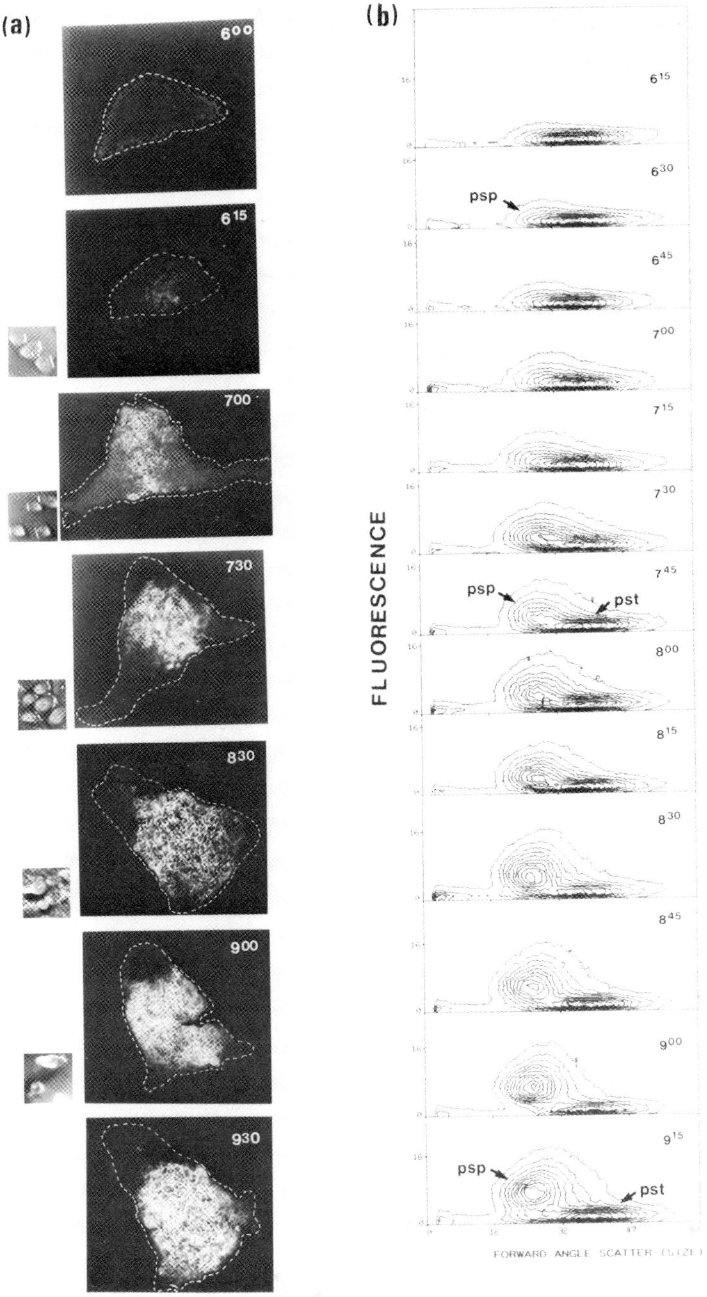

(a)
6 00
6 15
700
730
830
900
930
(b)
6 15
6 30
psp
6 45
7 00
7 15
7 30
7 45
psp
pst
8 00
8 15
8 30
8 45
9 00
9 15
psp
pst
FLUORESCENCE
FORWARD ANGLE SCATTER (SIZE)

Figure 1.

An important claim of this model is that in the aggregate there should be a 'salt and pepper' distribution of prespore and prestalk cells i.e. proportion regulation precedes pattern formation. We have not observed this in studies on frozen sections of early aggregates (12, Figure 1). Hence the model we favour proposes that polarity occurs first and that proportion regulation involves positional information i.e. cells differentiate according to their position in the aggregate. In this model proportioning and pattern formation occur concurrently and flow cytometry studies are consistent with this (12, Figure 1).

Our aim is to identify cell surface and matrix components which control and specify the different cell types and to determine their role in pattern formation. Tip formation is the most important time since at this stage both polarity and patterning are established. It is probable that contact molecules involved with cell-cell adhesion change at this stage. During aggregation an 80kd glycoprotein, contact site A, probably has an important role in cell adhesion (13) which may be taken over by a 95kd glycoprotein at tip formation (14). The slime sheath, which covers the cells of the multicellular stages of D.discoideum is synthesized at tip formation. These changes at the cell surface and the appearance of sheath undoubtably have a role in pattern formation, but its nature is quite unclear. This paper summarises our recent studies on cell surface and sheath antigens in the multicellular slug stage.

METHODS

Monoclonal antibodies have been used to identify cell surface and sheath antigens. Slime sheath antigens have been identified with an ELISA assay and characterized by Western blotting (15). Our strategy for isolating and characterizing cell surface antigens is outlined in Figure 2. D.discoideum slug cells were disaggregated mechanically and injected into 8 week old BALB/c mice. Monoclonal antibodies were raised according to standard techniques (16) using mouse myeloma cell line X63 Ag8.653 (17). The hybridoma supernatants were screened directly on disaggregated slug cells in a flow cytometer (18). This process not only allowed the direct recovery of monoclonal antibodies recognizing cell surface but also gave information concerning the cell type specificity of the monoclonal antibody (18). Quantitative

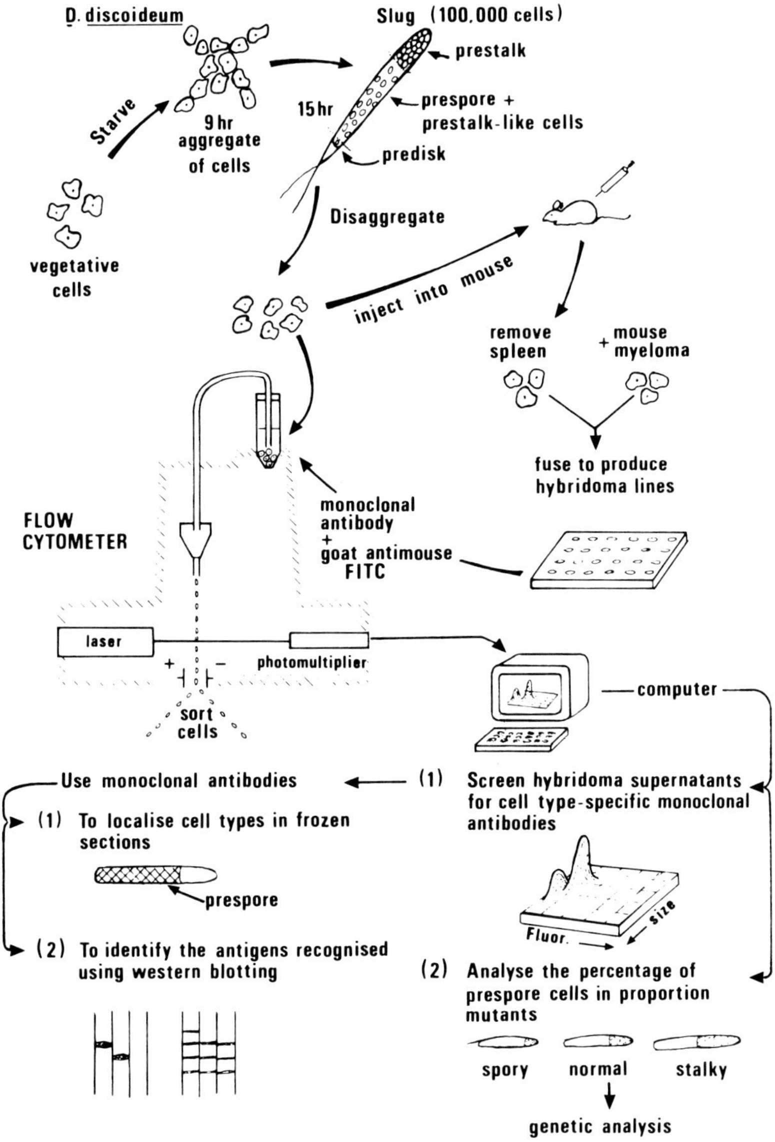

Figure 2. Techniques used to identify and study surface antigens of D. discoideum cells.

Figure 2.

techniques have been developed to allow rapid and accurate estimation of the proportions of different cell types in slugs (18) and this makes possible genetic analysis of proportion mutants (19). Simple techniques for obtaining frozen sections have been developed allowing the localization of monoclonal antibody to a specific class of cells in the slug (20). Western blotting techniques are used to identify the antigens recognised by the monoclonal antibodies (15,19,21). Genetic analysis involved mitotic crossingover and the use of the parasexual cycle as described elsewhere (6).

RESULTS AND DISCUSSION

The surface of D.discoideum slug cells. Studies using 2-D gel electrophoresis (22,23) or cDNA probes (24,25) have established that considerable changes occur at the time aggregating cells form the slug and the prespore/prestalk pattern is established. It is clear that many of the cell type specific proteins are not located at the cell surface, but are components of subcellular organelles in prespore or prestalk cells. A number of spore coat proteins are found in prespore vesicles whose contents are exocytosed at spore formation to form the bulk of the spore coat (10,26). On the other hand, two prestalk-specific monoclonal antibodies appear to identify autophagic vacuoles that are prominent in prestalk cells (10). There is an extensive literature on plasma membrane proteins of D.discoideum (reviewed in 14) but relatively few such proteins have been specifically identified with either prestalk or prespore cells. Exceptions are five glycoproteins, which bind wheat germ agglutinin, that are non-uniformly distributed in the slug (27) and the 95kd protein, which is probably a cell contact molecule in slugs, that is predominantly located on prespore cells (14). Monoclonal antibodies raised against the aggregation cell surface contact sites A glycoprotein (which disappears at slug formation), cross-react with other proteins that appear at slug formation (28,29). These proteins, with molecular weights of 25 kd and 90 kd, are prespore specific but their cell surface location is not yet confirmed (29).

The use of a flow cytometer in combination with monoclonal antibody techniques has revolutionised the study of lymphocyte cell surface differentiation (30). After developing techniques for disaggregating D.discoideum slugs to single cells (18), we have been able to use this approach to study D.discoideum pattern formation (19, Figure 2).

Already it is clear that there are substantial changes occurring at the cell surface during differentiation of prespore and prestalk cells. Only a handful of monoclonal antibodies has been characterised as described in Figure 2, but one predominantly prestalk specific and three prespore-specific determinants have been discovered.

One prespore specific monoclonal antibody, MUD1 (20) recognises a 32 kd glycoprotein PsA. PsA also carries the MUD50 determinant which is probably a carbohydrate moeity (see below). PsA is absent from vegetative and aggregative cells and it is first found inside prespore cells shortly before its appearance at the cell surface (12). PsA, as recognized by MUD1, remains on the surface of prespore cells in the slug stage and during culmination. It remains in mature spores, but it is not available for surface labelling (12). The loss of surface label in the transition from prespore to spore cells is very abrupt (Voet et al. In prep.) and is undoubtably due to the exocytosis of spore coat material found inside prespore vacuoles (26,31). It seems likely that PsA is identical to a 33 kd spore coat protein (Grant et al. In prep,32). Surprisingly PsA is also found in slime sheath and mature stalk cells (Grant et al. In prep.) but not in prestalk cells (20).

The above findings suggest that the PsA gene has complex developmental regulation. We have examined whether

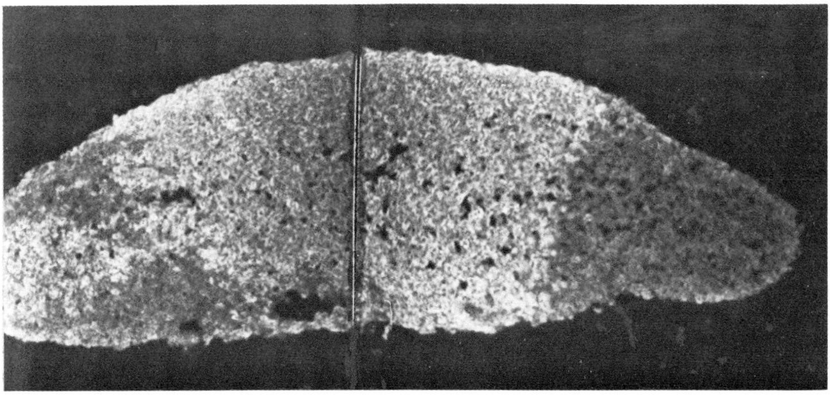

Figure 3. Immunofluorescence staining, using monoclonal antibody MUD50, of prespore cells in a 5 μm longitudinal frozen section of a slug of D.discoideum strain NP73. The slug is approximately 1 mm long. The tip (prestalk region) is on the right.

the PsA protein in its different locations is coded for by the same gene, using size polymorphisms detected by Western blotting of slug cells and slime sheath of several wild isolates of D.discoideum. Three size classes (30 kd, 32 kd and 34 kd) of the PsA protein have been identified and in every case the prespore and sheath PsA proteins have the same size. Using mitotic crossingover the PsA gene, pspA, has been mapped to linkage group I. Because heterozygous diploids express both size classes, the difference in size is unlikely to result from post-translational modification. It is tempting to suggest that the polymorphism results from PsA carrying variable numbers of a short repeated sequence of amino acids. At least two other cell surface proteins have repeated short sequences of amino acids. These are the malaria circumsporozoite protein which has twelve repeats of a 12 amino acid sequence (33) and the sea urchin protein bindin (three repeats of a 10 amino acid sequence) which is believed to be involved with sperm-egg adhesion (34).

Monoclonal antibody MUD50, which was raised against slime sheath, recognises prespore cells as determined by frozen sections (Figure 3) and flow cytometry (Grant et al. In prep.) The cell type specificity is somewhat surprising as the MUD50 antigenic determinant is found on several prespore surface proteins (15), including PsA. Another monoclonal antibody, MUD52, is similar to MUD50 in being prespore-specific and recognizing several prespore proteins, but it differs from MUD50 at least in its failure to recognize PsA. The basis for the cell type specificity of these monoclonal antibodies has not yet been determined, although it appears likely that only prespore cells carry the proteins which can be glycosylated to produce the MUD50 and MUD52 determinants.

A prestalk monoclonal antibody, MUD9, which was raised against whole slug cells, is not completely prestalk-specific. It is found on all aggregating cells but subsequently it is progressively lost from prespore cells. The current information about the developmental regulation of slug cell antigenic determinants is summarised in Figure 4.

Extracellular matrix, the slime sheath. Extracellular matrices are prominent features of developing organisms (e.g. sea urchin blastulation (35), mouse embryogenesis (36)). In the mouse, this matrix consists of collagenous and non-collagenous (fibronectin, laminin) glycoproteins and glycosaminoglycans. Collagen is undoubtably the extracellular backbone of the developing mouse embryo and each primary germ

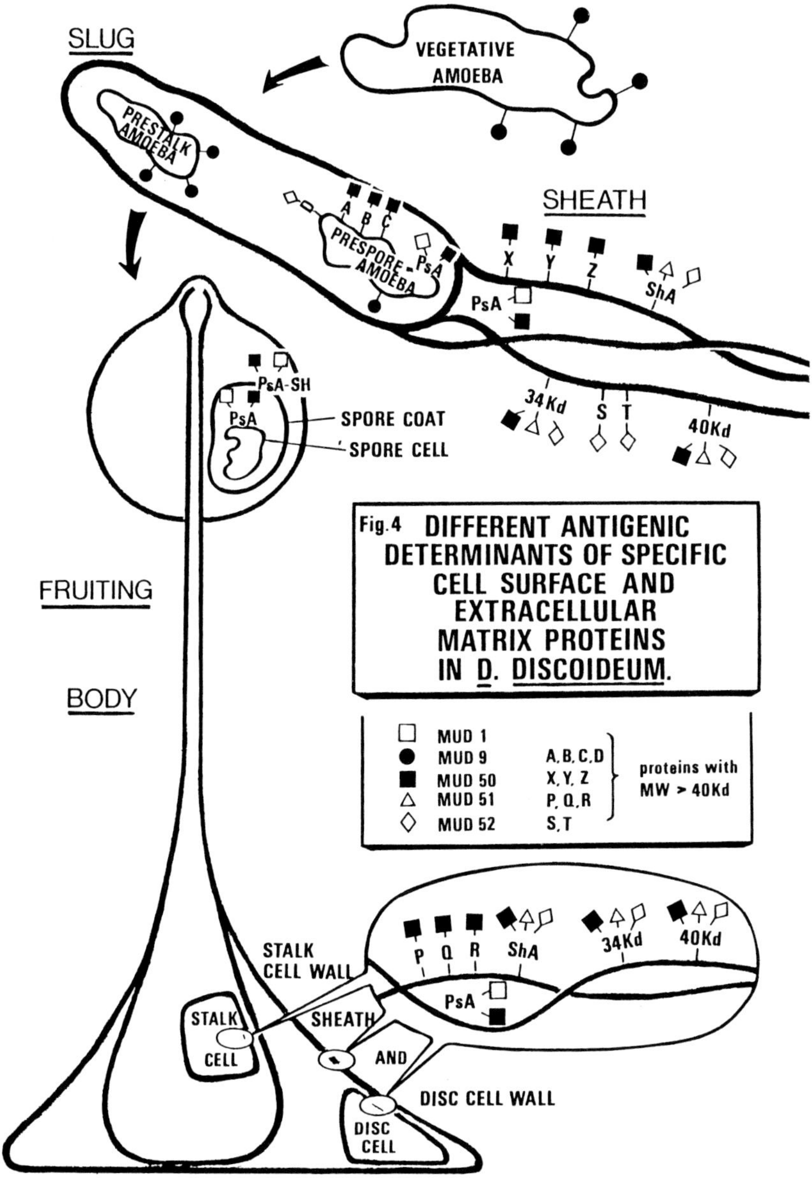

Fig. 4 DIFFERENT ANTIGENIC DETERMINANTS OF SPECIFIC CELL SURFACE AND EXTRACELLULAR MATRIX PROTEINS IN D. DISCOIDEUM.

layer and its derivatives is compartmentalized by matrix material. In addition to this skeletal function the matrix probably has other roles, such as guiding cell movements (36).

The slime sheath, which is secreted at the tip of the D.discoideum slug and forms a tube through which the organism migrates, is analogous to the extracellular matrix of more complex organisms. In contrast to animals, D.discoideum uses cellulose rather than collagen as the skeletal component of the sheath (37,38). The other components of the D.discoideum slime sheath are poorly characterized, although it is known that sheath comprises about half protein and half cellulose (37,39). Since the slime sheath is left behind as a trail, this provides a starting material for studies on sheath chemistry.

We have raised monoclonal antibodies to this material and, using Western blotting, discovered a number of sheath proteins (15). Four, named PsA (32 kd), ShA (32 kd), 34 kd and 40 kd, have been studied in some detail. The PsA protein is also found on prespore cells as has been discussed. The ShA, 34 kd and 40 kd glycoproteins share at least 3 different antigenic determinants recognized by monoclonal antibodies MUD50, MUD51 and MUD52. Two of these, recognized by MUD50 and MUD52, are probably carbohydrate determinants while MUD51 recognizes a pronase sensitive determinant (15). All of these antigens are developmentally regulated i.e. not found in vegetative cells. The MUD51 determinant is sheath-specific, while MUD50 and MUD52 also recognize cellular prespore proteins (but not ShA, 34 kd and 40 kd proteins which are probably sheath specific). MUD50 and MUD52 recognize several other sheath proteins (size 50 - 120 kd) in addition to the 32 kd, 34 kd and 40 kd glycoproteins (15, Figure 4).

Antigenic determinants are three dimensional structures often involving groups of residues well separated in the protein primary structure. In several systems common antigenic determinants, as defined by monoclonal antibodies, have been correlated with functional similarity of the proteins carrying the determinants (40,41). The fact that the ShA, 34 kd and 40 kd glycoproteins share three different antigenic determinants implies considerable structural similarity. It is possible that these glycoproteins have a similar function in that all are associated with the cellulose of the sheath. Unlike PsA, they can be released by cellulase treatment. The association is not covalent as all of the sheath proteins studied are released by SDS, urea or guanidine HCl treatment (15). Evidence for a possible interaction with cellulose is strengthened by the finding that the ShA, 34 kd and 40 kd glycoproteins are also found in stalk cells which have a cellulose wall that is similar to the sheath (15,42).

The above studies have revealed some basic features of the protein makeup of the slime sheath. While there is still much to be done in characterizing the high molecular weight sheath proteins, we can start to ask questions about sheath function in morphogenesis. Clearly an intact sheath is important for slug migration as mutants which don't make cellulose barely migrate (43). The sheath probably has a role in the slug/fruit switch (44), the decision to continue migration or to construct a fruiting body. This switch involves cellulose and/or the cellulase-releasable proteins as migration of slugs on cellulase-containing agar, which leads to the loss of ShA, 34 kd and 40 kd glycoproteins from the sheath, results in premature fruiting body construction in some strains (Grant, unpublished). Likewise, wild isolate strain WS584, which has altered and markedly reduced sheath proteins, migrates only for a short time. Some theories invoke differential permeability of the sheath (45,46) or sheath modification (47) as key elements in proportion regulation and pattern formation. We are now in a position to test such theories using the monoclonal antibodies as markers (15) and the rapid proportioning assay (19) to determine the effects on proportion regulation in sheath mutants such as WS584.

Antigenic and protein overlap between the sheath and slug cells of D.discoideum. From the preceding sections it is clear that monoclonal antibodies MUD1, MUD50 and MUD52 all identify determinants common to prespore (but not prestalk) cells and sheath in slugs of D.discoideum. This overlap between cell surface proteins and extracellular matrix (sheath) takes two forms. Firstly, the protein PsA is common to both prespore plasma membranes and sheath. There are several other proteins in the range 50-100 kd that may be common to prespore cells and slime sheath (15). Such overlap between cell surface and extracellular forms of the same protein is well established in other systems (e.g. fibronectin, (48)) and it may be involved with substrate-cell interactions involved with cell movement. The slime sheath is a primitive homeostatic organ in that it provides a constant substratum along which cells migrate. The second overlap involves the antigenic determinant but not necessarily the protein. The carbohydrate determinant recognized by MUD50 is found on a group of glycoproteins of 50-100 kd at least some of which are not common to sheath and prespore cells. The relationships between sheath and prespore proteins are summarized in Figure 4.

CONCLUSION

The studies reported here represent the start of a detailed program describing cell surface and extracellular matrix antigens during the development of D.discoideum, one of the simplest multicellular eukaryotes. It is already clear that prestalk and prespore cells have specific surfaces which differ from those expressed earlier on vegetative cells. In fact there is reason to suspect that the surface of differentiating D.discoideum cells is very dynamic, with new antigens appearing only to disappear a short time later. For example, a ~70 kd protein has been identified which appears about 3 hours after cells begin to starve only to disappear 8 hours later at early slug formation (49). A relationship between the extracellular sheath material and prespore cells is established here, but the function of this relationship is not at all clear. The approaches outlined in Figure 2 will allow us to fill out the details shown in Figure 4. It is to be expected that studies on mutants lacking specific cell surface antigens will contribute to our understanding of the function of these molecules, as will the ability to do genetic studies on mutants with altered proportions of different cell types (19).

From studies on the immunoglobulin family and other cell surface antigens of unknown function (e.g. Thy1, Qa, Tla) there is clear evidence that these proteins share common domains (50,51). It has been proposed that molecules corresponding to a single domain were first expressed on cell surfaces at an early stage of evolution of multicellular organisms. Their proposed role was initially in cell-cell recognition, but subsequently it is proposed that they were used to control cell interactions in tissue formation (51). At least a second family of progenitor cell surface molecules (identified by the W3/13 antigen on rat lymphocytes) has been proposed (51). As the surface molecules involved in D.discoideum pattern formation are discovered, the genes cloned and sequenced, it will be interesting to see if they also form a limited group of related molecules. The monoclonal antibodies so far obtained certainly show that groups of proteins identifying particular cell types or slime sheath share common antigenic determinants (Figure 4).

ACKNOWLEDGEMENTS

We thank Prof. J. Gregg (Uni. Florida) who participated in the studies on frozen sections and Shirley Hummelstad for typing the manuscript.

REFERENCES

1. Gerisch G (1982). Chemotaxis in Dictyostelium. Ann Rev Physiol 44:535-552.
2. Bonner JT (1967). "The Cellular Slime Molds," 2nd Ed. Princeton: Princeton Uni Press.
3. Loomis WF (1975). "Dictyostelium discoideum, a Developmental System." New York: Academic Press.
4. Stenhouse FO, Williams KL (1977). Patterning in Dictyostelium discoideum: the proportions of the three differentiated cell types (spore, stalk and basal disk) in the fruiting body. Dev Biol 59:140-152.
5. Newell PC (1982). Genetics. In Loomis WF (ed) "The Development of Dictyostelium discoideum". pp35-70. Acad. Press.
6. Welker DL, Williams KL (1982). A genetic map of Dictyostelium discoideum based on mitotic recombination. Genetics 102:691-710.
7. Hirth KP, Edwards CA, Firtel RA (1982). A DNA-mediated transformation system for Dictyostelium discoideum. Proc Natl Acad Sci USA 79:7356-7360.
8. Barclay SL, Meller E (1983). Efficient transformation of Dictyostelium discoideum amoebae. Mol Cell Biol 3:2117-2130.
9. Meinhardt H (1983). A model for the prestalk/prespore patterning in the slug of the slime mold Dictyostelium discoideum. Differentiation 24:191-202.
10. Tasaka M, Noce T, Takeuchi I (1983). Prestalk and prespore differentiation in Dictyostelium as detected by cell type-specific monoclonal antibodies. Proc Natl Acad Sci USA 80:5340-5344.
11. Sternfeld J, David CN (1981). Cell sorting during pattern formation in Dictyostelium. Differentiation 20:10-21.
12. Krefft M, Voet L, Gregg JH, Mairhofer H, Williams KL (1984). Evidence that positional information is used to establish the prestalk-prespore pattern in Dictyostelium discoideum aggregates. EMBO J 3: No.1.
13. Muller K, Gerisch G (1978). A specific glycoprotein as the target site of adhesion blocking Fab in aggregating Dictyostelium cells. Nature 274:445-449.

14. Parish RW (1983). Plasma membrane proteins in Dictyostelium. Mol Cell Biochem 50:75-95.
15. Grant WN, Williams KL (1983). Monoclonal antibody characterisation of slime sheath: the extracellular matrix of Dictyostelium discoideum. EMBO J 2:935-940.
16. de St. Groth SF, Scheidegger D (1980). Production of monoclonal antibodies: strategy and tactics. J Immunol Meth 35:1-21.
17. Kearney JF, Radbruch A, Liesgang B, Rajewsky K (1979). A new mouse myeloma cell line that has lost immunoglobulin expression but permits the construction of antibody-secreting hybrid cell lines. J Immunol 123: 1548-1555.
18. Voet L, Krefft M, Mairhofer H, Williams KL (1984). An assay for pattern formation in Dictyostelium discoideum using monoclonal antibodies, flow cytometry and subsequent data analysis. Cytometry. In Press.
19. Krefft M, Voet L, Mairhofer H, Williams KL (1983). Analysis of proportion regulation in slugs of Dictyostelium discoideum using a monoclonal antibody and a FACS IV. Exp Cell Res 147:235-239.
20. Gregg JH, Krefft M, Haas-Kraus A, Williams KL (1982). Antigenic differences detected between prespore cells of Dictyostelium discoideum and Dictyostelium mucoroides using monoclonal antibodies. Exp Cell Res 142:229-233.
21. Towbin M, Staehelin T, Gordon J (1979). Electrophoretic transfer of proteins from polyacrylamide gels to nitrocellulose sheets: procedure and some applications. Proc Natl Acad Sci USA 76:4350-4354.
22. Alton TH, Brenner M (1979). Comparison of proteins synthesized by anterior and posterior regions of Dictyostelium discoideum pseudoplasmodia. Dev Biol 71:1-7.
23. Borth W, Ratner D (1983) Different synthetic profiles and developmental fates of prespore versus prestalk proteins of Dictyostelium. Differentiation 24:213-219.
24. Barklis E, Lodish HF (1983). Regulation of Dictyostelium discoideum mRNAs specific for prespore or prestalk cells. Cell 32:1139-1148.
25. Mehdy M, Ratner D, Firtel RA (1983). Induction and modulation of cell-type-specific gene expression in Dictyostelium. Cell 32:763-771.
26. Devine KM, Bergmann JE, Loomis WF (1983). Spore coat proteins of Dictyostelium discoideum are packaged in prespore vesicles. Dev Biol 99:437-446.
27. West CM, McMahon D (1979). The axial distribution of plasma membrane molecules in pseudoplasmodia of the cellular slime mold Dictyostelium discoideum. Exp Cell Res 124:393-401.

28. Ochiai H, Schwarz H, Merkl R, Wagle G, Gerisch G (1982). Stage-specific antigens reacting with monoclonal antibodies against contact site A, a cell-surface glycoprotein of Dictyostelium discoideum. Cell Differentiation 11:1-13.
29. Murray BA, Niman HL, Loomis WF (1983). Monoclonal antibody recognizing gp80, a membrane glycoprotein implicated in intercellular adhesion of Dictyostelium discoideum. Mol Cell Biol 3:863-870.
30. Williams AF (1980). Cell-surface antigens of lymphocytes: markers and molecules. Biochem Soc Symp 45:27-50.
31. Hohl HR, Hamamoto ST (1969). Ultrastructure of spore differentiation in Dictyostelium: the prespore vacuole. J Ultrastruct Res 26:442-453.
32. Wilkinson DG, Hames BD (1983). Characterisation of the spore coat proteins of Dictyostelium discoideum. Eur J Biochem 129:637-643.
33. Godson GN, Ellis J, Svec P, Schlesinger DH, Nussenzweig V (1983). Identification and chemical synthesis of a tandemly repeated immunogenic region of Plasmodium knowlesi circumsporozoite protein. Nature 305:29-33.
34. Vacquier VD, Moy GD (1978). Macromolecules mediating sperm-egg recognition and adhesion during sea urchin fertilisation. ICN-UCLA Symp Mol Cell Biol 12:379-389.
35. McClay DR, Fink RD (1982). Sea urchin hyalin: appearance and function in development. Dev Biol 92:285-293.
36. Wartiovaara J, Leivo I, Vaheri A (1980). Matrix glycoproteins in early mouse development and in differentiation of teratocarcinoma cells. Symp Soc Dev Biol 38:305-324.
37. Hohl HR, Jehli J (1973). The presence of cellulose microfibrils in the proteinaceous slime track of Dictyostelium discoideum. Arch Microbiol 92:179-187.
38. Freeze H, Loomis WF (1977). Isolation and characterization of a component of the surface sheath of Dictyostelium discoideum. J Biol Chem 252:820-824.
39. Smith E, Williams KL (1979). Preparation of slime sheath from Dictyostelium discoideum. FEMS Microbiol Lett 6: 119-122.
40. Pruss RM, Mirsky R, Raff MC, Thorpe R, Dowding AJ, Anderton BH (1981). All classes of intermediate filaments share a common antigenic determinant defined by a monoclonal antibody. Cell 27:419-428.

41. Glenney JR, Glenney P, Weber K (1982). Erythroid spectrin, brain fodrin and intestinal brush border proteins (TW-260/240) are related molecules containing a common calmodulin-binding subunit bound to a variant cell type-specific subunit. Proc Natl Acad Sci USA 79:4002-4005.
42. Freeze H, Loomis WF (1978). Chemical analysis of stalk components of Dictyostelium discoideum. Biochim Biophys Acta 539:529-537.
43. Freeze H, Loomis WF (1977). The role of the fibrillar component of the surface sheath in the morphogenesis of Dictyostelium discoideum. Dev Biol 56:184-194.
44. Smith E, Williams KL (1980). Evidence for tip control of the slug/fruit switch in slugs of Dictyostelium discoideum. J Embryol Exp Morph 57:233-240.
45. Farnsworth P, Loomis WF (1975). A gradient in the thickness of the surface sheath in pseudoplasmodia of Dictyostelium discoideum. Dev Biol 46:349-357.
46. Sussman M, Schindler J (1978). A possible mechanism of morphogenetic regulation in Dictyostelium discoideum. Differentiation 10:1-5.
47. Ashworth JM (1971). Cell development in the cellular slime mould Dictyostelium discoideum. Symp Soc Exp Biol 25:27-47.
48. Furcht LT (1983). Structure and function of the adhesive glycoprotein fibronectin. Mod Cell Biol 1:53-117.
49. Brodie C, Klein C, Swierkosz J (1983). Monoclonal antibodies: use to detect developmentally regulated antigens on D.discoideum amoebae. Cell 32:1115-1123.
50. Williams AF (1984). The immunoglobulin family takes shape. Nature 308:12-13.
51. Williams AF (1982). Surface molecules and cell interactions. J Theor Biol 98:221-234.

II. GENE EXPRESSION DURING OOGENESIS AND EARLY DEVELOPMENT

Molecular Biology of Development, pages 93–108

GENE EXPRESSION DURING EMBRYOGENESIS IN XENOPUS LAEVIS

Jeffrey A. Winkles, Milan Jamrich, Erzsebet Jonas, Brian K. Kay, Seiji Miyatani, Thomas D. Sargent, and Igor B. Dawid

Laboratory of Molecular Genetics
National Institute of Child Health
and Human Development
National Institutes of Health
Bethesda, Maryland 20205

ABSTRACT A library of cloned cDNAs has been prepared that represents RNA molecules absent from the egg but present in the gastrula embryo of Xenopus laevis. These sequences, called DG sequences, are being studied in order to analyze the regulation and function of genes that are expressed in the early embryo. In this article we describe some general properties of the DG class of genes, and preliminary experiments on the detailed time course of accumulation and spatial distribution of several DG RNAs. Discussion is focused on a family of genes, the DG 42 family, which is activated in the blastula leading to very rapid accumulation of RNA through the gastrula and neurula stages. Thereafter, DG 42 RNA declines rapidly in abundance. The DG 42 family is represented by at least two genes which differ by about 20% in their nucleotide sequence; the two distinct members of the family are expressed coordinately.

INTRODUCTION

The early frog embryo derives its genetic information from two distinct sources: the maternal and the zygotic genomes. During oogenesis a prolonged period of active transcription leads to the accumulation of a large and complex population of maternal RNA which is inherited by the embryo (1,2,3). Although this RNA population undoubtedly contains many mRNAs, evidence has been adduced that a substantial fraction of the egg's poly(A)$^+$ RNA is not messenger (1,4,5). After fertilization, a period of rapid cleavage follows during which no nuclear RNA synthesis is detected; protein synthesis at this time is supported entirely by maternal mRNA. RNA synthesis is initiated during the blastula stage (6,7,8), at a time coinciding with the 13th cleavage (9,10). This point, named the midblastula transition (MBT), is also characterized by a loss of synchronous cell division and the beginning of membrane motility (9,10). By measuring RNA synthesis in blastula embryos dissociated into individual cells, it has been concluded that an amount of polyA$^+$ RNA equivalent to about one-third of that stored in the egg has been synthesized by gastrulation (8,11). Since the total quantity of poly(A)$^+$ RNA remains constant from egg to gastrula, this finding implies that much of the early RNA synthesis is balanced by turnover. Even more important than the quantitative considerations are the qualitative ones: since the RNA populations in eggs and gastrula embryos are very similar overall (3,12), it appears that many of the RNAs synthesized by the late blastula are homologous to the sequences originally present in the egg. Nevertheless, the accumulation of qualitatively new RNAs by the gastrula stage was suggested by the observation that cleavage and blastula development can occur in the absence of RNA transcription, but proper gastrulation cannot (see 3). In other words, the blastula can be formed utilizing information stored in the egg, but development of the gastrula and subsequent differentiation require the expression of the embryonic genome.

CONSTRUCTION OF THE DG cDNA LIBRARY

Our work was based on the hypothesis that qualitatively new RNA molecules accumulate in the Xenopus embryo by the gastrula stage. Such molecules should be of considerable developmental interest as they represent genes activated early in the embryonic developmental program. Genes whose products function specifically during gastrulation might be expected to be included among such early activated sequences. Our approach to the study of such genes was the isolation and cloning of DNA copies of RNA molecules that are present in Xenopus gastrulae, but absent from eggs. We have devised a differential cloning procedure which allows the production of a cDNA library that includes primarily, if not exclusively, sequences present in one, but absent from another RNA population. The DG (differentially expressed in gastrula) cDNA library was constructed as follows (13): cDNA was prepared from gastrula poly(A)$^+$RNA, hybridized with a 30-fold excess of ovarian poly(A)$^+$RNA, and the unhybridized single-stranded cDNA was isolated. After removal of short fragments, the greatly enriched cDNA was converted to double-stranded form and cloned in a plasmid vector by standard techniques.

The objective of the initial characterization of the DG cDNA library was to learn the extent to which the sequences in this library fulfill the desired criterion of absence from egg RNA. Figure 1 shows an experiment that addresses this question. Eighty-four cloned DG cDNAs were applied on each of four filters, together with positive and negative controls in the bottom row. Each filter was hybridized with a radioactive probe derived from cytoplasmic poly(A)$^+$RNA from a different developmental stage. As would be predicted from the method of library construction, the egg RNA probe did not react with any of the DG cDNAs; the same was true for the blastula RNA probe, consistent with the finding that no RNA synthesis occurs between fertilization and midblastula. In contrast, about half the DG cDNAs did react significantly with the gastrula probe. This result indicates that a number of qualitatively new RNA molecules accumulate to varying abundances between the blastula and gastrula stages. The last panel of Figure 1 shows that most DG RNAs decrease substantially in abundance from the gastrula to the three-day tadpole stage. This result supports the idea that many of the DG genes may function primarily during early development. This suggestion is further supported by the observation that most of the DG RNAs studied are polysomal at the gastrula stage (13).

The most abundant DG RNAs observed each constitute about 0.1 to 0.2% of the poly(A)$^+$RNA population at gastrula; the rarest detectable RNAs represent levels of about 0.005 to 0.01%. Many DG cDNA clones give no detectable reaction with gastrula RNA probes. These sequences probably represent very rare RNAs, since the negative clones contain Xenopus DNA, and the inserts have a length distribution similar to that of positive clones.

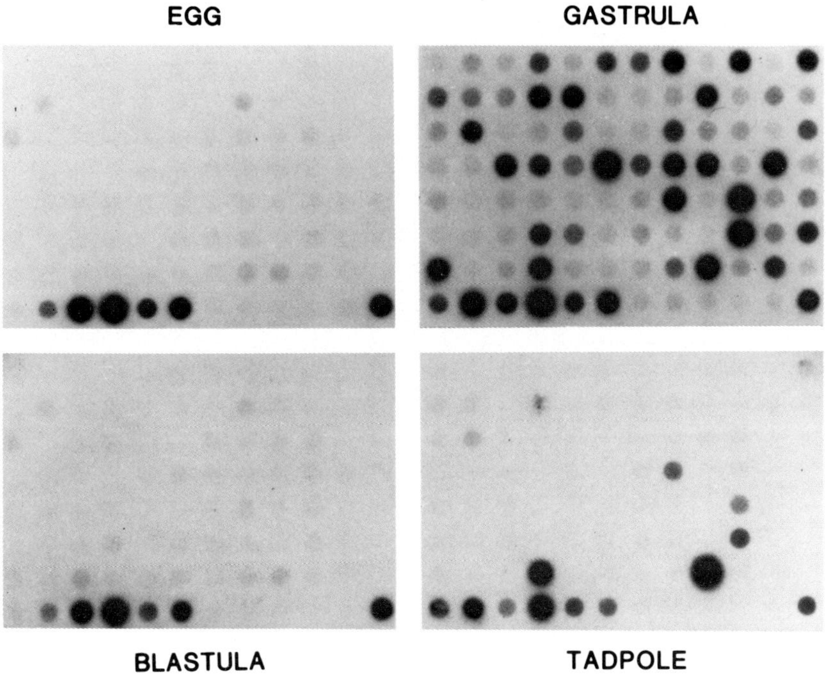

FIGURE 1. Autoradiograms of DG plasmid DNA dot blots hybridized to labeled probes derived from poly(A)$^+$ RNA of eggs, blastulae (about stage 8), gastrulae (stage 10 to 11) and tadpoles (stage 41). The four identical panels contain in their top seven rows arrays of 84 DG clones. Each bottom row contains, from left to right, six clones from a reference gastrula cDNA library (i.e., not enriched for differential expression), followed by five background dots, and lastly M13mp7 replicative form DNA, used as a hybridization control and abundance standard. The intensity of hybridization is approximately proportional to the abundance of the respective RNA in the poly(A)$^+$ RNA population. The RNAs of highest abundance represented in this grid have been estimated to account for 0.1 to 0.2% of the gastrula poly(A)$^+$ RNA population. The figure is reproduced from reference 13.

The total sequence complexity of the DG population is not known. Estimates depend critically on the problem of the "negative" clones: if these represent legitimate, but very rare DG RNAs, the total complexity of the DG population could be quite high. When only RNAs of rare to moderate abundance are considered their total number might be a few hundred. Some information regarding the distribution of the more abundant sequences in the DG library was obtained by testing for multiple occurrences of homologous sequences among the 84 cDNA clones that were studied (Figure 1). One family of sequences, whose prototype is DG 42, was found to be represented 7 times among the 84 clones. One other pair of homologous sequences was identified. These observations suggest that the 84 clones we have studied include a significant fraction, but certainly not all, representatives of the most abundant class of X. laevis DG RNAs.

ACCUMULATION OF DG RNAs DURING DEVELOPMENT

DG RNAs cannot be detected in total egg RNA, establishing the fact that de novo synthesis is responsible for the accumulation of these RNAs by gastrula rather than polyadenylation of pre-existing poly(A)$^-$RNA or release from a bound state (13). As mentioned before, no nuclear RNA synthesis can be detected in the Xenopus embryo during cleavage before the 13th division (6-11); thus, the accumulation of DG RNAs should occur after MBT. To achieve the substantial concentrations shown by some of the DG RNAs by gastrulation, there must be very rapid synthesis during the 4 to 6 hours between MBT and gastrula. Assuming that DG genes are not amplified at this stage -- a possibility that has yet to be checked -- we calculated that the accumulation per cell of DG 42 is similar to that of vitellogenin mRNA in the fully induced adult liver (approximately 9 molecules per gene per minute) (13). We have not measured actual rates of RNA synthesis but rather RNA accumulation; if processing is efficient so that every precursor molecule gives rise to a cytoplasmic mRNA and turnover is slow relative to synthesis these rates would be equivalent. An important question refers to the rates of RNA synthesis from those DG genes whose RNAs accumulate to much lower levels than DG 42. Are these genes turned on to the same high level but accumulate cytoplasmic RNA more slowly because of inefficient processing and rapid turnover, or are the rates of RNA synthesis on these genes inherently lower? We have no answer to this question at present.

A related but distinct question is whether all DG RNAs appear at MBT and then accumulate at parallel rates at least through gastrulation, or whether different and individual patterns of accumulation can be seen. To compare the temporal profiles of DG gene expression we have isolated RNA from synchronously developing embryos collected at short intervals from fertilization through neurulation, and measured DG RNA concentrations by RNA dot blot hybridization. The results suggest that there is substantial variation amongst DG RNAs in the time of onset of detectable accumulation and in the time of maximal abundance.

ARE DG RNAs DIFFERENTIALLY ACCUMULATED IN DIFFERENT REGIONS OF THE EMBRYO?

Differential gene expression accompanies and controls the process of differentiation. The total poly(A)$^+$ RNA populations of different tissues and developmental stages appear to be rather similar (14-15), perhaps because most mRNAs have functions which are not specific for any tissue but rather are of a "housekeeping" nature. In addition, however, different tissues and cell lineages contain a substantial number, from a few hundred to several thousand specific mRNAs, including mRNAs for abundant differentiated products such as globin, immunoglobulin, etc. When during the course of tissue differentiation do such specific mRNAs accumulate? The study of the accumulation of DG RNAs in the Xenopus embryo may provide some insights into this problem: DG genes are activated in the embryo before overt differentiation and cell migration initiates, and if there are genes whose expression leads differentiation, some DG RNAs might accumulate differentially in regions of the gastrula and neurula embryo. The localized appearance of DG RNAs might provide molecular markers for early differentiation of lineages and could suggest types of possible functions for some DG genes. On the other hand, the uniform accumulation of DG RNAs would suggest the stage specific requirement for new products in all parts of the rapidly differentiating embryo.

We have approached the problem of spatial specificity of DG gene expression by a two-step process. Xenopus embryos are quite large (about 1.1 mm diameter) and can be manually dissected. We are sectioning early neurulae through various planes, separating dorsal from ventral and anterior from posterior halves of the embryo. Dot blot analysis of RNA extracted from these fractions has led to the tentative conclusion that the majority of DG RNAs are distributed throughout the different regions of the embryo. However, we have evidence for localized expression in a few instances. The second phase of this analysis is in situ hybridization of tritium-labeled RNA probes to embryonic tissue sections (16,17,18). By this method RNAs of moderate to low abundance can be detected in the embryo, and thus many DG RNAs should be within the limits of sensitivity of this technique.

DG 42: A GENE FAMILY SPECIFICALLY EXPRESSED IN THE GASTRULA AND NEURULA STAGES

We have studied DG 42 in more detail than any other DG cDNA clone for several reasons. First, DG 42 RNA is one of the most abundant RNAs in its class. Second, its expression is highly stage specific: the RNA accumulates rapidly from MBT through gastrulation and peaks during neurulation, and then decays to undetectable levels two days later. Third, DG 42 represents a gene family, as will be described further below. And fourth, the fortuitous fact that the cDNA clone DG 42 is an almost full-length copy of the mRNA, facilitating analysis.

Evidence for the existence and expression of at least two genes in the DG 42 family was obtained from the study of cDNAs. In the group of 84 DG cDNAs that were studied (see Figure 1) there are six independent isolates that cross-hybridize with DG 42. It appears that the DG 42 family is over-represented in the 84 clones, since other DG RNAs of similar abundances in the gastrula are represented only once or twice in this set. The reason for this over-representation may be purely statistical, but it allowed us to study the DG RNA family in more detail. Figure 2 illustrates the structural relationships between six of the cDNAs in this family. This figure is based on the complete nucleotide sequence of DG 42 and partial sequences of the other clones. DG 42 is missing only a short section at the 5' end. An

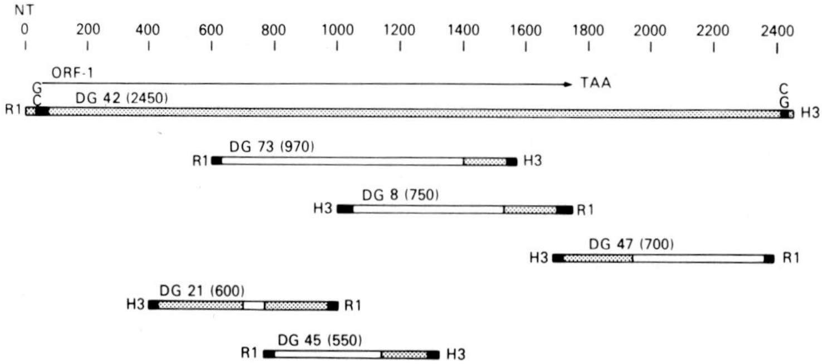

FIGURE 2. The DG 42 family of cDNAs. The maps represent six related cDNAs, aligned by sequence analysis. The cDNAs had been inserted with the aid of dG:dC tails into the Cla I site of pBR322 (13), and inserts were excised with Eco RI (R1) and Hind III (H3), leaving short regions of plasmid DNA attached. The black bars represent these short pBR322 sequences plus the dG:dC tails. The hatched bars represent sequenced cDNA regions, and open bars are segments that have not been sequenced. A scale in nucleotides is shown at the top, and the length of each cDNA clone is given in parentheses. The major open reading frame (ORF-1) of DG 42 is shown. DNA sequences were determined by the method of Maxam and Gilbert (25).

open reading frame starts upstream of the cloned region and continues for 1700 nucleotides; we believe this to be the reading frame that encodes the protein product of the DG 42 gene. While the other members of the cDNA family are shorter than DG 42 itself, they represent various regions of the mRNA. Three of the cDNAs analyzed match the sequence of DG 42 precisely. DG 47 has 2 substitutions and one 5-bp insertion, all in the 3' untranslated region. These slight differences might be due to polymorphism.

One member of the set, DG 21, is highly distinct. Whereas DG 21 is clearly related to DG 42, there are large differences leading to an overall homology level of only 80%. Nevertheless, DG 21 contains a protein coding sequence in the same open reading frame as DG 42, albeit with many changes in the amino acid sequence. Because of their considerable sequence difference it is easy to discriminate between the DG 42 and 21 RNAs by hybridization with the appropriate cDNA under stringent conditions. In this way we determined that the DG 42 and 21 genes are both expressed in the embryo, and that the two RNAs accumulate in parallel. These data suggest the existence in X. laevis of at least two distinct, non-allelic active genes in the family, one represented by cDNA 42, the other by cDNA 21. This conclusion is supported by Southern blot analysis of genomic DNA, as described below.

The relationship between the DG 42 and 21 genes and their products are reminiscent of that between the A and B genes of vitellogenin in the same animal: these genes have a homology of only about 80%, are non-allelic, and are both expressed at similar levels after estrogen induction in the adult liver (19). The situation is rather different in the case of the X. laevis calmodulin genes; two apparently non-allelic genes occur which encode the same protein in spite of a 5% difference in the nucleotide sequence of the protein-coding region (20). The frequent finding of gene duplications in X. laevis may be due to an apparent duplication of the entire genome in this species about 20 million years ago (21). It appears that X. laevis carries many duplicated genes which still fulfill the same function as the primordial gene. In the case of DG 42 and DG 21, the considerable differences in the nucleotide sequence and in the deduced sequence of the putative proteins suggest that functional differentiation between these two related genes may have occurred since their divergence began.

POLYMORPHISM IN DG GENES

Genomic Southern blots showed that DG 42 is homologous to several restriction fragments derived from the Xenopus genome (13), suggesting the presence of several genes. However, it was not clear to what extent the different restriction bands might be due to multiple genes or to polymorphism between the individuals which contributed DNA to the sample. We therefore performed Southern blot analysis with DNA from individual frogs (Fig. 3). Lanes A, B, and C show the blots obtained with DNA from 3 individual outbred animals. These patterns are very similar to each other and to the mixed pattern in reference 13. In contrast, lane D displays a much simpler pattern. This lane contains DNA from a homozygous diploid albino animal which has been produced and

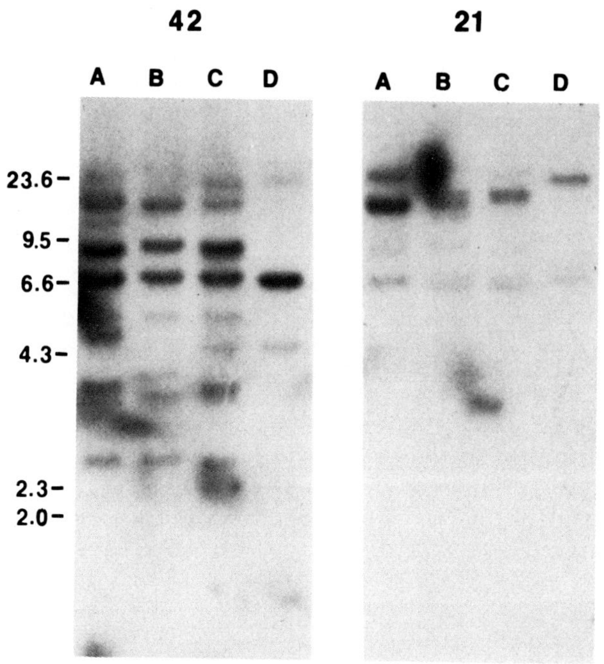

FIGURE 3. Genomic Southern blots with DG 42 and DG 21. Blood cell DNA from three individual outbred frogs (lane A, B, C), and from a diploid homozygous albino frog (lane D) was isolated, digested with Eco RI, separated by electrophoresis in agarose gels, blotted and hybridized with nick translated DG 42 or DG 21 under the conditions described in reference 13. Sizes of marker bands are indicated in kb.

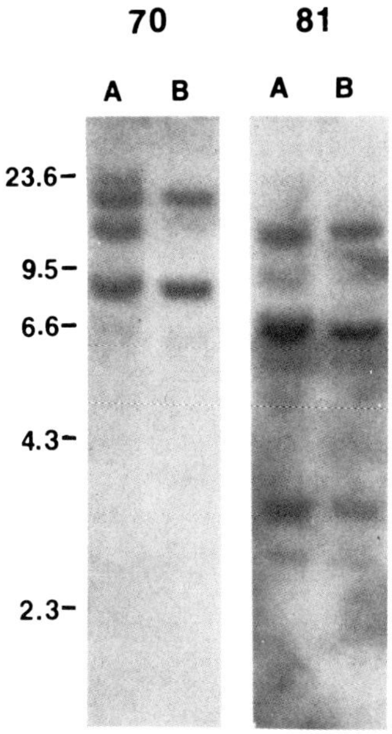

FIGURE 4. Genomic Southern blots with DG 70 and DG 81. DNA from a mixture of several individuals (lanes A), and the homozygous diploid frog (lanes B) was digested with Eco RI, blotted and hybridized with nick translated DG 70 or DG 81 as described in reference 13.

generously donated by Dr. Robert Tompkins. The right panel in Figure 3 shows the same genomic DNAs hybridized at high stringency with DG 21; under these conditions DG 42 and DG 21 do not cross-hybridize to a significant extent. The 3 individual outbred frogs show a similar pattern, although some polymorphic differences can be observed, and the albino individual again exhibits the simplest pattern. The relatively similar patterns found in the 3 outbred individuals by hybridization with DG 42 and DG 21 suggests that these genes are represented in a few copies in the genome, and that these copies are similarly organized in wild populations of X. laevis. However, the albino pattern indicates that considerable polymorphic differences may occur between different populations.

To test whether DG genes commonly show great structural differences between outbred individuals and the albino frog we carried out Southern blotting with two additional DG cDNAs (Figure 4). The patterns displayed by mixed DNA from outbred frogs are shown in the A lanes, while the albino patterns are presented in lanes B. While some differences between outbred and albino are evident, especially for DG 70, it is clear that the differences are minor when compared to the situation with DG 42. Thus, the DG 42 family may behave in an atypical manner, and other X. laevis genes may be more stable in their genomic arrangement in frog populations.

SUMMARY AND OUTLOOK

A major objective of modern developmental biology is to establish biochemical relationships between the regulated expression of individual genes and observable processes of morphogenesis. Genetics has been a useful tool in this pursuit, particularly with Drosophila melanogaster, for which a large number of mutations have been catalogued that affect development. Unfortunately, mutational analysis has been of limited use in studying the embryonic development of most other animals. Vertebrates have been especially intractable, mainly due to their long generation times and the impracticality of raising and screening large numbers of mutagenized individuals. The problem is that of identifying genes likely to be involved in coordinating differentiation without the aid of genetics.

Our approach has been an elaboration of that used by others: to isolate by molecular cloning genes that are specifically expressed at a time and place at which differentiation is known to occur. Xenopus gastrulation is an excellent subject for this type of analysis. There is a distinct separation between the beginning of embryogenesis at fertilization and the onset of nuclear control of differentiation at gastrulation. This separation is punctuated by several hours of transcriptional silence during cleavage and blastulation and makes it possible to compare the RNAs present in cells capable of further differentiation, i.e., the gastrula, to their immediate progenitor cells at blastula or earlier stages that lack this capability. The differences ought to include transcripts of genes that control morphogenesis. The selective cDNA cloning procedure we applied to this problem has resulted in the successful cloning of RNAs that fulfill the requirement of presence in gastrula and absence in earlier stages.

Considering the fact that the oocyte contributes a large and complex store of RNAs, one wonders why the early embryo should have a requirement for the accumulation of qualitatively new mRNAs, i.e., why there is a class of DG RNAs. Even though the protein product of these mRNAs may not be required -- or may even be unacceptable -- in the egg and blastula, the mRNAs might have been stored in an inactive state until needed, especially since translational regulation is known to occur in frog embryos (22,23). We do not know the biological reason for delaying the accumulation of DG RNAs until after the MBT, but it appears promising to look for general differences between DG RNAs and maternal mRNAs and their protein products.

Our present and future efforts center on two objectives: First, assuming that at least some of the cloned DG genes are important in controlling development, we wish to learn the biochemical functions of as many DG genes as practical. Information on the precise time and place of expression is of some use in inferring possible roles for these genes, but we recognize that it is usually not RNA that mediates gene function, but protein. One way to study DG proteins is to synthesize peptides based on the amino acid sequence predicted from the nucleotide sequence of the cDNA clone. Such peptides can be used to elicit antibodies capable of recognizing the cognate moiety of the native protein (24). Once such antibodies are available it should be possible to characterize the products of DG genes at the molecular level and to identify their localization in the embryo and in the cell.

Our second general objective is independent of the function of DG genes. Many of the DG RNAs we have studied are expressed at high levels only during gastrulation and neurulation. Most are present at peak levels at distinctive times and at least some are expressed in specific areas of the embryo. In other words, DG gene expression is regulated during early development, and we are investigating the molecular basis of this gene regulation. Transcriptional regulation is likely to be a component of this phenomenon, but posttranscriptional events may contribute in an important way. Our primary model is DG 42, and we have cloned several of the nuclear DG 42 genes from Xenopus genomic DNA libraries. These genomic clones include flanking regions which have been shown to harbor regulatory elements in other eukaryotic genes. Hybrid genes composed of DG 42 flanking regions and marker sequences can be introduced into fertilized egg by microinjection, and should allow detection of transcription products in the embryo, and eventually permit mutagenesis experiments to map sequences responsible for DG 42 regulation. Information obtained in this way will be of particular value as DG genes, unlike most other genes studied this way, are expressed during early development prior to terminal differentiation.

In summary, we have utilized selective cDNA cloning to demonstrate the existence of and to clone mRNA sequences that are present in gastrula and absent, or much rarer, at earlier stages. Gastrulation is a critical phase of animal development, but investigation of its biochemical basis has been impeded by the lack of molecular markers of gene expression during this period. We anticipate that many of the DG clones will prove useful in this regard.

REFERENCES

1. Smith LD, Richter JD (1984). Synthesis, accumulation, and utilization of maternal macromolecules during oogenesis and oocyte maturation. In Monroy A, Metz C (eds.): "Fertilization," New York: Academic Press, in press.
2. Davidson EH (1976). "Gene Activity in Early Development." New York: Academic Press.
3. Dawid IB, Kay BK, Sargent, TD (1983). Gene expression during Xenopus laevis development. Symp. Soc. Dev. Biol. 41:171.
4. Anderson DM, Richter JD, Chamberlin ME, Price DH, Britten RJ, Smith LD, Davidson EH (1982). Sequence organization of poly(A)RNA synthesized and accumulated in lampbrush stage Xenopus laevis oocytes. J. Mol. Biol. 155:281.
5. Richter JD, Anderson DM, Davidson EH, Smith LD (1984). Interspersed poly(A)RNAs of amphibian oocytes are not translatable. J. Mol. Biol. in press.
6. Brown DD, Littna E (1964). RNA synthesis during development of Xenopus laevis, the South African clawed toad. J. Mol. Biol. 8:669.
7. Bachvarova R, Davidson EH (1966). Nuclear activation at the onset of amphibian gastrulation. J. Exp. Zool. 163: 285.
8. Shiokawa K, Tashiro K, Misumi Y, Yamana K (1981). Non-coordinated synthesis of RNAs in pregastrular embryos of Xenopus laevis. Dev. Growth Differ. 23:589.
9. Newport J, Kirschner M (1982). A major developmental transition in early Xenopus embryos: I. Characterization and timing of cellular changes at the midblastula stage. Cell 30:675.
10. Newport J, Kirschner M (1982). A major developmental transition in early Xenopus embryos: II. Control of the onset of transcription. Cell 30:687.
11. Shiokawa K, Misumi Y, Yamana K (1981). Mobilization of newly synthesized RNAs into polysomes in Xenopus laevis embryos. Wilhelm Roux Arch. 190:103.
12. Dworkin MB, Dawid IB (1980). Construction of a cloned library of expressed embryonic gene sequences from Xenopus laevis. Develop. Biol. 76:475.
13. Sargent TD, Dawid IB (1983). Differential gene expression in the gastrula of Xenopus laevis. Science 222:135.

14. Hastie N, Bishop JO (1976). The expression of three abundance classes of messenger RNA in mouse tissues. Cell 9:761.
15. Perlman S, Rosbash M (1978). Analysis of Xenopus laevis ovary and somatic cell polyadenylated RNA by molecular hybridization. Develop. Biol. 63:197.
16. Angerer LM, Angerer RC (1981). Detection of poly(A)$^+$ RNA in sea urchin eggs by quantitative in situ hybridization. Nucl. Acids Res. 9:2819.
17. Hafen E, Levine M, Garber RL, Gehring WJ (1983). An improved in situ hybridization method for the detection of cellular RNAs in Drosophila tissue sections and its application for localizing transcripts of the homeotic Antennapedia gene complex. EMBO J. 2:617.
18. Jamrich M, Mahon KA, Gavis ER, Gall JG (1984). Histone RNA in amphibian oocytes visualized by in situ hybridization to methacrylate embedded tissue sections. Manuscript submitted for publication.
19. Wahli W, Dawid IB, Ryffel GU, Weber R (1981). Vitellogenesis and the vitellogenin gene family. Science 212:298.
20. Chien Y-H, Dawid IB (1984). Isolation and characterization of calmodulin genes from Xenopus laevis. Mol. Cell. Biol. 4:507.
21. Bisbee CA, Baker MA, Wilson AC, Hadji-Azimi I, Fischberg M (1977). Albumin phylogeny for clawed frogs (Xenopus). Science 195:785.
22. Woodland HR (1980). Histone synthesis during development of Xenopus. FEBS Lett. 121:1.
23. Dworkin MB, Hershey JWB (1981). Cellular titers and subcellular distributions of abundant polyadenylate-containing ribonucleic acid species during early development in the frog Xenopus laevis. Mol. Cell. Biol. 1:983.
24. Niman HL, Houghten RA, Walker LE, Reisfeld RA, Wilson IA, Hogle JM, Lerner RA. (1983). Generation of protein-reactive antibodies by short peptides is an event of high frequence: Implications for the structural basis of immune recognition. Proc. Natl. Acad. Sci. USA. 80:4949.
25. Maxam AM, Gilbert W (1980). Sequencing end-labeled DNA with base-specific chemical cleavages. Methods Enzymol. 65:400.

Molecular Biology of Development, pages 109–118

THE ACTIVATION OF ACTIN GENES IN EARLY XENOPUS DEVELOPMENT[1]

J.B. Gurdon, Sean Brennan,[2] Sharon Fairman, Nina Dathan, and T.J. Mohun

CRC Molecular Embryology Unit, Department of Zoology, Cambridge University, CB2 3EJ, England.

ABSTRACT Gene-specific cDNA probes in M13 have been prepared for three types of *Xenopus laevis* gene transcript, namely for γ-cytoskeletal, α-cardiac, and α-skeletal actin mRNA. We have used these probes to define the stage in development (end of gastrulation) when the transcription of these genes is first detected, and the region of an embryo (somites) where α-actin transcripts first appear. By culturing isolated pieces of mid-blastulae, we find that a commitment for α-actin gene activation has already been established in the mesoderm, but not in the ectoderm or endoderm, of embryos at this stage.

INTRODUCTION AND BACKGROUND

The injection of genes into amphibian oocytes and eggs has been used in several laboratories to analyze various aspects of gene activity (ref. 1, review). In most cases the genes injected have been from non-amphibian sources, and their transcription in *Xenopus* oocytes may not necessarily reflect normal mechanisms of regulation. For this reason, we have recently concentrated our attention on amphibian genes which are normally subject to regulation in amphibian oocytes or eggs.

[1]This work is supported by Cancer Research Campaign.
[2]Supported by NIH Fellowship No. GM 08810.

We have made substantial use of the family of genes which code for 5S RNA. There are two main classes of 5S genes (ref. 2, review). One class, the somatic type, is transcribed in all tissues, and appears to behave like constitutive genes; the other class, the oocyte type, is transcribed strongly in oocytes but not in most somatic cells. When isolated 5S genes or whole erythrocyte nuclei containing these genes in a non-transcribed state are injected into oocytes, they become active in transcription, as are the oocytes' own endogenous 5S genes. One interpretation of this effect is that oocytes contain a 5S gene activating substance which is needed to activate endogenous 5S genes and which is present in sufficient excess to also activate injected 5S genes. We now suspect that any such components are not gene-specific, but help to activate all genes transcribed by RNA polymerase III, and perhaps other genes also. This is because we recently discovered another family of genes, called OAX, which are also strongly activated when injected into *Xenopus* oocytes (3)(4), and are transcribed by polymerase III. Transfer RNA genes, transcribed by polymerase III, are very active in oocytes. Thus it seems that *Xenopus* oocytes may contain conditions or components which ensure the full activity of all genes transcribed by RNA polymerase III.

In the hope of finding a *gene-specific* regulatory mechanism, we investigated 5S gene transcription in early embryos. We found that endogenous 5S oocyte genes, as well as those in transplanted nuclei, are transcribed at a low level in late blastulae, but then become inactive during gastrulation, while the 5S somatic genes continue to be actively transcribed (5)(6). Early embryos therefore appear to possess a mechanism which differentially regulates these two closely related classes of genes.

Information that a study of 5S genes seems unable to provide is an insight into the mechanism by which embryos activate genes in a region- or tissue-specific way. As far as is known, 5S genes are expressed in all tissues, in contrast to genes active only in certain cell-types such as muscle. We have therefore initiated a study of the actin gene family and their regulation in *Xenopus* development. By analogy with 5S genes, actin genes fall into two classes. One, the cytoskeletal actins, appears to be expressed in all cell types. The other, tissue-specific class is expressed only in muscle cells.

METHODS

All are described or referred to in forthcoming publications (7)(8).

RESULTS AND DISCUSSION

The Preparation of Gene-Specific Probes.

cDNA clones were prepared from polyA$^+$ RNA of Xenopus embryos (mostly gastrulae and neurulae), and from Xenopus adult muscle. They were screened against a clone containing the coding region of a chicken β-actin cDNA, and transferred to M13 as subclones. A large number of subclones has been sequenced, and we now know the complete sequence of the 3' untranslated region, and at least half of the coding region, for 3 types of actin cDNA. After comparing the coding sequences of these cDNAs with the known coding sequences of other vertebrate actin cDNAs, we have classified our cDNAs as representing (i) a γ-cytoskeletal actin gene, (ii) a cardiac α-actin gene, and (iii) a skeletal muscle α-actin gene. We have used clones containing 200-300 bases of 3' untranslated region as gene-specific probes with which to recognize transcripts from these types of actin genes. When tested against RNA extracted from adult tissues, we have confirmed the identification of the three cDNAs as classified above. For example, the α-cardiac probe hybridizes to RNA from adult heart but not to RNA of skeletal muscle or to that of several other adult tissues. The preparation of the cDNA clones and the specificity of the M13 probes has been summarized elsewhere (7).

The Time of Activation of Actin Genes.

Using our gene-specific M13 probes we have determined the amounts of RNA complementary to the three kinds of genes we can moniter, using Northern and S1 nuclease protection assays. α-cardiac and α-skeletal transcripts are first detected just after the end of gastrulation (Xenopus stage 14, ref. 9); γ-cytoskeletal transcripts show a substantial increase at the same stage, but are present at a significant and constant level in all preceding stages (i.e. in half-sized oocytes, mature oocytes, eggs, and early embryos). An example of S1 analyses showing temporal gene activation is seen in Fig. 1 (see also ref. 7).

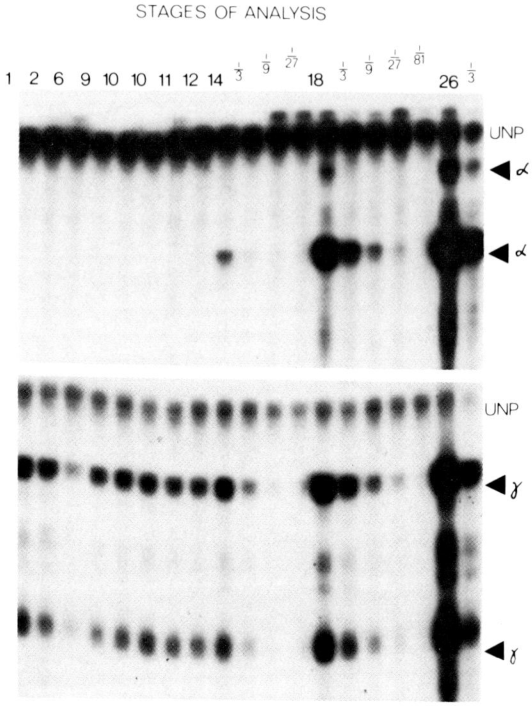

FIGURE 1. S1 nuclease protection analysis of RNA from Xenopus laevis embryos. The stages are those of Nieuwkoop and Faber (9). For each track, the RNA from half of one embryo was used. UNP, undigested probe. Arrows show protected DNA fragments characteristic of α-cardiac or γ-cytoskeletal actin mRNAs. The fractions show dilutions of the sample from the previous stage.

The time of actin gene activation deserves further comment. We may first ask whether stage 13-14 is a time of Xenopus development when many other genes are activated, or whether the mechanism responsible is specific to actin genes, and perhaps to a few others. A survey of previous reports shows that no identified genes are known to be first transcribed at stage 13 or 14. Ribosomal genes, transcribed by polymerase I, as well as 4S and 5S genes, transcribed by polymerase III, start to be transcribed before gastrulation (10,11). Genes transcribed by polymerase II before the end of gastrulation include those which code for histone H1 (12), for certain ribosomal proteins (13), for several proteins which differ in X.borealis and X.laevis (14), as well as genes whose transcripts appear during cleavage but disappear during gastrulation (15). Several proteins first appear at stages between neurula and tailbud in X.laevis - X.borealis hybrids (14). We therefore conclude that the time of activation of α-actin genes is not shared by other genes so far identified.

The time when α-actin transcripts first appear is followed almost immediately by the appearance of α-actin protein, which has been recorded at stage 14 (16,17). Therefore actin transcripts once synthesized seem to be immediately translated. At the other end of the time scale, we may ask whether actin gene transcription coincides with the time when cells become committed to differentiate as muscle. Embryological experiments in which pieces of gastrulae or blastulae are cut out, and either cultured in isolation or transplanted to other embryos, have established that in an early gastrula, and even in a mid-blastula, mesoderm cells are committed to form muscle (18,19) and to activate α-actin genes (see below). It is therefore clear that the time of commitment (stage 8), as judged by explant or transplant experiments, precedes by at least 10 hours the time of actual actin gene transcription (stage 13).

A further point worth mentioning comes from a comparison of actin transcripts recognized by the Northern and S1 nuclease procedures. A Northern analysis reveals mRNAs which hybridize to the probe only if they are of one or a few defined size classes. In contrast, our S1 probes, which contain only 200-300 nucleotides in the 3' untranslated region of actin mRNA, will recognize any complementary RNAs even if they differ in the positions of their 5' and 3' ends. The fact that both procedures first detect transcripts at a similar stage of development permits us to conclude that α-actin genes are at once transcribed into fully processed mRNAs. The temporal activation at stage 13-14 does not therefore reflect the processing of mRNAs which could have been transcribed some hours earlier, but rather a true activation at the transcriptional level.

The Region of Activation of Actin Genes.

An important question in development is whether a gene whose expression is tissue-specific in larval and adult stages is first activated in embryos with the same regional specificity. At stage 18, only a few hours after α-actin genes have started transcription, a *Xenopus* embryo contains a mesodermal layer of cells which is clearly separate from endoderm and ectoderm, and within the mesoderm itself a mid-dorsal region, the notochord, is clearly demarcated, and flanked on each side by somitic mesoderm. One possibility is that α-actin genes are first transcribed in all mesoderm cells but not in the endoderm or ectoderm, and that their expression is gradually narrowed down, in a series of subsequent steps, to the specialized (muscle) cells of which they are eventually characteristic. The other possibility is that, from the earliest time of their transcription, actin gene expression is already confined to those parts of the mesoderm from which muscle arises. The large size of amphibian embryos makes it possible to dissect out the notochord, somite and other nearby regions of a neurula, and the sensitivity of an S1 nuclease assay permits the recognition of transcripts in only a few hundred cells. The results have shown that, at stage 18 (neural folds), the notochord has no detectable α-actin transcripts, whereas an equivalent amount of somite tissue is intensely positive. The regions tested, and the results obtained, are shown diagrammatically in fig. 2. Details are given in reference 7.

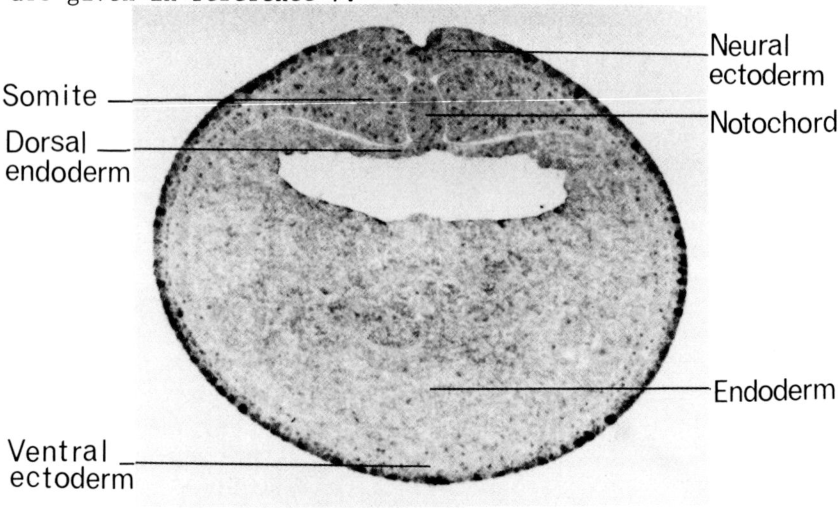

FIGURE 2. Transverse section through a stage 18 neural folds *Xenopus* embryo indicating the regions tested for the presence of α-actin transcripts.

We now have sufficient information to estimate the actual rate of transcription of α-actin genes in early development. The somite tissue in which these genes are strongly expressed at stage 18 constitutes about 5-10% of the whole embryo. If we assume that α-cardiac actin genes start to be transcribed at stage 13 (their transcripts being first detected at stage 14), we can calculate how many "gene-minutes" exist between stages 13 and 18. From published DNA values (20), we calculate that the total cell number per embryo increases from 50,000 at stage 13, to 67,000 at stage 18; at 23°C, this takes 5 hours. We calculate that 284,000 cell/hours, or 17×10^6 cell/mins, exist from stage 13 to stage 18. Assuming that there are two α-cardiac actin genes per diploid nucleus (Brennan *et al.*, unpubl.), there would be 4×10^7 α-cardiac gene/mins from the start of their transcription to stage 18. If only 5% of an embryo's cells have active α-actin genes, the total content of α-cardiac transcripts must be made in 2×10^6 gene/mins. If we assume a rate of gene transcripton of 10 transcripts/gene/min (15), then a stage 18 embryo would contain 2×10^7 transcripts.

One reason for believing that the abundance of transcripts estimated above may be correct comes from the following calculation. We estimate the specific activity of our α-cardiac probe to be about 10^8 c.p.m./µg, and about 50 c.p.m. of probe to be protected by the RNA of a stage 18 embryo. This works out to 5×10^7 α-cardiac actin transcripts per stage 18 embryo; this is a figure indistinguishable, in view of the assumptions and approximations above, from the value of 2×10^7 transcripts just reached. If these estimates are correct, we observe a rather simple pattern of α-actin gene transcription. It appears that, in about 5% of the cells of stage 13-18 embryos, cardiac α-actin genes start to be transcribed at the maximum rate. As soon as transcripts are formed, they are fully processed, translated and stable.

Commitment to α-Actin Gene Activation.

We have next asked whether the activation of α-actin genes at the end of gastrulation results from a commitment at an earlier stage of development, or whether it depends on the complicated movements and associated cell interactions that take place at gastrulation. We have inhibited normal gastrulation events by causing exogastrulae to be formed in a high salt medium, or by isolating pieces of mid-blastulae. Yet in both these cases, we observe α-actin transcription in mesodermal tissue at the normal time, but not in ectoderm or endoderm pieces. We have found that the cells contained in the mesoderm of a mid-blastula can be dissociated, mixed, and reaggregated, and still we observe normal α-actin gene activation. It seems clear from these results that a commitment to activate α-actin genes has already been established by the mid-blastula stage (stage 8, 4000-8000 cells).

A stage 8 mid-blastula embryo in *Xenopus* is formed about 5 hours after fertilization by a partitioning of egg cytoplasm. The most obvious cytoplasmic components of an egg, such as yolk and pigment, have much the same distribution in a blastula as in an egg. It is therefore possible that the commitment of mid-blastula mesoderm cells to activate their α-actin genes after gastrulation exists in the form of cytoplasmic determinant molecules originally localized in an egg. To find out whether this concept of gene activation in development is applicable to the muscle-specific actin genes of *Xenopus*, we are now separating blastomeres of very early embryos , and removing regions of cytoplasm from eggs. If it seems likely that determinants for α-actin gene activation exist in very early embryos, we plan to pursue this analysis by injecting genomic clones of *Xenopus* α-actin DNA into eggs, in the hope of identifying DNA sequences with which such determinants may interact.

ACKNOWLEDGEMENTS

We are indebted to W.G. Westley for the histological preparation shown in fig. 2.

REFERENCES

1. Gurdon JB, Melton DA (1981). Gene transfer in amphibian eggs and oocytes. Ann Rev Genet 15:189.
2. Korn LJ, Bogenhagen DF (1982). In "The Cell Nucleus," 12. Academic Press.

3. Wakefield L, Ackerman E, Gurdon JB (1983). The activation of RNA synthesis by somatic nuclei injected into Amphibian oocytes. Devel Biol 95:468.
4. Ackerman EJ (1983). Molecular cloning and sequencing of OAX DNA: an abundant gene family transcribed and activated in Xenopus oocytes. EMBO Jl 2:1417.
5. Wakefield L, Gurdon JB (1983). Cytoplasmic regulation of 5S RNA genes in nuclear-transplant embryos. EMBO Jl 2:1613.
6. Wormington WM, Brown DD (1983). Onset of 5S RNA gene regulation during Xenopus embryogenesis. Devel Biol 99:248.
7. Mohun et al., in preparation.
8. Gurdon et al., in preparation.
9. Nieuwkoop PD, Faber J (1956). Normal Table of Xenopus laevis (Daudin). North-Holland Publ Co Amsterdam.
10. Shiokawa K, Misumi Y, Yasuda Y, Nishio Y, Kurata S, Sameshima M, Yamana K (1979). Synthesis and transport of various RNA species in developing embryos of Xenopus laevis. Devel Biol 68:503.
11. Busby SJ, Reeder RH (1983). Spacer sequences regulate transcription of ribosomal gene plasmids injected into Xenopus embryos. Cell 34:989.
12. Woodland HR, Flynn JM, Wyllie AJ (1979). Utilization of stored mRNA in Xenopus embryos and its replacement by newly synthesized transcripts: histone H1 synthesis using interspecies hybrids. Cell 18:165.
13. Pierandrei-Amaldi P, Campioni, N. (1982). Expression of ribosomal-protein genes in Xenopus laevis development. Cell 30:163.
14. Woodland HR, Ballantine JEM (1980). Paternal gene expression in developing hybrid embryos of Xenopus laevis and Xenopus borealis. J Embryol exp Morph 60:359.
15. Sargent TD, Dawid IB (1983). Differential gene expression in the gastrula of Xenopus laevis. Science 222:135.
16. Ballantine JEM, Woodland HR, Sturgess EA (1979). Changes in protein synthesis during the development of Xenopus laevis. J Embryol exp Morph 51:137.
17. Sturgess EA, Ballantine JEM, Woodland HR, Mohun PR, Lane CD, Dimitriadis GJ (1980). Actin synthesis during the early development of Xenopus laevis. J Embryol exp Morph 58:303.
18. Slack JMW, Forman D (1980). An interaction between

dorsal and ventral regions of the marginal zone in early amphibian embryos. J Embryol exp Morph 56:283.

19. Nakamura O, Takasaki H, Mizohata T (1970). Differentiation during cleavage in Xenopus laevis. Proc Jap Acad 46:694.
20. Dawid IB (1965). Deoxyribonucleic acid in Amphibian eggs. J Mol Biol 12:581.

Molecular Biology of Development, pages 119–128

ISOLATION AND CHARACTERISATION OF A CELL[1] LINEAGE-SPECIFIC CYTOSKELETAL ACTIN GENE FAMILY OF *STRONGYLOCENTROTUS PURPURATUS*

Rosemary J. Shott-Akhurst,[2] Frank J. Calzone,
Roy J. Britten, and Eric H. Davidson

Division of Biology, California Institute of Technology
Pasadena, California 91125 USA

ABSTRACT We have isolated and characterised λ genomic isolates of the cytoskeletal actin gene subfamily, CyIII, which shows expression limited to the aboral ectoderm cell lineage of the early sea urchin embryo. This gene subfamily has three members, designated CyIIIa, b and c. Genes CyIIIa and b are both expressed early in embryogenesis, CyIIIa contributing the majority of actin transcripts at this stage. These two genes are linked 5'-3' CyIIIa-CyIIIb, separated by 6 kb and transcribed in the same direction. The third gene, CyIIIc, is a pseudogene which lacks sequences at the 5' end of the gene. It is not linked to the other two genes. We have sequenced the CyIIIa gene. Three introns are present whose positions are shared by other sea urchin cytoskeletal actin genes. A 5' non-coding region intron is located 26 bp upstream from the initiation codon, one intron occurs between amino acids 121 and 122 and a third intron is positioned between amino acids 203 and 204.

[1]This research was supported by the National Institute of Child Health and Human Development (HD 05753). R. J. S. was supported by a long-term EMBO fellowship and F. J. C. by an American Cancer Society Fellowship.

[2]Present address: St. Mary's Hospital Medical School, Department of Biochemistry, Paddington, London W2 1PG England.

INTRODUCTION

The genome of the sea urchin, ***Strongylocentrotus purpuratus***, contains eight non-allelic actin genes (1). In previous studies we demonstrated that at least six of these are actively transcribed at some stage in the life-cycle, five encoding cytoskeletal actins, CyI, CyIIa, CyIIb, CyIIIa and CyIIIb, and one encoding a muscle acting gene M (2). Each gene exhibits a unique temporal and spatial pattern of expression despite the extreme similarity between genes and, often, their juxtaposition within the genome (2, 3).

To date we have characterised the genomic organization of four of these genes, namely CyI, CyIIa and CyIIb, which are closely linked within the genome, and gene M, which is unlinked to other actin genes (1, 4). In this communication we report the isolation and characterisation of the lineage-specific cytoskeletal actin gene subfamily composed of the three CyIII actin genes. In the gastrula stage embryo, when CyIII transcript prevalence attains its highest values, over 50% of the total actin messenger RNA is transcribed from the CyIIIa actin gene, although none of these genes are active in any adult sea urchin tissues (2). *In situ* hybridisation experiments reveal that the expression of the CyIII actin genes is restricted to cells of the aboral ectoderm cell lineage (3, 5). This resembles the expression pattern of another set of genes that could be functionally related, the Spec genes, which encodes a family of troponin C-related calcium-binding proteins (6). An analysis of the structure and organization of this actin gene family should thus provide unique opportunity to probe both the mechanisms controlling differential expression of structurally-related actin genes and those coordinating regulation of genes which are functionally related.

RESULTS

a) DNA Probes Specific to the CyIII Actin Genes

Several subcloned DNA probes were utilised in the analysis of gene number and genomic organization of the CyIII actin genes. Probes specific to the 5' (amino acids 1-121) and 3' (amino acids 204-374) actin coding regions and to the 3' non-coding regions of genes, CyIIIa and CyIIIb were described previously (1-2).

Two independent 5' non-coding CyIIIa-specific gene probes were utilised in this study. The first was generated from a near full length cDNA clone, p9K2, complementary to the CyIIIa gene.

This cDNA was isolated from a gastrula stage cDNA library constructed by dG, dC tailing of double-stranded cDNA and ligation into the *Pst*I site of pBR322 (library provided by W. Klein). Digestion of this clone with *Pst*I and *Sal*I releases a 110 bp fragment containing the 5' non-coding DNA region plus 30 bp of the 5' coding region. Under the hybridisation conditions used this fragment hybridises specifically to CyIII genes, showing no cross-reaction with λ genomic isolates of CyI, IIa, IIb or M (data not shown). The second 5' non-coding CyIIIa-specific probe was an M13 subclone of the first exon of a CyIIIa genomic isolate, λBK23 (Fig. 3a) which contains 220 bp of 5' non-coding DNA sequences only.

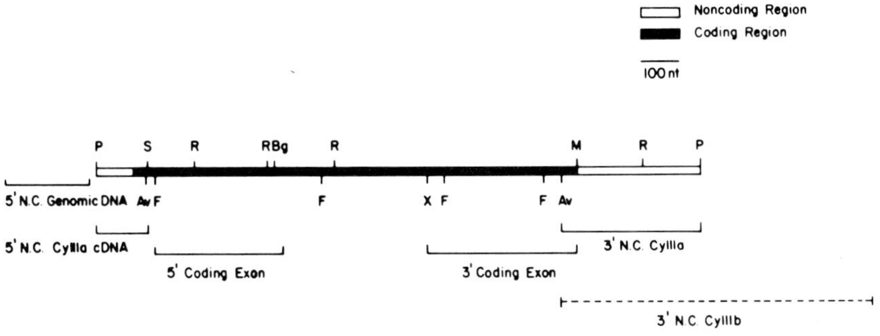

Figure 1. Restriction map of CyIIIa cDNA clone, p9K2. Restriction sites shown are *Sal*I (S), *Rsa*I (R), Hinf I (F), *Bgl*I (B), *Mbo*II (M), *Xho*I (X) and *Ava*II (A). Also shown are the position of the subcloned DNA probes used in this study. The cDNA insert is excised from the plasmid vector with *Pst*I (P).

A summary of positions of subcloned probes utilised relative to the CyIIIa DNA clone, p9K2, is shown in Figure 1.

b) CyIII Actin Gene Number

Genomic blots were performed at high stringency (0.75 M Na^+, 70°C hybridisation; 0.015 M Na^+, 70°C wash) with *Eco*RI-digested genomic DNA isolated from three sea urchin individuals, B, M and S (Fig. 2). Differences in the sizes of CyIII gene-containing *Eco*RI fragments revealed in DNA from different individuals is due to the extensive population sequence polymorphisms exhibited by the sea urchin genome (1).

The 3' non-coding CyIIIa and CyIIIb gene probes both hybridised to the same 6-7 restriction fragments within each individual's DNA, although the relative autoradiographic signal intensities of bands did differ depending on the probe used. (Compare Figs. 2a and 2b). These results demonstrate the conservation of 3' non-coding nucleotide sequences between members of the CyIII gene subfamily. DNA sequence analysis of genes CyIIIa and CyIIIb in fact shows only 10% nucleotide mismatch within the 3' non-coding regions of these genes. Individuals M and S contained six CyIII gene-containing *Eco*RI restriction fragments whereas individual B displayed an additional hybridising band (Fig. 2). Analysis of λ genomic isolates of the CyIII genes shows that the seventh band observed in individual B's DNA is due to the location of an *Eco*RI site within the 3' non-coding region of one allele of gene CyIIIc (Fig. 3). Thus, each diploid genome contains six CyIII alleles, or three genes.

The 3' non-coding CyIIIa-specific probe hybridises preferentially to two of the six bands (Fig. 2a), namely the two alleles of the CyIIIa gene. Conversely, the 3' non-coding CyIIIb-specific probe shows strong hybridisation to the remaining four allelic restriction fragments (five in individual B). These two genes have been designated CyIIIb, the homologous gene, and CyIIIc, a very closely related actin gene (<10% nucleotide divergence in the 3' non-coding regions).

Genomic blots were also performed using the 5' non-coding CyIIIa gene-specific probe (Fig. 2c). In this case, only four different hybridising *Eco*RI restriction fragments were observed, even using less stringent hybridisation conditions (0.75 Na^+, 60°C). It is therefore apparent that either the 5' non-coding nucleotide sequence of one of the CyIII genes must have diverged considerably from that of CyIIIa (>20%) or that one of the CyIII genes does not possess this region of the gene. Comparative analysis of restriction fragment sizes observed in the genome blots and of λ genomic isolates indicates that gene CyIIIc is in fact lacking a 5' non-coding region (see below).

c) Genomic Organization of CyIII Actin Genes

A λ genomic DNA library was constructed by partial *Mbo*I 3A digestion of DNA from individual B, followed by ligation of this DNA into the *Bam*HI site of the λ vector, EMBL3 (7). A portion of the library, representing 10 genome equivalents, was screened using the 3' non-coding CyIIIa- and CyIIIb-specific gene probes at a non-stringent hybridisation condition (60°C, 0.75 M Na^+). Thirty-six positive plaques were obtained of which 29 were plaque purified and characterised.

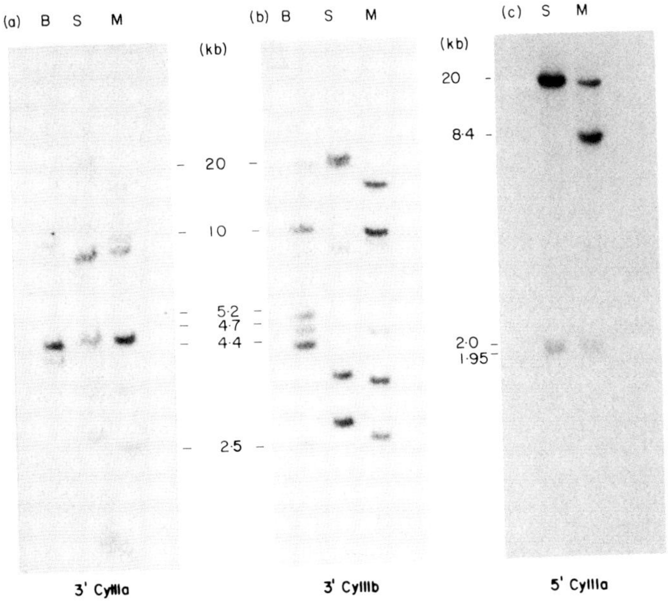

Figure 2. Genomic blot analysis of CyIII genes. Genomic DNAs from sea urchin individuals B, M and S were digested with *Eco*RI and subjected to genomic blotting using the following probes: a) 3' non-coding CyIIIa gene-specific probe, b) 3' non-coding CyIIIb gene-specific probe, c) 5' non-coding CyIIIa gene-specific probe. Gels a) and b) were run in parallel. In Figure 2b, individual B, seven bands were observed, two from gene CyIIIa (4.7 kb), two from gene CyIIIb (10 kb and 4.4 kb) and three from gene CyIIIc (20 kb and 5.2 kb) confirmed by analysis of genomic DNA isolates. In Figure 2c the 5' non-coding CyIIIa probe detects two alleles form gene CyIIIa (20 kb and 8.4 kb) and two alleles from gene CyIIIb (2 kb and 1.95 kb).

The 29 isolates (designated λBK's) were classified into gene types on the basis of restriction site differences within the 3' non-coding regions of the gene. Although there is strong nucleotide sequence conservation between the three genes, DNA sequencing of the 3' non-coding probes for CyIIIa and CyIIIb revealed the presence of an *Rsa*I site within this region of gene CyIIIa which was absent from gene CyIIIb (and CyIIIc). Also a *Dde*I site shared by genes CyIIIa and CyIIIb was found to be absent from gene CyIIIc. Thus, by performing double restriction enzyme digests with *Ava*II and *Rsa*I or *Dde*I followed by hybridisation with the 3' non-coding CyIII gene probes, it was possible to classify the 29 CyIII λ isolates as representatives of either gene CyIIIa, b or c. This classification was supported by the differential hybridisation characteristics of the λ genomic isolates with the 3' non-coding CyIIIa and CyIIIb probes (see Fig. 2) and by other differences in their restriction enzyme patterns (Fig. 3). Five of the isolates were classified as containing gene CyIIIa, eight contained gene CyIIIb and 16 gene CyIIIc.

Representative clones from all three genes were characterised in detail. The transcriptional orientation of each gene was determined by restriction mapping, using the 5' non-coding CyIIIa cDNA probe and the 3' non-coding CyIII gene probes. These results are summarised in Figure 3.

None of the 16 CyIIIc λ genomic isolates hybridised with either the 5' non-coding CyIIIa cDNA probe, or the 5' exon coding probe. These CyIIIc clones did, however, react with the 3' coding exon probe, thus allowing the transcriptional orientation of the gene to be assessed.

The 5' non-coding CyIIIa cDNA probe hybridised to two regions at either end of the CyIIIa genomic isolate, λBK22, although this isolate contained only one region hybridising with the actin coding region probes (Fig. 3). This suggests linkage of CyIII genes. Since the 5' non-coding CyIIIa DNA sequence is represented in only two genes (Fig. 2c) and CyIIIc lacks this 5' region, these data indicate that genes CyIIIa and CyIIIb are linked. This was confirmed by isolating and labelling the 2 kb *Sal*I-*Eco*RI fragment from the extreme 5' end of the CyIIIb genomic isolate, λBK17 (Fig. 3). As predicted, this probe cross-reacted with the corresponding region of the CyIIIa genomic isolate, λBK22 (data not shown). The two genes are thus linked in the same transcriptional orientation, 5' → 3', CyIIIa → CyIIIb. The two genes are separated by 6 kb.

d) CyIIIa Gene Structure

The primary structure of the major embryonic actin gene, CyIIIa, was determined by selective Sanger M13 dideoxy DNA sequencing of genomic clones, λBK22 and 23 (8), and comparison with that of the CyIIIa cDNA clone, p9K2 (Fig. 1). This sequence analysis revealed the presence of two introns within the coding region at amino acid positions 121/122 (900 nucleotides) and 203/204 (300 nucleotides) and a third 2 kb intron in the 5' non-coding region 26 nucleotides upstream from the initiation codon. Thus, the primary transcript size for the 1.8 kb CyIIIa mRNA is estimated at 5 kb.

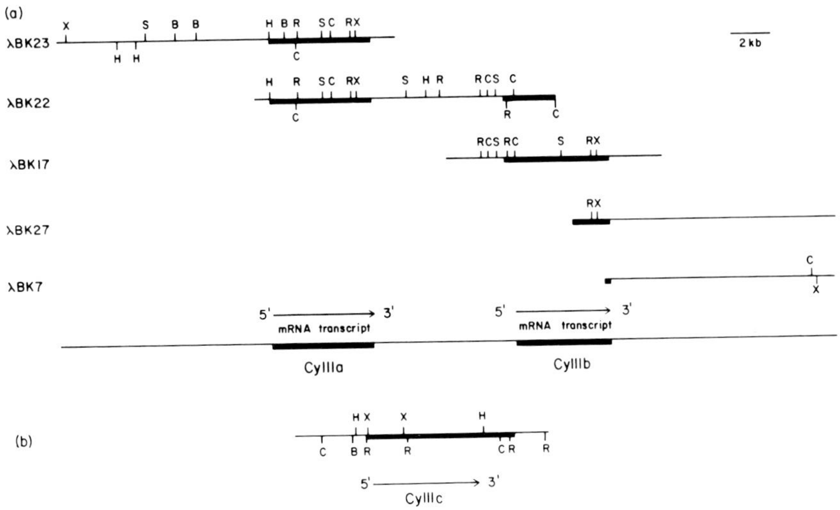

Figure 3. Restriction map of CyIII λ genomic isolates. Restriction enzymes shown are *Xho*I (X), *Sal*I (S), *Hin*dIII (H), *Bam*HI (B), *Sac*I (C), *Eco*RI (R). a) Maps of CyIIIa and CyIIIb genomic λ isolates. Primary transcript sizes are shown. b) Map of a CyIIIc genomic isolate, λBK1. 5' → 3' directionality is indicated. *Sal*I sites located within the polylinker of each recombinant EMBL3 phage vector may be used to release the genomic DNA inserts from the arms.

The coding region intron positions observed in gene CyIIIa are shared by all the sea urchin actin genes (4, 9; D. Durica, personal communication). In addition, at least genes CyI (9) and CyIIIb (D. Durica, personal communication) share the intron position at -26 nucleotides.

DISCUSSION

a) Actin Gene Structure and Evolution

The isolation and characterisation of λ genomic isolates representing members of the CyIII actin gene subfamily concludes a complete analysis of the gene number, structure and short range linkage of actin genes in the sea urchin genome. We have confirmed that the CyIII gene subfamily contains three members (1). Two of these are active genes, CyIIIa contributing the majority of actin transcripts (1.8 kb) during embryogenesis and CyIIIb supplying a minor portion of actin transcripts (2.1 kb) at this stage. The absence of a 5' coding exon region in gene CyIIIc suggests either the presence of a very large intron within the coding region (>3.5 kb) or that the CyIIIc gene is a pseudogene which lacks the first two or three exons of this gene subfamily. Genomic blotting experiments using the 5' non-coding CyIIIa-specific probe would support the latter hypothesis (Fig. 2c). Furthermore, preliminary sequence data (unpublished) shows a nucleotide insertion at codon 213 of gene CyIIIc which would cause a frame-shift mutation.

Assuming that gene CyIIc, like CyIIIc, is a pseudogene (1; J. J. Lee, personal communication) we can now divide the six active actin genes into three linkage groups. Genes CyI, CyIIa and CyIIb are linked with the transcriptional orientation in a 5' → 3' direction, (4, 9). Likewise genes CyIIIa and b are linked with the same transcriptional orientation in a 5' → 3' direction and the third gene, M, is unlinked to either of these groups, at least within a range of 10 kb.

It is now clear that all of the sea urchin cytoskeletal actin genes contain introns within their protein coding regions at amino acid positions 121/122 and 203/204 (4, 9). The muscle actin gene additionally possesses two further introns at amino acid positions 41/42 and 267/268 (4). As discussed by Scheller *et al.* (4) the evolutionary gene duplication event giving rise to the sea urchin actin multigene family probably occurred since divergence of the deuterostomes as none of these four intron positions are observed in protostomial organisms (4).

It is noteworthy, as discussed previously (e.g., 10-14), that all four of the intron positions in the sea urchin DNA actin gene are also observed in the vertebrate actin genes. This suggests that a gene bearing introns in these positions was present in the lower deuterostome common ancestor of the vertebrates and echinoderms.

b) Actin Gene Linkage and Expression

We have previously discussed the significance, if any, of the linkage of actin genes to the developmental regulation of their expression (1, 11). We concluded that linkage is most likely a consequence of the evolution of the gene family by gene duplication rather than providing any functional role in gene regulation (2). It could be argued that the localisation of the CyIII genes in close proximity to one another places the genes in a regulatory chromosomal "domain" the function of which is to limit the expression of these genes to the cells of the aboral ectoderm. However, we cannot, as yet, distinguish between this possibility and the alternative explanation that the regulatory DNA sequences of gene CyIIIb have not yet diverged sufficiently from those of gene CyIIIa to place the genes under separate controls, so that their expression in the same cell types simply follows from their common ancestors.

In view of the extreme sequence conservation and close linkage of these two CyIII genes it is probably more remarkable that the two genes do in fact show large differences both in the timing of onset of their expression during embryogenesis and in their absolute transcript prevalences within the embryo (2). DNA sequences in close proximity, possibly even within the body of the gene, must be involved in modulating these qualitative and qualitative levels of expression. A comparative analysis of the primary structure of these two genes should provide us with great insight as to the sequences essential in controlling these processes.

ACKNOWLEDGEMENTS

We thank Ted George, who was supported by a Summer Undergraduate Research Fellowship, for his assistance. Mr. James J. Lee provided invaluable assistance in provision of probes, clones and stocks, and in discussion of this ongoing work.

REFERENCES

1. Lee JJ, Shott RJ, Rose SJ III, Thomas TL, Britten RJ, Davidson EH (1984). J Mol Biol 172:149.
2. Shott RJ, Lee JJ, Britten RJ, Davidson EH (1984). Dev Biol 101:295.
3. Cox KH, Angerer LM, Lee JJ, Britten RJ, Davidson EH, Angerer RC (1984). In preparation.

4. Scheller RH, McAllister LB, Crain WR, Durica DS, Posakony JW, Thomas TL, Britten RJ, Davidson EH (1981). Mol Cell Biol 1:609.
5. Angerer RC, Davidson EH (1984). Science, in press.
6. Bruskin AM, Tyner AL, Wells DE, Showman RM, Klein WH (1981). Dev Biol 87:308.
7. Frischauf A-M, Lehrach H, Poustka A, Murray N (1983). J Mol Biol 170:827.
8. Sanger F, Nicklen S, Coulson AR (1977). Proc Natl Acad Sci USA 74:5463.
9. Schuler MA, McOsker P, Keller EB (1983). Mol Cell Biol 3:448
10. Buckingham ME, Minty AJ (1983). In Mackan N, Gregory SP, Flavell RA (eds): "Eukaryotic Genes; their Structure, Activity and Regulation," London: Butterworth.
11. Davidson EH, Thomas TL, Scheller RH, Britten RJ (1982). In Dover GA, Flavell RB (eds): "Genome Evolution," London: Academic Press.
12. Nudel U, Katcoff D, Zakut R, Sham M, Carmon Y, Finer M, Czosnek H (1982). Proc Natl Acad Sci USA 79:2763.
13. Fornwald JA, Kuncio G, Peng I, Ordahl CP (1982). Nucleic Acids Res 10:3861.
14. Cooper AD, Crain WR (1982). Nucleic Acids Res 10:4081.

Molecular Biology of Development, pages 129–143

REGULATION OF TRANSLATION DURING OOGENESIS[1]

L. Dennis Smith[+], J. D. Richter[‡], and M. A. Taylor

[+]Department of Biological Sciences, Purdue University, W. Lafayette, IN. 47907
[‡]Department of Biochemistry, University of Tennessee, Knoxville, Tenn. 37996-0840

ABSTRACT Growing oocytes characteristically synthesize and accumulate mRNA far in excess of their immediate needs. The mechanism(s) which limits the ability of oocytes to translate all of the putative mRNA remains unknown but several possibilities have been suggested including the idea that certain proteins associated with the maternal mRNA prevent translation. In this report, we present evidence that a class of oocyte-specific proteins extractable from native mRNPs may be reconstituted with mRNA in vitro to form particles which resemble native mRNPs. These proteins, but not proteins extracted from somatic cells, repress the translation of the mRNA when tested by injection into Xenopus oocytes. These proteins are present in greatest amount in small oocytes, and decrease as oogenesis progresses. We suggest that in small oocytes, in which only small amounts of mRNA are translated, most putative mRNA may be associated with proteins which prevent translation. As oogenesis progresses, more and more mRNA is required for protein synthesis and the need for proteins which sequester and/or prevent mRNA translation decreases. Thus the "unmasking" of maternal mRNA would be a gradual process and not a single event which occurs in response to a single stimulus.

[1]This work was supported by NIH grant HD 04229 (to LDS) and NIH grant GM 32559 (to JDR)

INTRODUCTION

Growing oocytes of all animal species synthesize and accumulate mRNA for use at some point later in development (1,2). In many organisms, translation of this maternal mRNA is initiated as a result of fertilization or parthenogentic activation. In amphibians, the induction of oocyte maturation results in increased protein synthesis which involves the translation of maternal mRNA (3,4). The mechanism(s) which regulates the translation of maternal mRNA in any species remains unknown. However, three possibilities have received the most attention in recent years. These are: (1) The rate of protein synthesis in oocytes (or unfertilized eggs) is limited by the availability of components of the translational apparatus other than mRNA; activation of maternal mRNA translation coincides with availability of the limiting components. (2) The structural organization of maternal mRNA prevents translation; activation requires structural modification. (3) Proteins associated with maternal mRNA prevent translation; activation involves removal or degradation of the repressor proteins.

The above possibilities are not mutually exclusive. For example, as much as 70% of the mass of maternal poly (A) RNA in sea urchin eggs as well as in Xenopus oocytes displays an interspersed sequence organization in which regions transcribed from single copy and repetitive DNA sequences are covalently attached (5, 6). Most of this interspersed RNA is not translatable (7). However, even considering this, the full-grown oocyte still contains much more bona fide translatable mRNA than is necessary to support observed rates of protein synthesis (3, 7).

Richter et al (4) have demonstrated that increased protein synthesis during oocyte maturation (3) must result from recruitment of mRNA from this pool. However, evidence obtained from injecting heterologous mRNA into oocytes has demonstrated that a component(s) of the translational machinery also is limiting in oocytes (8,9). Presumably, both the limiting component as well as mRNA availability must increase in order for protein synthesis to increase during oocyte maturation.

The idea that proteins associated with maternal mRNA suppress translation derives largely from studies on sea urchin eggs in which ribonucleoprotein particles from unfertilized eggs were reported not to translate in vitro

(10, 11). These results have been questioned in other studies which showed that nonpolysomal RNP from sea urchin eggs are translatable *in vitro* (12, 13). Similar studies have not been performed with amphibian oocytes, although the existence of oocyte-specific proteins associated with nontranslating poly(A)-containing mRNPs from *Xenopus laevis* oocytes has been reported (14, 15).

In this report, we discuss recent experiments in which the oocyte-specific proteins referred to above were tested for their ability to suppress translation of mRNA *in vivo*. We also present quantitative data on protein synthesis in oocytes at several stages of oogenesis which has allowed us to estimate the amount of maternal mRNA which is actually translated during oogenesis. These kinds of experiments have led us to suggest a dynamic model of translational regulation in which the role of putative suppressor proteins would change progressively over time.

RESULTS AND DISCUSSION

The Content of Translatable mRNA in Oocytes

Several studies have set a value of about 90 ng for the content of poly(A) RNA in full-grown stage 6 (16) *Xenopus* oocytes (17, 18). We have estimated that about 15% of the total poly(A) RNA represents mitochondrial transcripts (6). Furthermore, Richter et al have demonstrated that at least 90% of the interspersed poly(A) RNA, representing as much as 70% of the total poly(A) RNA, is translationally inactive when tested both *in vitro* and *in vivo*. After substracting this mass of poly(A) RNA, and the mitochondrial transcript, from the total, we estimate that the stage 6 oocyte contains about 20 ng of poly(A) RNA which is available for translation. Since the total content of poly(A) RNA is reached already by stage 2, and small oocytes also contain the interspersed poly(A) RNA, it is reasonable to suggest that the maternal mRNA "stockpile" is about 20 ng throughout most of oogenesis.

According to Woodland (19), about 2% of the ribosomes in stage 6 oocytes are found in polysomes. Our own studies summarized in Table 1 confirm this value very well. If one assumes that 4% of the polysomal RNA mass is mRNA, then stage 6 oocytes would contain a little more than 4 ng of the putative maternal mRNA on polysomes.

The data in Table 1 also include estimates of the content of ribosomes in polysomes for stage 3 and stage 1 oocytes. In stage 3 oocytes, the single value is not significantly different from that measured at stage 6, and it is not likely that the value obtained for stage 1 is significantly lower. However, both stage 3 and stage 1 oocytes contain substantially less total RNA. Hence, calculated estimates of the content of mRNA being translated are correspondingly lower. We estimate that stage 3 oocytes contain about 0.8 ng of mRNA on polysomes. The corresponding value for stage 1 oocytes is an order of magnitude lower.

Actual measurements of the rates of protein synthesis in oocytes at all stages of oogenesis (data not shown) accurately reflect the differences in calculated mRNA on polysomes listed in Table 1. In fact, from such data, and estimates of translational efficiency, one can obtain independent estimates of mRNA on polysomes. Those values agree very well with the numbers presented in Table 1. We conclude from such studies that the content of mRNA which oocytes translate during oogenesis progressively increases.

Protein synthesis in stage 6 oocytes occurs entirely on maternal templates. Based on the above calculations, at least 20% of the maternal stockpile would need to be translated in stage 6 oocytes. This value would be higher by about 2 fold if one examined stage 6 oocytes from hcG-stimulated females (3). We have not performed enucleation studies on smaller oocytes. Thus, it is not clear to what extent protein synthesis occurs on maternal templates in growing oocytes. However, if one considers that protein synthesis in stage 3 oocytes depends entirely on maternal mRNA, only a fraction of the putative mRNA pool would need to be translated. In fact, it is conceivable that none of the maternal stockpile would need to be translated since the rate of accumulation of newly synthesized heterogeneous RNA in stage 3 oocytes (20) would be sufficient to provide the mRNA found on polysomes within a day.

Since it is clear that translation in stage 6 oocytes is limited by components of the translational machinery, and a significant fraction of the maternal mRNA pool already is found on polysomes, it could be argued that "masking proteins" play a small role in regulating protein synthesis at the completion of oogenesis. Conversely,

Table 1

mRNA Content in Oocytes During Oogensis

Oocyte[a] Stage	RNA[b] Content (ug)	% ribosomes[c] in polysomes	mRNA in[d] polysomes (ng)
1	.15±.01 (2)	1.6±.5 (2)	.024
2	.35±.05 (2)		
3	1.00±.21 (3)	2.4± (1)	.816
4	2.26±.14 (3)		
5	3.14±.14 (3)		
6	4.83±.12 (3)	2.3±.7 (2)	4.00

[a] According to Dumont (16)

[b] Determined as described by Dolecki and Smith (18)

[c] Determined as described by Woodland (19)

[d] Calculated assuming that 25% of the total RNA in stage 1 oocytes is ribosomal (21), about 85% in stage 3 oocytes and about 90% in stage 6 oocytes is ribosomal (22), and that mRNA is 4% of the polysomal mass. Numbers in parentheses represent the number of independent experiments performed.

based on the quantitative data presented above, if such proteins do function in preventing the translation of maternal mRNA, one might expect to find them more concentrated in small oocytes in which little, or none, of the maternal message is being translated.

Developmentally Regulated RNA Binding Proteins During Oogenesis

Darnbrough and Ford (14) originally reported the existence of oocyte-specific proteins which were associated with poly(A) RNA in *Xenopus* oocytes, some of which were observed to decrease during oogenesis. They suggested that the oocyte-specific proteins might be involved in the sequestration and stabilization of stored RNA molecules in the oocyte cytoplasm.

Our own studies on oocyte RNA binding proteins began with a series of experiments in which mRNAs translated on free or membrane bound polysomes were injected into *Xenopus* oocytes. Both classes of mRNA competed with endogenous messages for translation. However, with the membrane bound messages, saturation of the translational machinery occurred at relatively low message concentrations, and injected mRNA accumulated as nontranslating mRNP (9). This mRNA could be recruited onto polysomes by proteins extracted from the rough endoplasmic reticulum (23), leading to the suggestion that cytosolic proteins might exist which could similarly regulate the translation of mRNAs on free polysomes. In order to study this possibility, we isolated proteins from oocytes at different stages of oogenesis and, after electrophoresis and electroblotting onto nitrocellulose paper, assayed for the ability of specific proteins to bind radioactive messages. This procedure enabled us to identify proteins which bound messages preferentially over other RNAs, but did not demonstrate proteins which discriminated between message classes (15). An example of this type of experiment is shown in Fig. 1.

Figure 1 shows the non-yolk proteins soluble in low salt extracted from oocyte stages 1-2, 3-4, and 5-6 after electrophoresis and blotting onto nitrocellulose paper. The three lanes on the left were stained with Coomassie blue and demonstrate that the same prevalent species of proteins were extracted from oocytes at each stage. However, when the same proteins were reacted with

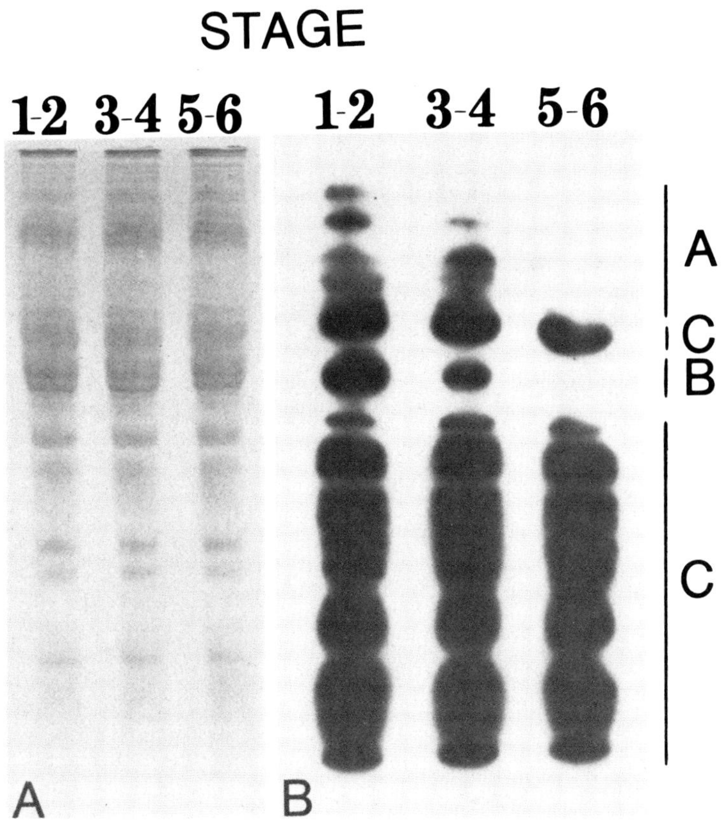

Fig. 1. Detection of RNA binding proteins in total soluble protein from oocytes of different stages. Oocytes of various stages were homogenized in 50 mM NaCl and 0.5 mM of phenyl methysufonyl fluoride and, after centrifugation and precipitation of the proteins in the supernatant the proteins were resuspended in SDS buffer for electrophoresis. After electrophoresis on a 12.5% polyacrylamide gel, the protein was electroblotted onto nitrocellulose paper and reacted with ^{125}I-globin mRNA. The Coomassie blue-stained protein(A) and autoradiograph (B) are shown. From Richter and Smith (15)

^{125}I-globin mRNA followed by autoradiography, a very different pattern was obtained. Some proteins present in small oocytes appeared to decrease as oogenesis progressed (regions A and B), while the remainder of the RNA-binding proteins were retained at a constant level (region C). Several somatic cells also contained RNA-binding proteins soluble in low salt, but these were different from those present in oocyte extracts; proteins in region A and B are oocyte specific.

If we assume that neither the recovery of the proteins from oocyte homogenates nor the binding affinities of the proteins for RNA change at different oocyte stages, the relative content of RNA binding protein at the different stages can be estimated from the level of radioactive globin mRNA bound to the proteins. Table 2 shows these values for the proteins delineated in Fig. 1 as regions A, B, and C. The actual amount of radioactivity in regions A and B decreased by about 95% between stages 1-2 and 5-6, while radioactivity in region C remains relatively constant. However, since each lane contains a constant amount of protein extracted from differing numbers of oocytes, the data in Table 2 also are listed as radioactivity in each region per oocyte equivalent relative to the amount in stage 1-2 oocytes. In this case, the proteins in region A still decrease by 78% from stage 1-2 through stage 5-6, those in region B decrease by 71%, while those in region C actually increase by almost 5 fold. Thus, the stage 1-2 oocytes contain two general classes of RNA binding proteins, those which decrease and those which increase during oogenesis.

Many of the proteins in region C also sediment with monosomes and polysomes when ribonucleoprotein particles from stage 1-2 oocytes are separated by sucrose gradient centrifugation and could include ribosomal proteins (15). However, the proteins in regions A and B are prominent within a region of the gradient corresponding to 40-60S RNPs, an area where non-translating RNPs would be expected to sediment. These basic proteins are also found in 40-60S particles which bind to oligo(dT)-cellulose. These and other experiments suggest that the developmentally regulated RNA binding proteins actually are bound to non-translating poly(A) RNAs in vivo.

Table 2. Relative Amount of RNA Binding Proteins During Oogenesis

STAGE	PROTEIN LOADED (μg)	NO. OOCYTES REPRESENTED	CPM AREA			AMOUNT OF BINDING PROTEIN RELATIVE TO THAT PRESENT IN STAGE 1–2 OOCYTES		
			A	B	C	A	B	C
1–2	100	19	1764	1541	15828	1	1	1
3–4	100	7	576	386	15464	0.88	0.68	2.65
5–6	100	4	82	96	16635	0.23	0.29	4.99

Nonyolk protein from oocytes of various stages was analyzed by electrophoresis on SDS 12.5% polyacrylamide gels. The protein was subsequently electroblotted onto nitrocellulose paper and was reacted with ^{125}I-globin mRNA. After autoradiography, the radioactivity in various regions of each lane (denoted as A, B, and C in Figure 1) was quantitated by liquid scintillation. From Richter, and Smith (15).

Do the Oocyte-Specific Proteins Suppress Translation?

In order to test the possible role of the oocyte-specific RNA-binding proteins in translation, we were influenced by several recent studies which demonstrated that RNAs and proteins may be reconstituted faithfully in vitro (24, 25, 26). Accordingly, we attempted to reconstitute a known message (globin mRNA) with the oocyte proteins, as well as several other proteins, followed by a test of the translatability of the reconstituted RNPs by injection into Xenopus oocytes.

When pure globin mRNA is injected into oocytes, the synthesis of globin is readily detected by extraction and electrophoresis of radioactive oocyte proteins (9). The actual level of radioactivity can be quantitated by cutting out the globin band from polyacrylamide gels and counting the band directly. In practice, this value is set at 100%. Globin mRNA mixed with basic proteins such as histone or ribosomal proteins also was translated after injection into oocytes. Furthermore, reconstitution of globin mRNA with basic proteins extracted from RNPs of liver, leg muscle, and heart in the same manner as the oocyte proteins did not suppress translation of the globin mRNA after injection to oocytes. In contrast, when globin mRNA was reconstituted with the oocyte-specific proteins extracted from stage 1-2 oocytes, synthesis of globin always was restricted after injection. These data are summarized in Table 3.

Trivial explanations for the inhibitory effects of oocyte proteins on translation of globin mRNA, such as mRNA degradation or structural modification, were eliminated by control experiments in which reconstituted mRNA was re-extracted and shown now to be translatable. Thus, the results of this study demonstrate that oocyte specific proteins extractable from native RNPs can reversibly repress the translation of mRNA in vivo (as well as in vitro). Based on these initial tests, these oocyte proteins, therefore, have the properties expected of proteins which can act to mask oocyte mRNA translation.

TABLE 3
Relative Translatabilities of Reconstituted Globin mRNA in Injected Oocytes

	Synthesis of globin relative to injected pure globin mRNA (%)						
	Experiment No.						
	1[a]	2[b]	3[c]	4[d]	5[e]	6[f]	x ± SDS
Pure globin mRNA	100	100	100	100	100	100	100 ± 0
Globin mRNA + histones	140	139	133	111	-	-	131 ± 12
Globin mRNA + ribosomal proteins	-	150	155	120	-	-	142 ± 15
Globin mRNA + basic liver proteins	-	105	137	101	-	-	114 ± 16
Globin mRNA + basic leg muscle proteins	-	-	-	-	-	97	97
Globin mRNA + basic heart proteins	-	-	-	-	-	96	96
Globin mRNA + oocyte RNA binding proteins	5	30	37	11	40	37	25 ± 17

[a]Four stage VI Xenopus oocytes were injected with 10 ng of pure globin mRNA, 10 ng of globin mRNA reconstituted with 10 ng calf-thymus histones, or 10 ng of oocyte binding proteins (preparation number 1) prepared as follows. Stage 1-2 Xenopus oocytes were homogenized and the postmitochondrial supernatant centrifuged through a sucrose gradient (16). The RNPs sedimenting to the 40-60s fraction were collected and applied to a phosphocellulose column. The proteins eluting in 1-2 M NaCl were used for reconstitution (16). Reconstitution conditions were 30-60 minutes on ice in 10 mM Tris-HCl, pH 7.5 and 50 mM NaCl. Oocytes were cultured for 12 hours in OR-2 medium (22) and then injected with ^{3}H-leucine (4). Oocytes then were cultured for 1 hour. The radioactive proteins were collected (4), electrophoresed on a SDS 12.5% polyacrylamide gel, and the radioactive globin band cut out, oxidized, and the radioactivity determined by liquid scintillation spectrometry (4).

[b,c,d,e,f]Five oocytes were injected with 3 ng pure globin mRNA, or globin mRNA reconstituted with 30 ng of calf thymus histones, Xenopus oocyte ribosomal proteins (i.e., Xenopus stage VI oocyte proteins extracted from RNP particles with high salt (2M NaCl) which sedimented at 80s and which had a molecular weight less than 60,000 as determined by gel filtration), Xenopus liver, leg muscle, and heart proteins prepared in an identical manner to the oocyte RNA binding proteins (16), or Xenopus stage 1-2 oocyte RNA binding proteins (16). Oocytes were cultured for 18 hours in modified Barth's saline (23) and then injected with ^{35}S-methionine (24). Oocytes were cultured for an additional one hour and the protein prepared as described (4). The relative amount of globin synthesis was determined by densitometry of autoradiograms or by direct quantitation of the radioactivity in the putative globin band. Part b used oocyte RNA binding protein preparation number 2, part c used preparation number 3, part d used preparation number 4 and part e used preparation number 5. The autoradiogram portrayed in Fig. 2 was taken from experiment number 4. From Richter, and L. D. Smith (27).

CONCLUDING COMMENTS

The induction of oocyte maturation with progesterone results in an increase in the rate of protein synthesis of at least two fold. It has been known for some time that this rate increase involves the translation of maternal mRNA. In recent studies we have demonstrated further that increased protein synthesis in maturing oocytes does not involve any changes in translational efficiency. Thus, increased protein synthesis must involve the translation of additional mRNAs. These mRNAs could derive from a stockpile of messages prevented from being translated by interaction with repressor proteins, by sequestration from the translational machinery, or by structural modification of the mRNA itself. Alternatively, since the capacity for translation in full-grown oocytes already is saturated, recruitment of additional mRNAs for translation may involve nothing more than an increase in the level of the rate-limiting component(s). A resolution among these alternatives is possible experimentally.

Protein synthesis in growing oocytes also could be regulated by the same processes discussed above. In that case, however, it is clear that substantially less mRNA is required for translation in the smaller oocytes (Table 1). In fact, it is not unreasonable to suggest that de novo transcriptional activity could account for the entire mRNA requirement in stage 1-2 oocytes. i.e., none of the "maternal mRNA stockpile" need be used for translation. Coincidentally, this is the point during oogenesis when the content of oocyte specific RNP proteins, which can suppress translation is highest.

We suggest that amphibian oocytes, like unfertilized sea urchin eggs, contain at least a portion of the so-called maternal mRNA in association with certain proteins which prevent translation. In amphibians, however, the times at which association with the proteins ("masking") and disassociation ("unmasking") occur appear to be rather protracted. Thus, we suggest that during the times at which maternal mRNA is synthesized (prior to stage 2), the message associates with proteins which not only prevents translation but function in guaranteeing long term survival. As oogenesis progresses and the demand for more messages to be translated increases, the proportion of mRNA protected by these proteins decreases. As oocytes approach the completion of oogenesis, the putative proteins would be associated with only a small proportion of the maternal mRNA and other limitations in the translational apparatus regulate the quantitative aspects of protein synthesis. In this view, so-called unmasking of mRNA is a progressive and continual process and not one activated in response to a specific stimulus such as fertilization or the induction of oocyte maturation.

REFERENCES

1. Davidson EH (1976). "Gene Activity in Early Development." New York: Academic Press.
2. Smith LD, Richter JD (1984). In Metz CB, Monroy A (eds): "The Biology of Fertilization." New York: Academic Press, in press.
3. Wasserman WJ, Richter JD, Smith LD (1982). Dev Biol 89:152.
4. Richter JD, Wasserman WJ, Smith LD (1982). Dev Biol 89:159.
5. Costantini FD, Britten RJ, Davidson EH (1980). Nature (London) 287:111.
6. Anderson DM, Richter JD, Chamberlin ME, Price DH, Britten RJ, Smith LD, Davidson EH (1982). J Mol Biol 155:281.
7. Richter JD, Smith LD, Anderson DM, Davidson EH (1984). J Mol Biol 173:227.
8. Laskey RA, Mills AD, Gurdon JB, Partington GA (1977). Cell 11:345.
9. Richter JD, Smith LD (1981). Cell 27:183.
10. Ilan J, Ilan J (1978). Dev Biol 66:375.
11. Jenkins NA, Kaumeyer JF, Young EM, Raff RA (1978). Dev Biol 63:279.
12. Moon RT, Danilchik MV, Hille MB (1982). Dev Biol 93:389.
13. Moon RT (1983). Differentiation 24:13.
14. Darnbrough C, Ford PJ (1981). Eur J Biochem 113:415.
15. Richter JD, Smith LD (1983). J Biol Chem 258:4864.
16. Dumont JN (1972). J Morphol 136:153.
17. Sagata N, Shiokawa K, Yamana K (1980). Dev Biol 77:431.
18. Dolecki GJ, Smith LD (1979). Dev Biol 69:217.
19. Woodland HR (1974). Dev Biol 40:90.
20. Anderson DM, Smith LD (1978). Dev Biol 67:274.
21. Mairy M, Denis H (1971). Dev Biol 24:143.
22. Scheer U (1973). Dev Biol 30:13.

23. Richter JD, Evers DC, Smith LD (1983). J Biol Chem 258:2614.
24. Pullman JM, Martin TE (1983). J Cell Biol 97:99.
25. Economidis IV, Pederson T (1983). Proc Natl Acad Sci (USA) 80:4296.
26. Baer BW, Kornberg RD (1983). J Cell Biol 96:717.
27. Richter JD, Smith LD (1984). Nature (London), in press.

Molecular Biology of Development, pages 145–163

mRNA LOCALIZATION IN THE MYOPLASM OF ASCIDIAN EMBRYOS[1]

W.R. Jeffery, C.R. Tomlinson, and R.D. Brodeur

Center for Developmental Biology, Department of Zoology, The University of Texas at Austin, Texas 78712

ABSTRACT *In situ* hybridization with poly(U) and cloned DNA probes suggests that the myoplasm of *Styela* eggs, although depleted in poly(A)$^+$RNA and containing the same concentration of histone mRNA as other cytoplasmic regions, is enriched in actin mRNA. Further information on the spatial distribution of mRNA was obtained by *in vitro* translation and product analysis of RNA derived from isolated myoplasmic crescents. Isolated crescents contained the normal myoplasmic constituents, at least 15 specific proteins, and about 20 different mRNAs that are enriched in this region relative to the other cytoplasmic regions of the egg. The uneven distribution of mRNA and its resistance to diffusion during ooplasmic segregation suggest that it is bound to a structural matrix. Eggs extracted with Triton X-100 exhibit a filamentous matrix, which is especially evident in the myoplasm. *In situ* hybridization shows this matrix to contain the same spatial distribution of poly(A)$^+$RNA, histone mRNA and actin mRNA as the intact egg. Depolymerization of actin filaments prior to extraction does not affect the distribution of mRNA. The results suggest that the pattern of mRNA distribution in *Styela* eggs is maintained by the association of mRNA with a non-actin, cytoskeletal matrix.

INTRODUCTION

About twenty individual muscle cells are present in bands on each side of the tail in tadpole larvae of the

[1]This work was supported by NIH grant HD-13970 and a grant from the Muscular Dystrophy Association.

ascidian Styela. The larval tail muscle and mesenchyme cells arise from 8 yellow-pigmented cells in the 64-cell embryo. The mesodermal lineage cells contain the cytoplasmic materials of the myoplasmic (or yellow) crescent (MC), a unique region in the vegetal hemisphere of the zygote formed from the cortical cytoplasm of the unfertilized egg during a spectacular episode of cytoplasmic rearrangements (ooplasmic segregation) after fertilization (1). The myoplasm is partitioned to 2 cells of the 2-cell embryo, 2 cells of the 4-cell embryo, 2 cells of the 8-cell embryo, 4 cells of the 16-cell embryo, 6 cells of the 32-cell embryo, and 8 cells of the 64-cell embryo. Myofibril differentiation, which appears to occur following further muscle cell proliferation between the neurula and tailbud stage (about 12 hours after fertilization at 16°C), is an intrinsic property of the cells containing myoplasm for it occurs in cleavage-arrested myoplasmic blastomeres in isolation from other embryonic cells (2). Cytoplasmic transfer experiments suggest that the myoplasm contains factors that promote the expression of muscle cell features in non-muscle lineage cells (3). The identity of these factors, often called cytoplasmic determinants, is still a mystery; but localized maternal RNA and protein molecules have often been advanced as the best candidates (4,5). Unfortunately, little is known about the spatial distribution of informational RNAs and proteins in eggs.

We have recently developed and employed in situ hybridization methods for mapping the spatial distribution of general and specific mRNAs in sections of Styela eggs and early embryos (6,7). Using these methods we have been able to follow the pattern of mRNA distribution in the myoplasm and other cytoplasmic regions during early development (8). We have found that histone mRNA is evenly distributed between the different cytoplasmic regions of the egg, but that actin mRNA is localized in the myoplasm and the ectoplasm (a cytoplasmic region that primarily enters the ectoplasmic cell lineages). The concentration of actin mRNA in the ectoplasm was expected because it contains more than 50% of the mass of maternal poly(A)$^+$RNA (6,8). The localization of actin mRNA in the myoplasm, however, is considered to be significant because this region is depleted in poly(A)$^+$RNA (8).

We have extended our in situ hybridization studies with a biochemical analysis of the composition of isolated MCs in the present investigation. This approach has permitted an assessment to be made of the extent of mRNA and protein localization in the myoplasm. We have also addressed the question of the cellular basis of mRNA localization in the myoplasm and other cytoplasmic regions of the Styela egg. The tenacious association of mRNA with specific cytoplasmic regions, even when they move extensively through the egg during ooplasmic segregation (6,8), suggests that mRNA molecules are bound to a regionalized structural matrix. It has recently been shown that Styela eggs extracted with the nonionic detergent Triton X-100 exhibit an elaborate filamentous matrix in the myoplasm (9). We have employed in situ hybridization of Triton X-100 extracted eggs and early embryos to determine whether this matrix is responsible for mRNA localization.

RESULTS

Isolation of Myoplasmic Crescents

The MC is thought to form by a process similar to capping (10). The recent isolation and characterization of concanavalin A-induced caps in Dictyostelium ameobae (11) encouraged us to attempt a similar isolation for the MC of Styela eggs. We were also aware of the enrichment of filamentous cytoskeletal elements (9), which are notoriously insoluble in media of high ionic strength (12), in the MC region. Eggs at the MC stage were collected by centrifugation and washed several times in an ice-cold medium containing 50mM Tris-HCl, 500mM NaCl, 10mM $MgCl_2$, 5mM $Cacl_2$, 10ug/ml leupeptin, and 10mM vanadyl ribonucleoside complex (pH 7.2) (isolation medium). The eggs were permeabilized by suspension in 20 volumes of isolation medium supplemented with Triton X-100 to 0.1% for 5 minutes at 4°C. The permeabilized eggs were centrifuged at 500xg for 1 minute to remove the follicle cells, which are separated from the surface of the chorion by this treatment, and resuspended in 20 volumes of isolation medium. The suspension was homogenized by 20 up and down strokes of a teflon pestle fitted into a Potter-Elvehjem glass homogenizer and centrifuged at 500xg for 3 minutes. The supernatant was decanted. The pellet was resuspended in 10 volumes of isolation medium, homogenized by 5 up and down strokes of the pestle and centrifuged. The pellet was examined by light microscopy. Cycles of homogenization, centrifugation, and microscopic examination as described above were continued until the pellet contained relatively clean MCs. The supernatants were pooled and analyzed as the nonmyoplasmic region of the egg. The major contaminants of the crude MC preparation were yolk particles and pieces of chorion. These could be removed by centrifuging the pellet (resuspended in 2 ml. of isolation medium) through a step gradient consisting of 10 ml of 0.2M sucrose over an 8 ml 1.9M sucrose pad. After centrifugation of the gradient for 45 minutes at 15,000xg (4°C), MCs banded at the interface between the sucrose solutions. The yield of crescents varied widely between different batches of eggs, but was sometimes as high as 80%.

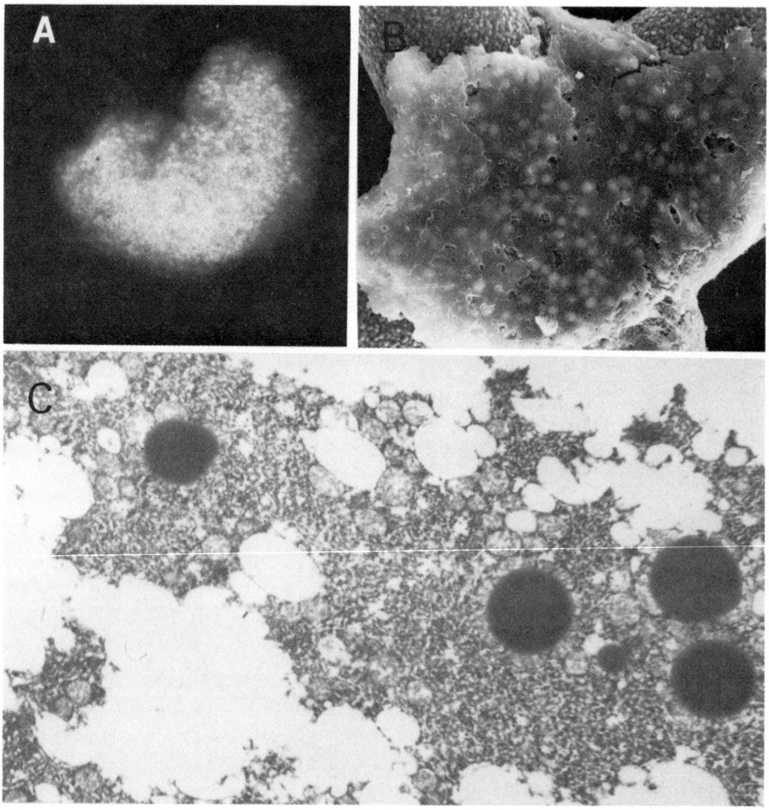

FIGURE 1. Isolated MCs as they appear when viewed by light microscopy (A), scanning electron microscopy (B), and transmission electron microscopy (C). Note in B the plasma membrane with small holes and pigment granules of the MC.

The quality of the isolated MCs was determined by light and electron microscopy. The myoplasm is known to contain large pigment granules associated with mitochondria, endoplasmic reticulum, and fine granular materials (13). Light microscopy showed the isolated crescents to consist of bright yellow pigment granules embedded in an opaque matrix (Fig. 1A). Scanning electron microscopy of whole mounts of MCs, fixed and processed as described previously for _Styela_ cytoskeletons (9), indicate that part of the plasma membrane, pocked with holes from the detergent treatment, remains in the isolated crescents (Fig. 1B). Thin sections of MCs examined by transmission electron microscopy showed pigment granules with associated mitochondria, internal membrane systems, and fine granular materials (Fig. 1C). These data indicate that the isolated MCs are structurally similar to MCs of intact eggs.

RNA and Protein Composition of Isolated Myoplasmic Crescents

The appropriate assays indicated that isolated MCs contain about 10% of the total egg protein and about 8% of the total RNA. These values correspond to 11%; the proportion of egg volume estimated to be allotted to myoplasm (8). Solution hybridization with [^{3}H]-poly(U) (14) of RNA extracted as described previously (8) from isolated MCs showed that the MCs contained only about 3% of the total egg poly(A) on a nucleotide basis. The low level of poly(A) in the isolated crescents confirms our earlier _in situ_ hybridization experiments that reported a paucity of poly(A)$^+$RNA in the myoplasm (8).

Two-dimensional gel electrophoresis was used to compare the proteins of isolated MCs and other regions of the egg. In these and the succeeding experiments that compare mRNA species in the myoplasm and the other cytoplasmic regions of the egg, the pooled supernatant fractions obtained during the preparation of the crescents is assumed to represent the contents of the ectoplasm and endoplasm; the other two major cytoplasmic regions of the _Styela_ egg. A number of differences were seen between the polypeptides of the supernatant and crescent fractions in silver nitrate-stained gels (Fig. 2). These differences are summarized in Table 1. Although many of the 133 detectible spots are present in both fractions, it is evident that the crescent fraction contains a significant number of polypeptides that are absent in the

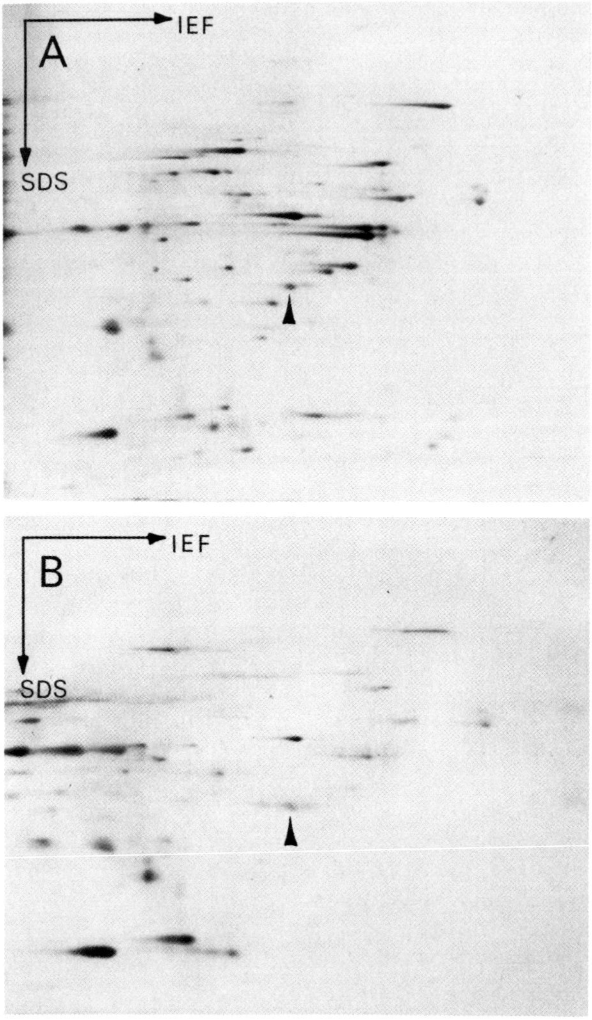

FIGURE 2. Two-dimensional gel electrophoresis of proteins present in isolated MCs (A) and the supernatant fraction corresponding to ectoplasm and endoplasm (B). Arrows mark the position of actin. Protein preparation, electrophoresis and silver staining was as described (9).

TABLE 1
SUMMARY OF POLYPEPTIDE DISTRIBUTION[a]

Class	Number	Percent total
Total	133	--
Shared equally	34	26
MC depleted	6	5
Non-MC	35	26
MC enriched	43	32
MC specific	15	11

[a]The number of polypeptides in each class was determined by visual inspection of the gels.

supernatant, and vice versa. The results suggest that the myoplasm contains a unique set of egg proteins.

RNA extracted from the crescent and supernatant fractions was translated in a message-dependent, wheat germ lysate at non-saturating levels as described previously (8) and the $[^{35}S]$-methionine labeled protein products were compared by two-dimensional gel electrophoresis and fluorography. This analysis is hampered by low levels of poly(A)$^{+}$RNA in the myoplasm and the difficulty of completely removing the vanadyl ribonucleoside, which is required to prevent RNA hydrolysis during MC isolation and reduces the efficiency of the *in vitro* translation. Consequently we are only able to detect about 100 polypeptide products in the gels (Fig. 3). Although almost every spot detected in the MC fraction can also be seen in the supernatant fraction, especially after extensive exposure of gels containing products from the latter, there are many quantitative differences between the fractions. The distribution of translation products between the MC and supernatant fractions is summarized in Table 2. The results suggest that most of the abundant messages present in the myoplasm are also represented in the other cytoplasmic regions of the *Styela* egg. The localization of these prevalent messages in the myoplasm appears to be quantitative rather than qualitative.

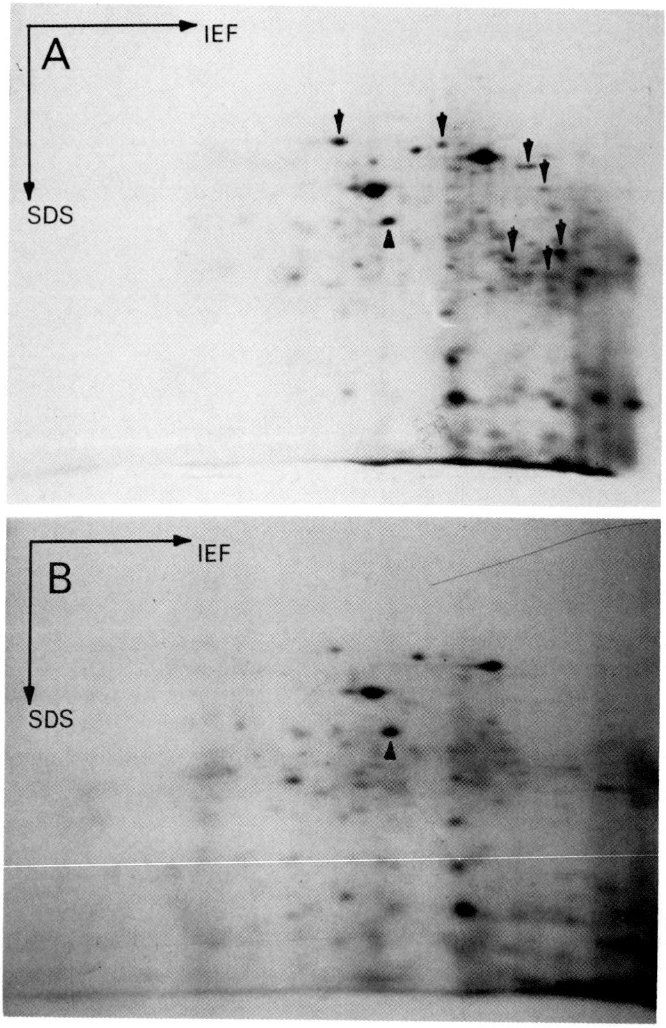

FIGURE 3. Two-dimensional gel electrophoresis and fluorography of translation products directed by equal quantities of RNA from the MC (A) and supernatant (B) fractions. Upward-pointing arrows indicate the positions of actin. Downward-pointing arrows show some of the products enriched in MC RNA.

TABLE 2
SUMMARY OF mRNA DISTRIBUTION[a]

Class	Number	Percent total
Total	109	--
Shared equally	86	79
MC depleted	3	3
MC enriched	20	18

[a]The number of translation products in each class was determined by visual inspection of the fluorographs.

Demonstration of Poly(A)$^+$RNA in Extracted Eggs

The uneven distribution of mRNA in the egg cytoplasm and the behavior of these molecules during ooplasmic segregation (6,8) suggest that they are bound to regionalized structural frameworks. Obvious candidates for these frameworks are cytoplasmic membrane systems, cytoskeletal elements, and their associated organelles. It is known that the myoplasm of *Styela* eggs, for instance, contains a localization of intracellular membranes (13) and cytoskeletal filaments (9). As a first attempt to distinguish between the possibility of membrane or cytoskeletal associations, we have extracted eggs with Triton X-100, a non-ionic detergent known to solubilize membrane lipids and free cellular proteins and nucleic acids, but not cytoskeletal materials and their associated components (15). The extraction conditions, which were optimized to yield a visibly intact cytoskeletal matrix, are outlined in reference 9. Although about 90% of the egg lipid, 80% of the protein, and 75% of the RNA was solubilized during the Triton X-100 treatment; more than 70% of the poly(A), as determined by solution hybridization with [^{3}H]-poly(U) (14), remained in the detergent-insoluble residue. The poly(A) in the detergent-insoluble fractions is retained at NaCl concentrations as high as 0.5M suggesting that it is not generated by artifactual interactions that are known to occur between nucleic acids and protein in low ionic strength media. The results suggest that poly(A)$^+$RNA is associated with a cytoskeletal matrix in *Styela* eggs.

Distribution of Poly(A)$^{+}$RNA in Extracted Eggs

If the cytoskeletal matrix is involved in maintaining the spatial distribution of mRNA, the pattern of mRNA localization observed in intact eggs might be expected to be reflected in detergent-extracted eggs. To test this possibility, sections of eggs that had been extracted with Triton X-100 at various stages of early development were subjected to *in situ* hybridization with poly(U) as described previously (16). The cytoplasmic regions of extracted eggs, like those of intact eggs (8), could be distinguished by differential staining (Fig. 4A). Earlier studies with intact

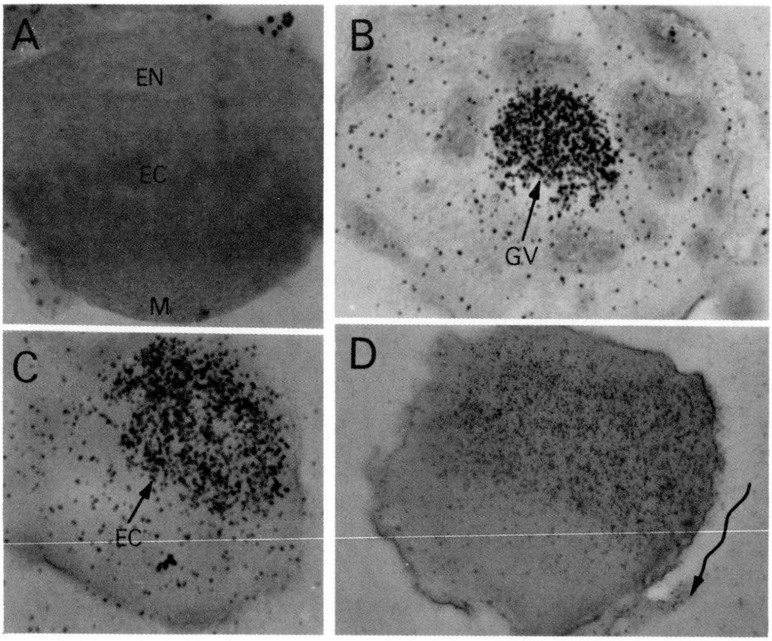

FIGURE 4. Distribution of poly(A) in sections of Triton X-100 extracted eggs as determined by *in situ* hybridization with poly(U) (16). A. Stained section of a fertilized egg showing endoplasm (EN), ectoplasm (EC), and myoplasm (M). B. Oocyte with grains in GV remnant. C. Fertilized egg with grains in ectoplasm. D. Centrifuged egg showing the stratified zones with grains in ectoplasmic zone. The arrow indicates the direction of centrifugal force.

cells showed that most of the poly(A)+RNA was localized in the germinal vesicle (GV) plasm of the oocyte and in its derivative, the ectoplasm, after maturation and during early development (6,8). The same situation was observed in extracted oocytes and eggs (Fig. 4B-C). The hybridization signal was concentrated in the remnant of the GV and in the area of the section corresponding to the ectoplasm. The heavily-labeled region shown in Fig. 4C was shown to be the ectoplasm by the following means. It is known that centrifugation displaces the cytoplasmic regions of Styela eggs from their normal positions and stratifies them into three major zones perpendicular to the axis of centrifugal force (17). From the centripetal to the centrifugal pole the zones are myoplasm, ectoplasm, and endoplasm. The locations of the three zones can be accurately defined in the extracted eggs because their positions are not affected by the detergent treatment. In situ hybridization showed that the heavily-labeled zone of extracted, centrifuged eggs was the intermediate or ectoplasmic zone (Fig. 4D). This experiment also shows that, as was previously described in intact eggs (8), the endoplasm and especially the myoplasm of extracted eggs are relatively deficient in poly(A)+RNA. Table 3 shows that the levels of hybridization in the cytoplasmic regions of extracted eggs do not markedly change between fertilization and the 2-cell stage. The results suggest that the spatial distribution of poly(A)+RNA characteristic of developing Styela eggs is preserved after detergent extraction. This result is consistent with the idea that the cytoskeletal matrix is involved in maintaining the normal distribution of mRNA.

TABLE 3
DISTRIBUTION OF POLY(A) IN EXTRACTED EGGS[a]

Stage	Ectoplasm	Myoplasm	Endoplasm
Unfert.	49.8±2.1	1.4	8.9±1.4
Fert.	61.1±3.8	0.7	10.1±1.1
2-cell	64.8±4.7	1.3	10.9±1.0

[a]Grain number is the mean±standard deviation of grain counts (8) in 25-50 400um^2, randomly-selected areas from sections of each stage.

Distribution of Histone and Actin mRNA in Extracted Eggs

Although the experiments described above suggest that the cytoskeletal matrix is involved in the maintenance of general poly(A)$^{+}$RNA distributions, they do not address the important question of whether the localization of specific messages is caused by their association with the matrix. The simplest hypothesis would be that localized messages, like those that encode actin, would be attached to a regionalized matrix whereas uniformly-distributed messages, like those encoding histones, would be present in the fluid cytoplasm. The presence of actin and histone mRNA in sections of detergent-extracted eggs was tested by in situ hybridization with the appropriate cloned DNA probes (8). The results show that both histone and actin mRNA are present in the extracted eggs and that their distributions

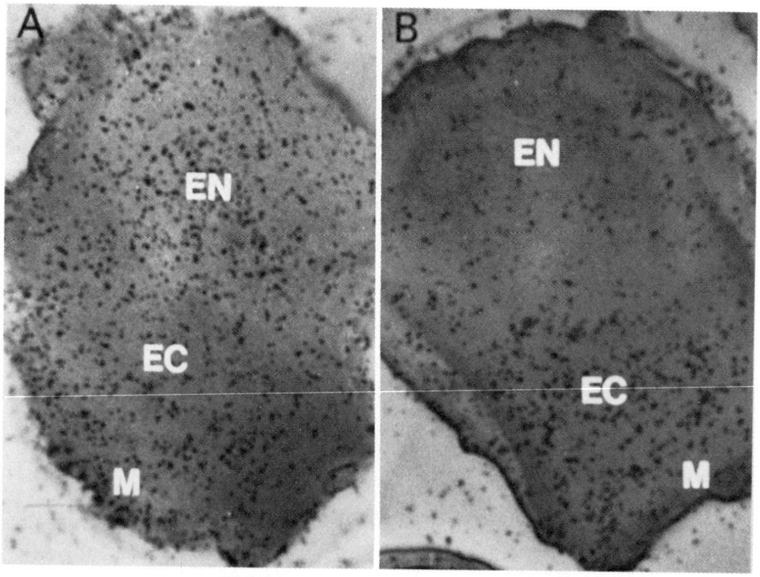

FIGURE 5. Distribution of histone mRNA (A) and actin mRNA (B) in sections of Triton X-100 extracted eggs as determined by in situ hybridization with cloned DNA probes (8). EN:Endoplasm. EC:Ectoplasm. M:Myoplasm.

are identical to those seen in intact eggs (Fig. 5). The data do not support the hypothesis that the cytoskeletal matrix contains only localized messages.

Nature of mRNA Binding Sites in Extracted Eggs

The retention of mRNA by Triton X-100 extracted eggs brings up the question of the identity of the mRNA binding sites in the cytoskeletal matrix. Wholesale binding of mRNA to microtubules appears to be eliminated in these experiments because they were carried out at 4°C. It has been suggested that mRNA or polyribosomes are associated with microfilaments in various somatic cells (16,18). The possibility that actin filaments are responsible for the association of mRNA with the cytoskeletal matrix of *Styela* eggs was tested in two ways. First, *in situ* hybridization signals from eggs treated with 10ug/ml cytochalasin B for 30 minutes prior to extraction with Triton X-100 were compared to those obtained from untreated, extracted eggs. We are certain that the drug entered the eggs because ooplasmic segregation and cleavage, processes known to be sensitive to cytochalasin B in ascidian eggs (19), were blocked in the treated eggs. Nevertheless, essentially equivalent signals were seen in

TABLE 4
THE EFFECT OF CYTOCHALASIN B AND DNase I ON THE RETENTION OF mRNA IN TRITON X-100 EXTRACTED EGGS[a]

Treatment	Probe	Percent control
Cytochalasin B	Poly(U)	112.4
	Histone	94.1
	Actin	86.5
DNase I	Poly(U)	96.5
	Histone	92.4
	Actin	114.2

[a]Percent control is the dividend of grain number from treated eggs divided by controls and multiplied by 100. Data is shown for the ectoplasm. The myoplasm and endoplasm behaved similarly.

sections of drug-treated eggs and controls hybridized in situ with poly(U) or with histone and actin DNA probes (Table 4). Second, in situ hybridization signals were compared between eggs extracted in media containing 1mg/ml DNase I and controls extracted for the same time in 1 mg/ml bovine serum albumin. DNase I, which is known to depolymerize F actin (20), specifically removes actin filaments from the cytoskeleton of Styela eggs (9). As shown in Table 4, the signals registered after in situ hybridization with poly(U), histone and actin probes were not markedly affected by DNase I treatment. These results suggest that the association of mRNA with the cytoskeletal matrix is not dependent on the integrity of actin filaments.

DISCUSSION

The purpose of this investigation was to obtain further information on the spatial distribution of mRNA in the myoplasm and other cytoplasmic regions of Styela eggs and to explore the cellular basis for mRNA localization. The first objective was faciliated by the development of an isolation method for the MC. This method is dependent on the relative insolubility of the myoplasm, a region of the Styela egg that is especially enriched in cytoskeletal filaments (9). The application of the isolation procedure described in this paper resulted in the preparation of MCs in fairly high yield and with relatively little contamination. The cytoskeletal domain (9) as well as the more soluble myoplasm surrounding it is probably isolated by this procedure because many more proteins than those known to exist in the cytoskeleton of Styela eggs (9) are present in the isolated MCs. Microscopic data show that there is a retention of the characteristic myoplasmic organelles in the isolated MCs. The isolated MCs are shown to contain pigment granules, mitochondria, endoplasmic reticulum, and even the portion of the plasma membrane that normally overlies the myoplasm. The retention of the plasma membrane and underlying elements in the isolated MCs is important because there is some evidence that the cortical region of eggs may be a repository of cytoplasmic determinants. It is planned to use the isolated MCs in developing microinjection assays for the muscle cell determinants.

The isolated MCs were found to contain a number of proteins that could not be detected in other cytoplasmic regions of the egg. This result is not surprising because others have shown that different regions of eggs contain a different spectrum of proteins (21-24). More relevant to the objectives of this investigation was the discovery of about 20 mRNAs which appear to be enriched in the myoplasm based on an approximation of the amount of protein product translated *in vitro*. It is significant, however, that we have not been able to detect mRNAs that are restricted to the myoplasm. All of the translation products we have designated as enriched in the myoplasm are also evident, albeit in lower amounts, in gels containing translation products encoded by mRNA from the non-myoplasmic portion of the egg. Thus myoplasmic localization appears to be quantitative rather than qualitative, at least for the most prevalent messages of *Styela* eggs. This conclusion is consistent with our *in situ* hybridization results which show actin mRNA in each cytoplasmic region, but more concentrated in the myoplasm and ectoplasm (8), and with the conclusions of others who have examined the distribution of mRNA in parts of *Ilyanassa* eggs by *in vitro* translation (25,26). More efficient translation of the mRNA from isolated MCs or solution hybridization experiments using cDNA probes of MC mRNA will be helpful in deciding whether any of the rarer messages of *Styela* eggs are qualitatively localized in the myoplasm.

One of the possible functions for mRNA molecules enriched in the myoplasm may be to code for muscle cell proteins after their segregation into the appropriate cell lineage. Although the well-studied muscle proteins are encoded by zygotic transcripts in other embryos (27,28), the exceedingly rapid development of ascidian embryos may call for a precocious accumulation of muscle-specific transcripts in the pool of maternal mRNA. This appears to be the case for alkaline phosphatase, which is synthesized in the endodermal cells under the direction of maternal transcripts (29). The storage of mRNA would of course be more efficient than the stockpiling of muscle proteins in the relatively small, yolk-laden egg of ascidians.

Our results suggest that the uneven distribution of RNA in *Styela* eggs and its remarkable resistance to diffusion during ooplasmic segregation is dependent on the association of mRNA with a Triton X-100 insoluble matrix. Earlier biochemical and morphological studies showed this matrix to contain at least two types of filamentous systems, one of which is composed of actin (9). Other cellular components are likely to be a part of the matrix, as evidenced by the retention of the myoplasmic pigment granules in Triton X-100 extracted Styela eggs (9). As in all other recent studies that have reported the association of mRNA with a detergent-insoluble cytoskeletal matrix, it is very difficult to rule out the possibility of artifact formation. The existence of similar patterns of mRNA distribution in intact and extracted eggs is not a strong argument against an artifactual association of messages with the matrix because localized mRNA molecules may develop an affinity for nearby cytoskeletal elements after the soluble materials of the cytoplasm have been removed. The only evidence against an artifactual origin of the mRNA-matrix interaction is the inability of high ionic strength media, which would be expected to reduce or eliminate electrostatic interactions between nucleic acids and proteins, to affect the retention of mRNA in Triton X-100 extracted eggs.

An important finding with regard to the possible cause of mRNA localization is that histone mRNA, which is evenly distributed in the Styela egg (8), is associated with the cytoskeletal matrix. The result is inconsistent with the hypothesis that localized messages are bound to the matrix and evenly-distributed messages are present in the fluid cytoplasm. All messages might bind to the matrix, but localized messages could also recognize a specific cytoskeletal domain. Different cytoskeletal domains are known to underlie the various cytoplasmic regions of Styela eggs (9). The domains could contain unique cytoskeletal systems, like the nuclear matrix discussed below, or specialized auxillary components, like the pigment granules known to be a part of the myoplasmic cytoskeletal domain (9). This possibility would also require that localized messages, or their associated protein moieties, contain complementary sites for recognizing the domains of the cytoskeletal matrix.

The identify of the cytoskeletal sites that bind mRNA is still an open question. The results of our cytochalasin B and DNase I experiments do not support the contention that actin filaments alone are responsible for mRNA binding (16, 18). The low temperature at which the extraction is carried out also eliminates microtubules. A good candidate for mRNA binding in the ectoplasm is the so-called nuclear matrix, since this region is almost entirely derived from the nucleoplasm that escapes from the GV at maturation. If mRNA, or its precursors, are bound to the nuclear matrix of the GV, it is possible that this structure (or some remnant of it) may be retained during early development as an mRNA binding framework that functions to segregate the ectoplasmic messages into the ectodermal cell lineages. There is also a non-actin cytoskeletal matrix in the myoplasm, probably composed of intermediate filaments, that could serve as an mRNA binding framework. Further progress in this area will require the identification of the cytoskeletal components that interact with mRNA.

ACKNOWLEDGEMENTS

We thank Dr. Stephen Meier for taking the micrograph shown in Figure 1B, Priscilla Kemp and Linda Wilson for expert technical assistance, Bonnie Brodeur for executing the figures, and Mary White for critical review of the manuscript. A portion of this work was conducted at the Marine Biological Laboratory, Woods Hole, MA.

REFERENCES

1. Conklin EG (1905). The organization and cell-lineage of the ascidian egg. J Acad Nat Sci Phila 13:1.
2. Crowther RJ, Whittaker JR (1983). Developmental autonomy of muscle fine structure in muscle lineage cells of ascidian embryos. Develop Biol 96:1.
3. Whittaker JR (1980). Acetylcholineasterase development in extra cells caused by changing the distribution of myoplasm in ascidian embryos. J Embryol Exp Morphol 55:343.
4. Davidson EH (1976). "Gene Activity in Early Development" New York:Academic, p 245.
5. Jeffery WR (1983). Maternal mRNA and the embryonic localization problem. In Siddiqui MAQ (ed): "Control of Embryonic Gene Expression," Boca Raton, FL: CRC, p 73.
6. Jeffery WR, Capco DG (1978). Differential accumulation and localization of poly(A)-containing RNA during early development of the ascidian Styela. Develop Biol 67:152.
7. Jeffery WR (1982). Messenger RNA in the cytoskeletal framework: Analysis by in situ hybridization. J Cell Biol 95:1.
8. Jeffery WR, Tomlinson CR, Brodeur RD (1983). Localization of actin messenger RNA during early ascidian development. Develop Biol 99:408.

9. Jeffery WR, Meier S (1983). A yellow crescent cytoskeletal domain in ascidian eggs and its role in early development. Develop Biol 96:125.
10. Jeffery WR (1984). Pattern formation by ooplasmic segregation in ascidian eggs. Biol Bull 166:277.
11. Condeelis J (1979). Isolation of concanavalin A caps during various stages of formation and their association with actin and myosin. J Cell Biol 80:751.
12. Cooke PA (1976). Filamentous cytoskeleton in vertebrate smooth muscle cells. J Cell Biol 68:538.
13. Berg WE, Humphreys WJ (1960). Electron microscopy of the four-cell stages of the ascidians _Ciona_ and _Styela_. Develop Biol 2:42.
14. Jeffery WR, Brawerman G (1974). Characterization of the steady-state population of messenger RNA and its polyadenylic acid segment in mammalian cells. Biochem 13:4633.
15. Lenk R, Ransom L, Kaufman Y, Penman S (1977). A cytoskeletal structure with associated polyribosomes obtained from HeLa cells. Cell 10:67.
16. Capco DG, Jeffery WR (1978). Differential distribution of poly(A)-containing RNA in the embryonic cells of _Oncopeltus fasciatus_: Analysis by _in situ_ hybridization with an [^{3}H]-poly(U) probe. Develop Biol 67:137.
17. Conklin EG (1931). The development of centrifuged eggs of ascidians. J Exp Zool 60:1.
18. Ramaekers FCS, Benedetti EL, Dunia I, Vorstenbosch P, Bloemendal H (1983). Polyribosomes associated with microfilaments in cultured lens cells. Biochim Biophys Acta 740:441.
19. Zalokar M (1974). Effect of colchicine and cytochalasin B on ooplasmic segregation of ascidian eggs. W Roux Arch 175:243.
20. Hitchcock SE, Carlson L, Lindberg U (1976). Depolymerization of F-actin by deoxyribonuclease I. Cell 7:531.
21. Moen TL, Namenwirth M (1977). The distribution of soluble proteins along the animal-vegetal axis of frog eggs. Develop Biol 58:1.
22. Gutzeit HO, Gehring WJ (1980). Localized synthesis of specific proteins during oogenesis and early embryogenesis in _Drosophila melanogaster_. W Roux Arch 187:151.

23. Schmidt O (1980). Insect egg cortex isolated by microsurgery: Specific protein pattern and uridine incorporation. W Roux Arch 188:23.

24. Jackle H, Eagleson GW (1980). Spatial distribution of abundant proteins in oocytes and fertilized eggs of the Mexican axolotl (Ambystoma mexicanum). Develop Biol 75:492.

25. Brandhorst BP, Newrock KM (1981). Post-transcriptional regulation of protein synthesis in Ilyanassa embryos and isolated polar lobes. Develop Biol 83: 250.

26. Collier JR, McCarthy ME (1981). Regulation of polypeptide synthesis during early embryogenesis of Ilyanassa obsoleta. Differentiation 19:31.

27. Sturgess EA, Ballantine JEM, Woodland HR, Mohun PR, Lane CD, Dimitridis GJ (1980). Actin synthesis during early development of Xenopus laevis. J Embryol Exp Morph 58:303.

28. Buckingham ME, Minty AJ, Robert B, Alonso S, Cohen A, Daubas P, Weydert A, Caravatti M (1982). Messengers coding for actins and myosins: Their accumulation during terminal differentiation in a mouse muscle cell line. In Pearson ML, Epstein HF (eds): "Muscle Development: Molecular and Cellular Control," Cold Spring Harbor, NY, p 201.

29. Whittaker JR (1977). Segregation during cleavage of a factor determining endodermal alkaline phosphatase development in ascidian embryos. J Exp Zool 202:139.

Molecular Biology of Development, pages 165–184

SPATIAL AND TEMPORAL APPEARANCE AND REDISTRIBUTION OF CELL SURFACE ANTIGENS DURING SEA URCHIN DEVELOPMENT[1]

David R McClay and Gary M Wessel

Department of Zoology, Duke University
Durham, NC 27706

ABSTRACT A number of monoclonal antibodies were identified that recognized specific regions of the sea urchin gastrula. The cell surface antigens identified by the antibodies appeared in a variety of spatial and temporal patterns. Some of the antigens became germ layer specific secondarily after being present on all germ layers. Other antigens appeared *de novo* and were located in a single germ layer from first transcription onward. Finally, some antigens were highly localized but not restricted to a single germ layer. These were expressed by portions of two germ layers. Thus, at gastrulation there is a dramatic appearance and redistribution of cell surface molecules at the time of origin of the three germ layers.

INTRODUCTION

In a series of studies by Gustafson and colleagues it was concluded that cell rearrangements in the sea urchin embryo at gastrulation involve two components: motility and cell surface recognition (1,2). They hypothesized that changes in the cell surface guide blastomeres in the reorganization from blastula to the gastrula with its three germ layers.

This theme of cell surface change has been a frequent component of models on pattern formation and cell recogni-

[1] This work was supported by USPHS HD 14483.

tion during morphogenesis (3-7). Implied by these changes is the appearance of new molecules at the cell surface. The evidence has been very indirect, however. For example, the sorting of cells within cell aggregates has demonstrated that cells (presumably through their cell surfaces) have the capacity to distinguish self from nonself (8, 9). Regeneration experiments and grafting experiments have demonstrated the probable existence of encoded patterns that are, in most models at least, thought to involve the cell surface (10, 11). Nevertheless, the components involved remain to be demonstrated. Recent progress with the chick embryo (12) and with the grasshopper (13, and this volume), offer encouragement by suggesting that adhesive specification mechanisms are beginning to be defined at the molecular level.

Until recently appropriate technologies were unavailable for direct examination of cell surfaces to ask whether pattern specifying components might exist. These components, if they existed, were thought to be rare and would require sensitive probes for detection. The development of the monoclonal antibody technology has provided a technical breakthrough for this field and already a number of cell surface components have been revealed that may participate in cell recognition, in pattern determination in the nervous system (14-16) and in early morphogenesis (17). The monoclonal technology has a strong experimental potential in that it may provide reagents to test directly the many models on pattern formation in development.

This report describes the pattern of expression of cell surface molecules during gastrulation of the sea urchin embryo and shows a series of spatial and temporal changes that occur on cells of all three germ layers. Monoclonal antibodies for the study were screened to select those that recognized antigens on specific germ layers then these antigens were mapped on the embryo by immunofluorescence and by electrophoresis from fertilization onward. When embryos were followed through developmental time several striking patterns were observed. The monoclonals revealed gradients, highly localized antigen distributions, spatial changes, temporal changes, de novo appearance and disappearance of antigens, and recompartmentalization of antigens. The existence of such a diverse group of antigens does not establish that the cell surface plays a central role in pattern formation but it does provide data for the prediction that a cell surface mosaic exists and changes with developmental time. It also provides encouragement for research because the antibodies may be useful probes for examining the molecular basis of patterns in morphogenesis. A preliminary report of this study has appeared elsewhere (17).

MATERIALS AND METHODS

Arbacia punctulata and *Lytechinus variegatus* were collected in Beaufort, NC. *Tripneustes esculentus*, *Echinometra lacunter* and *Eucidaris tribuloides* were collected in Bermuda. *Strongylocentrotus purpuratus* was obtained from supply companies on the West Coast of the U.S. Sections of *Dendraster excentricus* were the kind gift of Dr. Arthur Whiteley, The University of Washington, Seattle, WA.

Plasma membranes for injection and for monoclonal screens were obtained by douncing cells in 1 mM $NaHCO_3$ (pH 8) containing PMSF (phenylmethylsulfonyl fluoride, 0.01%). Membranes were prepared as described elsewhere (25).

Monoclonal antibodies were prepared and screened as described (17, 18). For histology, the embryos were fixed with Bouin's fixative, dehydrated in alcohol, and embedded in paraplast. Sections were examined by immunofluorescence by methods described previously (18).

Nitrocellulose (Western) blots were formed by transfer of SDS gel bands according to Towbin et al. (19). Antigens were identified on the blots using an immunoperoxidase protocol for detection of the primary monoclonal antibody (19).

RESULTS

The hybridomas and monoclonal antibodies were screened by ELISA and immunofluorescence to select surface antigens, and antigens that were germ layer specific. An ability to isolate ectoderm from endoderm (20) made it straightforward to identify more than 100 antibodies against antigens with a germ layer preference. The representative patterns displayed by this large group of monoclonals form the basis for the description of antigens given below. No attempt has been made to exhaustively describe every antibody. Instead, examples have been selected to show the diversity of patterns observed. The frequency with which a given pattern appears has little meaning in this kind of analysis because the number and variety of monoclonals one generates against multiple antigens depends upon many factors such as immunogenicity of an antigen, amount of antigen of a given specificity, serendipity, etc. The embryonic stages examined included the germinal vesicle, unfertilized egg, fertilized egg, cleavage stage, blastula, mesenchyme blastula, gastrula, prism, and pluteus larva.

ANTIGENS IN THE EGG THAT LATER BECOME CONFINED TO A SINGLE GERM LAYER

Although membranes of late gastrula stages were used for the immunization, a number of antigens could be traced back to the egg. When such antigens were found, they were intracellular prior to fertilization. They became associated with membranes or with the extracellular matrix at different times after fertilization. A number of antigens in the unfertilized egg later became associated with all plasma membranes without germ layer preference. The antigens described below, however, were present in the egg and became associated with, or compartmentalized to one germ layer secondarily.

Egg Antigens that Become Localized to Ectoderm.

At fertilization or shortly thereafter, presumptive ectodermal antigens are released to the surface of the embryo (Fig. 1)(the term "surface" is used because at the current level of resolution it is not possible to distinguish between antigens that are part of an extracellular matrix and antigens that are membrane-associated). An area in the egg cortex just beneath the surface loses fluorescent material, suggesting that the surface antigen originally comes from this area (Fig. 1e). A portion of each of the presumptive ectodermal antigens remains in intracellular compartments through early development. These are released, or turned over, at different rates, depending on the individual antigen. Some antigens are almost entirely extracellular by the 2-cell stage (Fig. 1c), while others are still detectable within blastula cells (Fig. 1d). At the gastrula stage some ectodermal antigens initially follow along the invaginating archenteron and then are lost secondarily from the endodermal surface to become associated predominantly or exclusively with the ectodermal surface (Fig. 1g, i).

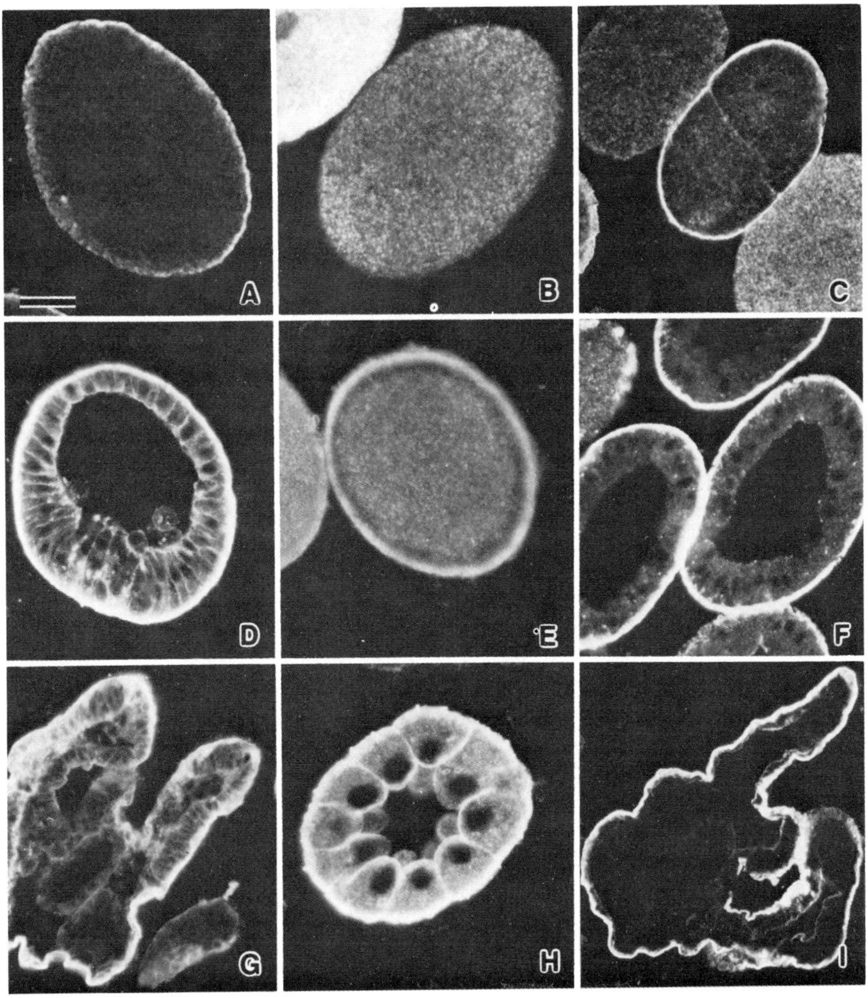

FIGURE 1 Ectodermal antigens originally found in the egg. Sections of Lytechinus were stained with two monoclonal antibodies to show the diversity of this group. (a) Polyclonal antibody to hyalin. (b,d,e,g,h) LL1c10. (c,f,i) De27e9. Both antigens are stored in the egg (a and b) as granules that are distinct from cortical granules (stained with anti-hyalin in g). 7e9 is released at fertilization and is predominantly on the cell surface by the two cell stage (c). At the blastula stage 7e9 surrounds the embryo (f). At the pluteus stage this antigen covers the embryo and the lining of the stomodaeum (i). Some of 1c10 is

released from the intracellular granules at fertilization (e), but this antigen continues to be within all cells through the blastula (h) and mesenchyme blastula stages (d). In the pluteus stage this antigen is found in the ectoderm; the staining is much reduced in endoderm, and the mesoderm is negative (g). All the embryos shown in Figs. 1-7 were fixed in the same way. Controls in each case included the following which were found to be negative for fluorescence: parent myeloma supernatant substituted for primary antibody, no primary antibody, and no secondary antibody. In addition, a polyclonal anti-sea urchin antibody was used as a positive control, and other monoclonal antibodies served as internal positive controls for nonspecific staining. All of the figures are at about the same magnification (bar in a= 25um). For reference, each egg has a diameter of ~120um.

Egg Antigens that Localize to the Basal Lamina.

The antigens described as "ectodermal" are first secreted toward the apical surface of all cells, then become associated only with ectoderm. The secretion of basal laminar antigens is first a function of all blastomeres. Later, some of the antigens are secreted only by the primary mesenchyme cells (Fig. 2). The antigens localized in the region of the basal lamina all come from the same class of egg granules (18). Many of these are released into the blastocoel early in cleavage others are released later at the mesenchyme blastula stage (18). As with the ectodermal antigens, the term "mesodermal" most likely results from continued synthesis of these antigens only in the mesoderm.

Egg Antigens that Later Become Endoderm-specific.

Like many other antigens in the egg these antigens segregate in a centrifugal field (18), in a pattern that is distinct from the segregation of yolk, the basal laminar antigens, ectodermal antigens, and of cortical granular antigens. At fertilization there is no obvious change in intracellular distribution of the presumptive endodermal antigens. In fact through the blastula stage until the late gastrula-prism stage, many of these antigens often appear to be equally distributed in intracellular compartments in all blastomeres (Fig. 3b). As the pluteus stage approaches these antigens become increasingly confined to the endoderm, often to a small segment of the gut (Fig. 3c,3d).

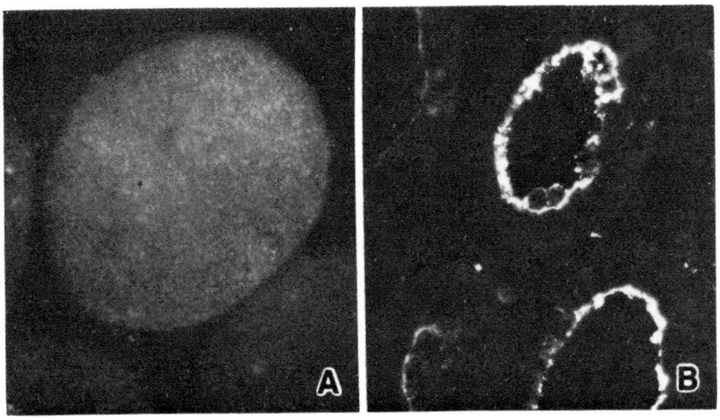

FIGURE 2. An antigen in the egg that becomes localized to the basal lamina. In (a) antigen LL1b10 was detected but at reduced amounts when compared to the lining of the blastocoel at the mesenchyme blastula stage (b).

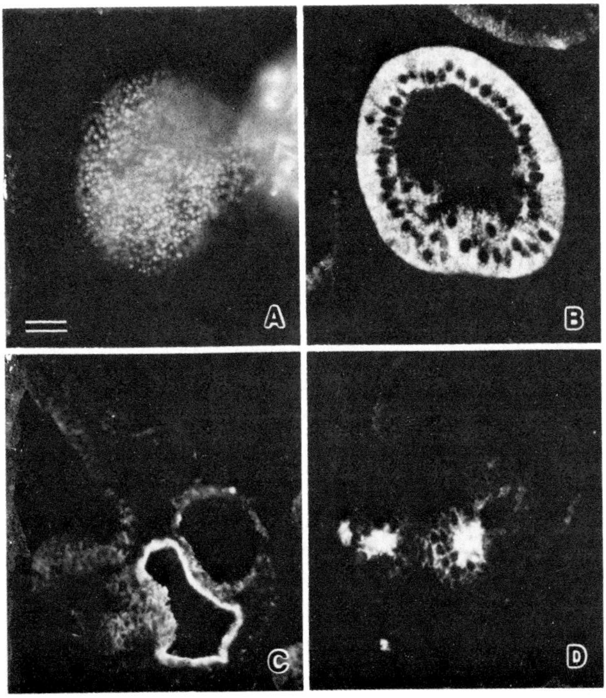

FIGURE 3. Antigens in the egg that later become endodermal. (a-c) AA1a3, (d) LL5f7. (a) shows a section of an

unfertilized egg to demonstrate the intracellular granules characteristic of antigens in the egg. Bar = 10um. In the mesenchyme blastula the antigen remains intracellular and in all blastomeres (b), but by the pluteus stage the antigen is concentrated in the midgut and absent or reduced elsewhere (c). Another antigen in this class becomes confined to cell surfaces in the midgut and hindgut (d). (b-d) magnification same as in Fig. 1.

It is possible that some of the antigens that are being described as "endoderm" are yolk proteins, but four facts argue against that possibility. First, yolk proteins are abundant, yet the endodermal antigens, as seen on Western blots, are not associated with gel bands of abundant proteins. Second, the molecular weights do not compare (21). Third, the antigens ultimately are localized at the cell surface which is not a location where yolk is usually thought to be. And fourth, an antibody against yolk does not stain in a pattern that resembles any of the antibodies in this class.

ANTIGENS THAT APPEAR *DE NOVO* AT GASTRULATION

The term *de novo* is used to describe antigens that appear for the first time at some time during development. Without additional information it cannot be assumed that these antigens are newly synthesized, since their antigenicity could be a secondary modification of a preexisting protein. Our screens have picked up *de novo* antigens in the mesoderm and endoderm and these will be described below. Elsewhere Carpenter, et al., (22, and this volume), have detected several ectodermal antigens that appear *de novo*.

Mesodermal Antigens.

One very striking pattern is seen by a group of monoclonals, all of which begin to stain the embryo at the same developmental time, but by Western blot analysis, consist of a heterogeneous group of antigens. Of those AA1g8 is the best characterized and will be described here. The AA1g8 antigen is not present prior to the mesenchyme blastula stage (Fig. 4a). At the mesenchyme blastula stage the antigen appears at a very precise time. This embryonic stage is defined by the delamination of primary mesenchyme cells from the blastula wall. Cells located at the vegetal pole move into the blastocoel one by one until a certain

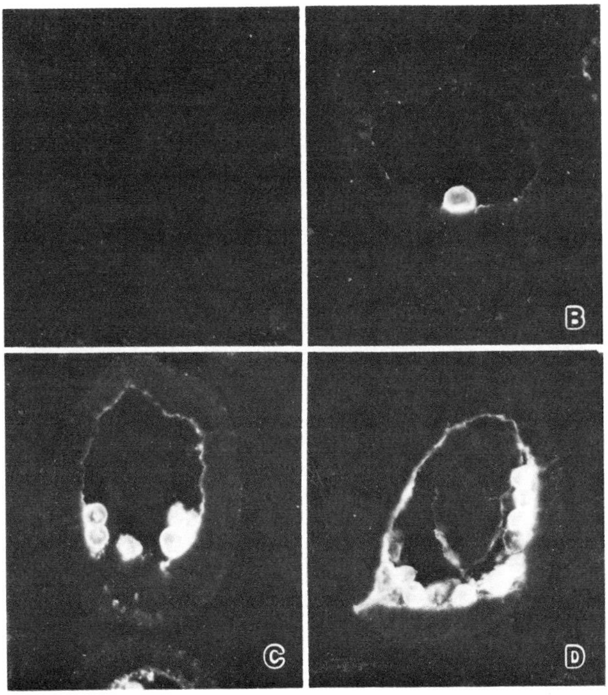

FIGURE 4. Mesodermal antigen, AA1g8, that appears at the mesenchyme blastula stage. The antigen is not present at the blastula stage (a). It first appears on the first of the primary mesenchyme cells to complete delamination (b). As each primary mesenchyme cell completes its migration through the basal lamina, it becomes positive for AA1g8 (c). The antigen is deposited along the basal lamina from the vegetal to the animal pole. When spicules appear, two projections of the antigen pass through the ectoderm wall immediately beneath the area of spicule origin (d).

specific number come to lie on the floor of the blastocoel. The AA1g8 antigen appears just at the time a cell completes delamination (Fig. 4b, 4c). A number of sections capture a situation in which one or more cells have completed the delamination process and others are still in the process. Only the cells that have completed delamination express AA1g8. Each of the cells expressing AA1g8 has an intracellular spot of stain, suggesting perhaps a site in the synthetic pathway at which antigenicity is first attained.

The antigen covers the primary mesenchyme cells and then is deposited along the blastocoelic surface, first at the vegetal pole and then toward the animal pole (Fig. 4b, 4c). At the prism stage the antigen penetrates the ectodermal covering immediately beneath the site of spicule synthesis (Fig. 4d). The antigen also covers the spicule envelope (24). The specificity of 1g8 has been explored in different species to ask whether it recognizes primary mesenchyme cells in a spectrum of echinoderm species. The species examined were *Lytechinus variegatus*, *Tripneustes esculentus*, *Strongylocentrotus purpuratus*, *Eucidaris tribuloides*, and *Dendraster excentricus*, a sand dollar. The antigen was present and specific for primary mesenchyme cells in all of these species except *Eucidaris*, which does not have primary mesenchyme cells. Spicules in *Eucidaris* originate from secondary mesenchyme cells, and these cells also did not express the 1g8 antigen. In one case there was a difference in the time of expression; in *Dendraster* primary mesenchyme cells released the antigen but not until the midgastrula stage.

Endoderm Antigen

An antigen that is specific for endoderm was observed to appear on cells of the vegetal plate at the beginning of invagination of the archenteron (Fig. 5a, 5b). The antigen is present on the apical surface of the cells and less so on the basal surface (Fig. 5b, 5c). At the midgastrula stage the antigen is found on cells of the lower 2/3 of the archenteron. Secondary mesenchyme cells and anterior endodermal cells do not express the antigen.

Antigens that Assume Precise Spatial Localizations Without Regard to Germ Layer Boundaries.

Many of the monoclonal antibodies identify antigens that appear in a very confined area of the embryo. Two antigens, in particular, serve as examples for a highly restricted distribution. The first, shown in Figure 6, is an antigen that is restricted to the stomodael opening. The cells that are positive for this antigen are both of ectodermal and of endodermal origin.

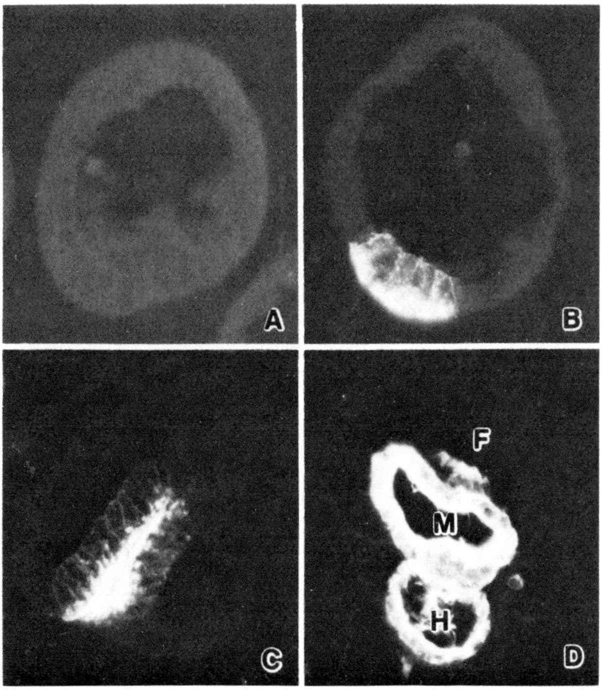

FIGURE 5. An endodermal antigen that appears early in the gastrula stage. At the mesenchyme blastula stage (a) the antigen is not present. The antigen, De25c7, first appears in and on cells of the vegetal plate shortly before the first evidence of archenteron invagination (b). At the midgastrula stage (c) the antigen is on the apical surface of each presumptive endodermal cell with the exception of cells in the esophageal region. Finally, in the pluteus stage the antigen is present in the midgut and hindgut but not in the foregut (d). (Fig. 5c). The boundary between cells that express the antigen and those that do not is striking both at the midgastrula stage where there is no apparent anatomical marker delineating a boundary and in the pluteus where the boundary is defined by the constriction between the foregut and the midgut (Fig. 5d).

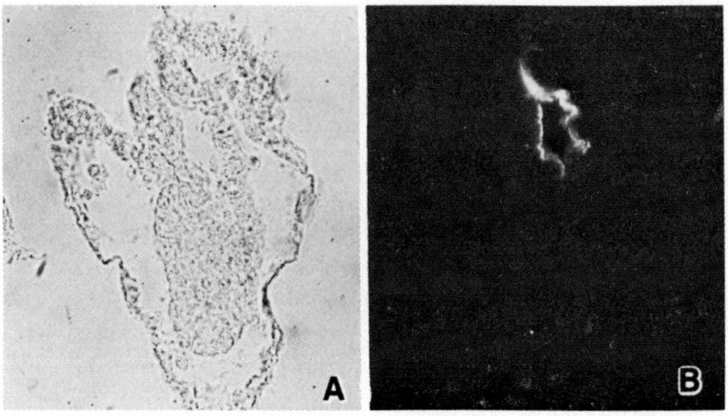

FIGURE 6. Antigen restricted to the stomodael opening. At the pluteus stage 'Lips' is detected only on cells lining the stomodaeum. (a) is a brightfield of the fluorescent section in (b).

Figure 7 shows the developmental history of another antigen with a novel distribution. This antigen is secreted at the blastula stage and as a signal of that secretion there is an intracellular spot of antigen within each blastomere. The antigen eventually covers the surface of the gastrula in a gradient of antigen increasing in concentration from the animal to the vegetal pole (Fig. 7b). The antigen persists through development on the surface of the ectoderm surrounding the original blastopore and on the surface of the hindgut (Fig. 7b). Thus the ectoderm and endoderm, that were originally at the vegetal pole, retain the antigen and it is lost from other tissues.

MOLECULAR IDENTIFICATION OF ANTIGENS

The antigens selected for this study were screened on Western blots to determine molecular weight(s). Also it was important to learn whether a monoclonal antibody was following the same antigen throughout development. In Table 1 it can be seen that some of the antigens are on multiple gel bands even though the antibody was known to be monoclonal. A striking example of this is the antigen found in a gradient over the surface of the gastrula (Fig. 7). The antibody

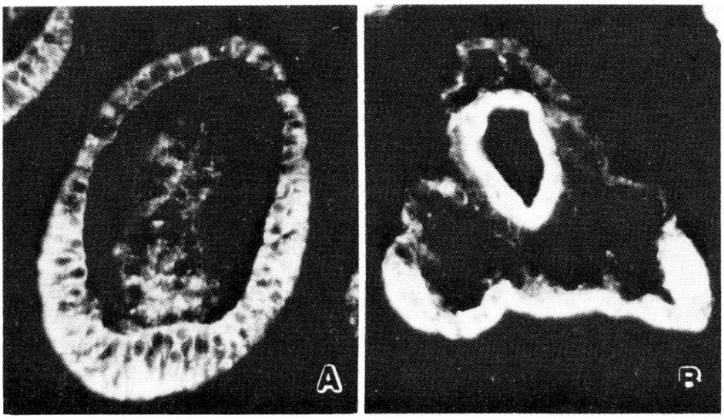

FIGURE 7. An antigen that is expressed at the vegetal pole. In the gastrula 'Cheeky' is found at the vegetal pole with a decreased intensity of staining toward the animal pole (a). In the pluteus cells of the hindgut and of the ectoderm surrounding the anus stain intensely (b) while other cells have little, if any, of the antigen.

recognizes ~12 bands on the gel. Some of these could be degradation products, but it is also likely that this antigen is a component present on several macromolecules. Most of the antigens including all of those in Table 1 are relatively rare components in that the bands were identified on Western blots using immunochemical probes and these bands did not correspond to bands that were stained with Coomassie blue. The de novo proteins were not picked up by blotting prior to their appearance by immunofluorescence.

For each of the antigens tested where there was positive immunofluorescence at several embryonic stages, the Western blot analysis indicated that the antigen was of the same molecular weight(s) at each stage. Thus, it would appear that the same antigens were being followed throughout development in this study.

DISCUSSION

Cell surface antigens are described using a collection of monoclonal antibodies against sea urchin embryonic cell membranes or extracellular matrix. The group described in this report were selected for their germ layer preference in the gastrula or pluteus stage. When this group was examined

ANTIBODY	GERM LAYER RECOGNIZED	MEMBRANE ASSOCIATED	APPARENT MOL WT
LL1c10	Ectoderm	M	210,000, 110,000
De27e9	Ectoderm, Stomodaem	H	190,000, 140,000
LL1b10	Basal lamina	M,H	90,000, 115,000, 130,000
LL1a3	Endoderm	M	24,000
AA1g8	Mesoderm	M,H	120,000
De25c7	Endoderm	M	170,000
Lips	Stomodaem	M	37,000, 45,000
Cheeky	Ectoderm, Endoderm	M	22,000, 45,000, 66,000+[2]

TABLE 1. Apparent molecular weight of antigens recognized by monoclonal antibodies used in this report. (1) The antigen is associate with a membrane fraction (M), or with an embryo homogenate (H). (2) Cheeky detects about 12 bands total, some of which may be molecular aggregates.

by immunofluorescence it revealed a wide diversity of patterns. Earlier approaches with polyclonal antisera (25) revealed expression of germlayer-specific antigens but the spatial and temporal precision of antigen localization was not appreciated until monoclonal antibodies were available. Before discussing the present results the limitations of such immunofluorescent and electrophoretic studies should be reviewed. First, a study that follows an antigen through development may not detect the same antigen at every stage. If two proteins were to share the same antigenic determinant, their coordinate or sequential expression could confuse a description based on immunofluorescent data alone. Western blot analysis of early and late stages showed that, for each antigen observed in this study, molecular weights remained constant at every stage suggesting that the antigen detected at one stage was the same antigen throughout. Second, immunofluorescence has a limited resolution. "First detection" or "site restriction" observations may not reveal the true nature of an antigen if it is expressed at levels below the resolution of the technique. Most of the antibodies used in this study had a very high titer predicting a low threshold of detection, but even so there are detection limits for any antibody. Third, neither the immunofluorescence nor the monoclonal technique reveals good quantitative information on the expression of an antigen. A rare antigen could be highly immunogenic, antibody production levels are variable, and antibody binding strength is variable. The variables restrict the interpretation of any pattern unless biochemical data are available to provide additional information. As an example of how the immunofluorescence has limitations, consider the antigen in Figure 7. It appears to be expressed in an animal-vegetal gradient on the gastrula and it becomes restricted to the hindgut and the ventral ectoderm surrounding the larval anus at the pluteus stage (Fig. 7). Gel analysis reveals that ~12 bands are recognized by the monoclonal antibody (Table 1). A number of possibilities could account for the presence of the gradient. For example, each of the 12 proteins could be expressed in a gradient, or all of the antigens might be present at the vegetal pole and qualitatively fewer of them represented in cells toward the animal pole. It is also possible the gradient could be in the expression or the activity of a single enzyme that supplies a group of glycoproteins with a common oligosaccharide side chain. Thus, while immunofluorescence reveals a number of patterns, many could be open to multiple interpretations in the absence of biochemical information. The two general conclusions that can be made from this study, however, are that there is both a precise spatial and a precise temporal distribution of cell surface antigens that accompany differentiation in early development.

The precision in time of expression of an antigen was most striking for the mesodermal antigens expressed at the mesenchyme blastula stage (Fig. 4). Primary mesenchyme cells are direct descendants of the micromeres of the 16-cell stage embryo. The mesodermal antigens are expressed precisely when the cells complete their delamination from the blastula cell wall. There was no evidence that the presumptive primary mesenchyme cells expressed the antigen until delamination was completed. The expression was programmed to micromeres by the 16-cell stage since micromeres, isolated at the 16-cell stage and grown in culture, expressed the antigen at a time equivalent to the mesenchyme blastula stage (24, 26). One cannot conclude necessarily from this data, however, that the synthesis of the mesodermal antigens are due to the _de novo_ expression of an embryonic gene. It has been shown that a number of maternal transcripts are not translated until relatively late in development (27). Thus *de novo* antigen synthesis does not predict *de novo* transcription.

The patterns and the restricted distribution of antigens provide a new concept to the data accumulated by those working on gene expression and protein synthesis during early development. Studies on expression of new mRNA suggest that fewer than 10% of the transcripts expressed during development are new(27, 28). It has been calculated that there are around 15,000 different transcripts expressed in the gastrula stage of which only a small number are new and most of these are in the "rare" class. The data are supported by studies on protein synthesis at the gastrula stage where of about 1,000 proteins detected on 2-D gels, only about 1% changed in synthesis more than 100 fold, and an additional 10% showed quantitative changes of about tenfold between the blastula and gastrula stages (28, 29). The monoclonal data reveal several patterns that add to that data. First, the calculations of "rare" transcripts were made by dividing the amount of an mRNA species by the number of cells in the embryo. This calculation is inaccurate if a protein is restricted to a few cells as seen with the stomodael antigens. Thus, a so-called "rare" transcript could be far more common, locally, if expressed by only a few cells.

A second pattern not seen in the biochemical studies is the shift in location of an antigen during developmental time. A number of the antigens followed initially were found everywhere in the embryo. These later were restricted to an area that encompassed parts of two germ layers. Thus the monoclonal antibodies reveal spatial distributions with precise boundaries that sometimes change with developmental time without necessarily being confined to a single germ layer.

A third pattern that would escape most biochemical studies is the pattern of secretion and deposition of the extracellular matrix. Many of the antigens were stored in the unfertilized egg, and those that later became part of the outer, or extraembryonic matrix were released at fertilization. As more of these antigens accumulated on the embryo surface there was a corresponding decrease in the amount of antigen remaining in blastomeres until the only trace of intracellular deposition of antigen was in vesicles at the apical end of the cells. A second extracellular group of antigens were destined for the basal laminar region. Many of these antigens were also in the egg but their release was delayed, presumably until the blastocoel developed. The construction of the basal lamina was not by a synchronous deposition of matrix components. Some of the antigens were released early in the blastula stage, others were released much later. Antigens from both the ectoderm and the mesoderm became a part of the basal lamina, including antigens that appeared de novo during development (18). Thus, in both the egg and in ectoderm antigens somehow sort out to be released independently in strictly polarized directions. Evidence exists in other systems for such polarized release mechanisms (30) so it will probably be observed frequently as specific probes become available. It is not enough to postulate different intracellular membrane receptors for the sorting out because it then becomes a problem to explain how the receptors sort out. However it occurs, the polarized release of antigen adds an additional dimension contributing to protein expression in embryos.

A number of the antigens found in the egg were assigned to a germ layer based on their eventual fate. This assumes the antigen in the egg is the same antigen that is present later in development. The best indication that this assumption is true is that Western blots of an antigen in eggs and in later stages generally identify the same antigen bands. Several unknowns remain, however. It is not known, for example, whether the antigens that become restricted to a germ layer are the same molecules that were stored in the egg, or whether these antigens are lost to be replaced by others during development. Perhaps the simplest explanation of the antigen compartmentalization is that an antigen that is present in the egg eventually is lost from cells that do not synthesize it, and is retained by tissues that do synthesize it.

Cell movement at the mesenchyme blastula stage and archenteron formation at the gastrula stage are among the most dramatic events in morphogenesis. It has been shown that primary mesenchyme cells undergo three cell recognition changes during the approximate 30 minutes it takes for these cells to invade the blastocoel (28, 35). Coincident with the delamination event we have observed the primary mesenchyme cells to begin expressing several antigens similar in pattern to that shown in Figure 4. Other new antigens, specific for the endoderm cell surface, appear at the beginning of archenteron formation. The monoclonal antibodies can provide probes not only for the genes that are expressed at precise times but can also be used as probes to define function of the newly expressed proteins.

The patterns seen in this study almost certainly are

not the complete set of cell surface patterns expressed by sea urchin embryos at gastrulation. Nevertheless, the diversity shown in this limited data set provides the impression that the cell surface is an ever changing mosaic. This demonstration is of value as support for models that predict cell surface change in morphogenesis. In addition, the antibodies have the potential as probes for actual functions in morphogenesis.

REFERENCES

1. Gustafson T, Wolpert L (1963). The cellular basis morphogenesis and sea urchin development. Int Rev Cytol 15:139.
2. Gustafson T, Wolpert L (1967). Cellular movement and contact in sea urchin development. Biol Rev 42:442.
3. Weiss P (1961). Guiding principles in cell locomotion and cell aggregation. Exp Cell Res, Suppl 8:260.
4. Moscona AA (1962). Analysis of cell recombinations in experimental synthesis of tissues in vitro. J Cell Comp Physiol 60:Suppl 1:65.
5. Sperry RW (1965). Embryogenesis of behavioral nerve nets. In DeHaan R, Ursprung H (eds) "Organogenesis" New York: Holt, Rinehart, Winston, p 161.
6. Nardi JB (1981). Epithelial invagination: Adhesive properties of cells can govern position and directionality of epithelial folding. Differentiation 20:97.
7. French V, Bryant PJ, Bryant SV (1976). Pattern regulation in epimorphic fields. Science 193:969.
8. Townes PL, Holtfreter J (1955). Directed movements and selective adhesion of embryonic amphibian cells. J Exp Zool 128:53.
9. Moscona AA (1957). The development in vitro of chimeric aggregates of dissociated embryonic chick and mouse cells. Proc Natl Acad Sci US 43:184.
10. Sperry RW (1963). Chemoaffinity in the orderly growth of nerve fiber patterns and connections. Proc Natl Acad Sci US 50:703.
11. Bryant SV, Holder N, Tank PW (1982). Cell-cell interactions and distal outgrowth in amphibian limbs. Amer Zool 22:143.
12. Edelman GM, Gallin WJ, Delouvee A, Cunningham B, Thiery JP (1983). Early epochal maps of two different cell

adhesion molecules. Proc Natl Acad Sci US 80:4384.
13. Goodman CS, Raper JA, Ho RK, Chang S (1982). Pathfinding by neuronal growth cones in grasshopper embryos. In Subtelny S, Green PB (eds) "Developmental Order: Its Origin and Regulation" New York: AR Liss, p 275.
14. Trisler GD, Schneider MD, Nirenberg M (1981). A topographic gradient of molecules in retina can be used to identify neuron position. Proc Natl Acad Sci US 78:2145.
15. Barnstable CJ (1980). Monoclonal antibodies which recognize different cell types in the rat retina. Nature 286:231.
16. Chang S, Ho R, Raper JA, Goodman CS (1981). A monoclonal antibody which stains pioneer neurons and early axonal pathways in grasshopper embryos. Soc Neurosci 7:347.
17. McClay DR, Cannon GW, Wessel GM, Fink RD, Marchase RB (1983). Patterns of antigenic expression in early sea urchin development. In Jeffries WR, Raff RA (eds): "Time, Space, and Pattern in Embryonic Development," New York, AR Liss, p 157.
18. Wessel GM, Marchase RB, McClay DR (1984). Ontogeny of the basal lamina in the sea urchin. Dev Biol. In Press.
19. Towbin H, Staehelin T, Gordon J (1979). Electrophoretic transfer of proteins from polyacrylamide gels to nitrocellulose sheets: Procedure and some applications. Proc Natl Acad Sci US 76:4350.
20. McClay DR, Marchase RB (1979). Separation of ectoderm and endoderm from sea urchin pluteus larvae and demonstration of germ layer-specific antigens. Dev Biol 79:289.
21. Ozaki H (1982). Vitellogenesis in the sand dollar _Dendraster excentricus_. Cell Diff 11:315.
22. Carpenter CD, Bruskin AM, Hardin PM, Keast MJ, Anstrom J, Tyner AL, Brandhorst BP, Klein WH (1984). Novel proteins belonging to the troponin c superfamily are encoded by a set of mRNAs in sea urchin embryos. Cell 36:663.
23. Wessel GM, McClay DR, In Preparation.
24. Fink RD, McClay DR (1984). Three cell recognition changes accompany the ingression of sea urchin primary mesenchyme cells. Submitted.
25. McClay DR, Chambers AF (1978). Identification of four classes of cell surface antigens appearing at gastrulation in sea urchin embryos. Dev Biol 63:179.
26. Okazaki K (1975). Spicule formation by isolated micromeres of the sea urchin embryo. Amer Zool 15:567.
27. Davidson EH, Hough-Evans BR, Britten RJ (1982).Molecular biology of the sea urchin embryo. Science 217:17.
28. Brandhorst BP, Bedard PA, Tufaro F (1983). Patterns of

protein metabolism and the role of maternal mRNA in sea urchin embryos. In Jeffrey WR, Raff RA (eds) "Time, Space and Pattern in Embryonic Development," New York: AR Liss, p 29.

29. Bedard PA, Brandhorst BP (1983). Patterns of protein synthesis and metabolism during sea urchin embryogenesis. Dev Biol 96:74.
30. Rodriguez Boulan E, Sabatini DD (1978). Asymmetric budding of viruses in epithelial monolayers. Proc Natl Acad Sci US 75:5071.

Molecular Biology of Development, pages 185–197

GENETIC APPROACH TO EARLY DEVELOPMENT

A. P. Mahowald, K. Konrad,
L. Engstrom, and N. Perrimon

Department of Developmental Genetics and Anatomy
Case Western Reserve University
Cleveland, Ohio 44106

ABSTRACT

Extensive mutagenesis of the Drosophila genome indicates that approximately 10% of the genome can be mutated to female sterility. However, most of these loci have only one female sterile allele and are probably weak mutations of essential genes. There is only a limited number of loci that produce a true maternal effect on early development. Detailed analysis of the phenotype of some of the X chromosomal loci indicates that they specifically affect the ability of the blastoderm to gastrulate properly. Utilizing the dominant female sterile technique, systematic searches of the lethal class of mutations on the X chromosome has led to a new estimate of the number of lethal mutations that are also active during oogenesis and whose deficiency produces a distinct maternal effect on development. The pupal lethal class is especially rich in these loci. The importance of this group of genetic loci in understanding embryogenesis is stressed.

INTRODUCTION

Many approaches have documented the complexity of the genetic control of early development. Purely molecular

This work was supported by grants from the NIH (HD-17607,-17608) and the Cleveland Foundation.

analyses have characterized the rich store of maternal information found in the egg in the form of messenger RNA and have shown that much of this complexity continues during early development (1). Recently molecular analyses of some sequences within the maternal store have been undertaken by recombinant DNA technology (2). However, in most instances the function of these sequences is not known.

A second approach for understanding early development is to identify by mutational analysis the set of loci that are required for normal development. Saturation mutagenesis screens for female sterility mutations on the X-chromosome have indicated that approximately 10% of the genome can mutate to female sterility (3,4,5). In addition, the analysis of germ line clones of lethal mutations (6) indicate that at least 70% of the lethals are needed during oogenesis. Thus, there is considerable genetic evidence for the complexity of gene action in egg production.

On the basis of frequency of alleles for loci mutating to female sterility, a number of calculations have been made to estimate the total number of loci that are in the female sterile class (3,4,7). Many complementation groups are only represented by one allele (Table 1) and this group is thought to represent the N1 or unique class in the Poisson distribution analysis. Complementation analysis between representatives of different female sterile screens indicate that about 75% of the complementation groups are represented by only one mutation (3,4,5, unpublished observations). On the other hand, among female sterile loci some have many alleles, among which some have the properties of true amorphs [e.g., gastrulation defective (see below)]. While it is difficult to exclude all the complementation groups with only one or two alleles from the true female sterile (fs) class, we wish to propose another potential classification:

Class 1: true fs loci (i.e., amorph is fs).

Class 2: hypomorph of lethal loci.

Class 3: neomorph or gain of function loci.

In this classification the number of alleles recovered in screens for female sterility mutations is the basis for catagorizing the locus. Among class 1 loci, each locus is represented by many alleles and the frequency of obtaining alleles reflects the mutability of the gene. In Class 2 each locus possesses one or more alleles and the number reflects the inherent mutagenicity of the gene and the level of zygotic expression required for viability. Finally, in class 3 each locus will ordinarily be represented by only one female sterile allele.

TABLE 1: Mutagenesis screens for female steriles on the X-chromosome

Source	# of loci	# with 1 allele	# with 2 alleles	# with >2 alleles
Gans(3)	63	48	8	7
Mohler(4,and unpublished)	114	69	17	28

There are a number of arguments in support of this reclassification of female sterility loci. Many examples of female sterile mutations have effects on adult morphology and/or viability, e.g., diminutive, tiny, fused. Recently, we have discovered that the ovarian phenotype of germ line clones of many lethal loci resembles the ovarian abnormalities of female steriles of class 2 (6; unpublished observations). Finally, the fact that new sets of loci with only one example continually arise in screens forfemale sterility mutations suggest that these are not found more often because their usual phenotype is lethal.

Some of the true female sterility loci fit the characteristics of "luxury" genes, i.e., genes specialized for a specific role in ovarian development and function. Thus, for example, at least one third of the female sterile loci affect the functions of non-germline cells (9,27) such as for egg shell production (10) and another large group is needed for proper differentiation of the nurse cell-oocyte cluster (e.g., otu, 11). There are only a few that appear to be required for the development of the future embryo itself. In these instances the oocytes produced by homogeneous mothers show a true maternal lethal effect on embryonic development.

PURELY MATERNAL EFFECT ON EARLY EMBRYOGENESIS

We have completed a detailed analysis of the gastrulation-defective (fs(1)gd) locus at 11A3 (12) The locus shows no effect at any other time during development. By means of temperature shifts with temperature sensitive alleles we have shown that the genetic information is needed during the terminal portions of oogenesis and the first 60 minutes of embryogenesis. The developmental abnormality appears after the blastoderm stage (210 min) when the movements of gastrulation fail to occur. Thus, there is no invagination of the mesoderm along the mid-ventral surface and both anterior and posterior invagination of endoderm are absent. Furthermore, there is only limited formation of neuroblasts and these form mostly in the anteriormost region of the embryo. The locus is not required for cell viability since the embryo remains alive and will even differentiate a variety of cell types, e.g., an amorphous neural mass, some gut tissue, and even an occasional contractile muscle cell. The cuticle forms from the hypoderm and appears to have only structures resembling the dorsal portion of the embryo.

Thus, this locus appears to be specific for early development and must play an essential role in preparing the blastoderm for gastrulation. There are a number of similar loci in the genome. Nusslein-Volhard (13) has described dorsal on the second chromosome with very similar properties, and recently a number of loci on the third chromosome with similar phenotypes have been found (14). Consequently, it appears that one of the principal developmental events in the embryo, prepared for by oogenetic information, is gastrulation.

A similarly large set of maternal effect loci has been found which produces a shortened or torso embryo (15; unpublished observations). In the case of two loci on the X chromosome that we have studied, the first detectable change in development occurs during blastoderm formation at the posterior tip. Pole cells form normally but then a series of abnormal events ensues. There is a wave of secondary pole cell formation and an apparently continuous budding off of small cytoplasmic blebs below the pole cells. Then, at the time of blastoderm cellularization, furrows do not form between the nuclei at the posterior tip so that a small restricted region is left free of cells. The result is a small hole in the blastoderm (hence the name pole hole for this locus). The second mutation is allelic to fs(1)Nas (16,unpublished observations). It is possible that the cause of this defect is related to the fact that the centrosome region of each blastoderm nucleus, which normally is located at the peripheral edge of the nucleus at the time of furrowing (17,18), is found at other locations in embryos produced by mutations producing the torso phenotype.

Following gastrulation of embryos with the pole hole defect, developmental abnormalities affect all structures formed from the posterior 15% to 20% of the blastoderm (about 1,000 cells) and possibly parts of the anterior region. The posterior midgut invagination does not take place and no structure characteristic of abdominal segment 8 or the analia form. The cells of the posterior 20% of the blastoderm apparently do not become assimilated into other components of the embryo, but are lost. This set of loci are especially intriguing because they are needed for establishing specific regions of the pre-gastrula pattern of cell determination.

RESCUABLE MATERNAL EFFECT MUTATIONS

An interesting class of maternal effect loci are those which support normal development either by action during oogenesis or by paternally provided gene activity in the embryo. Thus, normal development ensues for a mutant embryo provided the mother has the wild type gene; however, a homozygous mutant mother is sterile unless the zygote receives the wild type allele in the sperm (for review, see 19). Some of these loci have been identified as genes required for metabolic functions such as pyrimidine biosynthesis (20). Others, however, are essential for the proper establishment of the pattern of cell determination at the blastoderm. For example, almondex and pecanex gene activity is needed for proper establishment of the ventral ectoderm. In the absence of either gene excess numbers of neuroblasts form at the expense of the ectoderm (21,22). The critical point regarding these mutations is that the activity can be provided either during oogenesis or in the brief period prior to the blastoderm. Consequently, these gene products which are essential for proper determination do not fit either within the maternal repertoire or with required zygotic activity. The gene product of the extra sex comb (esc) locus is another example of a gene product required for the normal expression of genes prerequisite for segmental determination which can be provided maternally or zygotically (23,24). These examples suggested to us that there may be other examples of this type of essential maternal gene activity which would also be loci essential for viability (ergo lethal loci).

MATERNAL EFFECTS OF LETHAL LOCI

There have been a number of indications that many loci are active in oogenesis as well as critical for essential functions at other times during the life cycle of *Drosophila* (25,26). We (6) have recently investigated the maternal role of 48 lethal loci on the X chromosome utilizing the dominant female sterile technique (27) (Figure 1).

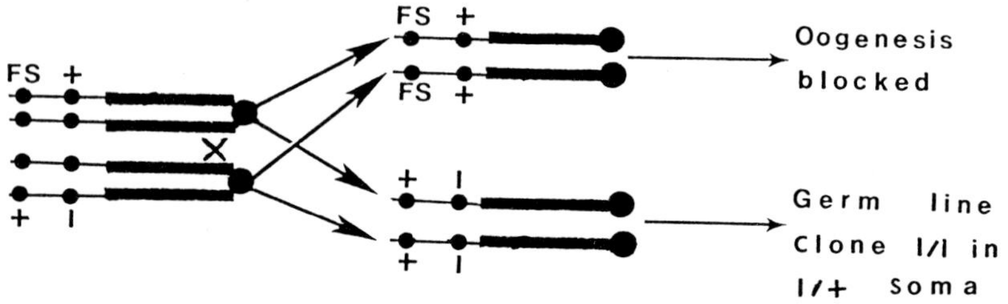

Figure 1. Somatic recombination, induced by X-ray irradiation, produces germ line clones homozygous for lethal alleles (l) and lacking the dominant female sterile allele (Fs). Any flies laying eggs have germ line clones. By utilizing this approach, it is possible to screen relatively large numbers of lethal mutations for any role these loci have in oogenesis.

While approximately 30% of the loci showed no change in the lethal phase of mutant embryos derived from germ line clones homozygous for the lethal mutation, the remainder showed clear maternal effects: 30% appeared to be cell lethals, 35% severely affected the process of oogenesis, and 5% produced normal appearing eggs which subsequently showed maternally dependent developmental abnormalities.

We have analyzed in considerable detail one representative of this last set of lethal loci. lethal (1)

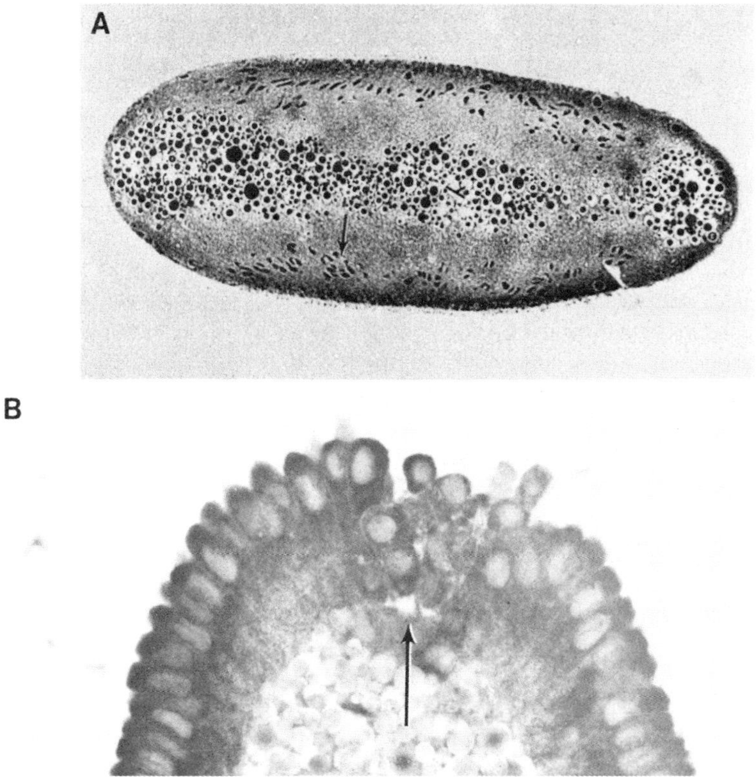

Figure 2. Phenotype of blastoderm stage embryos lacking all genetic activity of l(1)ph (A) or having paternally provided genetic activity (B). The arrows designate the multilayered nuclei in (A) and the defective posterior tip in (B).

pole hole (l(1)ph) is a discless pupal lethal, mapped via deficiencies to 2F6 on the X-chromosome (28). The class of discless lethals represent loci essential for cell cycle function (29,30). Germ line clones of extreme or amorphic alleles of l(1)ph produce embryos with two phenotypes (31): embryos without wild type gene function develop syncytial blastoderms with multiple layers of nuclei (Figure 2A); embryos which receive the wild type gene paternally show a normal blastoderm except that the posterior cells below the pole cells fail to form (Figure 2B). This resembles closely the pole hole embryo found in torso- like female sterile mutations. Subsequently, the two classes of embryos produce dramatically different phenotypes. The first class produces a poorly differentiated cuticle with rare cuticular differentiations (Figure 3A). The pole hole class produces the torso phenotype (Figure 3B).

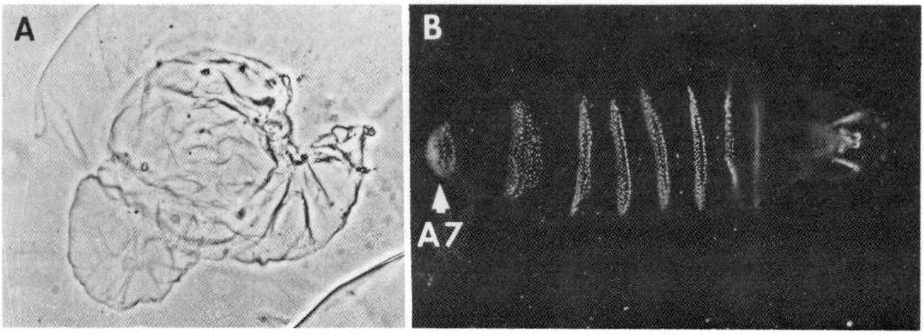

Figure 3. Phenotypes of late embryos without (A) and with paternally provided (B) l(1)ph gene activity. The terminal segment in the torso-like embryo is abdominal 7 (A7).

Loci similar to l(1)ph would appear to be especially important for understanding the genetic control of embryonic development. These genetic functions are essential for cellular reproduction not only in imaginal disk tissue, but also in embryonic development. Baker and others (30) have already shown that early pupal lethals are rich in mutations affecting the cell cycle. Since the process of metamorphosis, in which the adult form is produced from diploid imaginal cell anlagen, is amazingly similar in dynamics to the developmental processes occurring in embryonic life, it is reasonable to suggest that the same gene products are required. Thus, clusters or polyclones of cells proliferate and then integrate with each other to produce the new form that characterizes the post-metamorphic imago stage. Loci which have lethal phase during pupation may be especially rich sources of loci with key roles during oogenesis and embryonic development.

DISCUSSION

The mutational approach to understanding the complexity of the oocyte's contribution to early development is obviously very powerful. The number of loci that are specialized for purely an oogenetic component is smaller than might have been predicted from the size of the maternal complexity determined by purely molecular analyses. However, most of this molecular complexity has been shown to be common to other stages of development. In a similar fashion, we have now shown that a large fraction of the key components of the maternal contribution to early development are also critical at some other point in the life cycle. Before we can expect to understand the genetic control of egg structure and its contribution to early embryogenesis, we will have to explore in much greater detail this category of genetic loci affecting egg structure.

We have begun an analysis of this type of loci by screening a set of 80 pupal lethals for those which show a maternal contribution. A number of points deserve attention. In our initial sample of 48 lethals scattered along the whole X chromosome and having a lethal phase varying from embryonic to pupal, 30% of the lethals affected oocyte viability. In the set of 80 pupal lethals, only 10% showed autonomous cell lethality. In this latter set of lethals, we also found a major set of loci that produced specific developmental abnormalities. Interestingly, some of the phenotypes are distinctly different from those found among the purely female sterile category, indicating that there will be other loci that affect early development in new ways scattered throughout the genome.

In order to utilize the dominant female sterile technique with the autosomes, other loci will have to be found on each major chromosome arm with similar properties to Fs(1)K1237. Alternatively, it should be possible to translocate Fs(1)K1237 to the needed positions on the autosomes. Then it will be possible to systematically search the genome for lethal loci that have key roles in the maternal preparation for early development.

The phenotype of l(1)ph is especially interesting because of its relationship to the common maternal effect phenotype called torso, produced by a number of female sterile loci. We have examined the defect in blastoderm formation at the posterior tip of an allele of fs(1)Nasrat that we found which is 100% penetrant for the pole hole phenotype. In embryos produced by these flies, there is a disruption of normal blastoderm formation which is associated with the dislocation of the centrosome region of the blastema nuclei. It is probable that a similar aberration occurs in heterozygous l(1)ph embryos produced by germ line clones since these embryos show an identical phenotype both at the blastoderm and at the end of the embryonic period. The phenotype of homozygous mutant embryos produced by germ line clones may be due to a more extreme display of this defect in cell formation.

The question of centrosome location in relationship to cell division has been previously noticed in two instances relative to the maintenance of continued cell division. After the formation of the 16-cell cyst in the germarium of *Drosophila*, the two centrioles become located away from the nuclear envelope and separate from each other prior to their movement into the future oocyte. Later, these cystocytes become polyploid and the centrioles in the oocyte disappear (32). In a similar fashion, the centrioles of follicle cells become dissociated from the nuclei at the end of the mitotic divisions and subsequently disappear (33). It is possible that this change in configuration of the centrosome relative to the nucleus is related to the retention of continued mitotic division, characteristic of imaginal cells. The events occurring at the blastoderm of pole hole embryos may be a localized occurrence of this dissociation of centrioles. This abnormal feature of cell organization may then spread to the posterior 20% of the blastoderm prior to the completion of embryogenesis, leading to the loss of the this region of the embryo. In homozygous l(1)ph embryos, derived from heterozygous mothers, the defect is only seen in those cell types that must continue to divide during the larval period, i.e., the imaginal disk cells and the neuroblasts. Thus, this class of mutations may be especially important in the continued maintenance of cell division during the life cycle of the fly.

ACKNOWLEDGEMENTS

We wish to especially acknowledge Dr. C. Nusslein-Volhard for giving us the torso mutation and Dr. T. Schupbach for providing us examples of new torso- like female sterile loci on the second chromosome. We are grateful to Dr. J. Dawson Mohler for sharing many alleles and for extensive discussions, and to Drs. Paul Hardy and Heidi Degelmann for their assistance in recent studies on pole hole embryos.

REFERENCES

1. Brandhorst BP, Bedard P-A, Tufaro F (1983). Patterns of protein metabolism and the role of maternal RNA in sea urchin embryos. In Jeffrey W, Raff R (eds): "Time, Space, and Pattern in Embryonic Development," New York: Alan R. Liss, p 29.
2. Angerer LM, DeLeon DV, Angerer RC, Showman RM, Wells DE, Raff RA (1984). Delayed accumulation of maternal histone mRNA during sea urchin oogenesis. Dev Biol 101,477.
3. Gans M, Audit C, Masson M (1975). Isolation and characterization of sex-linked female sterile mutants in Drosophila melanogaster. Genetics 81:683.
4. Mohler JD (1977). Developmental genetics of the Drosophila egg. I. Identification of 59 sex-linked cistrons with maternal effects on embryonic development. Genetics 85:259.
5. Komitopoulou K, Gans M, Margaritis LH, Kafatos FC (1983). Isolation and characterization of sex-linked female-sterile mutants in Drosophila melanogaster with special attention to eggshell mutants. Genetics 105:897.
6. Perrimon N, Engstrom L, Mahowald AP (1984) Analysis of the effects of zygotic lethal mutations on germ line functions in Drosophila. (Submitted for publication).
7. King RC, Mohler JD (1975) The genetic analysis of oogenesis in Drosophila melanogaster. In King RC (ed): "Handbood of Genetics." New York: Plenum Press, Vol 3, p757.
8. Lindsley DL, Grell EL (1968). "Genetic variations of Drosophila melanogaster." Carnegie Inst. Wash., Publ. No. 627.
9. Wieschaus E, Audit C, Masson M (1981). A clonal analysis of the roles of somatic and germ line during oogenesis in Drosophila. Develop Biol 88:92.
10. Underwood EM, Mahowald AP (1980). The chorion defect of the ocelliless mutation is caused by abnormal follicle cell function. Dev Gen 1:247.
11. King RC, Riley SF (1982). Ovarian pathologies generated by various alleles of the otu locus in Drosophila melanogaster. Develop Genet 3:69.
12. Konrad K, Goralski TJ, Mahowald AP (1984). Developmental

analysis of the gastrulation-defective locus in Drosophila melanogaster. (in preparation).
13. Nusslein-Volhard C (1979). Maternal effect mutations that alter the spatial coordinates of the embryos of Drosophila melanogaster. In Subtelny S, Konisberg IR (eds): "Determinants of Spatial Organization," New York: Academic Press, p 185.
14. Anderson KV, Nusslein-Volhard C (1983), Genetic analysis of dorsal-ventral embryonic pattern in Drosophila. In Malacinski G, Bryant S (eds): "Primers in Developmental Biology," New York: Macmillan, (in press).
15. Nusslein-Volhard C, Wieschaus E, Jurgens G (1982). Segmentierung bei Drosophila - Eine genetische Analyse. Verh Dtsch Zool Ges 1982:91.
16. Counce SJ, Ede DA (1957) The effect on embryogenesis of a sex-linked female sterility factor in Drosophila melanogaster. J Embryol exp Morph 5:404.
17. Mahowald AP (1963). Electron microscopy of the formation of the cellular blastoderm in Drosophila melanogaster. Exp Cell Res 32:457.
18. Fullilove SL, Jacobson AG (1971) Nuclear elongation and cytokinesis in Drosophila montana. Dev Biol 26:560.
19. Konrad KD, Engstrom L, Perrimon N, Mahowald AP (1984). Genetic analysis of oogenesis and the role of maternal expression in early development. In Browder L (ed): "Developmental Biology: A Comprehensive Synthesis, Oogenesis." Plenum, New York: (in press).
20. Norby S (1973). The biochemical genetics of rudimentary mutants of Drosophila melanogaster. Hereditas 73:11.
21. Jimenez G, Campos-Ortega JA (1982). Maternal effects of zygotic mutants affecting early neurogenesis in Drosophila. Wilhelm Roux' Arch 191:191.
22. LaBonne S, Mahowald AP (unpublished observations).
23. Struhl G (1981). A gene product required for correct initiation of segmental determination in Drosophila. Nature 293:36.
24. Struhl G (1983). Role of the esc+ gene product in ensuring the selective expression of segment-specific homeotic genes in Drosophila. J Embryol exp Morph 76:297.
25. Garcia-Bellido A, Moscoso Del Pardo J (1979) Genetic analysis of maternal information in Drosophila. Nature 278:346.
26. Robbins LG (1983). Maternal zygotic lethal interactions in Drosophila melanogaster: zeste-white region single-cistron mutations. Genetics 103:633.

27. Perrimon N, Gans M (1983). Clonal analysis of the tissue specificity of recessive female sterile mutations of Drosophila melanogaster using a dominant female sterile mutation, Fs(1)K1237. Dev Biol 100:365.
28. Perrimon N, Engstrom L, Mahowald AP (1984) Developmental genetics of the 2E-F region of the Drosophila X-chromosome: a region rich in "developmentally important" genes. (Submitted for publication).
29. Szabad J, Bryant PJ (1982). The mode of action of "discless" mutations in Drosophila melanogaster. Dev Biol 93:240.
30. Baker GS, Smith DA, Gatti M (1982) Region-specific effects on chromosome integrity of mutations at essential loci in Drosophila melanogaster. Proc Natl Acad Sci USA 79:1205.
31. Perrimon N, Engstrom L, Mahowald AP (1984). A pupal lethal mutation with a partially rescuable maternal effect on embryonic development in Drosophila melanogaster. (in preparation).
32. Mahowald AP, Strassheim JM (1970). Intercellular migration of centrioles in the germarium of Drosophila melanogaster. J Cell Biol 45:306.
33. Mahowald AP, Caulton JH, Edwards MK, Floyd AD (1979). Loss of centrioles and polyploidization in follicle cells of Drosophila melanogaster. Exp Cell Res 118:404.

Molecular Biology of Development, pages 199–212

NUCLEOLAR DOMINANCE AND THE DEVELOPMENTAL REGULATION OF RNA POLYMERASE I PROMOTERS IN XENOPUS[1]

Ronald H. Reeder, Sharon Busby[2], Marietta Dunaway, Steven C. Pruitt, Garry Morgan[3], Paul Labhart, Aimee H. Bakken[4], and Barbara Sollner-Webb[5].

Division of Basic Sciences, Fred Hutchinson Cancer Research Center, Seattle, Washington 98104

ABSTRACT DNA sequence elements that regulate transcription of the ribosomal genes in Xenopus laevis have been mapped. These include a 140 bp promoter at the 5' end of each gene and repetitive elements in the spacer which act as enhancers for the promoter. The enhancers are part of repeated sequences that are 60 or 81 bp long and they act in relative independence of position or orientation. The enhancers appear to attract a transcription factor(s) that is also attracted by the promoter and both enhancers and promoters function only on closed circular plasmids. Ribosomal genes injected into Xenopus embryos come under developmental control and only begin transcription after the mid-blastula transition. The level of transcription is proportional to the number of enhancers adjacent to the promoter.
In F1 hybrids between X. laevis and X. borealis the laevis ribosomal genes are dominant in early development regardless of which species furnished the egg (no maternal effect). The dominance phenomenon can be re-created by injecting competing plasmids into

[1]This work was supported by NIH grants GM26624 to RHR and GM 28905 to AHB.
[2]Zymogenetics, Seattle, WA 98103
[3]Department of Genetics, University of Nottingham, England
[4]Zoology Department, Univ. of Wash., Seattle, WA 98195
[5]Department of Physiological Chemistry, Johns Hopkins Medical School, Baltimore, MD 21205

oocytes of either species. Laevis is dominant primarily because its spacer on average has more enhancers (seventeen or more) than does a borealis spacer (four in the two spacers that have been studied).
The X. laevis spacer also contains two or more duplications of the gene promoter. These spacer promoters differ from the gene promoter in 14 or 15 nucleotides out of 140 and are usually silent or very weakly active in their chromosomal location. When cloned ribosomal genes are injected into oocyte nuclei, however, the spacer promoters are equal in activity to the gene promoter. Co-injection of an intercalating agent such as ethidium bromide or chloroquine differentially represses the spacer promoters with little effect on the gene promoter. We propose that the spacer promoters are sensitive to changes in DNA supercoiling and that changes in the torsional stress of the chromosome determines whether they are active or not.

INTRODUCTION

In the frog, Xenopus, the 18S, 5.8S and 28S RNA's of ribosomes are transcribed as part of a 7.5 kb precursor RNA. The genes for this 7.5 kb precursor (ribosomal genes) are tandemly repeated at a single chromosomal site in each haploid chromosome set and are present in about 500 copies per site. The genes are separated from each other by spacer regions of variable length which are usually (but not always) transcriptionally silent. The structure of a typical spacer region is shown diagramatically in Figure 1. The genes themselves all appear to be identical and are transcribed by RNA polymerase I, a polymerase which seems to have no other function in the cell. The ribosomal genes begin transcription early in development, shortly after the mid-blastula transition, and their product is required by all cells. For such a housekeeping gene one might therefore expect a relatively simple means of transcriptional regulation. However, the regulatory machinery for the ribosomal genes gives promise of being as complex as that for any differentiated polymerase II gene. Recent work has shown that the polymerase I promoter is relatively large (about 140 bp) and contains several functional domains. The frequency with which this promoter activates is influenced by enhancer sequences far upstream of the promoter, and these enhancers provide a mechanistic explanation for the phenomenon of nucleolar dominance in Xenopus. Certain polymerase I promoters also have the potential to be regulated by the degree of supercoiling of the chromosome. These findings are summarized in the following article.

STRUCTURE OF THE GENE PROMOTER

The Xenopus laevis ribosomal genes contain two types of RNA polymerase I promoters. The gene promoter is located at the 5' end of each gene and regulates transcription of the 7.5 kb rRNA precursor. Duplications of the gene promoter are also present and are located at intervals further out in the spacer (see Figure 1). These spacer promoters (also called Bam islands, see reference 1) differ from the gene promoter in 14 or 15 nucleotides and, as we will discuss in a later section, can be regulated independently of the gene promoter.

The sequences required for activity of the gene promoter lie between -142 and +6 relative to the site of transcription initiation (2,3). Initial experiments showed that when deletion mutants were assayed by injection into oocyte nuclei, a much smaller region, from -7 to +6 would suffice for accurate initiation (3). More extensive work suggests, however, that the oocytes that gave this result were in some way unusual since we now routinely see a requirement for the full -142 to +6 promoter in oocyte injection experiments. Transcription in homogenates of isolated oocyte nuclei also suggested the existence of at least two promoter domains (3). In nuclear homogenates from some oocytes 5' deletions closer than -142 abolished all transcription. In other homogenates, however, a reduced but significant level of transcript was seen until the 5' deletions went beyond -7 at which point all transcription ceased.

Further evidence for the presence of domains in the gene promoter comes from DNaseI footprinting experiments in which end-labeled DNA containing the promoter is incubated in whole oocyte homogenates and then lightly digested with DNaseI. At least three DNaseI protection regions are seen in these experiments. Protected region I covers the site of initiation and coincides well with the

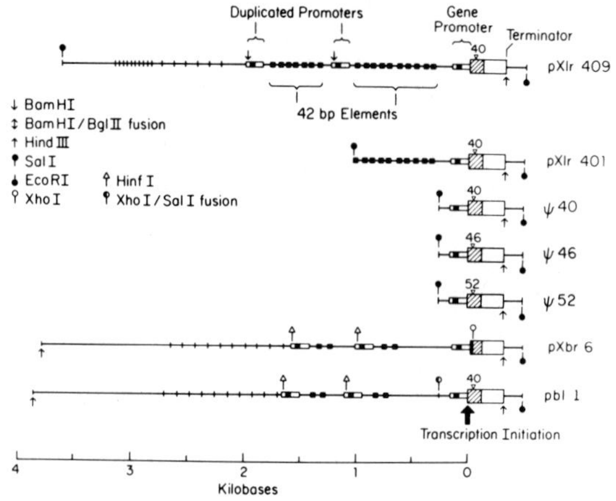

FIGURE 1. Structure of ribosomal gene plasmids derived from X. laevis and X. borealis.

pXlr409 This plasmid has a complete spacer attached to a 315 bp minigene body and is derived from X. laevis ribosomal DNA (8). It contains a gene promoter at the 5' end of the gene region, two duplicated spacer promoters, and seventeen enhancer elements (represented by small black boxes). In addition it has 40 bp of linker DNA inserted at position +31 inside the gene so that its transcript can be distinguished from endogenous laevis precursor rRNA by primer extension assay (7). All constructs shown here are in pBR322 and can be asayed by primer extension using the same 88 bp primer (7).

pXlr401 Same as pXlr409 except the spacer has been truncated down to -980.

40 Same as 40 except that 46 bp of linker DNA have been inserted at +31.

52 Same as 40 except that 52 bp of linker DNA have been inserted at +31.

pXbr6 A borealis ribosomal gene containing a full length spacer was cut at a unique StuI site 9 nucleotides inside the gene, an XhoI linker was added, and borealis spacer plus gene promoter was fused to the XhoI site in the 40 bp linker region of the laevis minigene, 40. This allows the borealis promoter to be assayed using the same primer as was used for all the laevis plasmids.

pBl 1 The entire borealis spacer (down to -200) was removed from pXbr6 and attached to the SalI site of 40.

"small" promoter seen in some oocyte injection experiments. Region II extends from about -70 to -100 and correlates well with a 42 bp block of sequence (from -72 to -114) that is also present in each one of the multiple 60/81 bp repeating elements further out in the spacer. As we will describe in the next section, these 60/81 bp elements act as enhancers for the gene promoter and we hypothesize that the 42 bp homology region is the sequence actually responsible for enhancer activity. Footprinting also reveals a region III, from -120 to -140 whose possible function is still unknown.

Both the gene and spacer promoters appear to function as on-off switches. Electron microscope evidence indicates that, in the presence of excess RNA polymerase, these promoters are either completely silent or else they are active and load polymerase at a density that probably represents the maximum possible packing (one polymerase for every 60 or 70 bp of DNA; this argument is summarized in reference 4). We think the gene promoter itself has the ability to load polymerase at this density since removal of spacer sequences down to -320 does not diminish polymerase packing on active genes (5).

RNA POLYMERASE I ENHANCERS

When a ribosomal gene bearing a truncated spacer is injected into oocyte nuclei in competition with a gene bearing a full spacer, the gene bearing the full spacer is dominant (6,7). The regions of the spacer responsible for this competition effect are the 60/81 bp repetitive elements (7). An example of a competition experiment between three different genes injected into oocytes is shown in Figure 2. The 60/81 bp repeats function in relative independence of position or orientation. A single block of ten of these repeats on a plasmid with three gene promoters will bring all three promoters under its influence (8). We hypothesize that the 60/81 bp repeats act as attraction sites for binding of a transcription factor(s) whose eventual function is to activate the gene promoter. When factor is limiting plasmids with more of these repeats are larger targets for binding and therefore out-compete plasmids with fewer repeats. After binding, the factor must in some way cause rapid activation of the gene promoter. We favor the notion that it does so by moving along the DNA until it binds stabley to the gene promoter. However, other models are also possible and have not as yet been ruled out. A block of 60/81 bp repeats on one plasmid (with no other gene or promoter) will still effectively compete against a promoter on a separate plasmid. This supports the idea that the 60/81 bp repeats and the promoter bind the same factor(s) (8).

As we noted earlier, the gene promoter behaves as an on-off switch and is capable of loading polymerase at maximum density in the complete absence of upstream spacer sequences (5). Therefore, we conclude that the 60/81 bp repeats influence transcription by increasing the frequency of activation of the gene promoter rather than

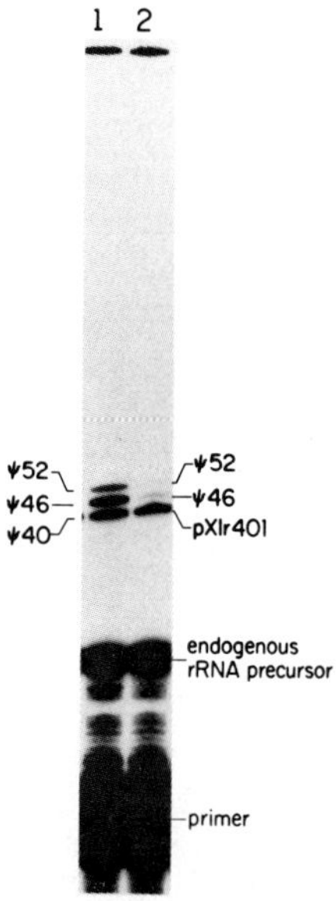

FIGURE 2. Ribosomal gene enhancer assay by oocyte injection.

Lane 1 200 pg each of 40, 46 and 52 were co-injected into the same laevis oocyte nuclei and transcription was assayed the next day by primer extension (see reference 7 for the method). Lane 2, same as lane 1 except 40 was replaced with pXlr401, a plasmid that carries ten enhancer elements in addition to the gene promoter.

by altering the frequency of polymerase loading once the promoter is activated.

We can also detect enhancer function of the 60/81 bp repeats by injecting plasmids into fertilized eggs and assaying for transcription at late gastrula (after the genes have begun transcription at mid-blastula; 9). In this assay the transcription signal observed is directly proportional to the number of 60/81 bp repeats adjacent to the promoter. An example of an embryo experiment is shown in Figure 3.

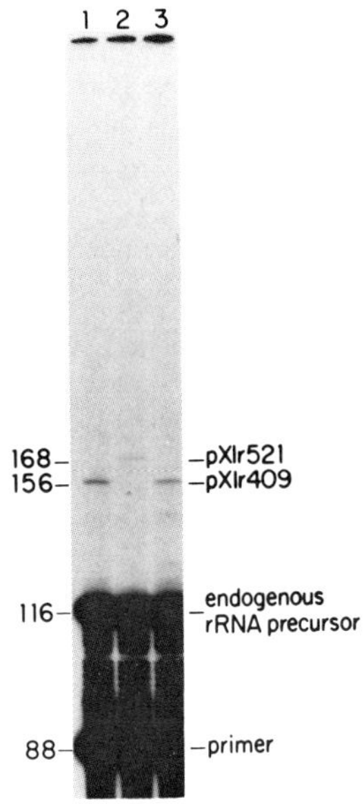

FIGURE 3. Ribosomal gene enhancer assay by embryo injection.

Various laevis ribosomal gene plasmids were injected either singly or together into fertilized laevis embryos prior to first cleavage. Transcription was assayed by primer extension when the embryos had reached late gastrula. Plasmids injected were: Lane 1, pXlr409; Lane 2, pXlr521 (same as pXlr401 except it has a 52 bp linker insert); Lane 3, pXlr409 and pXlr521 together.

POLYMERASE I ENHANCERS AND NUCLEOLAR DOMINANCE

When two closely related species are crossed it is often observed that the ribosomal genes of one species are transcriptionally dominant in the F_1 hybrid. This phenomenon was first observed in several genera of plants (reviewed in reference 10) and has also been seen in crosses between Xenopus laevis and Xenopus borealis (11,12; X. borealis was mistakenly identified as X. mulleri in some earlier publications). The most striking feature of this type of nucleolar dominance is the overall lack of a maternal effect. For example, laevis ribosomal genes are dominant regardless of which species furnished the egg or the sperm. The suppressed ribosomal genes are not lost or damaged in any way, and, in fact, significant borealis transcription often reappears in later development stages.

We have found that we can duplicate the basic phenomena of nucleolar dominance by injecting laevis and

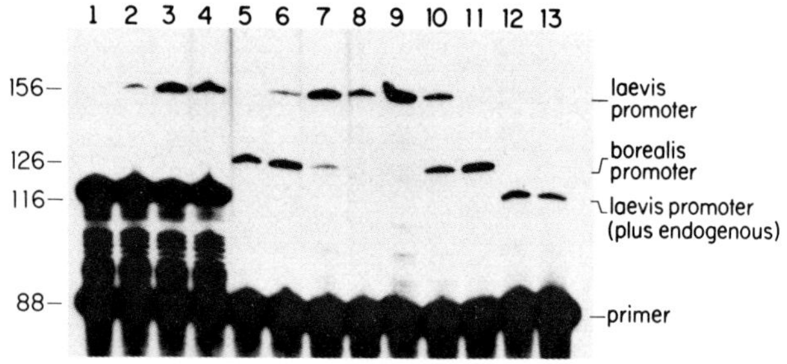

FIGURE 4. Nucleolar dominance analyzed by injection of laevis and borealis ribosomal genes into oocyte of either species.

200 pg of various ribosomal gene plasmids were injected either singly or in combination into oocyte nuclei and assayed the next day by primer extension. Lanes 1 to 4, laevis oocytes; lanes 5 to 13, borealis oocytes. Plasmids injected were: Lane 1, pXbr6; lane 2, pXbr6 plus 40; lane 3, pXbr6 plus pXlr401; lane 4, pXbr6 plus pXlr409; lane 5, pXbr6; lane 6, pXbr6 plus 40; lane 7, pXbr6 plus pXlr401; lane 8, pXbr6 plus pXlr409; lane 9, pbl 1; lane 10, pbl 1 plus pXbr6; lane 11, pXbr6; lane 12, pbl 1 plus pXlr980 (same as pXlr401 but lacking the 40 bp linker insert); lane 13, pbl 1 plus pXlr209 (same as pXlr409 except lacking the 40 bp linker insert).

borealis ribosomal gene plasmids into oocytes of either species. The ribosomal gene constructs that were used in this work are shown in Figure 1. The result of injecting competing pairs of these plasmids into either laevis or borealis oocytes is shown in Figure 4. The main conclusion to emerge from this work is that laevis is dominant because its spacer has more enhancer elements than does borealis. For example, the shortest naturally occurring laevis spacer (such as the spacer on pXlr409, Figure 1) has seventeen enhancer elements. In contrast, borealis spacers on average have only four enhancer elements (pXbr6, Figure 1). (The borealis spacer does not have 60/81 bp repeats but does have four copies of the 42 bp homology element that is the central core of the laevis 60/81 bp repeats. For this reason we expect that the 42 bp elements in borealis function as enhancers.) When this pair is injected in competition with each other, either in laevis oocytes (Figure 4, lane 4) or in borealis oocytes (Figure 4, lane 8) the laevis gene is strongly dominant. In fact, for any of the competing pairs injected in Figure 4, one can accurately predict which will be dominant by simply counting the number of enhancers adjacent to the promoter. To emphasize this fact we put the borealis spacer onto the laevis promoter (pbl 1) and competed pbl 1 against various laevis plasmids. Figure 4, lanes 9 through 13, shows that dominance relations correlate with the spacer (and enhancer number) but not with the species of the promoter.

A second factor adding to the dominance of laevis ribosomal genes is the fact that the laevis promoter functions well with borealis transcription machinery (Figure 4, lane 8) but the borealis promoter functions very poorly with laevis machinery (Fig. 4, lane 1).

A second type of nucleolar dominance has been described in somatic cell hybrids between distantly related species such as mouse and human (13,14). In this case the species are so far apart evolutionarily that their transcription machineries do not cross react at all (15,16). In mouse-human somatic cell hybrids the chromosomes of one species are preferentially lost and this is invariably the species whose ribosomal genes are suppressed. We and others have proposed that this type of nucleolar dominance is due to the loss or inactivation of a gene for an essential transcription factor from the dominated genome (17,18,19).

PROMOTER REGULATION BY SUPERCOILING

The X. laevis spacer contains two or more promoters which differ from the gene promoter in 14 to 15 positions. The activity of these spacer promoters can range from completely silent to as active as the gene promoter. At first we thought it likely that the differences in activity would be due to sequence differences in the spacer promoters themselves. We identified a frog that showed abundant spacer transcription (4), and cloned and sequenced its spacer promoters. The sequence of a spacer promoter from the "active" frog (164a) is compared with all other known spacer promoter sequences in Figure 5. The striking fact is that all the spacer promoters are nearly identical and the few single base changes do not correlate well with their activity in vivo.

We then attached several spacer promoters to ribosomal minigene bodies so that we could co-inject them into oocytes with the gene promoter and measure the activity of both using the same primer extension probe. We found that all spacer promoters are equally active as the gene promoter when injected into oocytes and assayed by primer extension, a result that confirmed a similar report by Moss (6). Apparently oocyte injection does not mimic the true in vivo situation where these promoters are often silent.

A possible solution to this puzzle appeared when we tried co-injecting an intercalating agent along with the ribosomal gene plasmids (Figure 6). At low doses of ethidium bromide the spacer promoter is almost completely repressed while at higher doses it comes on again.

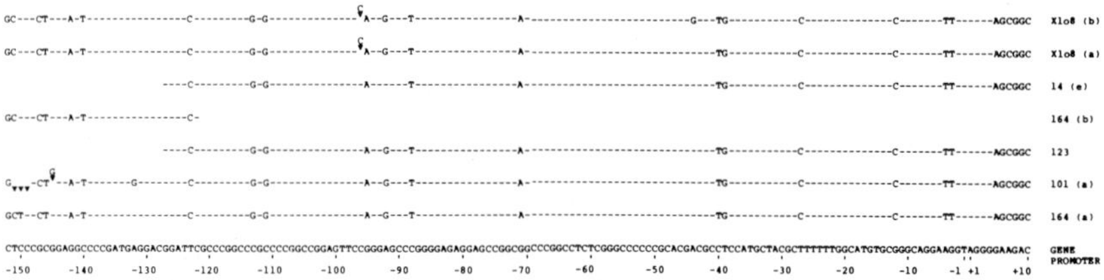

FIGURE 5. Comparison of gene and spacer promoter sequences.

The gene promoter (shown on the bottom line) is within the region from -142 to +6 (3). Shown above are six independent spacer promoters that have been sequenced. Xlo8 a and b are from (1), 14e is from (21) and the rest are unpublished sequences of G. Morgan.

Throughout this range the gene promoter is either unaffected or slightly stimulated until, at the highest doses both promoters are repressed. The same effect can be obtained with chloroquine. The intercalator can be added with the DNA or injected six hours later with similar results. The line of X. laevis kidney cells which we maintain has an abnormally high level of spacer transcription. Treatment of these intact cells with chloroquine can completely repress the spacer transcription without affecting the gene promoter.

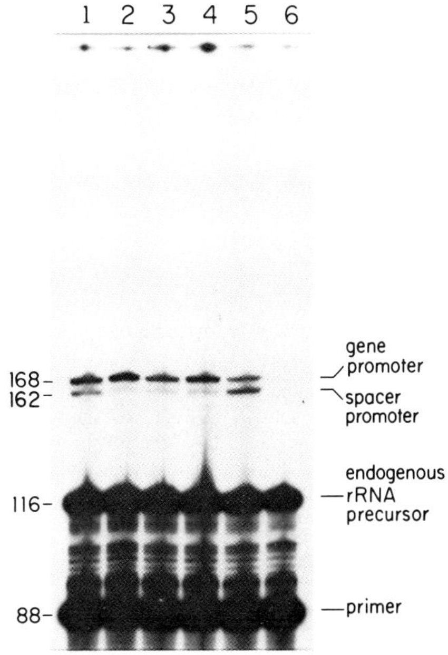

FIGURE 6. Effect of ethidium bromide on the activity of gene and spacer promoters injected into oocyte nuclei.
A 300 bp fragment containing a spacer promoter from pXlr409 was ligated to a Bam HI site in the 40 bp linker of 40 to make pXlr14G. pXlr14G was mixed with an equi-molar amount of 52 and injected into oocyte nuclei along with various amounts of ethidium bromide and assayed for transcription 12 hours later by primer extension. 20 nl of ethidium bromide was co-injected at the following concentrations: lane 1, 0 ug/ml; lane 2, 5 ug/ml; lane 3, 20 ug/ml; lane 4, 100 ug/ml; lane 5, 500 ug/ml; lane 6, 2000 ug/ml.

The fact that in oocyte injections low doses of intercalator inhibits spacer promoters while higher doses turns them on again argues strongly that the effect is not due to simple steric interference due to the intercalating molecules. Instead, it seems likely that the effect is due to changes in the super helicity of the promoter. This correlates well with previous observations that the gene promoter must be on a closed circular molecule in order to function (20). And it strongly implies that, despite the abundant topoisomerases in the cell, a promoter can be effectively put under different degrees of superhelical stress.

From these results we propose that in the chromosome, spacer promoters are normally under a degree of superhelical stress that keeps them repressed without affecting the gene promoter. When plasmids are injected into oocytes, however, they assemble into chromatin with a degree of super helicity that is not truly physiological and which allows abnormal expression of the spacer promoters. Addition of an intercalator modifies this superhelicity and, at low doses, phenocopies the normal chromosomal state.

We still do not know why promoters on some spacers are active in vivo while others are not. But we suspect it is due to other spacer sequences that cause an altered degree of supercoiling in those particular spacers.

PROSPECTS FOR THE FUTURE

From the results summarized in this article we can begin to understand how two mechanisms, enhancers and supercoiling, can act to regulate the activity of RNA polymerase I promoters. Although more work is needed to completely define the sequences involved in these regulatory events, the more important challenge now is to isolate and characterize the proteins that interact with these regulatory sequences. It is also worth pointing out that the physiological significance of these regulatory mechanisms is very poorly understood at present. Thus, it is clear that considerable more work is necessary before we can truly claim to understand the developmental regulation of the Xenopus ribosomal genes.

REFERENCES

1. Boseley P, Moss T, Machler M, Portmann R, and Birnstiel M (1979). Sequence organization of the spacer DNA in a ribosomal gene unit of X. laevis. Cell 17:19.
2. Moss T (1982). Transcription of cloned Xenopus laevis ribosomal DNA microinjected into Xenopus oocytes, and the identification of an RNA polymerase I promoter. Cell 30:835.
3. Sollner-Webb B, Wilkinson JAK, Roan J, and Reeder RH (1983). Nested control regions promote Xenopus ribosomal RNA synthesis by RNA polymerase I. Cell 35:199.
4. Morgan GT, Reeder RH, and Bakken AH (1983). Transcription in cloned spacers of Xenopus laevis ribosomal DNA. Proc. Nat. Acad. Sci. USA 80:6490.
5. Bakken AH, Morgan G, Sollner-Webb B, Roan J, Busby S, and Reeder RH (1982). Mapping of transcription initiation and termination signals on Xenopus laevis ribosomal DNA. Proc. Nat. Acad. Sci. USA 79:56.
6. Moss T (1983). A transcriptional function for the repetitive ribosomal spacer in Xenopus laevis. Nature 302:221.
7. Reeder RH, Roan JG, and Dunaway M (1983). Spacer regulation of Xenopus ribosomal gene transcription: Competition in oocytes. Cell 35:449.
8. Labhart P, and Reeder RH (1984). Enhancer-like properties of the 60/81 elements in the ribosomal gene spacer of Xenopus laevis. Cell in press.
9. Busby SJ, and Reeder RH (1983). Spacer sequences regulate transcription of ribosomal gene plasmids injected into Xenopus embryos. Cell 34:989.
10. Wallace H, and Langridge WHR (1971). Differential amphiplasty and the control of ribosomal RNA synthesis. Heredity 27:1.
11. Honjo T, and Reeder RH (1973). Preferential transcription of Xenopus laevis ribosomal RNA in interspecies hybrids between Xenopus laevis and Xenopus mulleri. J. Mol. Biol. 80:217.
12. Cassidy DM, and Blackler AW (1974). Repression of nucleolar organizer activity in an interspecific hybrid of the genus Xenopus. Dev. Biol. 41:84.

13. Croce CM, Talavera A, Basilico C, and Miller OJ (1977). Suppression of production of mouse 28S ribosomal RNA in mouse-human hybrids segregating mouse chromosomes. Proc. Nat. Acad. Sci. USA 74:694.
14. Elicieri GL, and Green H (1969). Ribosomal RNA synthesis in human-mouse hybrid cells. J. Mol. Biol. 41:253.
15. Grummt I, Roth E, and Paule MR (1982). Ribosomal RNA transcription in vitro is species specific. Nature 296:173.
16. Mishima Y, Financsek I, Kominami R, and Muramatsu M (1982). Fractionation and reconstitution of factors required for accurate transcription of mammalian ribosomal RNA genes: identification of a species-dependent initiation factor. Nucleic Acids Res. 10: 6659.
17. Onishi T, Berglund C, and Reeder RH (1984). On the mechanism of nucleolar dominance in mouse-human somatic cell hybrids. Proc. Nat. Acad. Sci. USA 81:484.
18. McStay B, and Bird A (1983). The origin of the rRNA precursor from Xenopus borealis, analyzed in vivo and in vitro. Nucleic Acids Res. 11:8167.
19. Meisfeld R, and Arnheim N (1984). Species-specific rDNA transcription is due to promoter-specific binding factors. Molec. Cell. Biol. 4:221.
20. Pruitt SC, and Reeder RA (1984). Effect of topological constraint on transcription of rDNA in Xenopus oocytes: Comparison of plasmid and endogenous genes. J. Mol. Biol. in press.
21. Sollner-Webb B, and Reeder RH (1979). The nucleotide sequence of the initiation and termination sites for ribosomal RNA transcription in X. laevis. Cell 18:485.

Molecular Biology of Development, pages 213–225

HOW MOUSE EGGS PUT ON AND TAKE OFF THEIR EXTRACELLULAR COAT[1]

Paul M. Wassarman, Jeffrey M. Greve, Rosario M. Perona, Richard J. Roller, and George S. Salzmann

Department of Biological Chemistry
Harvard Medical School
Boston, MA 02115

ABSTRACT During a two to three week period, as mouse oocytes grow from 12 to 85 μm in diameter, a relatively thick (7 μm) extracellular coat appears that completely surrounds the oocyte's plasma membrane. This extracellular coat, called the zona pellucida, serves important functions during fertilization and early development, and is not shed by embryos until just prior to implantation (day-5 of embryonic development). In mice, the zona pellucida is composed of three different glycoproteins, designated ZP1 (200 kd), ZP2 (120 kd), and ZP3 (83 kd). Since synthesis of these relatively abundant glycoproteins is stringently regulated with respect to both cell type and stage of development, they are attractive candidates for studies of the regulation of gene expression during early mammalian development. Furthermore, since hatching of mouse embryos from the zona pellucida involves a specific differentiative event at a particular time during early embryogenesis, this too represents an attractive system for study. Various aspects of these topics are presented here.

[1]This work was supported in part by the National Institute of Child Health and Human Development.

INTRODUCTION

Once a mouse oocyte enters its growth phase, a period that lasts about two weeks, its fate is determined. The fully-grown oocyte will be ovulated to become an unfertilized egg and the egg will either be fertilized or undergo atresia. Therefore, oocyte growth represents a final stage in the complex process of oogenesis and, as such, the last opportunity for oocytes to prepare for ovulation and meiotic maturation (conversion to an egg), fertilization, and early embryogenesis. Consequently, it is not surprising that oocyte growth is characterized not only by a tremendous increase in cell size (about 350-fold in volume), but also by substantial alterations in oocyte morphology, ultrastructure, and metabolism (1,2).

An obvious feature of mouse oocyte growth is the appearance of an extracellular coat, the zona pellucida, that increases in width concomitant with increasing oocyte diameter. Whereas a nongrowing oocyte (about 12 μm in diameter) apparently does not have a zona pellucida, a fully-grown oocyte (about 85 μm in diameter) has a relatively thick coat (about 7 μm in width) completely surrounding its plasma membrane. Production of a zona pellucida is extremely important since it contains species-specific sperm receptors essential for fertilization of ovulated eggs, participates in a secondary block to polyspermy following fertilization, and maintains the integrity of preimplantation embryos as they slowly make their way to the uterus. Finally, the preimplantation embryo must divest itself of its zona pellucida in order to implant in the uterus and continue through normal development.

For several years, our research has focused on various aspects of the zona pellucida. Among our specific interests are the (i) regulation of expression and organization of genes that code for glycoproteins that constitute the zona pellucida, (ii) structure of the zona pellucida, (iii) nature and mechanism of action of sperm receptors present in the zona pellucida, and (iv) molecular basis of "hatching", whereby the blastocyst escapes from its zona pellucida. Progress made in our laboratory in three of these four areas of investigation will be summarized here. Information about our research on the mouse egg's sperm receptor can be found elsewhere (3-6).

RESULTS

Structure of the Mouse Zona Pellucida

Like many other extracellular coats, the zona pellucida is a porous matrix of fibers and granules, permeable to large macromolecules and even to small viruses. Whereas transmission electron micrographs of fixed and stained thin sections of zonae pellucidae reveal a loose network of intertwined fibers (Fig. 1), it has not been possible to obtain a great deal of structural information using preparations of this sort. Consequently, we have recently examined mouse oocyte zonae pellucidae dissolved under a variety of conditions (e.g., low pH, enzymatically, or in the presence of reducing agents), sprayed onto a substrate, rotary shadowed or negatively stained, and subjected to electron microscopy (7, J. Greve and P. Wassarman, unpublished results). In general, dissolution of zonae pellucidae yielded fragments small enough so as to be amenable to analysis by electron microscopy, but large enough so as to retain structural features of the three-dimensional matrix. Such analyses have revealed that, overall, the zona pellucida can be considered as a three-dimensional array of interconnected filaments, with each filament consisting of "beads" having a mean diameter of about 10 nm and spaced regularly along the axis of the filament about every 17 nm (Fig. 1). Results of preliminary experiments further suggest that ZP1 serves as the crosslinker of zona pellucida filaments.

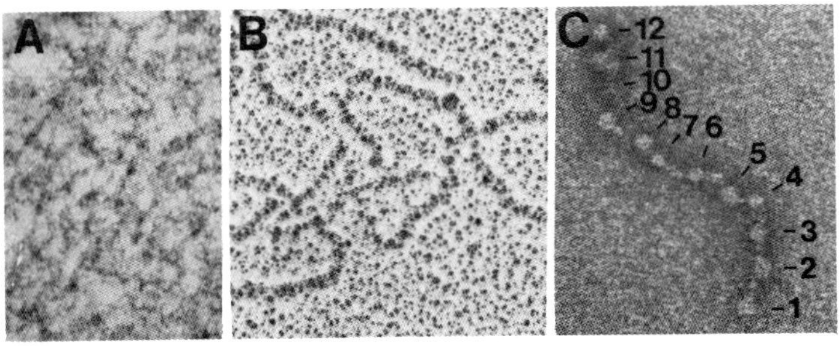

FIGURE 1. Electron microscopy of the zona pellucida. A. Negatively stained thin section; B. Rotary shadowed filaments; C. Negatively stained filament.

Composition of the Mouse Zona Pellucida

Mouse zonae pellucidae isolated from fully-grown oocytes each contain about 3 ng of protein, or about 10% of the oocyte's total protein. Surprisingly, we found that this 3 ng is distributed among only three different species of glycoproteins having molecular weights of 200, 120, and 83 kd, and referred to these as ZP1, ZP2, and ZP3, respectively (8). All three zona pellucida glycoproteins are acidic (isoelectric points less than 5.5) and only ZP1 consists of more than one polypeptide chain (linked by intermolecular disulfides). These are the "mature" forms of the glycoproteins found in the zona pellucida itself.

Cellular Origin of Mouse Zona Pellucida Glycoproteins

Each growing mouse oocyte is surrounded by follicle cells that increase in number as the oocyte grows; these cells eventually comprise the corpus luteum that functions as an endocrine gland following ovulation and supports post-implantation development. By using growing oocytes, completely denuded of surrounding follicle cells and metabolically radiolabeled with either methionine or fucose, we were able to demonstrate that only the growing oocyte itself synthesizes and secretes ZP1, 2, and 3 (9-12). The presence of surrounding follicle cells had no significant effect on the synthesis of zona pellucida glycoproteins by growing oocytes, and follicle cells themselves did not synthesize ZP1, 2, or 3. While we have not been able to detect any synthesis of ZP1, 2, and 3 in nongrowing oocytes, all three glycoproteins are synthesized and secreted by growing and fully-grown oocytes, but not by ovulated eggs or cleavage stage embryos (Fig. 2).

Therefore, the three zona pellucida glycoproteins, ZP1, 2, and 3, are: (i) synthesized by a single cell type, the oocyte; (ii) synthesized only during a two to three week period of development; and (iii) major products, representing 10% or more of total protein synthesis. Because of these and other characteristics, we feel that zona pellucida biosynthesis represents an attractive system for study of differential gene expression during mammalian development. Accordingly, we have recently constructed a recombinant cDNA library from individually isolated growing mouse oocytes using the expression vector lambda gt11. The library has been screened using antisera directed against zona pellucida glycoproteins and positive clones characterized. These cDNAs will be used to examine both expression and organization of the zona pellucida genes.

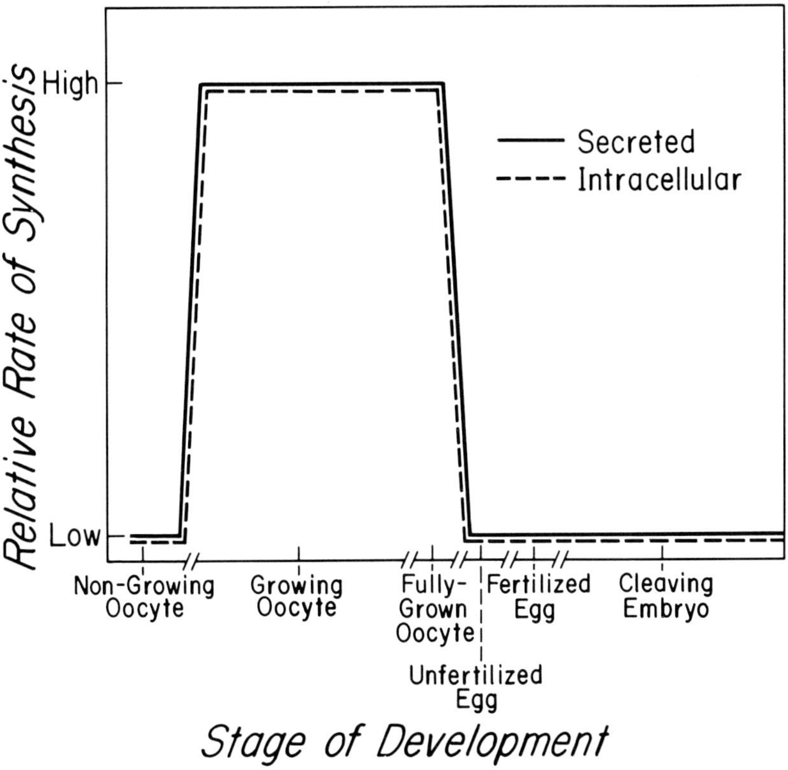

FIGURE 2. Diagram of the relative rates of zona pellucida glycoprotein synthesis during mouse development. The solid line represents measurements of mature ZP1, 2, and 3 found in the zona pellucida (secreted). The dashed line represents measurements of precursors of ZP1, 2, and 3 found in the oocyte (intracellular). A designation of "low" indicates that synthesis was not detected, whereas "high" indicates that zona pellucida glycoprotein synthesis represented about 10% of total protein synthesis.

Biosynthesis of Mouse Zona Pellucida Glycoproteins

Like many other secreted glycoproteins, mature ZP1, 2, and 3 contain asparagine-linked (N-linked) oligosaccharides. By using polyclonal antisera directed against the zona pellucida glycoproteins, together with metabolic radiolabelling of intracellular precursors of these glycoproteins, we have been able to learn a great deal about the biosynthetic pathways of ZP1, 2, and 3 (10-12, G. Salzmann and P. Wassarman, unpublished results). For example, pulse-chase and other types of experiments have revealed that the immediate precursors of mature ZP1, 2, and 3 have molecular weights of 150, 91, and 53/56 kd, respectively, and are considerably more basic than the mature species. Furthermore, experiments utilizing either endoglycosidases H and F (Endo H and F) to remove N-linked oligosaccharides or the antibiotic tunicamycin to prevent N-linked glycosylation of nascent proteins, have provided information about the nature of the precursors of zona pellucida glycoproteins.

ZP1 (200 kd) is synthesized as a 135 kd molecular weight dimer held together by intermolecular disulfides, to which high mannose-type oligosaccharides are added, presumably cotranslationally, giving rise to a 150 kd molecular weight intermediate (Fig. 3).

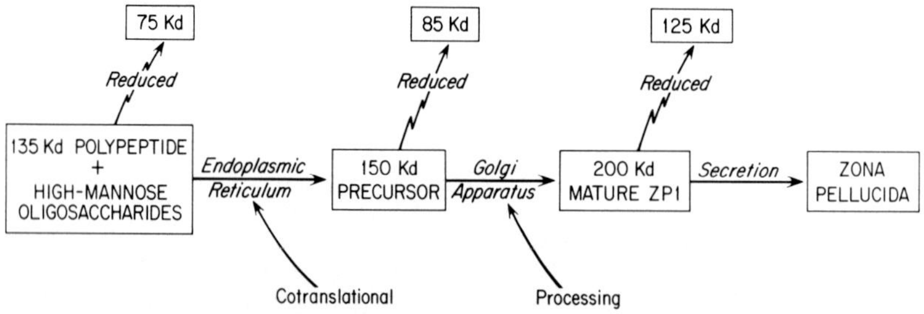

FIGURE 3. Biosynthetic pathway for ZP1.

ZP2 (120 kd) is synthesized as an 81 kd molecular weight polypeptide chain to which six high mannose-type oligosaccharides are added, presumably cotranslationally, giving rise to a 91 kd molecular weight intermediate (Fig. 4).

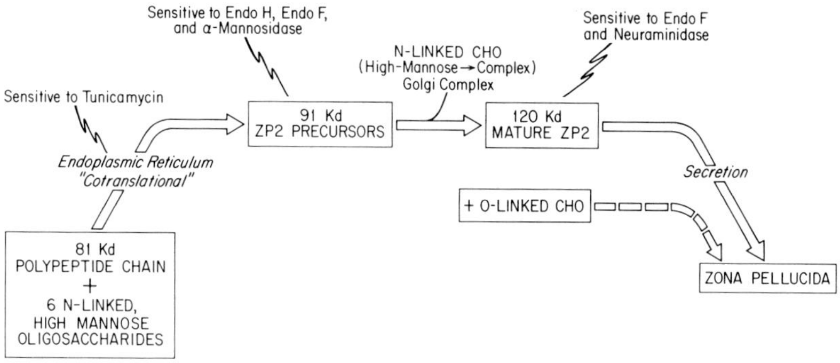

FIGURE 4. Biosynthetic pathway for ZP2.

ZP3 (83 kd) is synthesized as a 44 kd molecular weight polypeptide chain to which either three or four high mannose-type oligosaccharides are added, presumably cotranslationally, giving rise to 53 and 56 kd molecular weight intermediates, respectively (Fig. 5). It is not clear whether the presence of two precursors means (i) that our in vitro system is inefficient at adding the fourth N-linked oligosaccharide chain, or (ii) that there are two different protein species, one with three and another with four acceptor sites.

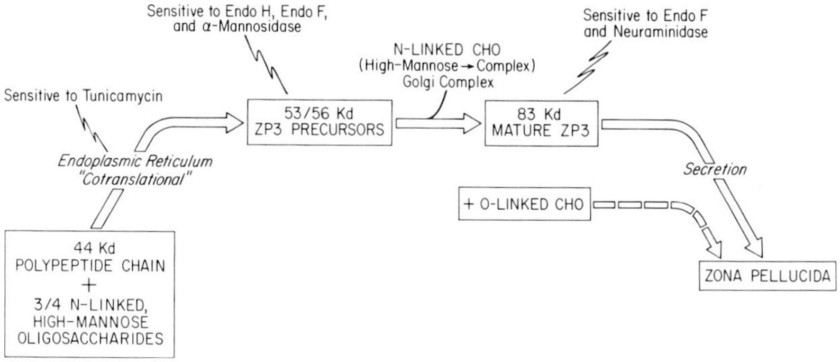

FIGURE 5. Biosynthetic pathway for ZP3.

In all cases (ZP1, 2, and 3), the N-linked oligosaccharides are processed to complex-type, presumably in the Golgi, prior to secretion of the mature glycoproteins into the zona pellucida (Figs. 3-5). Furthermore, at least in the cases of ZP2 and 3, serine/threonine-linked (O-linked) carbohydrate is added to the glycoproteins prior to secretion. The O-linked carbohydrate plays a vital role in the sperm receptor activity of ZP3 (5,6, H. Florman and P. Wassarman, unpublished results), whereas the N-linked oligosaccharides apparently play a role in secretion of ZP2 (12).

How Mouse Eggs Take Off Their Extracellular Coat

In order to be able to implant in the uterus and continue normal development, a mouse blastocyst must first "hatch" from its zona pellucida. Hatching occurs even during culture of mouse embryos in vitro, resulting in empty zonae pellucidae, each with a small hole through which the embryo has escaped, and embryos that attempt to implant in the culture dish. We have recently investigated several aspects of hatching in vitro and discovered the involvement of a trypsin-like proteinase, that we call "strypsin", in this process. Furthermore, we found that the appearance of strypsin activity occurred during early cleavage and was restricted at first to no more than one or two cells at about the time of compaction of the 8-16-cell embryo, and later to a group of cells in the mural trophectoderm of the expanded blastocyst (R. Perona and P. Wassarman, unpublished results).

Hatching of a mouse embryo from its zona pellucida occurs at the expanded blastocyst stage of development (more than 100 cells), when the embryo consists of inner cell mass (ICM; gives rise to the fetus) and trophectoderm (gives rise to the placenta)(Fig. 6). We have found that blastocysts fail to hatch from their zona pellucida when cultured in the presence of various serine proteinase inhibitors (e.g., alpha-1 antitrypsin, PMSF, aprotinin, leupeptin, and antipain), whereas exopeptidase (e.g., bestatin and amastatin) and metalloproteinase (e.g., phosphoramidon) inhibitors do not prevent hatching. Furthermore, since soybean trypsin inhibitor prevents hatching, but chymostatin and elastin do not, it is likely that a trypsin-like proteinase (strypsin) is involved. Using a fluorometric assay to detect strypsin activity in early embryo homogenates

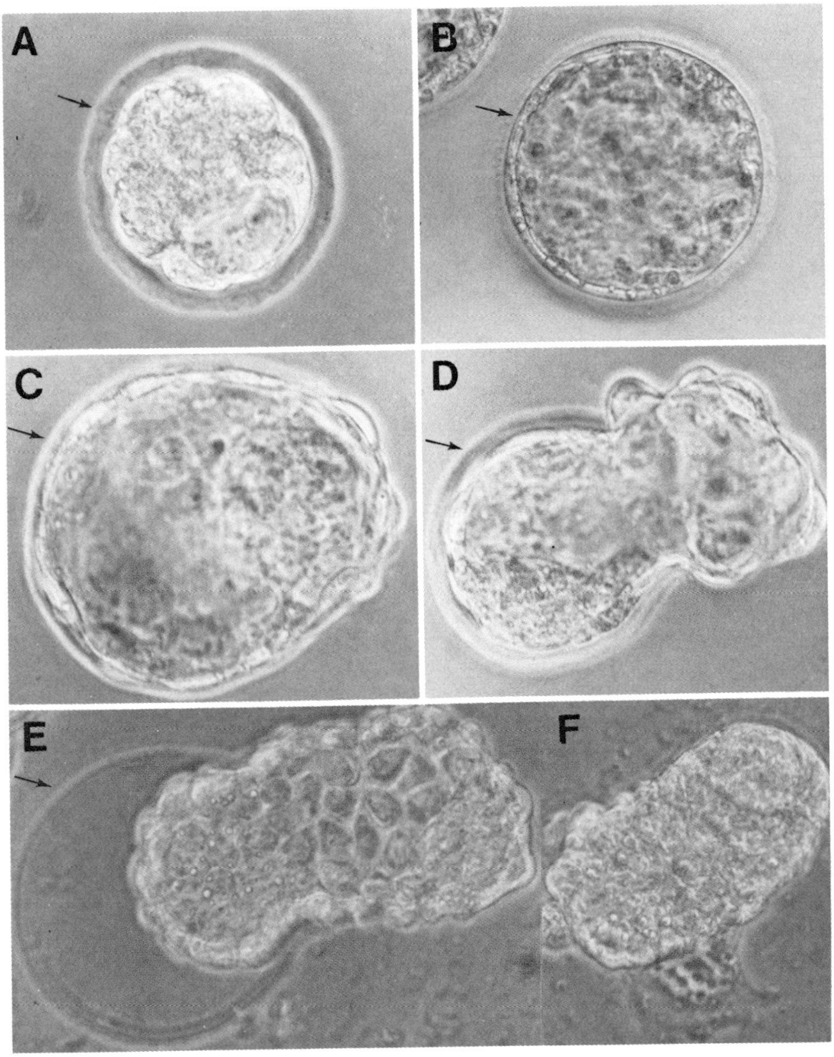

FIGURE 6. Selected stages of early embryogenesis in the mouse. Shown are light photomicrographs of morula (A), blastocyst (B), hatching blastocyst (C, D, and E), and hatched blastocyst (F) stage embryos cultured in vitro. In each case the zona pellucida is indicated by an arrow.

([N-CBZ-glycyl-glycyl-argininamido]-methyl-coumarin, as substrate), we found that the activity first appeared at the compacted 8-16-cell stage (morula), increased to a maximum level when about 50% of the embryos had hatched, and then decreased to background levels. Electrophoretic analyses have revealed that strypsin has a molecular weight of about 74 kd and various properties of the enzyme strongly suggest that strypsin is a cell surface proteinase.

In order to determine whether or not strypsin is localized to a particular region of the early mouse embryo, we have used a histochemical procedure that involves lightly fixed cells and is based on proteolysis of the substrate, benzoyl-arginine-methoxy-beta-naphthylamide, in the presence of a diazo dye. Examples of results obtained using this procedure are presented in Fig. 7.

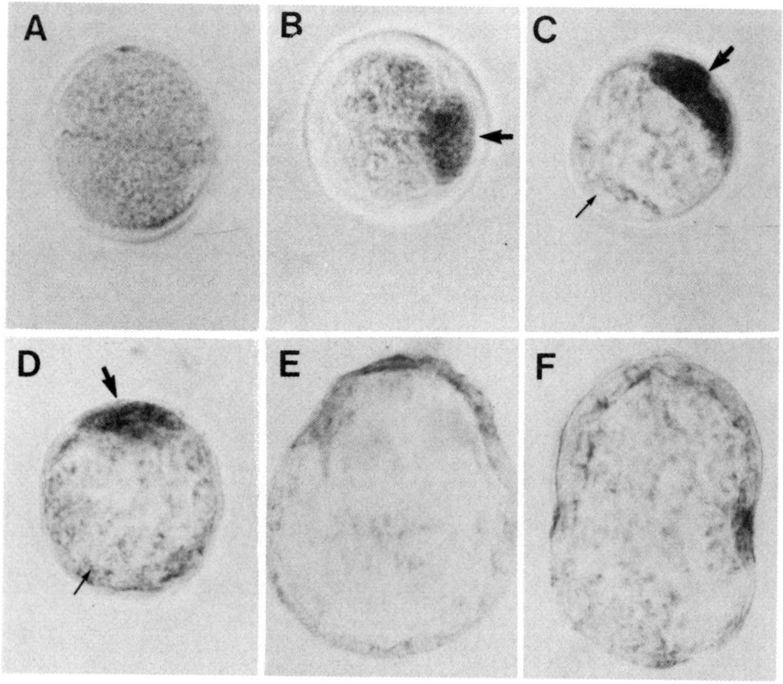

FIGURE 7. Light photomicrographs of mouse embryos subjected to histochemistry. Shown are 2-cell (A) and 8-cell (B) embryos, blastocysts (C) and expanded blastocysts (D), and hatching blastocysts (E,F). Strypsin staining (thick arrow) and ICM (thin arrow) are indicated.

Histochemistry of early mouse embryos has revealed that: (i) strypsin activity first appears at the 8-16-cell stage of development, in only one or two cells of the embryo; (ii) strypsin activity is always associated with the trophectoderm, never with the inner cell mass; and (iii) strypsin activity is always associated with mural, not polar (overlying the inner cell mass), trophectoderm. Finally, experiments of the type diagrammed in Fig. 8, have demonstrated that appearance of strypsin activity correlates with compaction of the 8-cell embryo in vitro; in the absence of compaction (achieved by removal of calcium from the culture medium), strypsin activity cannot be detected.

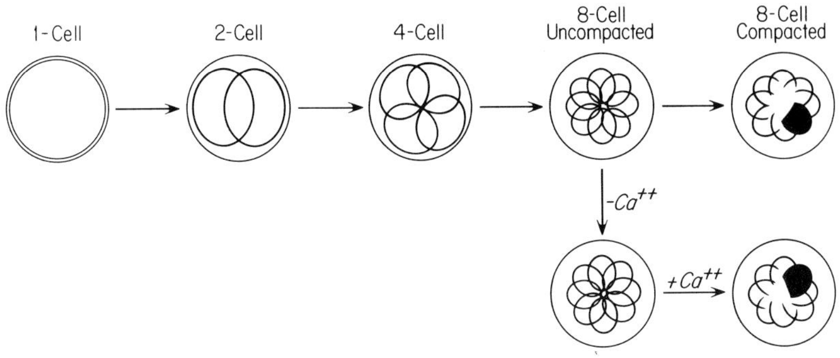

FIGURE 8. Diagrammatic representation of experiments implicating compaction of the 8-cell embryo with appearance of strypsin activity.

SUMMARY

Here, we have reviewed some of our progress in understanding how mouse eggs put on and take off their extracellular coat, the zona pellucida. Our studies of various aspects of this simple organelle have revealed a stringently regulated developmental program that begins during oocyte growth and ends with hatching of blastocysts. We have found that production of the zona pellucida involves expression of genes encoding only three different glycoproteins for a relatively short period of time in a single cell type, the oocyte. The three glycoproteins are assembled into filaments possessing a structural repeat and these filaments are interconnected giving rise to a three-dimensional matrix. The completed zona pellucida performs its functions during fertilization and preimplantation development and, finally, must be discarded. Accordingly, at the blastocyst stage, embryos hatch from the zona pellucida using a trypsin-like proteinase we call "strypsin". Appearance of strypsin activity correlates with compaction of the 8-cell stage embryo in vitro, and the proteinase appears to be restricted initially to a single cell. Presumably, this cell gives rise to the group of cells possessing strypsin activity found in the mural trophectoderm of blastocysts.

ACKNOWLEDGMENTS

J.M.G. and R.M.P. are postdoctoral fellows supported in part by the Rockefeller Foundation and Juan March Foundation, respectively. G.S.S. and R.J.R. are predoctoral fellows supported in part by National Research Service Awards. Finally, we are grateful to the other members of our laboratory group for advice and constructive criticism throughout the course of this research.

REFERENCES

1. Wassarman PM, Josefowicz WJ (1978). Oocyte development in the mouse: an ultrastructural comparison of oocytes isolated at various stages of growth and meiotic competence. J Morphol 156:209.
2. Wassarman PM (1983). Oogenesis: synthetic events in the developing mammalian egg. In Hartmann JH (ed): "Mechanism and Control of Animal Fertilization," New York: Academic Press, p 1.
3. Bleil JD, Wassarman PM (1980). Mammalian sperm-egg interaction: identification of a glycoprotein in mouse egg zonae pellucidae possessing receptor activity for sperm. Cell 20:873.
4. Bleil JD, Wassarman PM (1983). Sperm-egg interactions in the mouse: sequence of events and induction of the acrosome reaction by a zona pellucida glycoprotein. Dev Biol 95:317.

5. Wassarman PM (1982). Fertilization. In Yamada K (ed): "Cell Interactions and Development: Molecular Mechanisms," New York: Wiley, p 1.
Wassarman PM, Bleil JD (1982). The role of zona pellucida glycoproteins as regulators of sperm-egg interactions in the mouse. In Frazier WA, Glaser L, Gottlieb DI (eds): "Cellular Recognition," New York: Alan R. Liss, p 845.
Wassarman PM, Florman HM, Greve JM (1984). Receptor mediated sperm-egg interactions in mammals. In Metz CB, Monroy A (eds): "Biology of Fertilization," New York: Academic Press, in press.
6. Florman HM, Wassarman PM (1983). The mouse egg's receptor for sperm: involvement of O-linked carbohydrate. J Cell Biol 97:26a.
Florman HM, Bechtol KB, Wassarman PM (1984). Enzymatic dissection of the functions of the mouse egg's receptor for sperm. Dev Biol, in press.
7. Greve JM, Wassarman PM (1983). The mouse egg's extracellular coat: organization of a glycoprotein matrix. J Cell Biol 97:229a.
8. Bleil JD, Wassarman PM (1978). Identification and characterization of the proteins of the zona pellucida. J Cell Biol 79:173a.
Bleil JD, Wassarman PM (1980). Structure and function of the zona pellucida: identification and characterization of the proteins of the mouse oocyte's zona pellucida. Dev Biol 76:185.
9. Bleil JD, Wassarman PM (1980). Synthesis of zona pellucida proteins by denuded and follicle-enclosed mouse oocytes during culture in vitro. Proc Nat Acad Sci USA 77:1029.
10. Greve JM, Salzmann GS, Roller RJ, Wassarman PM (1982). Biosynthesis of the major zona pellucida glycoprotein secreted by oocytes during mammalian oogenesis. Cell 31:749.
11. Salzmann GS, Greve JM, Roller RJ, Wassarman PM (1983). Biosynthesis of the sperm receptor during oogenesis in the mouse. EMBO J 2:1451.
12. Roller RJ, Wassarman PM (1983). Role of asparagine-linked oligosaccharides in secretion of glycoproteins of the mouse egg's extracellular coat. J Biol Chem 258:13243.

Molecular Biology of Development, pages 227–239

THE EXPRESSION OF REPETITIVE SEQUENCES ON AMPHIBIAN LAMPBRUSH CHROMOSOMES[1]

Kathleen A. Mahon[2] and Joseph G. Gall[3]

Department of Biology, Yale University
New Haven, CT 06511

ABSTRACT We have studied two highly reiterated satellite DNA sequences from the newt *Nothophthalmus viridescens* which have been shown to be transcribed on lampbrush chromosomes by *in situ* hybridization. Satellite 1, a 222 bp tandemly repeated sequence, is located in the pericentric heterochromatin of all chromosomes and at two sites on the arms of chromosomes 2 and 6. Both strands of satellite 1 are transcribed at the non-centromeric loci, apparently due to readthrough transcription from adjacent histone gene promotors. This pattern of transcription, in which there is a failure of termination at the 3' end of active genes, is probably a general feature of the lampbrush chromosome stage and can explain the extremely long transcription units present on the lampbrush loops. Transcripts containing satellite 1 are almost exclusively nuclear and are not stably accumulated. Satellite 2 is an abundant 330 bp sequence which differs from satellite 1 in both its genomic organization and in its pattern of expression in the oocyte. In contrast to the predominantly heterochromatic location of of satellite 1, tandemly repeated copies of satellite 2 occur in clusters dispersed throughout the genome. Transcripts homologous to satellite 2 are stable and appear to accumulate in the oocyte.

[1] Supported by grant GM 12427 from the NIH.
[2] Present address: Laboratory of Molecular Genetics, NIH, Bethesda, MD 20205.
[3] Present address: Department of Embryology, Carnegie Institute of Washington, Baltimore, MD 21210.

INTRODUCTION

In many animals, a substantial proportion of the development of the oocyte takes place during a period of suspended diplotene known as the lampbrush chromosome stage. At this time, the chromosomes appear as axes of relatively condensed chromatin from which loops of extended, transcriptionally active DNA periodically project. Morphological data on the length and polarity of ribonucleoprotein matrix material on the loops (1,2,3), and ultrastructural analysis of lampbrush chromatin by the Miller spreading technique (4,5), indicate that the transcription units (TUs) on the loops are extremely long. The length of the loops and their TUs is proportional to the DNA content of the organism. In urodele amphibians, with their very high C values, these TUs range from 10->200 um in length (2). Given the interspersion of repetitive sequences in the eukaryotic genome, the great length of the TUs on lampbrush chromosomes is an indication that many of these repeated sequences may be transcribed to some extent. Indeed, repetitive sequences are extensively represented in oocyte RNA (6).

Taking advantage of the large size of amphibian lampbrush loops and the high density of nascent RNA transcripts associated with them, a number of investigators have used the technique of *in situ* hybridization to study the expression of repetitive sequences in the oocyte nucleus (7-16). This approach has proven invaluable for the detection of sequences which are expressed, but also has the unique advantage of simultaneously identifying the chromosomal location of synthesis. The later property is of particular utility in studying the transcription of repetitive sequences, since it is possible to distinguish members of a repeated family which are transcriptionally active from those which are not.

Using *in situ* hybridization to nascent lampbrush transcripts coupled with a molecular analysis of transcription, we have studied the expression of two satellite DNAs--sequences traditionally thought to be transcriptionally silent--on the lampbrush chromosomes of the newt *Notophthalmus viridescens*. Both of these satellite sequences are abundant, tandemly reiterated components of the newt genome. Sequence data have indicated that neither satellite has protein coding capacity, yet we have found that both are transcribed on the loops of lampbrush chromosomes. Despite these similarities, the two satellites, designated satellte 1 and 2, differ from each other in their distribution in the newt genome and in their transcriptonal properties.

CHARACTERIZATION OF NEWT SATELLITES

When genomic DNA of Notophthalmus viridescens was digested with restriction enzymes and electrophoresed on an agarose gel it was apparent from the ethidium bromide staining pattern that many enzymes cleave the DNA periodically, generating visible ladders of fragments of multimeric lengths (Figure 1b). The ladder-like pattern is diagnostic of tandemly repeated satellite DNA, and results from slight sequence divergence among the repeats causing a gain or loss of restriction sites in members of the repeat family. Closer inspection indicated that the banding pattern was actually composed of two multimeric ladders of fragments superimposed: one with a periodicity of 222 bp, and the other with a periodicity of 330 bp.

To study these repeated sequences further, the monomer and dimer bands of the 222 bp series and the monomer fragments of the 330 bp series were isolated from an agarose gel containing Bgl II digested N. viridescens DNA and cloned into the Bam HI site of pBR322. Recombinant plasmids pNv11 and pNv15, containing the monomer and dimer of the 222 bp series, respectively, and pNv13, containing the monomer of the 330 bp series, were selected for further analysis. Radiolabeled pNv11 and pNv13 were hybridized to genomic Southern blots containing Bgl II digested newt DNA. The monomer clone pNv11 hybridized exclusively to the 222 bp oligomeric ladder (Figure 1a), whereas pNv13 had homology to only the 330 bp multiple series (Figure 1c), indicating that these are members of distinct repetitive sequence families. The newt sequences in the 222 bp series complementary to pNv11 were designated satellite 1 and the family of sequences represented by the insert in pNv13 were called satellite 2.

Satellite 1

In situ hybridization of ^{3}H-labeled satellite 1 probes to the DNA in newt mitotic and lampbrush chromosomes

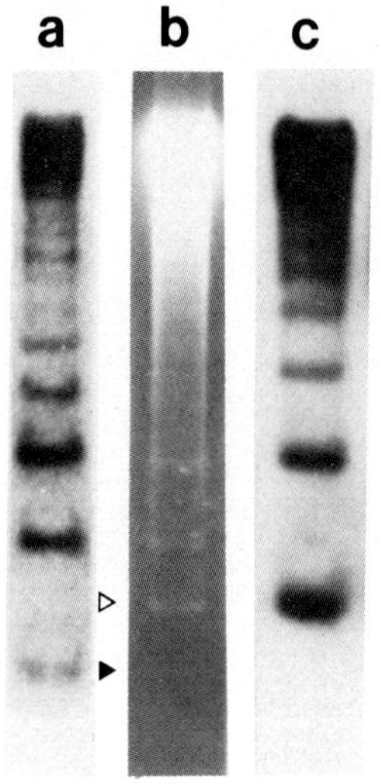

Figure 1. Genomic organization of newt satellite DNAs.
Notophthalmus viridescens testis DNA was digested with Bgl II and electrophoresed on a 2% agarose gel. The ladder-like pattern discernible in the ethidium bromide stained gel (lane b) is comprised of two multimeric series of restriction fragments: one with a basic repeat length of 222 bp (closed arrow), the other with a repeat length of 330 bp (open arrow). The autoradiograms of filters containing Bgl II digested DNA hybridized to ^{32}P-labeled pNv11, a cloned 222 bp monomer (lane a), and pNv13, a cloned 330 bp monomer (lane c), show that satellite 1 and satellite 2 are distinct repeated sequences.

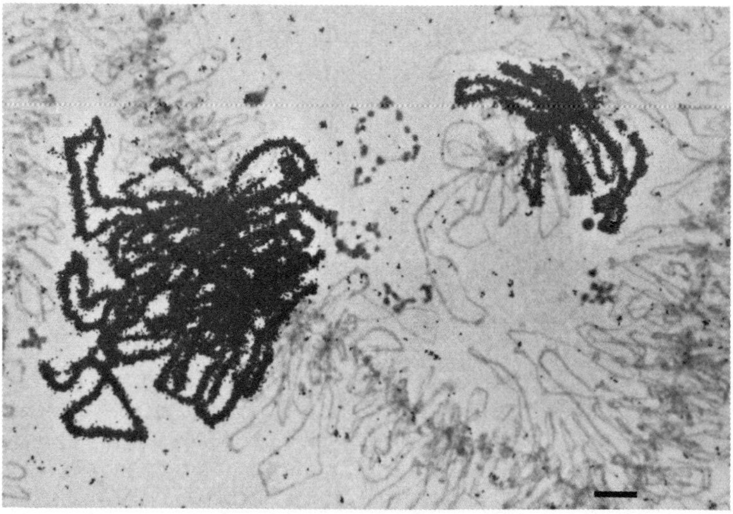

Figure 2. *In situ* hybridization of satellite 1 sequences to lampbrush chromosome transcripts.
Shown are loops at the sphere loci of chromosomes 2 and 6 labeled after hybridization to ^{3}H-cRNA to mNv15-47, a single-stranded M13 clone containing a dimer of satellite 1. Spec. act. of probe: 9 X 10^7 dpm/μg. Exp. 4 days. Coomassie blue stain. Bar = 10μm.

revealed that satellite 1 sequences are situated in the pericentric heterochromatin of all chromosomes and on two loci on the arms of chromosomes 2 and 6 (17). On lampbrush chromosomes, these non-centromeric sites are demarcated by proteinaceous bodies known as spheres. The newt insert of pNv11 was sequenced and found to contain no open translational reading frames (16). The heterochromatic localization of satellite 1, its lack of obvious coding potential, and the hybridization pattern apparent on Southern blots, suggested that in virtually every respect of its genomic organization, satellite 1 resembled a typical satellite DNA sequence.

When satellite 1 probes were hybridized to the nascent RNA transcripts on non-denatured lampbrush chromosomes, intense labeling was seen over loops at both sphere loci, but only very rarely at the centromeric location (Figure 2), indicating a difference in expression at these two loci. By hybridizing single-stranded probes synthesized from M13 subclones of pNv15 to lampbrush chromosomes, it was found that both strands of satellite 1 are transcribed on the sphere loci loops (16).

Transcripts of satellite 1 are primarily limited to the nucleus of the developing oocyte (Mahon et al, manuscript in prep.). Considered together with the lack of open reading frames in the sequence of the satellite, it is doubtful that satellite 1 transcripts are involved in translational processes and suggests that these transcripts are unstable and degrade after synthesis.

The reason for the transcription of satellite 1 became clear when it was found that very long tandem arrays of satellite 1 sequence separate the 600-800 histone gene clusters of the newt (18). This association is depicted schematically in Fig. 3. Single-stranded histone probes hybridized to loops at the same cytologically identifiable sphere loci (17). Because the TUs labeled by the histone probes were much longer than the histone coding sequences alone, it was proposed that transcription at the sphere loci initiates at one of the five histone gene promotors, fails to terminate at the 3' end of the gene, and continues into the adjacent satellite sequence, generating the extremely long transcription units seen (16). According to this model, the existence of transcripts from both strands of the satellite can be accounted for by the presence of histone gene promotors on both strands of the cluster.

Although it has not yet been established unequivocally

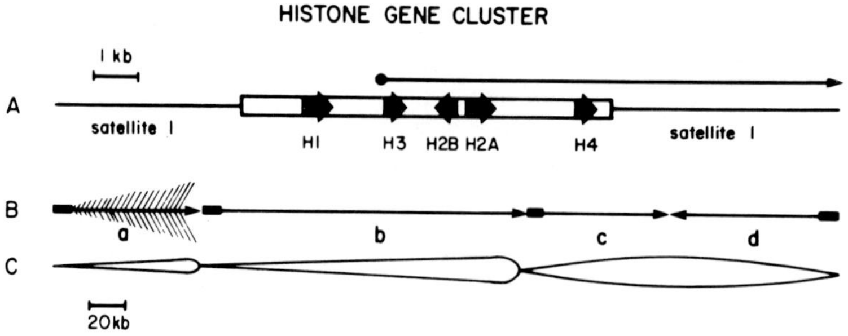

Figure 3. A map of the histone gene cluster and an interpretation of its transcription.
A. The arrangement of the histone genes within the histone gene cluster. Each cluster is separated by long arrays of tandemly repeated satellite 1 sequence. The direction of transcription is indicated by the arrows. Transcription initiates on histone gene promotors and proceeds without interruption into the flanking satellite 1 sequences. Transcription begining at the H3 promotor is indicated by the long arrow.
B. Four consecutive histone gene clusters (black boxes) separated by variable lengths of satellite 1 tracts (thin lines). Arrows indicate four possible directions of transcription, giving rise to four TUs (a,b,c,d). a,b, and c are examples of TUs arranged head to tail; c and d are examples of convergent TUs. Nascent transcripts are depicted on transcription unit a.
C. A diagram of how these TUs would appear in the light microscope.

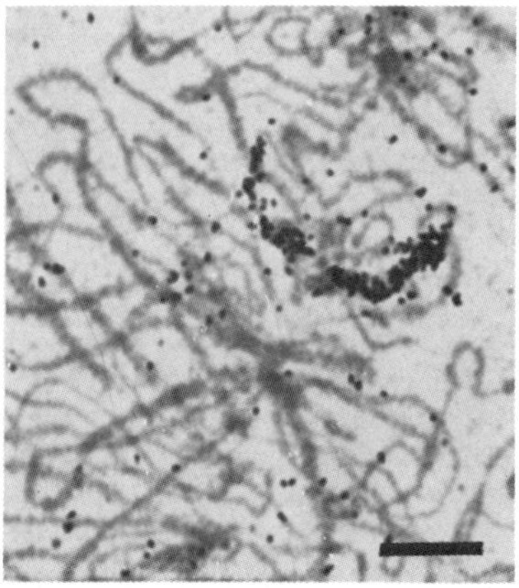

Figure 4. _In situ_ hybridization of satellite 2 to lampbrush loops.
Loops labeled after hybridization to ^{3}H-labeled cRNA to mNv13-08, a single-stranded M13 clone containing a monomer repeat of satellite 2. The hybridization pattern seen on these loops is representative of other loops labeled with satellite 2 probes. Spec. act. of probe 1.2×10^7 dpm/ug. Exp. 3 weeks. Coomassie blue stain. Bar = 10μ m.

that the histone and satellite 1 RNAs are covalently linked on the same primary transcript, a great deal of indirect data support the model of transcription posed above. A detailed cytogenetic and biochemical study of transcription on the sphere loci loops using single-stranded probes specific to various spacer sequences in the histone gene cluster has substantiated various predictions of the model (19; Diaz and Gall, manuscript in prep; O'Rear and Gall, manuscript in prep.).

Satellite 2

The data from our analysis of the 330 bp repeated sequence, satellite 2, will be published elsewhere, but the results of these studies will be briefly summarized here for the purposes of discussion. The hybridization pattern of satellite 2 probes to newt genomic DNA digests (Figure 1c) indicated that this sequence is an abundant, tandemly repeated component of the newt genome. However, <u>in situ</u> localization of satellite 2 probes to the DNA on <u>mitotic</u> chromosomes revealed that, unlike satellite 1, labeling was not sequestered in long blocks in the heterochromatic regions of the chromosomes, but was scattered over the entire chromosome complement. It therefore appears that satellite 2 sequences occur in tandemly repeated clusters dispersed throughout the genome. Transcripts complementary to satellite 2 are also synthesized on lampbrush chromosomes. Loops hybridizing to satellite 2 probes are generally few (5-7) and sparsely labeled (Figure 4). Despite the abundance and interspersion of this satellite, very little is actually present in the long TUs of lampbrush chromosomes--much less than would be expected if there is random readthrough transcription on the loops. Furthermore, although both strands of satellite 2 are transcribed on lampbrush chromosomes, small discrete transcripts (330-660 nt) complementary to only one strand accumulate in the oocyte. The identity and significance of these transcripts are currently under investigation.

DISCUSSION

It has long been known that many interspersed

repetitive sequences are transcribed in the oocyte (reviewed in 20). This repetitive component of the total RNA population suffers many fates: some of the transcripts appear to turn over in the nucleus with the bulk of the hnRNA, while others are found in the cytoplasmic poly A+ RNA of the oocyte and early embryo (6, 21, 22). The functional significance of repetitive transcripts of either class is not clear.

In situ hybridization has fascilitated the identification of many repeated sequences which are transcribed on lampbrush chromosomes. These sequences include the repeated genes for 5S RNA (7; Jamrich et al, manuscript in prep.), rRNA (15), and histones (9, 10), as well as middle repetitive DNA (cot 0.2-50) (11), highly repetitive DNA (cot 0.0-0.2) (12), and the simple sequence polymers of poly(dCT):poly(dAG) and poly(dC): poly(dG) (9, 10). The hybridization of cloned satellite DNA sequences of the newt to lampbrush chromosomes (13, 14) clearly demonstrated that sequences of this type were not as transcriptionally inert as had always been assumed. Varley et al (14) postulated that a general lack of transcription termination could account for both the abundance of repetitive transcripts produced as well as the length of the TUs seen on lampbrush chromosomes.

However, it was only after the detailed study of transcription at the histone locus that the true nature of a transcription unit containing satellite DNA sequences could be dissected in detail. First, the hybridization of single-stranded probes specific for histone and satellite 1 sequences to lampbrush chromosomes demonstrated that the unit of labeling consisted of one thin-to-thick gradient of RNP matrix on the loop. This provided the first biochemical evidence that these regions of morphological polarity, as had long been suspected, corresponded to transcription units. Defining the morphology of the transcription unit at the level of the light microscope was critical for the interpretations of in situ hybrids that followed. Secondly, it was possible to integrate a molecular analysis of the organization of the histone gene complex and its surrounding satellite 1 sequences with data obtained from cytohybridization. The results clearly showed that satellite 1 and histone sequences are coordinately expressed in the long TUs on the sphere loci loops, apparently because of a failure of transcription termination at the 3' end of the histone coding regions.

We propose that transcription can initiate on any one of the five histone gene promotors within a cluster, and that it continues through the rest of the cluster, even through other histone genes, into the flanking satellite sequences (16). Since long TUs like those of the sphere loci loops are typically observed on lampbrush chromosomes, this pattern of transcription, in which the termination signals presumably utilized in somatic cells are ignored, may be a general and ubiquitous characteristic of the lampbrush chromosome stage.

The expression of repeated sequences on lampbrush chromosomes by apparent readthrough transcription is not restricted to the newt, with its extremely long transcription units, but has been documented in another amphibian, Xenopus, which has a much smaller genome and correspondingly smaller loops and TUs (23). As is the case with satellite 1 and 2 of Notophthalmus, both strands of each of the three repeat families examined by these authors are represented in transcripts on Xenopus lampbrush chromosomes. Transcription of both strands of a repetitive sequence is to be expected if the analogy to the TUs at the sphere loci is of universal applicability to lampbrush transcripts, since statistically at least a few members of a repeat family will be found in either orientation downstream from an active gene promotor.

It is clear from the study of Xenopus lampbrush transcription and the analysis of Notophthalmus satellite 1 and 2 outlined here that not all copies of a particular repeat sequence are transcriptionally active in the lampbrush chromosome stage. Only the satellite 1 sequences at the sphere loci are expressed; the centromeric copies are transcriptionally quiescent. Satellite 2 is widely distributed throughout the genome, yet loops containing satellite 2 transcripts are few. This suggests that either transcription is not under the control of the repeat sequence itself, which appears to be the case with satellite 1, or that there are qualitative differences among the individual members of the repetitive sequence families, some being capable of regulating their own transcription. The two possibilities are not mutually exclusive. It is conceivable that certain members of a given repeat family are able to promote RNA synthesis, while other members are transcribed simply because they lie fortuitously 3' to an actively transcribing gene. The 5S RNA and rRNA genes of the newt appear to be transcribed in this way on lampbrush chromosomes (15; Jamrich et al, manuscript in prep.). Both strands of satellite 2 are transcribed on the loops of lampbrush chromosomes, which, given the abundance and widespread interspersion of this sequence, is consistent with the model of readthrough transcription.

The accumulation of small, discrete transcripts of only one strand of satellite 2 in the oocyte suggests that at least a subset of the repeats sustain regulated transcription either in the long transcription units seen on lampbrush chromosomes, or perhaps in very short TUs on the axis. Short, repeat length transcription units may escape detection on lampbrush chromosomes by in situ hybridization because they would only be long enough to contain a few transcripts and would probably not be located on the loops. For example, it is clear that bona fide 5S RNA transcripts are synthesized on such small TUs on the chromosome axis (Pukkila, 1975), and it is possible that most, if not all, Pol III transcripts are similarly transcribed. It is currently not certain which polymerase is responsible for the generation of the small satellite 2 transcripts, although the nascent transcripts hybridizing to satellite 2 probes (Fig. 4) are almost certainly due to Pol II since transcription on all of the loops is sensitive to low levels of α-amanitin (Bucci et al, 1971; Shultz et al, 1981). This question is currently under investigation.

Finally, our analysis of the transcription of satellite 1 and 2 of *Notophthalmus* clearly illustrates both the power and the limitation of our approach to defining sequences expressed during oogenesis. The study of transcription at the sphere locus, while providing a very clear picture of the pattern of transcription on lampbrush chromosomes which would not have been possible by any other means, has also introduced a certain amount of reservation into the interpretation of lampbrush chromosome in situ hybrids. It cannot be tacitly assumed, as we once naively believed, that positive hybridization of a particular sequence to lampbrush chromosome loops is an automatic indication that these transcripts have significance to the development of the oocyte, or that they contribute to the stored reserve of maternal message. If the interpretation of transcription at the sphere locus has general validity, then much of the RNA synthesized may have no significance. It is possible that even structural genes can be transcribed coincidentally by readthrough if positioned downstream from an active gene.

Conversely, as our studies of satellite 2 indicate, it cannot be assumed that all transcription units hybridizing to a particular sequence actually give rise to the transcripts which are present in the cytoplasm of the oocyte. There is a gap in the chain of evidence which necessarily qualifies any definitive statements equating primary transcript with mRNA product. However, our studies have given us greater insight into transcription during oogenesis, and it is our hope that as more regions of the amphibian genome are subjected to the kind of molecular and cytogenetic scrutiny given to the histone gene complex of the newt, combined with a continued biochemical analysis of transcription and RNA metabolism throughout oogenesis, we will ultimately have a much clearer idea of the role that lampbrush chromosomes play in these processes.

ACKNOWLEDGMENTS

We thank Dr. Milan Jamrich and Dr. Mary Ellen Digan for helpful comments on this manuscript.

REFERENCES

1) Callan HG (1955). Recent work on the structure of cell nuclei. Int Union Biol Sci B 21:89.
2) Gall JG (1956). On the submicroscopic structure of chromosomes. Brookhaven Symp Biol 8:17.
3) Callan HG, Lloyd L (1960) Lampbrush chromosomes of crested newts Triturus cristatus (Laurenti). Phil Trans Roy Soc B. 243:135.
4) Miller OL, Hamkalo BA (1972). Visualization of RNA synthesis on chromosomes. Int Rev Cytol 33:1.
5) Hamkalo BA, Miller OL (1973) Electronmicroscopy of genetic activity. Ann Rev Biochem 42:379.
6) Anderson DM, Richter JD, Chamberlin ME, Price DH, Britten RJ, Smith LD, Davidson EH (1982). Sequence organization of the poly (A) RNA synthesized and accumulated in lampbrush chromosome stage Xenopus laevis oocytes. J Mol Biol 155:281.
7) Pukkila PJ (1975). Identification of the lampbrush chromosome loops which transcribe 5S ribosomal RNA in Notophthalmus (Triturus) viridescens. Chromosoma 53:71.

8) Old RW, Callan HG, Gross KW (1977). Localization of histone gene transcripts in newt lampbrush chromosomes by in situ hybridization. J Cell Sci 27:57.
9) Callan HG, Old RW (1980). In situ hybridization to lampbrush chromosomes: a potential source of error exposed. J Cell Sci 41:115.
10) Callan HG, Old RW, Gross KW (1980). Problems exposed by the results of in situ hybridization to lampbrush chromosomes. Eur J Cell Biol 22:21.
11) Macgregor HC, Andrews C (1977). The arrangement and transcription of "middle repetitive" DNA sequences on lampbrush chromosomes of Triturus. Chromosoma 63:109.
12) Macgregor HC (1979). In situ hybridization of highly repetitive DNA to chromosomes of Triturus cristatus. Chromosoma 71:57.
13) Varley JM, Macgregor HC, Erba HP (1980a). Satellite DNA is transcribed on lampbrush chromosomes. Nature 283:686.
14) Varley JM, Macgregor HC, Nardi I, Andrews C, Erba HP (1980b). Cytological evidence of transcription of highly repeated DNA sequences during the lampbrush chromosome stage in Triturus cristatus carnifex. Chromosoma 80:289.
15) Morgan G, Macgregor HC, Colman A (1980). Multiple ribosomal gene sites revealed by in situ hybridization of Xenopus rDNA to Triturus lampbrush chromosomes. Chromosoma 80:309.
16) Diaz MO, Barsacchi-Pilone G, Mahon KA, Gall JG (1981). Transcripts from both strands of a satellite DNA occur on lampbrush chromosome loops of the newt Notophthalmus. Cell 24:649.
17) Gall JG, Stephenson EC, Erba HP, Diaz MO, Barsacchi-Pilone G (1981). Histone genes are located at the sphere loci of newt lampbrush chromosomes. Chromosoma 84:159.
18) Stephenson EC, Erba HP, Gall JG (1981). Histone gene clusters of the newt Notophthalmus are separated by long tracts of satellite DNA. Cell 24:639.
19) Gall JG, Diaz MO, Stephenson EC, Mahon KA (1983). The transcription unit of lampbrush chromosomes. In Subtelny S, Kafatos F (eds): "Gene Structure and Regulation in Development", New York: Alan R. Liss, p 137.
20) Davidson E (1976). "Gene Activity in Early Development". New York: Academic Press.

21) Davidson EH, Posakony JW (1982). Repetitive sequence transcripts in development. Nature 297:633.
22) Spohr G, Reith W, Sures I (1981). Organization and sequence analysis of a cluster of repetitive DNA elements from *Xenopus laevis*. J Mol Biol 151:573.
23) Jamrich M, Warrior R, Steele R, Gall JG (1983). Transcription of repetitive sequences on *Xenopus* lampbrush chromosomes. Proc Nat Acad Sci 80:3364.
24) Bucci S, Nardi I, Macino G, Flume L (1971). Incorporation of tritiated uridine in nuclei of *Triturus* oocytes treated with α-amanitin. Exp Cell Res 69:462.
25) Shultz LD, Kay BK, Gall JG (1981). *In vitro* RNA synthesis in oocyte nuclei of the newt *Notophthalmus*. Chromosoma 82:171.

Molecular Biology of Development, pages 241–252

A METHOD FOR ANALYZING OOCYTE MESSENGER RNAS WHICH PERSIST IN THE EMBRYO OF DROSOPHILA MELANOGASTER[1]

William H. Phillips, Jeffrey A. Winkles[2], and Robert M. Grainger

Department of Biology, University of Virginia
Charlottesville, Virginia 22901

ABSTRACT In order to follow the fate of oocyte mRNAs during *Drosophila* embryogenesis, a pulse-chase scheme using density labelling was devised. This method permits us to collect embryos that contain oocyte RNAs which are density labelled, while RNAs synthesized during embryogenesis are of a light buoyant density. These heavy and light RNAs are preparatively separable on Na/KI equilibrium gradients, and a sufficient amount of oocyte and embryonic mRNAs can be recovered to stimulate translation *in vitro*. In embryos we have detected and identified stable oocyte mRNAs which are transcribed sometime between early oogenesis and the beginning of vitellogenesis. This experimental approach permits the first systematic analysis of oocyte mRNAs which persist in the embryo and which may contribute to such important embryonic processes as axis formation and cell determination.

INTRODUCTION

It has been shown in many organisms that RNAs synthesized in oogenesis persist in the embryo and are probably sufficient to permit normal development to proceed, in some cases, through the blastula stage (1,2).

[1]This work was supported by NIH Grant GM 26139.
[2]Present address: Laboratory of Molecular Genetics, National Institute of Child Health and Human Development, NIH, Bethesda, MD 20205.

In organisms which permit extensive genetic analysis, such as *Drosophila*, it has been possible to demonstrate, with maternal effect mutants, that oocyte gene products are critical for embryonic cell determination and axis formation (3,4). There has been no effective general approach for studying the fate of these oocyte gene products in the embryo; we describe here a methodology which permits us to isolate these oocyte RNAs which persist in *Drosophila* embryos. In this way we hope that the function of such RNAs in embryonic processes can be clarified. The method employs stable isotopes to density label oocyte RNAs, which can then be separated from unlabelled embryonically-synthesized RNAs. Using this approach, we have shown that a fraction of the dense RNA transcribed during early and middle oogenesis persists in the young embryo and that there are stable oocyte mRNAs in the RNA population which can be translated *in vitro*.

THE OOCYTE PULSE-CHASE METHOD FOR DENSITY LABELLING OOCYTE RNA

It is feasible to label RNAs synthesized during *Drosophila* oogenesis, because oocyte development is short and oocytes proceed single-file through development at a constant rate in the ovariole. The complete period of oogenesis is about 5 days (5). During the first two days, active stem cell division takes place, generating an oocyte and its interconnected nurse cells. The next 60 hours comprise the early and middle oogenetic stages during which the oocyte develops with minimal growth. During vitellogenesis, which spans the next 20 hours, there is an increase in volume and a rapid growth, largely a result of nurse cell contributions and yolk uptake into the oocyte. Finally, during the last 4 hours of oogenesis, the oocyte stops growing and completes maturation, whereupon it is fertilized internally and laid as an embryo.

The labelling scheme is unavoidably complicated by the fact that it is not possible to label all oocyte RNAs without also labelling embryonic RNAs, since the large reserve of metabolites accumulated in the later stages of oogenesis are the only source of nutrients for the developing embryo until it hatches as a first instar larvae. If we were to density label oocytes throughout oogenesis, embryonic RNAs would also become density

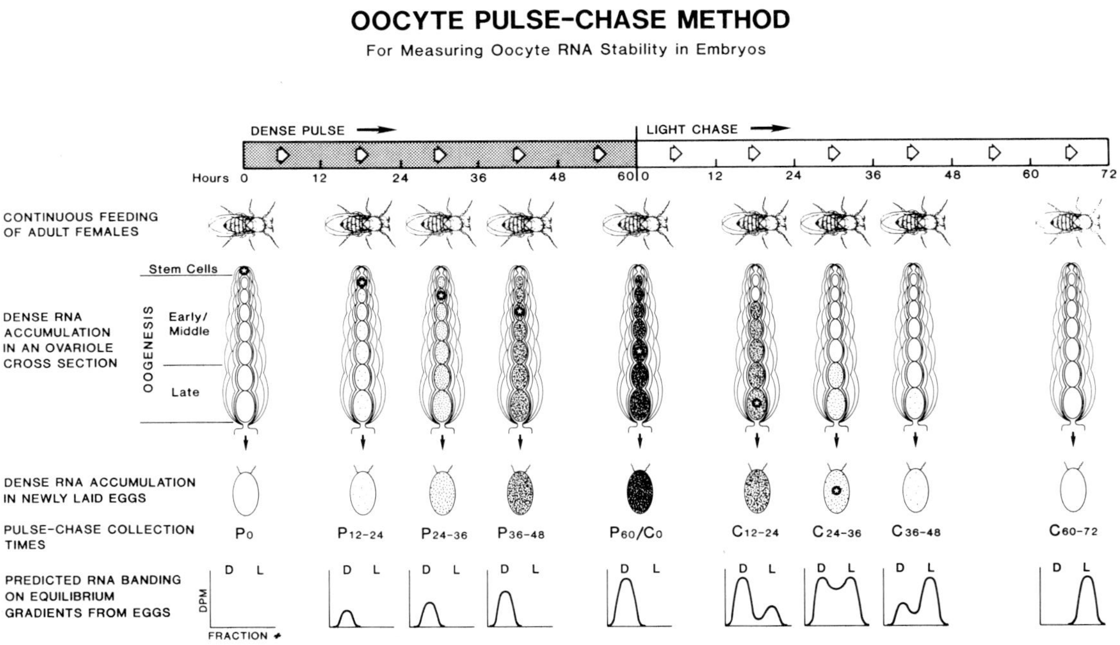

Figure 1. Schematic diagram of the oocyte pulse-chase method for measuring oocyte RNA stability in embryos. Adult female flies are fed on ^{13}C-^{15}N-50%^{2}H-labelled algae paste containing ^{3}H-uridine for 60 hours (pulse). The flies are then fed on a "light" yeast paste for the remainder of the experiment (chase). The stippling indicates the accumulation of dense RNA in a developing oocyte. The predicted RNA banding pattern is shown for embryos collected every twelve hours during the pulse and chase. The star identifies a particular oocyte as it develops during the pulse labelling and chase.

labelled. We have found that it is possible to label oocyte components specifically by a pulse-chase procedure in which oocytes are density labelled during the previtellogenic stages of oogenesis, but accumulate unlabelled components during the later stages of oogenesis. In this way, a large pool of unlabelled (light) precursors is accumulated for embryonic synthesis and RNAs synthesized during embryogenesis have a light buoyant density. These RNAs can be separated from those dense RNAs synthesized during early and middle oogenesis which have persisted in the embryo. During this procedure there is turnover of some dense RNA, with release of some dense precursors; however, because of the large accumulation of light precursors, this label is diluted extensively and reincorporated into macromolecules which have essentially a light buoyant density.

The oocyte pulse-chase method is depicted in Figure 1 and is carried out as follows: adult female *Drosophila melanogaster* are fed on an isotopically labelled algae paste (^{13}C-^{15}N-50%^{2}H-substituted) along with ^{3}H-uridine to monitor the incorporation of the density label. Embryos are collected at 12 hour intervals during the experiment. The dense pulse is maintained for a total of 60 hours with no toxic effects. Shortly after the beginning of the density labelling, embryos that are laid will contain density labelled RNA synthesized during late or middle oogenesis (e.g. P_{0-12} or P_{12-24}). Embryos collected from a later pulse of 36 to 48 hours (P_{36-48}) contain dense RNA synthesized not only from middle and late oogenesis, but early oogenesis as well. By 60 hours of the dense pulse (P_{60}) newly laid embryos consist almost entirely of dense components synthesized throughout all but the earliest stages of oogenesis.

It is at this time that a "light" chase is initiated in order to obtain embryos in which only those RNAs synthesized during early and middle oogenesis are density labelled (Figure 1). Flies are switched from dense algae paste to an unlabelled food source, a yeast paste, (time C_0). In the first collection (C_{0-12}) embryos are comprised of dense RNA transcripts, except for a small amount of light RNA synthesized at the very end of oogenesis. Embryos collected from 12 to 24 hours of the chase (C_{12-24}) have dense transcripts synthesized during early and middle oogenesis including the earliest part of vitellogenesis. In these embryos, RNA synthesized during embryogenesis would be expected to be of light buoyant density; by this time in the chase, most precursors of embryonic synthesis are derived from either the store of light metabolites accumulated during late oogenesis, or from the turnover of light RNAs synthesized during this period. Some dense precursors will also be available for embryonic transcription due to the turnover of density labelled oocyte RNAs, as well as any uptake of residual dense precursors remaining in the adult female fly during the chase. However, the contribution of these dense components is minimal, and embryonic RNA will be close the light buoyant density position as seen in Figure 1. Embryos gathered from later chase points will contain dense RNA transcripts originating from successively earlier stages of oogenesis. Therefore, as the chase progresses, the fraction of RNAs in the embryo which are dense decreases with a corresponding increase in light RNAs. By 60 to 72 hours (C_{60-72}) embryos will once again be composed of light RNA from all stages of oogenesis.

A single oocyte is starred in Figure 1 to show its approximate location in the ovariole during the course of the pulse-chase scheme. At the end of the pulse, P_{60}/C_0, this oocyte has been density labelled only during its development through early and middle oogenesis. During the chase, the oocyte proceeds through late oogenesis synthesizing light buoyant density RNA, as well as accumulating a large store of unlabelled yolk and metabolic components and is finally collected at C_{24-36}.

ANALYSIS OF DENSITY DISTRIBUTIONS OF RNAS DURING AN OOCYTE PULSE-CHASE EXPERIMENT

An oocyte pulse-chase experiment was performed as outlined above. The gradient system used to separate dense and light RNAs consists of a mixture of NaI and KI salts; we developed this system to obtain maximum resolution and minimal damage to the RNA for subsequent *in vitro* translation (6). The density distribution of labelled RNA isolated from newly laid embryos during the course of the chase is depicted in Figure 2. To demonstrate this separation of dense and light RNA in this system, RNA from a C_{0-12} collection (dense) was mixed with a light marker RNA in Figure 2B. As the scheme would predict, eggs laid at the very end of the pulse

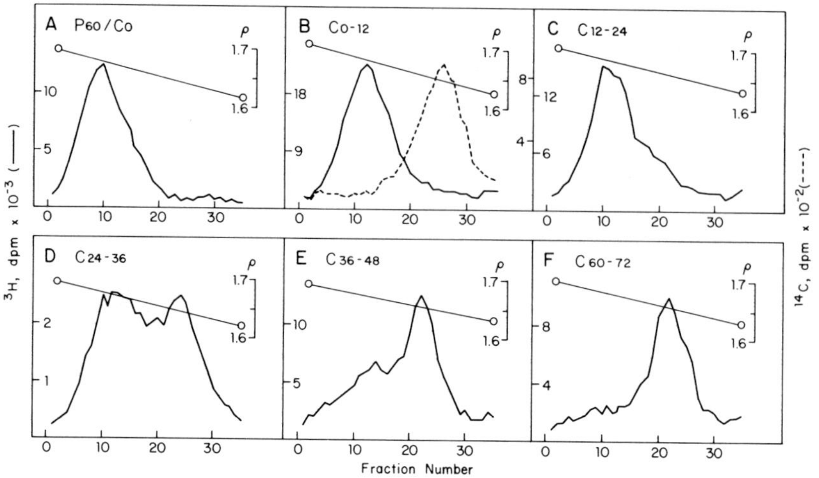

Figure 2. Density distribution of RNAs from embryos collected during an oocyte pulse-chase experiment. An oocyte pulse-chase experiment was performed as in Figure 1. Embryos were collected every 12 hours during the chase. RNA was extracted from embryos at each timepoint and centrifuged to equilibrium in a 4.5ml NaI/KI gradient (Beckman TI 80 rotor, 48,000 rpm, 26°C, 72 hr). Fractions were collected and the radioactivity was determined by liquid scintillation counting.
(______) RNA from chase embryos.
(-----) ^{14}C light buoyant density RNA marker

(P_{60}/C_0) are comprised of RNA which is completely dense (Figure 2A). Eggs that are gathered from successively later chase points show decreasing amounts of label in the dense region of the gradient and increasing amounts in the light region. By 60 to 72 hours of the chase, we can only detect light buoyant density RNA in embryos (Figure 2F).

Although the pattern of incorporation of label is consistent with our conception of how the pulse-chase should occur, it is essential to show that, in embryos collected during the chase, all embryonically-synthesized RNAs have a light buoyant density. It is known that under certain conditions adult female flies may retain fertilized eggs for a more prolonged period than normal. It should be remembered that C_{0-12} embryos are density labelled during late oogenesis and, hence, synthesize dense embryonic RNAs from a dense precursor pool. If C_{0-12} eggs were retained and laid along with embryos from a later chase collection, there would be a contaminating subset of embryonic transcripts which are dense. We have monitored the reincorporation from precursor pools of C_{12-24} and C_{24-36} embryos (data not shown; see ref. 6). The results have indicated that any dense RNA remaining in these chase embryos must be stable oocyte RNA, while any embryonically-synthesized RNA is of light buoyant density.

While the distribution of RNAs in NaI/KI gradients shown in Figure 2 suggests that dense and light RNA are almost completely separable, the problem is more complex. Detection of a tritiated light peak during the chase is due to the decay of the dense oocyte ^{3}H-RNAs. These turnover products are diluted out by the large reserve of unlabelled metabolites from late oogenesis and are then reincorporated into newly synthesized embryonic RNA. Therefore, the specific activity of the tritium label must be substantially lower in the light region of the gradient; the relative proportion of light RNA must be greater than is represented by its radioactive incorporation. If we consider a Gaussian distribution of the mass of the dense oocyte and light embryonic RNAs in a gradient, then there is cause to consider cross-contamination between the two. To ensure that we can recover a pure population of dense and light RNAs for further analysis, we rerun the dense and light regions of a single gradient on two separated gradients .

IN VITRO TRANSLATION OF OOCYTE RNAS WHICH PERSIST IN THE EMBRYO

Dense and light RNAs from embryos collected at C_{12-24}, C_{24-36}, and C_{36-48} chase points were isolated and samples were translated in a rabbit reticulocyte lysate containing ^{35}S-methionine; the overall incorporation of ^{35}S-methionine per microgram of total RNA was then measured. Results are presented in Table 1. Light RNA synthesized during late oogenesis and early embryogenesis retrieved from C_{12-24}, C_{24-36}, and

TABLE 1
TRANSLATIONAL EFFICIENCY OF RNA RECOVERED FROM NaI/KI EQUILIBRIUM GRADIENTS

RNA SAMPLE		^{35}S-DPM INCORPORATION PER MICROGRAM TOTAL RNA
C_{12-24}	Dense	35,740
C_{12-24}	Light	54,028
C_{24-36}	Dense	9,480
C_{24-36}	Light	30,348
C_{36-48}	Dense	8,180
C_{36-48}	Light	36,087
Total embryo RNA		179,088

An oocyte pulse-chase experiment was performed as described in the text. Extracted RNA from the inidicated timepoints was centrifuged on NaI/KI equilibrium gradients as described in Figure 2. Separated dense and light RNA fractions were rerun on a second set of gradients to insure that there would be negligible cross-contamination of dense and light RNAs. RNA samples were added to 6 microliters of a rabbit reticulocyte lysate containing ^{35}S-methionine. Radiolabel incorporation was determined by TCA precipitation. Total embryo RNA was extracted from 12 hour unlabelled embryos and translated directly.

C_{36-48} embryos, all stimulate the translation system, but incorporation is at least 3 to 4 times less than that of unlabelled, total embryo RNA that was never run on a salt gradient. This is probably the result of degradation of RNA in NaI/KI gradients. Dense oocyte RNAs from C_{24-36} and C_{36-48} embryos show translational activity which is only slightly above background; however, a substantial level of incorporation occurs with C_{12-24} dense oocyte RNA. Therefore, it is possible to detect mRNAs synthesized during oogenesis which persist in the embryo.

Figure 3 shows the translation products of dense and light mRNAs from various chase points which have been displayed on a 10% SDS-polyacrylamide gel. Light RNA from all chase points examined show identical translation products. Only dense oocyte RNA from C_{12-24} embryos has

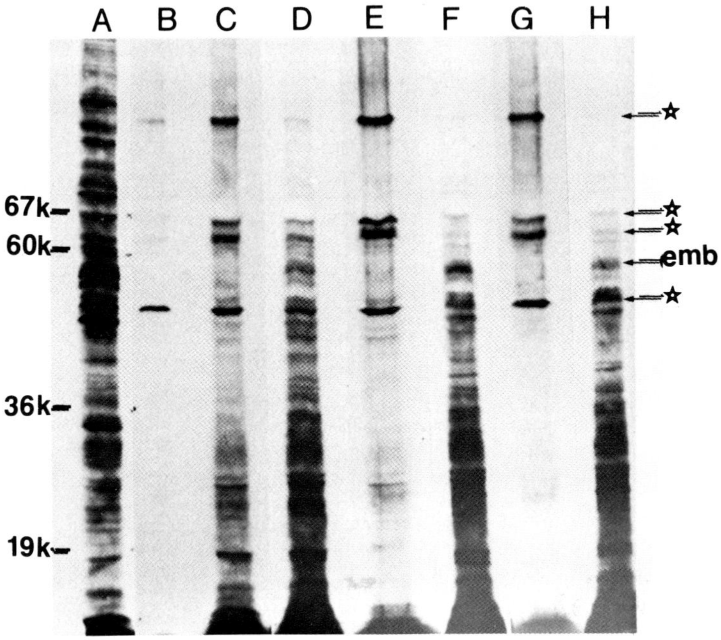

Figure 3. Translation products of dense and light RNAs from an oocyte pulse-chase experiment. From C_{12-24}, C_{24-36}, and C_{36-48} embryo collections, dense and light RNAs were separated and translated <u>in vitro</u> as discussed in Table 1. Translation products were electrophoresed on 10% SDS-polyacrylamide gels and then prepared for fluorography. A composite of different exposures was made from the same gel so that each lane represents approximately the same number of ^{35}S-disintegrations. Lane A.) 12 hr unlabelled, total embryo RNA, B.) peptides endogenous to the reticulocyte lysate, C.) C_{12-24} dense RNA, D.) C_{12-24} light RNA, E.) C_{24-36} dense RNA, F.) C_{24-36} light RNA, G.) C_{36-48} dense RNA, H.) C_{36-48} light RNA. The "emb" arrow points to an embryo-specific product and the starred arrows to the endogenous bands from the rabbit reticulocyte system.

a high enough incorporation to see peptides bands clearly, other than those endogenous to the reticulocyte system. The translation products from the dense C_{12-24} oocyte mRNAs appear to be very similar to those of the light embryonic RNAs, although there is one prominent band in light RNA (shown with "emb" labelled arrow) which is not seen in dense RNA. Since we do see at least this one clearcut difference between dense and light RNA, this indicates that the separation of dense and light RNA must have been complete. It should also be pointed out the unlabelled, total embryonic RNA has more high molecular weight translation products than are seen in RNAs retrieved from NaI/KI gradients, again indicating that there is probably some mRNA degradation during the isolation procedure.

DISCUSSION

We have devised a pulse-chase experiment whereby RNA which is transcribed during *Drosophila* oogenesis and persists into the embryonic period, can be isolated for further analysis. We have shown that such stable oocyte RNAs exist, and by *in vitro* translation, that some mRNAs must persist in the embryo that were transcribed in early or middle oogenesis (from C_{12-24} embryos). We cannot, however, detect stable RNAs from very early periods of oogenesis (as indicated by results with C_{24-36} and C_{36-48} embryos). There are very few differences between dense and light mRNA populations that can be discerned by analysis of one-dimensional SDS-polyacrylamide gels; we are currently in the process of a more extensive analysis of these translation products by a two-dimensional gel analysis to discern whether there are more subtle differences in these RNA populations.

Our separation scheme makes it possible to follow the fate of density labelled oocyte RNAs at later embryonic or larval stages as well as in young embryos. We hope to determine whether a subset of oocyte mRNAs is preferentially stable at these later stages, and may therefore have an important function in the embryo.

Although our density labelling technique is limited to analyzing the stability of only the early oogenetic RNAs, this is still extremely useful. These transcripts are likely to persist in only very small quantities in the embryo and almost certainly could not be discerned in any way but by our purification scheme. Also, this early period is particularly interesting from a developmental viewpoint, because temperature sensitivity studies argue that gene products are required at this time to specify embryonic axis formation (7).

We hope to apply our technology to determine whether any oocyte mRNAs are localized in embryos, and may therefore be important in such processes as embryonic axis formation and cell determination. Many studies have reported localization of maternal RNAs which act as morphogenetic determinants during development. For example, in Smittia, another dipteran, there exists a maternal mRNA at the anterior end of the egg which is responsible for proper anterior-posterior axis formation (8). We plan to use our pulse-chase scheme to identify any stable oocyte RNAs from early oogenesis which persist and become localized in either the anterior or posterior halves of embryos.

Although we have only mentioned the use of the oocyte pulse-chase experiment in terms of identifying oocyte RNA, it is likely that other molecules of oocyte origin may be localized and functionally interesting during embryogenesis. An intriguing study on sea urchin embryo nuclear proteins shows that many non-histone proteins which are found in embryonic nuclei must arise from a specific maternal store of proteins (9). Because of the likely significance of maternal proteins in embryonic processes, we are currently characterizing the early oocyte proteins of Drosophila and following their fate during embryogenesis and larval development using methods similar to those presented here for RNA analysis (10).

ACKNOWLEDGEMENTS

W.H.P. wishes to thank Timothy S. Charlebois for his critical comments and many helpful discussions. We also would like to express our appreciation to Lindsay Catlin for the illustrations and figures in this manuscript.

REFERENCES

1. Denney PC, Tyler A (1964). Activation of protein biosynthesis in non-nucleate fragments of sea urchin eggs. Biochem Biophys Res Comm 14:245.
2. Gross PR, Cousineau GH (1963). Effects of actinomycin-D on macromolecular synthesis and early development of sea urchin eggs. Biochem Biophys Res Comm 10:321.
3. Nüsslein-Volhard C (1979). Maternal effect mutations that alter the spatial coordinates of the embryo of Drosophila melanogaster. In Subtelny S, Konigsberg I (eds): "Determinants of Spatial Organization," New York: Academic Press p 185.
4. Santamaria P, Nüsslein-Volhard C (1983). Partial rescue of dorsal, a maternal effect mutation. EMBO J 2:1695.
5. King RC (1970). "Ovarian Development in Drosophila melanogaster." New York: Academic Press.
6. Winkles JA (1983). "Stability of Ribosomal RNA and Messenger RNA During Drosophila melanogaster." Ph.D. Dissertation University of Virginia, Charlottesville.
7. Nüsslein-Volhard C (1977). Genetic analysis of pattern-formation in the embryo of Drosophila melanogaster. Wilhem Roux Arch 183:249.
8. Kandler-Singer I, Kalthoff K (1976). RNase sensitivity of an anterior morphogenetic determinant in an insect egg (Smittia sp., Chironomidae, Diptera). Proc Natl Acad Sci USA 73:3739.
9. Kuhn O, Wilt FH (1981). Chromatin proteins of sea urchin embryos: dual origin from an oogenetic reservoir and new synthesis. Dev Biol 85:416.
10. Phillips WH, Grainger RM (1982). Fate of oocyte proteins during Drosophila melanogaster development. Amer Zool 22:927.

Molecular Biology of Development, pages 253–262

ISOLATION AND CHARACTERIZATION OF MOUSE GENOMIC DNA CLONES OF AN EARLY DIFFERENTIATION MARKER : ENDO A[1]

M. Vasseur, P. Duprey, C. Marle, P. Brûlet and F. Jacob

Unité de Génétique cellulaire
du Collège de France et de l'Institut Pasteur
25 rue du Dr. Roux, 75724 Paris Cedex 15

ABSTRACT. The expression of a cytoskeletal protein, Endo A, is differentially regulated during mouse early embryogenesis and embryonal carcinoma cell differentiation. The recombinant cDNA Rec XVI, which hybridizes to the Endo A mRNA, detects two specific bands on a Southern blot of genomic mouse DNA digested with EcoRI. We have screened a mouse genomic library and isolated two classes of genomic clones : type $\alpha1$ contains a 9.5 Kb gene composed of 8 exons and 7 introns while type $\alpha2$ contains a 1.6 Kb gene devoid of introns. These two genes are closely associated with short repetitive interspersed sequences which are flancking the 1.6 Kb gene and located within the third intron of the 9.5 Kb gene.

INTRODUCTION

The cytoskeleton of vertebrates contains various elements such as microfilaments, microtubules and intermediate sized filaments.

Intermediate filaments are formed in different cell types by different protein constituents, providing specific

[1]This work was supported by grants from the Centre National de la Recherche Scientifique (LA 269, ATP 95153, ATP 95189), the Institut National de la Santé et de la Recherche Médicale, the Ligue Nationale Française contre le Cancer and the André Meyer Foundation.

markers for cell lineages and differentiation (1, 2). Such markers have been widely used to probe differentiation of mouse embryonal carcinoma cells (3, 4) and early embryonic development (5, 6). In order to obtain stage specific probes, monoclonal antibodies raised against intermediate filament proteins extracted from a trophoblastoma cell line (TDM-1) have been prepared (7). These antibodies label filament network in trophectoderm but not in inner cell mass cells of mouse blastocyst (8) and neither in F9 embryonal carcinoma cells.

One of the monoclonal antibodies, referred to as TROMA 1, recognizes a 55 K protein called Endo A which has been described as a major cytoskeletal protein found in murine endodermal cell lines (9). A cDNA clone has been isolated from a trophoblastoma cDNA library. This cDNA detects the specific mRNA encoding the 55 K protein which is present in trophoblastoma cells but not in F9 cells (10).

We have used this cDNA as a probe to isolate the gene encoding Endo A from a genomic mouse library. We describe here the isolation and the characterization of the genomic clones.

METHODS

Cells.

Teratocarcinoma cell lines were cultured under standard procedures. TDM 1 is a trophoblastoma cell line (11).

Screening of a Mouse Genomic Library.

Recombinant Charon 4A phages carrying the mouse gene library were screened by using the in situ plaque hybridization technique (12). Hybridization with the ^{32}P-labeled Rec XVI cDNA was performed in 5 X standard sodium citrate, 5 X Denhardt's buffer and 50 % formamide at 42°C for 16 hours.

Restriction Endonuclease Digests and Gel Electrophoresis.

Restriction endonuclease was from Bethesda Research Laboratories and Boehringer. Analytical as well as preparative agarose gel electrophoreses were carried out in a Tris acetate buffer as described (13).

Subcloning of Endo A Genes Fragments in Plasmid pBR 322.

EcoRI digested phage DNA (1 γ) was mixed with 0.1 µg of EcoRI digested and phosphatased pBR 322 DNA. T4 DNA ligase was aded and buffer was adjusted to 66mM Tris HCl pH 7.5/6.6 mM $MgCl_2$/66 mM ATP/10 mM OTT. After incubation for 16 hours at 10°C, this ligated DNA was used to transform $CaCl_2$ treated E.coli C600.

Plasmid DNA was prepared by the alkaline lysis procedure (14) and was further purified by centrifugation to equilibrium in cesium chloride ethidium bromide density gradient.

Cellular DNA and RNA Preparation.

DNA and RNA extraction were performed as described in (15).

RESULTS

Cloning of the Endo A Gene.

We have hybridized the cDNA Rec XVI (10) with a Southern blot of EcoRI digested mouse DNA. The cDNA hybridized to two specific bands, of respectively 2.5 and 2.3 Kb (Fig. 1 A). Since no sites for EcoRI are present in the cDNA, one should expect that two different Endo A genes have to be found in the genomic library.

Of the 6.10^5 recombinant plaques screened, 7 clones hybridized with the ^{32}P-labeled cDNA. The DNA of these 7 clones was digested with EcoRI, run on agarose gel, and transferred onto nitrocellulose sheets. Hybridization with the cDNA probe revealed that the clones fall into two groups. The first group contains the 2.5 Kb EcoRI fragment

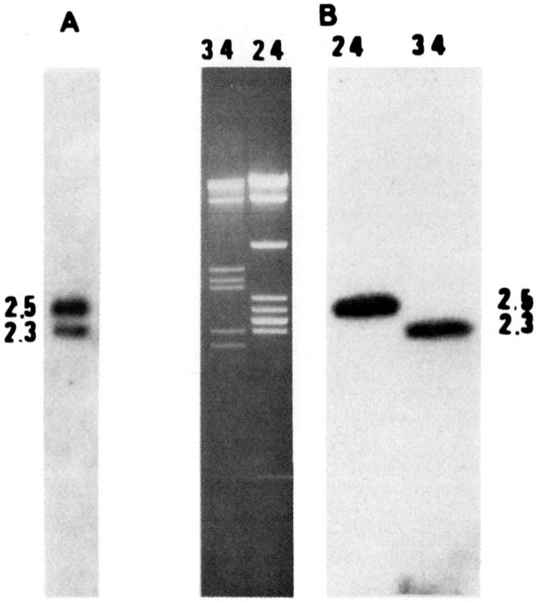

FIGURE 1. A. Southern blot analysis of EcoRI digested mouse DNA probe with the ^{32}P-labeled cDNA Rec XVI. B. Pattern of EcoRI digested DNA of the two types of recombinant clones : λ24 referred to as α1 and λ34 referred to as α2 in the text (left). Southern blot of such patterns hybridized with the ^{32}P-labeled cDNA Rec XVI (right).

(this type will be referred to as α1), the second type contains the 2.3 Kb fragment (α2) (Fig. 1 B). Digestion of the two types of clones with several other restriction endonucleases gave rise to different patterns. This result indicates that the two types of genes obtained from the library are different.

Characterization and Mapping of the Endo A Genes.

In order to analyze the relationship between the two genes encoding Endo A, we have performed cross hybridization between the two recombinant phages (α1 and α2) representing the two classes of genes. We have found that while

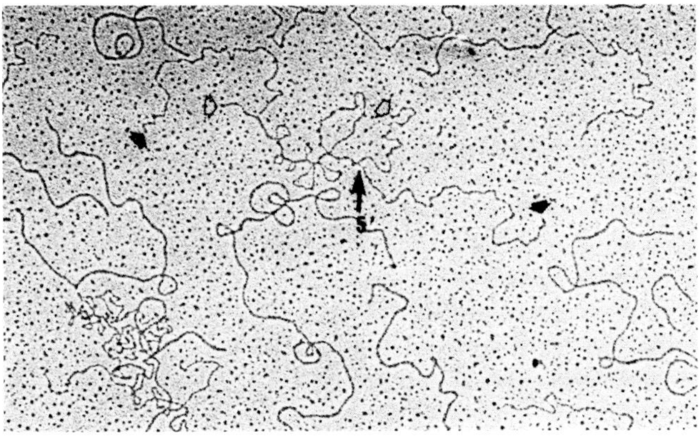

FIGURE 2. Electron microscopy picture of a typical heteroduplex between p1H1, a plasmid containing the full length of the Endo A gene (black arrows) and p2R14, a subclone of α2 containing the 2.3 Kb EcoRI fragment which encompass the pseudogene (open arrows).

sequences contained in α1 hybridize only to the 2.3 Kb band of α2, α2 sequences hybridize with 4 bands of EcoRI digested α1. Subcloning of the two types of recombinants have confirmed that sequences from the 2.3 Kb fragment of α2 were able to hybridize with sequences spread along 10 Kb in α1 (data not shown).

It thus appears that the mouse genome contains two different Endo A genes in which sequence organization is considerably different. The comparison of sequence organization was performed by electron microscopy of heteroduplex between the two genes. We have hybridized the RcoRI 2.3 Kb subclone isolated from α2 (referred to as p2R14) to a Hind III subclone of 14 Kb long, originating from α1 (referred to as p1H1), which totally encompasses the sequences able to hybridize to such an heteroduplex. Such an heteroduplex is presented in Fig. 2.

The gene contained in α1 is interrupted by seven intervening sequences while the DNA of the other type seems to be devoid of any intervening sequence. The total length of the double stranded part of the heteroduplex was estimated to 1650 ± 65 bp. The simplest interpretation of such results is to assume that the gene for Endo A is comprised

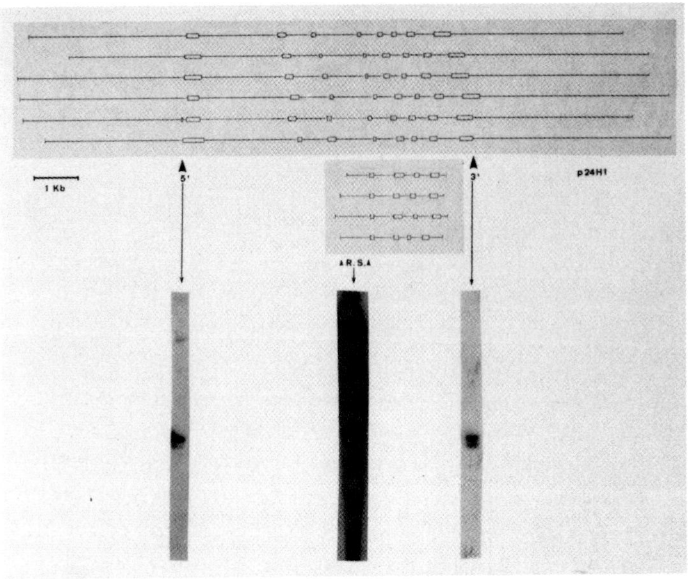

FIGURE 3. Pattern of computer analysis of some heteroduplexes between the full length gene, or a part of it, and the pseudogene. The repetitive sequence (R.S.) located into the third intron yield a smear when used as a probe on a Southern blot of mouse DNA. The first and the 8th exon hybridize only to the 2.5 and 2.3 Kb EcoRI fragments already detected by Rec XVI.

of 8 coding segments, separated by 7 intervening sequences, which cover a distance of approximately 9.5 Kb. The other gene may be considered as a pseudogene which has been derived from a cDNA copy of Endo A mRNA. This hypothesis has been confirmed by other results, including DNA/RNA heteroduplex analysis and S1 mapping (data not shown).

Repetitive Sequence associated with Endo A.

Both recombinant clones α1 and α2 contain repetitive sequences. Mapping of these sequences shows that they are flanking the pseudogene, in clone α2, while in clone α1, a 200 bp repeat unit is localized into the third intron of the gene (Fig. 3). No cross hybridizations have been detected between these different repetitive sequences.

The 200 bp repetitive sequence hybridizes with low molecular weight, poly A^+ RNA (0.6 and 0.7 Kb) present in large amount in F9 cells, less represented in TDM 1 and undetectable in 3T3 (data not shown). Such patterns have been recently described : small cytoplasmic RNAs encoded by the B2 family of repetitive sequences (16) have been found in SV40 transformed mouse cells (17) and other tumor cells including embryonal carcinoma cells (8). We have prepared a B2 type probe of 163 bp long from a cloned H_2D cDNA (19) and found, by hybridization experiments, that the repetitive sequence included in the third intron of Endo A gene belongs to this family.

DISCUSSION

Molecular analysis of early embryonic cell determination requires the availability of stage specific probes which can be used to monitor the evolution of the pattern of gene expression. The cytokeratin Endo A appears during the process of compaction (20) but its expression is restricted to trophectoderm cells in the early blastocyst (8). It is a valuable marker for the study of the first embryonic differentiation.

We have cloned the genes encoding Endo A in order to (i) analyze the structure of the gene(s), (ii) study the non coding sequences flanking the gene and (iii) obtain a probe larger than the cDNA Rec XVI. We have found that Endo A was encoded by a 9.5 Kb-long gene which contains 7 introns. We have also found a pseudogene of 1.65 Kb devoid of introns. The pseudogene which contains all the coding sequences of Endo A may be a useful probe to isolate the other cytokeratin genes, since it is known that all keratins share in common the central domain of the molecule (21).

A short B2 type repetitive sequence has been detected in the third intron of the gene. The presence of this type of repetitive element into intervening sequences has already been described for different genes including mouse albumin and alpha-foetoprotein (22), and an H_2D pseudogene (23). A repetitive sequence of 83 bp long has been also found among introns of genes specifically expressed in the brain of rats (24). It has been proposed that dispersed repetitive elements are "hot spots" for genetic recombination and may be responsible for rearrangements of genes and exons (25). Since Endo A belongs to the broad family of cytokeratin, it would be of interest to search for similar type of repetitive elements in the other genes of the family. Another hypothesis is that repetitive sequences may play a role in the regulation of genetic expression (26, 27), either at the level of transcription or mRNA maturation. In contrast to their widely dispersed pattern, some families of interspersed repetitive elements are expressed at high level in particular tissues and at very much lower level in others. This is the case for the small RNAs encoded by B2 sequences which seem to be associated with the transformed or embryonic phenotype. Work is in progress to analyze the expression of this class of sequence in the early steps of embryogenesis, together with the expression of Endo A.

Acknowledgements.

We thank Odile Croissant for her help in the electron microscopy analysis of heteroduplexes.

REFERENCES

1. Fuchs E, Hanugoklu I (1983). Unraveling the structure of the intermediate filaments. Cell 34:332.
2. Roop DR, Hawley-Nelson P, Cheng CK, Yuspa SH (1983). keratin gene expression in mouse epidermis and cultured epidermal cells. Proc Natl Acad Sci USA 80:716.
3. Paulin D, Forest N, Perreau J (1980). Cytoskeletal proteins used as marker of differentiation in mouse teratocarcinoma cells. J Mol Biol 144:95.
4. Paulin D, Jakob H, Jacob F, Weber K, Osborn M (1982). In vitro differentiation of mouse teratocarcinoma cells monitored by intermediate filament expression. Differentiation 22:90.
5. Jackson BW, Grund C, Schmid E, Bürki K, Franke WW, Illmensee K (1980). Formation of cytoskeletal elements during mouse embryogenesis. Intermediate filaments of the cytokeratin type and desmosomes in preimplantation embryos. Differentiation 17:161.
6. Paulin D, Babinet C, Weber K, Osborn M (1980). Antibodies as probes of cellular differentiation and cytoskeletal organization in the mouse blastocyst. Exp

Cell Res 130:297.
7. Brûlet P, Babinet C, Kemler R, Jacob F (1980). Monoclonal antibodies against trophectoderm-specific markers during mouse blastocyst formation. Proc Natl Acad Sci USA 77:4113.
8. Kemler R, Brûlet P, Schnebelen MT, Gaillard J, Jacob F (1981). Reactivity of monoclonal antibodies against intermediate filament proteins during embryonic development. J Embryol Exp Morph 64:45.
9. Oshima RG (1982). Developmental expression of murine extra-embryonic endodermal cytoskeletal proteins. J Biol Chem 257:3414.
10. Brûlet P, Jacob F (1982). Molecular cloning of a cDNA sequence encoding a trophectoderm-specific marker during mouse blastocyst formation. Proc Natl Acad Sci USA 79:2328.
11. Nicolas JF, Jakob H, Jacob F (1981). Teratocarcinoma-derived cell lines and their use in the study of differentiation. In Sato G (ed): "Functionally differentiated cell lines," New York: Alan R. Liss, p 185.
12. Benton W, Danis RW (1977). Using of phage immunity in molecular cloning experiments. Science 196:180.
13. Vasseur M, Kress C, Montreau N, Blangy D (1980). Isolation and characterization of polyoma mutants able to develop in murine embryonal carcinoma cells. Proc Natl Acad Sci USA 77:1068.
14. Birboim H, Doly J (1979). A rapid alkaline extraction procedure for screening recombinant plasmid DNA. Nucl Ac Res 7:1513.
15. Blin N, Stafford D (1976). A general method for isolation of high molecular weight DNA from eucaryotes. Nucl Ac Res 3:2303.
16. Kramerov DA, Grigoryan AA, Ryskov AP, Georgiev GP (1979). Long double-stranded sequences (dsRNA-B) of nuclear pre-mRNA consist of a few highly abundant classes of sequences : evidence from DNA cloning experiments. Nucl Ac Res 6:697.
17. Murphy D, Brickell PM, Latchman DS, Willison K, Rigby PWJ (1983). Transcripts regulated during normal embryonic development and oncogenic transformation share a repetitive element. Cell 35:865.
18. Brickell PM, Latchman DS, Murphy D, Willison K, Rigby PWJ (1983). Activation of a Qa/Tla class I major histocompatibility antigen gene is a general feature of oncogenesis in the mouse. Nature 306:756.

19. Lalanne JL, Bregegere F, Delarbre C, Abastado JP, Gachelin G, Kourilsky P (1982). Comparison of nucleotide sequences of mRNAs belonging to the mouse H-2 multigene family. Nucl Ac Res 10:1039.
20. Oshima RG, Howe WE, Klier FG, Adamson ED, Shevinsky LF (1983). Intermediate filament protein synthesis in preimplantation murine embryos. Dev Biol 99:447.
21. Hanukoglu I, Fuchs E (1982). The cDNA sequence of a human epidermal keratin : divergence of sequence but conservation of structure among intermediate filament proteins. Cell 31:243.
22. Kioussis D, Eiferman F, van de Rijn P, Gorin MB, Ingram RS, Tilghman SM (1981). The evolution of alpha-fetoprotein and albumin. II. The structures of the alpha-fetoprotein and albumin genes in the mouse. J Biol Chem 25:1960.
23. Steinmetz M, Moore KW, Frelinger JG, Sher BT, Shen FW, Hood L (1981). A pseudogene homologous to mouse transplantation antigens: Transplantation antigens are encoded by eight exons that correlate with protein domains. Cell 25:683.
24. Milner RJ, Bloom FE, Lai C, Lerner RA, Sutcliffe JG (1984). Brain-specific genes have identifier sequences in their introns. Proc Natl Acad Sci USA 81:713.
25. Gilbert W (1978). Why genes in pieces ? Nature 271: 501.
26. Davidson EH, Britten RJ (1979). Regulation of gene expression : Possible role of repetitive sequences. Science 204:1052.
27. Davidson EH, Posakony JW (1982). Repetitive sequence transcripts in development. Nature 297:693.

Molecular Biology of Development, pages 263–272

INTERMEDIATE FILAMENTS IN TISSUE CULTURE CELLS AND EARLY EMBRYOS OF DROSOPHILA MELANOGASTER

Marika F. Walter and Bruce M. Alberts

Department of Biochemistry and Biophysics
University of California, San Francisco
San Francisco, California 94143

ABSTRACT The presence of an intermediate-sized filament cytoskeleton in embryonic cell lines of *Drosophila melanogaster* has been described. This cytoskeleton exhibits features characteristic of those found for the vertebrate vimentin cytoskeleton, and it is stained by a monoclonal antibody directed against a vimentin-like *Drosophila* protein. As judged by a protein blotting assay, a set of intermediate filament proteins that remains unchanged through early embryonic development is recognized by the monoclonal antibody in *Drosophila* embryos. The location of these proteins has been determined by immunofluorescent staining of whole embryos at various stages of development. In early embryos, where all of the nuclei are located in the interior of the embryo (nuclear cycles 1 to 9), the cortical cytoplasm at the periphery is stained in a diffuse manner that seems to reveal a meshwork of filaments all over the embryo. Later, when the nuclei migrate out to form a monolayer just beneath the surface (nuclear cycles 10 to 14), the intermediate filament proteins become confined to the cytoplasm that immediately surrounds each nucleus. For cycle 11 embryos, this results in a patchy distribution of the stained intermediate filament proteins since areas of the cortical cytoplasm that are far from any nucleus stain poorly. At later stages, the nuclei increase in number and become more tightly packed together, giving rise to a honeycomb-like pattern of intermediate filament proteins between them.

INTRODUCTION

An intermediate filament (10 nm) cytoskeleton that contains vimentin has been well characterized in vertebrate cells (1). Although relatively little is known about intermediate filaments in invertebrates, a monoclonal antibody directed against a major 46,000 dalton protein from Drosophila cross-reacts with vimentin from vertebrates (2,3). By using this antibody for cytological staining, one can show that a cytoskeleton of intermediate-sized filaments is present in tissue culture cells of Drosophila melanogaster that exhibits features characteristic of the vertebrate vimentin cytoskeleton (4).

The function of the intermediate filament cytoskeleton is largely unknown. It has been suggested that it may be involved in maintaining structural organization within a cell, such as by anchoring the nucleus (5), by providing attachment sites for polyribosomes (6), or by maintaining the proper structure of the cytoplasm during mitosis (7). However, the injection of antibodies can be used to collapse the intermediate filaments in vertebrate tissue culture cells without a detectable effect on cellular functions (8). In order to obtain a better insight into the function of this type of cytoskeleton, it seems desirable to study it in a developing system. We have therefore begun to study the presence and distribution of intermediate filaments in Drosophila embryos. By protein blotting and indirect immunofluorescence techniques, we demonstrate that intermediate filament proteins are present in the cortex of early embryos, before the onset of zygotic transcription. As these embryos develop, they remain as giant single cells for 13 near-synchronous nuclear divisions (9,10). In this report, we describe how the organization of intermediate filaments in the cortex of this syncytial embryo changes as development progresses.

METHODS

The monoclonal antibody Ah6/5/9 (henceforth abbreviated Ah6), which was made by standard hybridoma technology against a gel-purified 46,000 dalton protein from Drosophila Kc cells (3), was used in all experiments described here. This 46,000 dalton protein appears to be the main Drosophila intermediate filament cytoskeletal protein. Unless otherwise stated, all methods used in this study have been described elsewhere (2-4,10-12).

RESULTS

The Cytoplasmic Distribution of Vimentin-Like Proteins in Drosophila Tissue Culture Cells.

In order to examine the organization of the intermediate filaments in Drosophila cells, tissue culture cells (Schneider's line 2) were grown on coverslips, fixed, and prepared for indirect immunofluorescence, as described elsewhere (2). When stained with the monoclonal antibody Ah6, which recognizes the major Drosophila intermediate filament protein, the cytoskeleton appears as a very fine cytoplasmic meshwork that is dense around the nucleus and radiates to the periphery of the cell (Figure 1). It differs slightly from the vimentin cytoskeleton described in vertebrate cells in that no thick filament bundles are detected. The same antibody cross-reacts with all vertebrate cells tested, and the characteristic vertebrate staining pattern is found when cells of vertebrate origin (CHO, BHK) are stained with the Ah6 antibody (2).

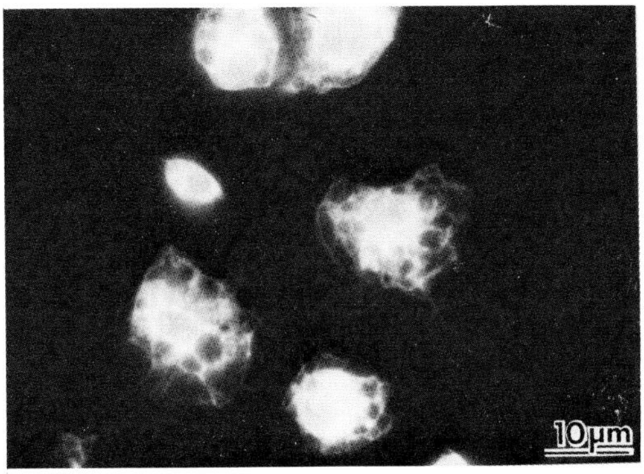

Figure 1. Indirect immunofluorescence of an embryonic cell line of Drosophila melanogaster (Schneider's line 2) fixed with formaldehyde and stained with monoclonal antibody Ah6 followed by a FITC-conjugated secondary antibody.

A study of the filamentous meshwork stained by the Ah6 antibody in Drosophila cells using immuno-electron microscopy reveals filaments with a diameter of 10 nm which have a very similar appearance to that of vertebrate intermediate filaments (4). We conclude that these Drosophila tissue culture cells contain a vimentin-like intermediate filament cytoskeleton that closely resembles analogous structures found in vertebrate cells, except for the fact that few, if any, large bundles of intermediate filaments are formed.

The Presence of Intermediate Filaments in Drosophila Embryos.

To test for the presence of intermediate filament proteins in Drosophila embryos, two approaches were used. First, crude detergent-solubilized extracts of embryos were fractionated by gel electrophoresis and then examined for reaction with the Ah6 monoclonal antibody by protein blotting (11). In addition, fixed embryos at various stages of development were stained with the Ah6 antibody and visualized by indirect immunofluorescence.

In order to search for intermediate filament proteins by the first approach, protein extracts of batches of embryos of different ages were separated by electrophoresis through sodium dodecyl sulfate (SDS) polyacrylamide gels (lanes 2 to 8) and then transferred onto nitrocellulose paper. The paper was reacted with the Ah6 antibody, washed and then treated with [^{125}I]-conjugated secondary antibody. The autoradiogram in Figure 2 shows the position of protein bands on this protein blot that react with the Ah6 antibody. For comparison, a similar analysis of the total proteins present in Drosophila Kc tissue culture cells is shown (lane 1). Note that all of the proteins that are recognized by the Ah6 antibody in Kc cells are also present in the preblastoderm and older embryos, except for the highest molecular weight protein (110,000 daltons). Furthermore, since the patterns in lanes 2 to 8 are identical, it seems that the protein species recognized by the Ah6 antibody are present in nearly the same relative amounts at all times of early development. This conclusion is supported by additional experiments of this type in which batches of embryos of other ages have been analyzed (from 30 min to 12 hours after egg deposition; data not shown).

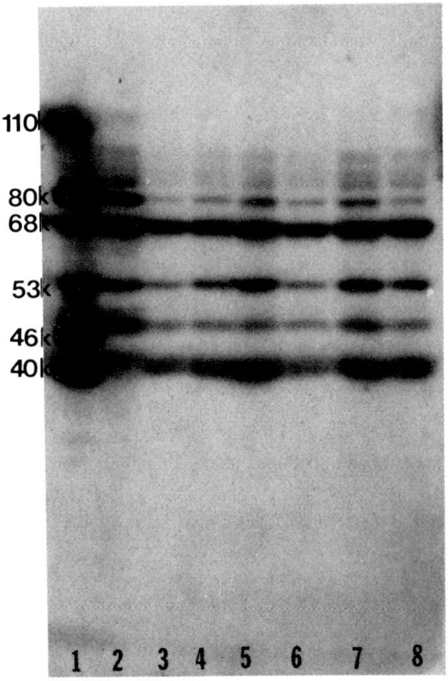

Figure 2. Analysis of the protein bands fractionated by SDS polyacrylamide gel electrophoresis that react with the Ah6 antibody against intermediate filament protein. Autoradiograms of protein blots are shown. Lane 1: Drosophila Kc tissue culture cell extracts showing reaction of the antibody with six protein bands. Lanes 2-8 are Drosophila embryo extracts: lane 2, overnight egg deposition; lane 3, 1 hr egg deposition fractionated immediately; lanes 4 to 8, same embryos as in lane 3, but aged for a further 1,2,3,4, and 5 hr, respectively. To prevent proteolysis, fresh embryos were disrupted by sonication in 1% SDS containing EGTA and immediately heated to 100°C.

The experiment in Figure 2 demonstrates that the intermediate filament proteins are present even in very young embryos. Since there are only traces of nuclear transcription prior to stage 9 in these embryos (13,14), we conclude that these proteins have been encoded by maternal genes.

The Spatial Distribution of Intermediate Filament Proteins in Drosophila Embryos.

We have also studied the intermediate filament cytoskeleton in these early embryos by indirect immunofluorescence. For this purpose, embryos between 30 min and 2.5 hr old were collected from population cages, permeabilized with hexane, and then fixed briefly with formaldehyde using minor modifications of previous techniques (12,16). The embryos were then treated with the Ah6 antibody and stained with a RITC-conjugated secondary antibody. DNA was visualized in the same embryos by staining with Hoechst dye 33258. The stained embryos were visualized by epifluorescent illumination, photographed, and staged as described elsewhere (10). Immunofluorescent staining of the intermediate filament proteins shows that in very early embryos (cycles 1 to 9) there is a cytoplasmic meshwork of these proteins distributed rather evenly near the surface of the embryo (Figure 3a). At these times of development, the nuclei are located in the central core of the embryo. When the nuclei migrate to the periphery at nuclear cycle 10, the antibody staining near the embryo surface begins to reorganize and become more distinct. By cycle 11, the intermediate filament proteins become confined to the cytoplasm that surrounds each nucleus (Figure 3b). This gives rise to a patchy distribution of the stained proteins, since neither the interior of the nucleus nor areas of the cortical cytoplasm that are far from any nucleus seem to contain intermediate filaments.

As the number of nuclei increases in the surface monolayer at later stages, the nuclei become more and more closely packed together. In this process, the amount of cytoplasm between nuclei decreases, and at cycle 13 the stained intermediate filament proteins are squeezed into the small amount of space that separates the largely hexagonally-packed nuclei, giving rise to a honeycomb pattern of staining (Figures 3c and 3d).

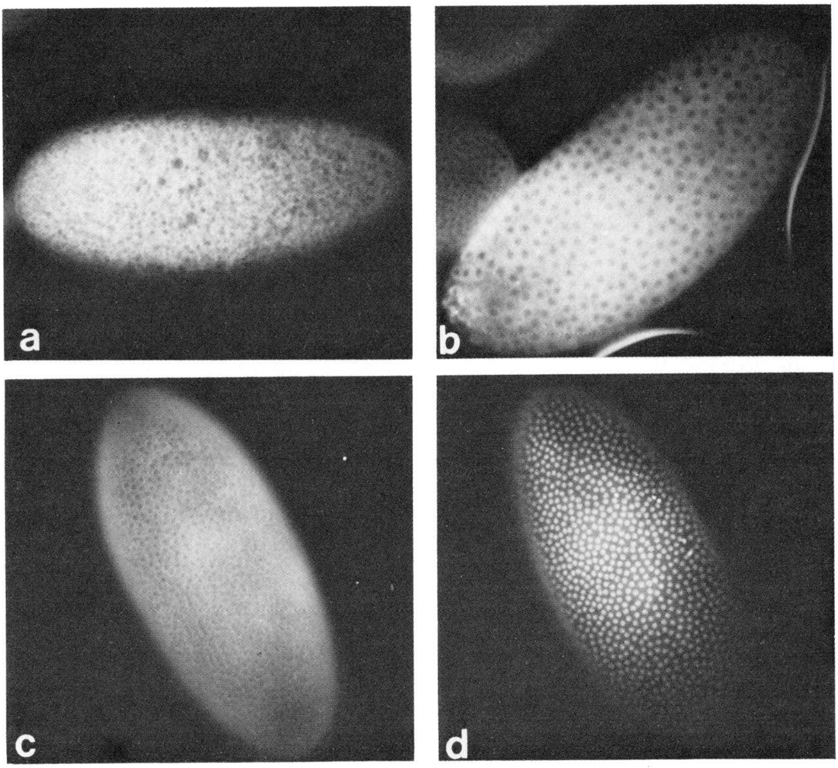

Figure 3. Drosophila embryos of different ages viewed after immunofluorescent staining.

(a): A cycle 4 embryo stained for intermediate filament proteins with the Ah6 antibody. At this time, all of the nuclei are deep in the core of the embryo, while the intermediate filament proteins that are visible are near the embryo surface. (b): A cycle 11 embryo stained for intermediate filament proteins with the Ah6 antibody. The nuclei have finished their outward migration to form a monolayer just under the embryo plasma membrane. Plasma membranes will form between these nuclei to form individual cells only much later during nuclear cycle 14; thus, at cycle 11 the embryo is a syncytium, consisting of a single giant cell that contains about 1000 nuclei which share the same cytoplasm. (c) and (d): A cycle 13 embryo stained for intermediate filament proteins with Ah6 antibody (c) and for DNA with Hoechst dye 33258 (d). The embryo is still a syncytium at this stage. All of the photographs in this figure are surface views.

DISCUSSION

By using a monoclonal antibody prepared against a 46,000 dalton vimentin-like protein from *Drosophila*, we have been able to demonstrate the presence of an intermediate filament cytoskeleton in invertebrate cells. In *Drosophila* tissue culture cells, the intermediate filament cytoskeleton consists of fine fibers that form a dense network around the nucleus and radiate outward to the cell periphery (Figure 1). Except for the thicker bundles of intermediate filaments that can usually be seen in invertebrate cells, the organization in *Drosophila* is similar to that observed in vertebrates, consistent with the notion that this type of cytoskeleton functions in part to help anchor the nucleus in the cell (5).

In early preblastoderm *Drosophila* embryos, the intermediate filament proteins are mainly localized in the cortical cytoplasm near the egg plasma membrane (Figure 3a). At these times of development, no nuclei are present at the periphery of the embryos. We detect intermediate filament proteins as early as nuclear cycle 2. Since there is very little transcription of the zygotic nuclei before nuclear migration (13,14), these proteins must have been encoded by maternal genes. Other cytoskeletal proteins are also present in the cortical cytoplasm of *Drosophila* embryos before nuclear migration, including actin (15,16) and tubulin (16). It remains to be determined whether these three cytoskeletal systems occupy different layers in the cortex or whether they instead form an interwoven meshwork to which all of the filamentous structures are attached.

After nuclear migration to the periphery of the embryo at nuclear cycle 10, the distribution of the intermediate filaments changes. Now the Ah6 antibody stains the cytoplasmic area around each nucleus, resulting in a honeycomb-like pattern (Figures 3b and 3c). Since protein blotting reveals no change in the intermediate filament proteins during these early stages of development (Figure 2), it appears that pre-existing intermediate filaments are reorganized after the nuclei migrate to the embryo surface.

Intermediate filaments have recently been detected in vertebrate embryos. In preimplantation mouse embryos, intermediate filaments of the cytokeratin type are found in the late morula and blastocyst stage (17), as well as in earlier cleavage stages (18). Cytokeratin filaments are also present in *Xenopus* oocytes and cleavage stages (19).

In both of these vertebrate embryos, vimentin intermediate filaments appear only much later in development. In contrast, we find that vimentin-like intermediate filaments are present from the earliest times of Drosophila development.

It is tempting to speculate that the intermediate filament cytoskeleton plays an important structural role in developing Drosophila embryos, serving to help organize the nuclei in the syncytial cytoplasm after they have migrated out to the embryo periphery. This hypothesis can be tested by examining the effects of an experimental disruption of the intermediate filament cytoskeleton in living embryos. This report represents an initial characterization of the unperturbed cytoskeleton, which is required for such a study.

REFERENCES

1. Lazarides E (1980). Intermediate filaments as mechanical integrators of cellular space. Nature (London) 283:249.
2. Falkner FG, Saumweber H, Biessmann H (1981). Two Drosophila melanogaster proteins related to intermediate filament proteins of vertebrate cells. J Cell Biol 91:175.
3. Walter MF, Biessmann H (1984). Cross reaction of a monoclonal antibody with intermediate filament like proteins from vertebrate and nonvertebrate organisms. Mol Cell Biochem, in press.
4. Walter MF, Biessmann H (1984). Intermediate sized filaments in Drosophila tissue culture cells. J Cell Biol, submitted.
5. Wang E, Chopin CW (1981). Effects of vanadate on intracellular distribution and function of 10nm filaments. Proc Natl Acad Sci USA 78:2363.
6. Fulton AB, Wan KM, Penman S (1980). The spatial distribution of polysomes in 3T3 cells and the associated assembly of proteins into the skeletal framework. Cell 20:849.
7. Zieve GW, Heidemann SR, McIntosh JR (1980). Isolation and partial characterization of a cage of filaments that surrounds the mammalian mitotic spindle. J Cell Biol 87:160.

8. Klymkowsky MW (1981). Intermediate filaments in 3T3 cells collapse after intracellular injection of a monoclonal anit-intermediate filament antibody. Nature 291:249.
9. Zalokar M, Erk I (1976). Division and migration of nuclei during early embryogenesis of Drosophila melanogaster. J Micro Biol Cell 25:97.
10. Foe VE, Alberts BM (1983). Studies of nuclear and cytoplasmic behaviour during the five mitotic cycles that precede gastrulation in Drosophila embryogenesis. J Cell Sci 61:31.
11. Risau W, Saumweber H, Symmons P (1981). Monoclonal antibodies against membrane protein of Drosophila. Localization by indirect immunofluorescence and detection of antigen using a new protein blotting procedure. Exp Cell Res 133:147.
12. Mitchison T, Sedat J (1983). Localization of antigenic determinants in whole Drosophila embryos. Develop Biol 99:261.
13. Lamb MM, Laird CD (1976). Increase in nuclear polyA-containing RNA at syncytial blastoderm in Drosophila melanogaster embryos. Develop Biol 52:31.
14. Zalokar M (1976). Autoradiographic study of protein and RNA formation during early development of Drosophila eggs. Develop Biol 49:425.
15. Warn RM, Magrath R, Wess S (1984). Distribution of F-actin during cleavage of the Drosophila syncytial blastoderm. J Cell Biol 98:156.
16. Karr T, Alberts BM (1984). Manuscript in preparation.
17. Jackson BW, Grund C, Schmid E, Burki K, Franke WW, Ilmensee K (1980). Formation of cytoskeletal elements during mouse embryogenesis. Intermediate filaments of the cytokeratin type and desmosomes in preimplantation embryos. Differentiation 17:161.
18. Lehtonen E, Lehto VP, Vartio T, Badley RA, Virtanen I (1983). Expression of cytokeratin polypeptides in mouse oocytes and preimplantation embryos. Devel Biol 100:158.
19. Franz JK, Gall L, Williams MA, Picheral B, Franke WW (1983). Intermediate-size filaments in a germ cell: Expression of cytokeratins in oocytes and eggs of the frog Xenopus. Proc Natl Acad Sci USA 80:6254.

III. DEVELOPMENTAL EXPRESSION OF GENE FAMILIES

Molecular Biology of Development, pages 275–292

ACTIN AND MYOSIN GENES, AND THEIR EXPRESSION DURING MYOGENESIS IN THE MOUSE[1]

Margaret E. Buckingham, Serge Alonso, Paul Barton
Gabriele Bugaisky[2], Arlette Cohen, Philippe Daubas,
Ian Garner, Adrian Minty[3], Benoît Robert, André Weydert

Department of Molecular Biology, Pasteur Institute
25 rue du Dr Roux, 75015 Paris, France

ABSTRACT During the initial formation of skeletal muscle fibres in the mammalian foetus, and during their subsequent maturation, different isoforms of the contractile proteins accumulate. Using recombinant probes, the transcripts coding for actins, myosin heavy chains, and myosin alkali light chains have been characterized both at different stages of foetal muscle development in the mouse *in vivo*, and from the onset of fibre formation in differentiating muscle cell lines. In the case of the actin and myosin light chain multigene families transcripts of a gene expressed as a major species in an adult cardiac tissue accumulate during skeletal muscle development, although this process follows different kinetics in the two families. In contrast, the myosin heavy chain family has developmental isoforms specific to skeletal muscle; and transcripts of a foetal and an adult gene accumulate sequentially. The strategy of gene expression during development is clearly different for each of these multigene families. The structure of a promotor region of a

[1]This work was supported by grants from the C.N.R.S., I.N.S.E.R.M., the M.I.R., and the M.D.A. of America. S.A. was financed by a research fellowship from Roussel, France, G.B. from the A.M.F., I.G. from the S.E.R.C/Royal Society of Great Britain, and P.B. from the M.D.A. of America.
[2]Present address: Department of Medicine, University of Chicago, Chicago, Illinois 60637
[3]Present address: Veterans Administration Medical Center, Palo Alto, California 94304.

cardiac actin gene, and of the gene encoding two myosin light chains LC_{1F} and LC_{3F} is reported. The organization and chromosomal localization of actin and myosin genes has been investigated, using a genetic approach in the mouse. With the possible exception of the myosin heavy chain genes expressed specifically during the development of a striated muscle, genes within the myosin and actin multigene families are not linked. Genes expressed in the same phenotype are also not linked, and indeed have been mapped to different chromosomes. This therefore precludes any models of cis-acting regulation of grouped actin and myosin genes during the establishment of a phenotype.

INTRODUCTION

The formation and maturation of skeletal muscle represents an interesting biological system in which to study the establishment of a differentiated phenotype and its subsequent modification during foetal and post-natal development. When dividing myoblasts fuse together to form multinucleated muscle fibres quantitative and qualitative changes take place in the proteins synthesized. Membrane proteins such as the acetylcholine receptor appear on the cell surface, the enzymic constituents of the cell are modified to accommodate the energy requirements of muscle metabolism, and muscle specific contractile proteins accumulate in organized sarcomeric structures (for review see (1)). Different isoforms of the contractile proteins with distinct primary structures are present in different types of muscle and in non-muscle tissues. They are therefore encoded by multigene families (see (2)). Our particular interest is centered on the regulation of gene expression within the actin and myosin families during mammalian (mouse) myogenesis. In striated muscles characteristic isoforms are present in different types of adult fibres as indicated in Table I, and in addition, in the case of the myosin heavy chain, isoforms have been identified in skeletal and cardiac tissues which are specific to a given stage of foetal or neonatal development (3,4). For the actins and myosin light chains no distinct developmental isoforms have been detected; instead some proteins accumulated in adult cardiac tissue are present in developing skeletal muscle (5,6) and vice versa (7,8).

Tissue	Actins (α, β, γ)	Myosin Heavy Chains (MHC)	Myosin Light Chains alkali (LC_1, LC_3)
Skeletal Muscle			
Fast foetal	—	MHC_{emb}	LC_{1emb} (LC_{1F})
Fast new born	—	MHC_{NB}	(LC_{1emb}) LC_{1F} (LC_{3F})
Fast adult	α_{sk}	$2MHC_F$ MHC_{SF}	LC_{1F} LC_{3F}
Slow foetal	—	MHC_{emb}	LC_{1emb} (LC_{1F})
Slow new born	—	MHC_{NB} MHC_F	(LC_{1emb}) LC_{1F} (LC_{3F}) LC_{1S} LC'_{1S}
Slow adult	α_{sk}	MHC_S	LC_{1S} LC'_{1S}
Cardiac Muscle			
Atria foetal	—	—	LC_{1A}
Atria new born	—	—	—
Atria adult	α_c	MHC_A (? = MHC_{V1})	LC_{1A} (? ≡ LC_{1emb})
Ventricles foetal	—	MHC_{V3}	LC_{1A} LC_{1V}
Ventricles new born	—	(MHC_{V3}) MHC_{V1}	LC_{1V}
Ventricles adult	α_c	(MHC_{V1}) MHC_{V3}	LC_{1V} (? = LC_{1S})
Smooth Muscle	α_{sm} α_{sm}	MHC_{sm}	LC_{1nm}
Non Muscle	β γ	$2MHC_{nm}$	LC_{1nm}
Minimum Gene Number	6	$\geq$ 11	$\geq$6

TABLE I. Actin and myosin isoforms in different muscle and non-muscle tissues (based on protein data for mammals, 2/84)

The initial formation of muscle fibres is thus characterized by a transition from the synthesis of non-muscle actins and myosins to those characteristic of the earliest foetal phenotype (9). This process is first detectable *in vivo* in the mouse at the 10th/11th day when limb buds are just beginning to form from the mesodermally derived somites of the embryo (see (10)). During later foetal and neonatal development the muscle mass increases, both in terms of fibre density and fibre size ; further transitions take place in the types of muscle myosin, and actins present (see (2)). An aspect of myogenesis which distinguishes it from some other systems (e.g. erythropoiesis (11)) which have been taken as models for developmental regulation, is that many of these changes are taking place within the same fibres, or cells. It has been suggested that the earliest myotubes may be generated from a distinct population of myoblasts (12), but later fibres are probably derived from a common precursor cell population, and the increase in muscle mass during later foetal life is due to the fusion of these myoblasts into new and into existing fibres in which contractile proteins continue to accumulate (see (10)). The physiological factors responsible for changes in the types of isoforms expressed in a fibre are not well understood, both circulating levels of thyroid hormone and neuronal influences are probably implicated (13). The developing muscle undergoes changes in the type of innervation from polyneuronal to mono-neuronal (see (10)), and certainly after birth the type of nerve (e.g. fast or slow) determines the phenotype of a fibre, to the extent that this can be reversed by changing the pattern of electrical stimulation of an adult fibre (14). An experimentally useful attribute of muscle cells is that they will form muscle fibres spontaneously in tissue culture, either in primary cultures of foetal muscle or in permanent myoblast cell lines. Myotubes cultured in the absence of nerve will undergo at least some of the modifications in isoform expression, suggesting that changing humoral or neuronal input is not essential for all modulation of the muscle phenotype (see (1)).

Recombinant DNA technology has made it possible to develope probes for the mRNAs and genes coding for contractile protein isoforms, and thus to look directly at their regulation in muscle. We have examined changes in the expression of actin and myosin genes during muscle development in the mouse *in vivo*, and compared this with the situation in muscle cell lines. We also report here some structural features of the isolated genes, and a genetic analysis of linkage relationships between them, both with reference to other members of the same multigene family, and to other muscle protein genes expressed in the same phenotype.

THE EXPRESSION OF GENES ENCODING DIFFERENT ACTIN AND MYOSIN ISOFORMS

Members of the actin or myosin multigene families in mammals demonstrate considerable sequence homology, for example skeletal and cardiac actin proteins differ by 4/375 amino acids and have about 14% divergence in their nucleic acid sequences due to differences in codon usage. In order to distinguish between mRNAs or genes of the same family, it is necessary either to employ a probe derived from a non-coding part of the sequence of the mRNA or gene, which is then isoform specific, or to use a coding sequence probe and stringent hybridization washing conditions (e.g. 0.1 x SSC, 70°C). At low stringency (e.g. 0.1 x SSC, 45°C) most sequences in the family are detected (15).

Experiments *in vivo*

Studies on mRNA accumulation in skeletal muscle tissue from mice at different stages of foetal and neonatal development demonstrate that the three multigene families, actins, myosin heavy chains (MHC) and alkali myosin light chains (MLC) undergo independent developmental transitions. S_1 protection experiments were performed using a probe, derived from a sequence encoding myosin heavy chain in adult fast skeletal muscle, containing a 3' non-coding sequence specific for this isoform and a -COOH terminal coding sequence which has homologies with other myosin heavy chain genes. In late foetal mouse muscle a major mRNA species is present which protected part of the coding region of the labelled probe. At birth there is a rapid transition from this foetal MHC sequence to accumulation of the homologous adult MHC mRNA (16). In contrast the alkali myosin light chain family shows co-accumulation of foetal and adult isoforms in late foetal life. Unlike the myosin heavy chain, expression of the foetal MLC is not confined to developing skeletal muscle; two dimensional gel analysis of the proteins already suggested that it is the major light chain of adult heart atria (LC_{1A})(17). We have isolated the mRNA and gene encoding LC_{1emb} and have shown that the same gene is indeed transcribed in both tissues (18).

The adult myosin light chain gene has also been isolated. This encodes two proteins LC_{1F} and LC_{3F} (see later). Results derived, for example, from Northern blotting experiments with these probes, where the different MLC mRNAs can be distinguished on the basis of size as well as nucleotide sequence, demonstrate that the major MLC mRNA in late foetal muscle is that coding for LC_{1F}. The mRNAs for LC_{3F} and LC_{1emb} are present in approximately equal, lesser, amounts. Preliminary experiments, based on dot blots alone, suggest that LC_{1emb} is the major alkali myosin light chain in earlier foetal muscle ; the adult mRNAs are not detectable. After birth in the mouse equal amounts of the mRNAs for LC_{1F} and LC_{3F} rapidly accumulate (19).

In the case of the actins no developmental isoform has been described at the protein level. However we isolated recombinants from new born skeletal muscle which contained a cardiac type actin coding sequence. S_1 protection experiments, and genetic mapping confirmed that transcripts from the same cardiac actin gene are present in the adult heart and in developing skeletal muscle (5). Recently in collaboration with J. Vandekerckhove we have shown that the cardiac actin protein accumulates with skeletal actin. This situation therefore resembles that for the alkali myosin light chain LC_{1emb}. However in some important respects it is different. In the case of the actins skeletal actin transcripts are also found in developing cardiac tissue (7, 8), and indeed low level expression of the cardiac or skeletal gene is also found in the other adult tissue. No such reciprocal situation exists for the myosin light chains. The adult skeletal genes LC_{1F} and LC_{3F} are only expressed in skeletal muscle, and the cardiac LC_{1V} is similarly heart specific. The LC_{1emb} gene is expressed in both developing skeletal muscle and cardiac ventricles, but is found only in cardiac atrial tissue in the normal adult animal. The developmental regulation of the two multigene families is also different. Cardiac actin sequences co-accumulate with those of skeletal actin and represent about 30-40% of striated muscle actin mRNA in late foetal muscle (5). After birth cardiac actin mRNA is still present in significant amounts (about 20% in 5-10 day skeletal muscle). In early foetal muscle, dot blotting experiments suggest that not cardiac, but skeletal actin transcripts predominate. Maximum accumulation of cardiac actin mRNA takes place in 17/18 day foetal muscle (19). This corresponds to the time when maximum increase in muscle volume is taking place (10). Cardiac and skeletal actin proteins only differ by 4/375 amino acids and the functional significance of these differences has not yet been demonstrated. It has been suggested that multiple genes such as those for the actins, may have developed in response to a regulatory requirement (see 20), such that the genes rather than the proteins expressed in one phenotype be distinct from those in another. However there is as yet no direct evidence for two genes encoding the <u>same</u> protein expressed in different tissues or at different stages of development. An alternative point of view is that the actin isoforms do have functionally significant differences. The down-modulation of one in favor of the other in the adult tissue would then correspond to a functional fine turning and the presence of both proteins in developing muscle a response to a requirement for the rapid production of large quantities of striated muscle actin. In the case of the myosin heavy chains, it is clear that at least the adult isoforms have physiologically different properties (21). The role of the myosin light chains is more problematic (22) - establishment of the LC_{1emb} sequence, for example, from that of the gene will indicate how structurally different it is from the adult isoforms. It appears to be most closely related to LC_{1V}, the major cardiac ventricular light chain (18). In the immediate future the introduction of genes for different contractile protein isoforms into mammalian cells, should distinguish the functional role of some at least of the actins and myosins.

In conclusion, different developmental strategies can be distinguished within the actin and myosin multigene families during the formation of mammalian skeletal muscle. For the myosin heavy chains there is sequential accumulation of skeletal muscle specific transcripts, with an abrupt transition from foetal to adult mRNAs at birth. In contrast, no developmental specific isoforms of the actins and myosin light chains have been identified, products of genes expressed in adult cardiac tissue accumulate in developing skeletal muscle. However the nature and the developmental timing of this phenomenon differ between these two gene families (see Table II).

Experiments with Muscle Cell Lines

Muscle cell lines provide model systems in which to look at the initial formation of muscle fibres which takes place in culture more rapidly and synchronously than *in vivo* (see (1)). They are also potentially important for cell transformation experiments directed at asking questions about the regulation of muscle gene transcription. It is now well established that mRNAs encoding muscle contractile proteins accumulate rapidly at the time when myoblasts begin to fuse to form muscle fibres ; there is concomitant loss of the non-muscle mRNAs (e.g. 9). However the types of developmental isoform present are less well characterized. We have investigated this for the actins

TABLE II: ACCUMULATION OF mRNAs DURING SKELETAL MUSCLE DEVELOPMENT

Skeletal muscle	Foetal		New Born			Adult	Fused muscle cell cultures T984-C110		L6
Age (days)	12/15	18	1/2	5/6	11/12	90	2	11	2
Myosin Heavy Chains									
foetal MHC	ND	+++	++				ND	ND	ND
adult MHC			+	+++	+++	+++			
Alkali Myosin Light Chains									
foetal $LC1_{emb}$	++	+					+++	+++	+++
adult $LC1_F$	(+)	+++	+++	+++	+++	+++	++	++	
$LC3_F$			++	+++	+++	+++	+	++	
Actins									
foetal/cardiac α_c	(+)	++	++	++	(+)	(+)	+++	+	
adult skeletal α_{sk}	++	+++	+++	+++	+++	+++	+++	+++	+++

and myosin light chains in a mouse cell line T984-C110 isolated in the Pasteur (23), and in a rat line, L6, isolated by D. Yaffé (24). Analysis of both the proteins and the RNAs demonstrates that T984-C110 expresses the foetal and adult light chain genes LC_{1emb}, LC_{1F}, and LC_{3F} from the onset of fibre formation. About twice as much of the foetal light chain is synthesized. Fibres cultured for several weeks continue to synthesize the three light chains in the same proportions (25). Both cardiac and skeletal actin mRNAs and proteins also accumulate as the first myotubes form. However the relative proportion of cardiac actin transcripts is down-modulated in older myotubes to 25% of their original level (5,25). In contrast the rat cell line, L6, expresses only the LC_{1emb} gene and that of skeletal muscle actin (25). The two cell lines therefore have different muscle phenotypes. It is possible that the L6 line is analogous to the early myoblast population, whereas the fibres of T984-C110 more closely resemble those in later foetal muscle (Table II). Differences in cell lines such as these may account for some of the apparent discrepencies in the extent of regulation seen with exogenous muscle genes re-introduced into the cells (see this symposium).

STRUCTURAL ASPECTS OF ACTIN AND MYOSIN GENES

The Cardiac Actin Gene

In some inbred mice lines (e.g. Balb C) the 5' end of the cardiac actin gene is reduplicated (26). In other lines (e.g. C3H) this is not the case and the gene is nevertheless expressed both in cardiac tissue and in developing skeletal muscle. The re-duplication is therefore not functionally significant in this respect. It nevertheless poses some interesting problems. The reduplication corresponds to 3.5 Kb, at the 5' end of the actin gene including the second exon (Fig.1). No mutation in the coding sequence has been observed which would affect processing or translation. In the promotor region, the conventional signals TATA, CAAT etc. are conserved, and there is no reason to suppose that the promotor is not functional. It is not clear what happens to RNA initiated at this promotor - whether transcription of any read-through messenger will be terminated ; we are currently investigating the nature of the transcripts. Interestingly at a position 370 bp upstream from the CAP site there is an alternating stretch of 24 purine/pyrimidine residues, which can thus potentially form a left handed DNA structure of the " Z" type. Such a structure may have regulatory significance in modulating the transcriptional accessibility of the cardiac actin gene. Reintroduction of the promotor into muscle cells with and without this sequence should define its importance.

```
         10        20        30        40        50        60
GAATTCTGGCTTTCCATCTAGATGAGCATAGTCCCAGCTCAGGCTAAAGGAGACCAGAAG
         70        30        90       100       110       120
ATGCACTACCCAGGTNCCCCTTTTGCATCCAGGTACTCAAATATCCTGTCAAGCTGCTCT
        130       140       150       160       170       180
TTGTGCCTATGACTCTGCTCCTCTGTCTCCCAGTACTGGAAGATGAGCAAGCTGCTGTCT
        190       200       210       220       230       240
GCCCCTGCAGCCCCAGCCCAGCTGTCAGGACCCTTCTCCAAGGGCAGGGCTAGCCAAGTC
        250       260       270       280       290       300
TTCCGGCATGTGTGTGTGTGTGTGTGTGTGTGTGTGTGTGTGTGTGTGTGTGTGTGGACT
        310       320       330       340       350       360
CATTGCCCTTAGTTTTTGGAAGGGCTGAAGAGCAATAAGCCCACTCCACAACTAGGGAGC
        370       380       390       400       410       420
TCCCCCACCCAAGGGGCGCATGGCATGAGATAGCCTTCCCCGCCCCCACCCCTTGCTGGC
        430       440       450       460       470       480
CTGCCCCTCCCATACCTCCCTATATGGCCATGCTCTGACTGCCCCCTCCCCTTCCTTACA
        490       500       510       520       530       540
TGGTCTGGGAGCCCCCTGGCTGATCCTCTACCCTGCCCTTGGCTTCCAAGAATGGCCTCA
        550       560       570       580       590       600
CGGTCCTAGATGGTGCTAAGGCGACCAAATAAGGCAAGGTGGCAGATCAGGGGCCCCCCA
        610       620       630       640       650       660
CCCCTGCCCCCGGCTGCTCCAACTGACCCCGTCCATCAGAGAGCTATAAAGCTGCGCTCC
        670       680       690       700       710       720
AGGCGACTGACACCCAGTGCCTGCCACCAGCGCCAGCCCAGCTGAATCCAGCCGCCCCTA
        730       740       750       760       770       780
GCACGGTGAGTCCCAGCCTTGCTCCCTGCAGGACCTTGTCAGCACTGTGCTTTTGTGCTG

TTGGATCC
```

FIGURE 1. Mouse cardiac actin gene 5' promotor region (duplication)

The Gene Encoding the two Fast Light Chains LC_{1F} and LC_{3F} of adult mouse skeletal muscle.

The two alkali myosin light chains LC_{1F} and LC_{3F} have a common -COOH terminal sequence of 141 amino acids. They differ at the $-NH_2$ terminal, LC_{1F} and LC_{3F} have 50 and 8 distinct amino acids, respectively (27). The conservation of species variation in sequence between the -COOH sequence of the two light chains led to the suggestion that they may be encoded by a common genomic fragment (28). Isolation of cDNA clones with a common nucleotide sequence in the 3' coding and non-coding parts of the mRNAs was in keeping with this view (e.g. 29). In the laboratory mouse two

FIG.2: THE MYOSIN FAST ALKALI LIGHT CHAIN GENE IN THE MOUSE

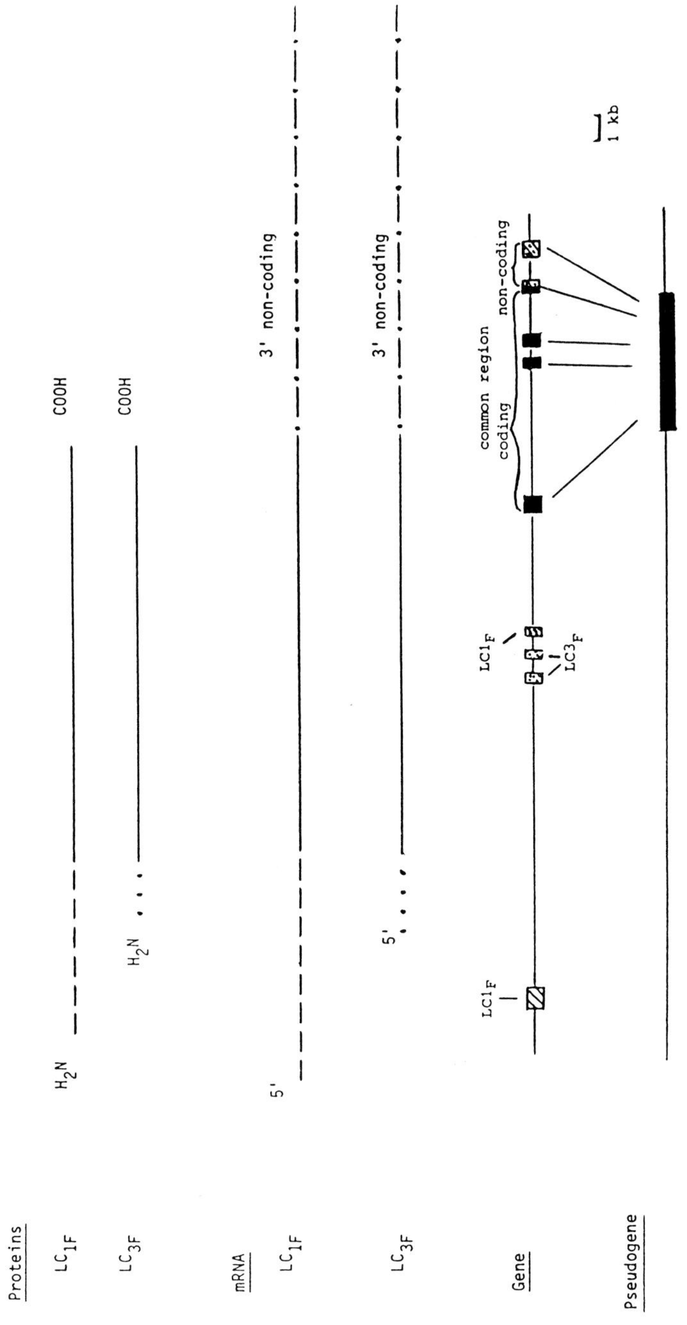

genetic loci for this sequence are present, although in other mouse species there is only one, suggesting that the second may be a pseudogene. Isolation and characterization of recombinant phage containing these DNA sequences confirm that one locus corresponds to an intronless pseudogene containing specifically the common coding region of LC_{1F} and LC_{3F}. The other locus corresponds to the functional gene (Fig.2). This spans 21 kb. The specific 5' exons are separated from the common coding region by large introns. A complex intron/exon organization places exons for the LC_{3F} specific coding sequences within an intron in the LC_{1F} specific region. DNA sequencing and S_1 protection experiments demonstrate two functional promotors. The production of the two mRNAs for LC_{1F} and LC_{3F}, which are present in equal amounts in adult muscle, but not in foetal skeletal muscle where LC_{1F} accumulates first, probably involves differential initiation of transcription at one promotor rather than the other, followed by differential splicing to give the appropriate transcript (30).

This phenomenon, whereby two or more different mRNAs and proteins are generated from the same genomic fragment is not unique to the myosin light chains. In mammals there is evidence that isoforms of α-tropomyosin (31) and troponin T (32) are encoded by alternating distinct and common exons of the same α-tropomyosin or troponin T gene. In Drosophila this has been demonstrated for the myosin heavy chain where differential splicing at the 3' end of the gene results in different terminal exons (33,34). In fact, in order to conserve the highly ordered structure of a muscle sarcomere the interaction between contractile proteins in certain regions of their molecules is probably best kept constant. Divergence of other parts of the molecule allows for functional variations, in interactions with regulatory proteins, for example. The generation of diversity in a contractile protein multigene family by a strategy which differentially splices some exons while keeping others constant satisfies such a requirement.

THE GENOMIC ORGANIZATION OF ACTIN AND MYOSIN GENES, IN THE MOUSE

In attempting to understand how the expression of a muscle phenotype is regulated at the level of gene transcription, it is important to know how the genes concerned are organized. The simplest "operon-type" model, for example, might be based on a battery of genes grouped together and regulated in "cis". The fact that developmental changes in the level of mRNAs encoding different isoforms of the actins and myosins are not co-ordinated would tend to argue against such a model. We have used mouse genetics to look at linkage relationships between different actin and myosin genes. The approach is based on the identification of a restriction fragment length polymorphism for a given gene, detected by Southern blotting, between two parental mouse lines. The segregation of this polymorphism in the progeny can be compared with that of other genes (e.g. (15)). If recombinant inbred lines are used where the progeny has been inbred to homozygosity genetic distances of about 1 cM are screened. If, on the other hand, back-crosses between different inbred mouse species (e.g. (DBA/2 x Spretus) x DBA/2) are used, these are either heterozygous with respect to the polymorphism of one parent (Spretus/DBA/2) or homozygous (DBA/2 / DBA/2), and much larger chromosomal distances (c. 30 cM) are screened. It is this latter approach that we have used initially. Precise localization of a gene in terms of co-segregation with mapped chromosomal markers in that region has been confirmed using recombinant inbred lines. The results are presented in Table III.

We conclude that those genes which we have examined (for which we have specific probes) within the same multigene family (actins, MLC, MHC), are not linked. The MLC genes,for example, map to different mouse chromosomes (35). Cardiac and skeletal muscle myosin heavy chain genes expressed in the adult tissues also map to different chromosomes (38). However there is evidence from structural analysis of the foetal and adult cardiac MHC genes (V_1 and V_3) that these are within a few Kb of each other in the genome (37). There is also a suggestion that the different MHC genes expressed during the development of skeletal muscle are on the same chromosome (38,39). It is possible that this developmental series of genes is also linked, suggesting that, as in the case of the β globin genes for example, that genes which are expressed sequentially in a given tissue may be subject to a regulatory restraint which keeps them together on the chromosome. In general, genes in a multigene family, formed initially by reduplication and consequently organized in tandem, are more likely to diverge into distinct coding units and hence to be retained in the rare instances where they are functionally useful, if they are rapidly dispersed. Gene conversion events which will tend to re-homogenize similar sequences are 10^3 more frequent

Table III

NON CO-SEGREGATION OF ACTIN AND MYOSIN STRIATED MUSCLE GENES IN DBA/2 : MUS SPRETUS BACK CROSSES

	1	2	3	4	5	6	7	8	9	10	11	12	13	14	15	16	17	18	19	20	Chr.
LC_{1F}/LC_{3F}	-	+	+	+	+	-	-	-	-	+	+	-	-	+	+	+	+	-	-	-	1
Skeletal actin	+	+	-	+	-	-	+	-	+	-	+	+	-	-	-	+	+	-	+	-	
Adult skeletal MHC	-	+	+	-	+	-	+	-	+	+	+	+	+	-	-	+	-	+	+	-	? 11
LC_{1A}	-	+	-	-	-	+	-	-	+	-	+	+	+	+	-	+	-	+	+	-	11
Cardiac actin	+	-	-	-	+	-	+	-	+	-	+	+	-	+	+	+	-	-	+	-	
Cardiac MHC	-	-	-	-	+	+	-	-	+	+	+	-	-	+	+	-	+	+	+	-	14
LC_{IV}	-	-	+	-	+	+	+	+	-	+	+	-	-	-	-	-	+	+	-	+	9

(at least in yeast) between neighbouring genes, intrachromosally (40). The myosin light chain gene LC_{1emb} occurs only in mammals and probably evolved relatively recently by reduplication from the LC_{1V} gene ; nevertheless it is on another chromosome (35).

Genes expressed in the same phenotype are also dispersed (35). No linkage is detected over 20-30 cM (1 cM = approx. 1000 Kb) for MHC, MLC or actin genes expressed in adult skeletal or adult cardiac muscle. Genes of the same family expressed together during the development of these tissues (e.g. skeletal and cardiac actins, foetal and adult light chains) are also not linked. Analysis of the chromosomal content of hybrid cell lines, has also indicated that muscle genes expressed in the same phenotype are on different chromosomes (38). The conclusion therefore is that simple models, evoking cis-acting regulation of a group of actin and myosin genes are not tenable. These differing patterns of mRNA accumulation of each multigene family during skeletal muscle maturation indicate that they respond differently to any trans-acting regulatory factors resulting from changing neuronal, hormonal or other influences during development of the animal. The microdissection of these mechanisms for a given gene is now feasible, using gene transformation technology. Understanding the integral regulation of the different multigene families remains more difficult.

REFERENCES

1. Buckingham M.E (1984). Actin and myosin multigene families : their expression during the formation of skeletal muscle. In Essays in Biochemistry Vol 20, ed. P.N. Campbell & R.D. Marshall, pub. Academic Press. In press.
2. Buckingham M E & Minty A J (1983). Contractile protein genes. In Eukaryotic genes, ed. N. MacLean & R.A. Flavell, pub. Butterworths 21:365.
3. Whalen R G, Sell S M, Butler-Browne G S, Schwartz K, Bouveret P & Pinset I (1981) Three myosin heavy chain isozymes appear sequentially in developing rat muscle. Nature 292:805.
4. Lompré AM, Mercardier JJ, Wisnewsky C, Bouveret P, Pantaloni C, d'Albis A , Schwartz K (1981) Species and age dependent changes in the relative amounts of cardiac myosin isoenzymes in mammals. Develop. Biol 84:286.

5. Minty AJ, Alonso S, Caravatti M, & Buckingham ME (1982) A fetal skeletal muscle actin mRNA in the mouse, and its identity with cardiac actin mRNA. Cell 30:185.
6. Whalen RG, Butler-Browne GS & Gros F (1978) Identification of a novel form of myosin light chain present in embryonic muscle tissue and cultured muscle cells. J. Mol. Biol. 126:415.
7. Gunning P, Ponte P, Blau H & Kedes L (1983) Alpha-skeletal and alpha-cardiac actin genes are co-expressed in adult human skeletal muscle and heart. Mol. Cell Biol. 3:1985.
8. Mayer Y, Czosnek H, Zeelon PE, Yaffé D & Nudel U (1984) Expression of the genes coding for the skeletal muscle and cardiac actins in the heart. Nucl. Acids Res. 12:1087.
9. Caravatti M, Minty AJ, Robert B, Montarras D, Weydert A, Cohen A, Daubas P & Buckingham ME (1982) Regulation of muscle gene expression : the accumulation of mRNAs coding for muscle specific proteins during myogenesis in a mouse cell line. J. Mol. Biol. 160:59.
10. Ontell M (1982) The growth and metabolism of developing muscle. In "Biochemical Development of the foetus and neonate" ed. C.T. Jones, pub. Elsevier Biochemical Press p. 213.
11. Marks PA & Rifkind RA (1978) Erythroleukemic differentiation. Ann. Rev. Biochem. 47:419.
12. Rutz R & Hauschka S (1982) Clonal analysis of vertebrate myogenesis. Dev. Biol. 91:103.
13. Gambke B, Lyons GE, Haselgrove J, Kelly AM & Rubinstein NA (1983) Thyroidal and neural control of myosin transitions during development of rat fast and slow muscles. FEBS Lett. 156:335.
14. Weeds AG, Trentham DR, Kean CJC, Buller AJ (1974) Myosin from cross-reinnervated cat muscles. Nature 247:135.
15. Minty AJ, Alonso S, Guénet JL & Buckingham ME (1983) Number and organization of actin related sequences in the mouse genome. J. Mol. Biol. 167:77.
16. Weydert A, Daubas P, Caravatti M, Minty A, Bugaisky G, Cohen A, Robert B & Buckingham M (1983) Sequential accumulation of mRNAs encoding different myosin heavy chain isoforms during skeletal muscle development *in vivo* detected with a recombinant plasmid identified as coding for an adult fast myosin heavy chain from mouse skeletal muscle. J. Biol. Cell. 258:13867.

17. Whalen RG, Sell SM, Erikson A & Thornell LE (1982) Myosin subunit types in skeletal and cardiac tissues and their developmental distribution. Dev. Biol. 91:478.
18. Barton P, Robert B, Fiszman M, Leder D & Buckingham M (1984) The same alkali myosin light chain gene is expressed in adult atria and foetal skeletal muscle. Submitted for publication.
19. Barton P, Bugaisky G, Buckingham M et al. In preparation.
20. Davidson EH, Thomas TL, Scheller RH & Britten J (1982) The sea urchin actin genes and a speculation on the evolutionary significance of small gene families. In "Genome Evolution" . Ed. G.A. Dover & R.A. Flavell, Pub. Academic Press, New York. p. 177.
21. Streter FA, Balint M & Gergely J (1975) Structural and functional changes of myosin during development. Dev. Biol. 46:317.
22. Wagner PD & Giniger E (1981) Hydrolysis of ATP and reversible binding to F-actin by myosin heavy chains free of all light chains. Nature 292:560.
23. Jakob H, Buckingham ME, Cohen A, Dupont L, Fiszman M & Jacob F (1978) A skeletal muscle cell line isolated from a mouse teratocarcinoma undergoes apparently normal terminal differentiation *in vitro*. Exp. Cell Res. 114:403.
24. Yaffé D (1969) Cellular aspects of muscle differentiation *in vitro*. In "Current Topics in Developmental Biology". Academic Press, New York. Vol 4, p.37.
25. Bugaisky G, Buckingham ME et al. In preparation.
26. Garner I, Minty AJ, Alonso S, Buckingham M. In preparation.
27. Frank G & Weeds AG (1974) The amino acid sequence of the alkali light chain of rabbit skeletal muscle myosin Eur. J. Biochem. 44:317.
28. Matsuda G, Maita T & Umegane T (1981) The primary structure of L1light chain of chicken fast skeletal muscle myosin and its genetic implication. FEBS Lett. 126:111.
29. Robert B, Weydert A, Caravatti M, Minty A, Cohen A, Daubas P, Gros F & Buckingham ME (1982) cDNA recombinant plasmid complementary to mRNAs for light chains 1 and 3 of mouse skeletal muscle myosin. Proc. Natl. Acad. Sci. USA. 79:2437.
30. Robert B, Daubas P, Akimenko MA, Cohen A, Guénet JL & Buckingham ME (1984) A single locus in the mouse encodes both myosin light chains 1 and 3, a second locus corresponds to a related pseudogene. Submitted.

31. Ruiz-Opazo N, Weinberger J & Nadal-Ginard B (1983) Different tissue specific forms of α-tropomyosin encoded by the same gene. J. Cell Biol. 97:329a.
32. Wilkinson JM, Moir AJG & Waterfield MD (1984) The expression of multiple forms of troponin T in chicken fast skeletal muscle. Eur. J. Biochem. In press.
33. Rozek CE & Davidson N (1983) Drosophila has one myosin heavy chain gene with three developmentally regulated transcripts. Cell 32:23.
34. Bernstein SI, Mogami K, Donady JJ & Emerson CP (1983) Drosophila muscle myosin heavy chain is encoded by a single gene located in a cluster of muscle mutations. Nature 302:393.
35. Robert B, Barton P, Minty A, Daubas P, Weydert A, Bonhomme F, Guénet JL, Buckingham M (1984). In preparation.
36. Weydert A, Daubas P, Bonhomme F, Guénet JL & Buckingham M (1984) In preparation.
37. Wydro RM, Nguyen HT, Gubits RM & Nadal-Ginard B (1983) Characterization of sarcomeric myosin heavy chain genes. J. Biol. Chem. 258:670.
38. Czosnek H, Nudel U, Shani M, Barker PC, Pravtcheva DD, Ruddle FH & Yaffé D (1982) The genescoding for the muscle contractile proteins, myosin heavy chain, myosin light chain 2 and skeletal muscle actin are located on three different mouse chromosomes. EMBO J. 1:1299.
39. Leinwand LA, Fournier REK, Nadal-Ginard B & Shows TB (1983) Multigene family for sarcomeric myosin heavy chain in mouse and human DNA : localization on a single chromosome. Science 221:766.
40. Dubay, M. & Petes, T.D. (1982). Recombination between genes located on non-homologous chromosomes in S. Cerevisiae. Genetics, 101:369

Molecular Biology of Development, pages 293–308

THE REGULATION OF CELL-TYPE-SPECIFIC GENES IN DICTYOSTELIUM

Mona C. Mehdy, Charles L. Saxe III, and Richard A. Firtel

Department of Biology
University of California, San Diego
La Jolla, California 92093

ABSTRACT We are using cloned probes of genes which are preferentially expressed in prestalk or prespore cells to examine regulation of cell differentiation in Dictyostelium. Members of each cell-type-specific group of genes that we are studying show coordinate regulation. The prestalk specific mRNAs are not detectable in vegetatively growing wild-type cells, accumulate during late aggregation, and require cAMP and a diffusible factor for their accumulation in culture. In contrast to the prestalk specific mRNAs, the prespore specific mRNAs accumulate much later in development and their accumulation may require cell surface interactions in addition to cAMP and the diffusible factor. The diffusible factor has different effects on gene expression than the previously reported regulatory molecules, DIF (Differentiation Inducing Factor) and NH_3. Its effect is selective for these developmental genes; no effect is observed on the mRNA levels of actin and another gene transcribed in vegetative cells and early development. Later in development, both groups of cell-type-specific genes require only cAMP for their continued expression although the mechanisms of regulation are different. Dissociation of aggregates results in the rapid loss of both sets of mRNAs in the absence of exogenous cAMP but not the cell-type nonspecific gene. Addition of cAMP to disaggregated cells deprived of cAMP for several hours results in the rapid reaccumulation of the cell-type-specific mRNAs. The structures of the coordinately regulated prespore and prestalk genes are presently being examined. It is hoped that the complementary studies of gene structure and regulation by physiological inducers will clarify the structural features of genes involved in their differential regulation during development.

INTRODUCTION

The cellular slime mold, Dictyostelium discoideum, has been widely investigated as a model developmental system. Starvation induces cells to aggregate into loose mounds over a period of approximately 9 hours. After a defined series of morphological stages, a mature fruiting body is formed after 24-26 hours of development. This terminal structure is composed of two basic cell types: approximately 80% spore cells and 20% stalk cells (1). Differentiated precursor cells have been distinguished at the tip formation stage (12-13 hours; 2,3).

The structures and expression of numerous developmentally regulated genes have been analyzed in a variety of organisms including Dictyostelium. However, the physiological mechanisms controlling cell-type specific gene expression and cell differentiation are not well understood in most systems. Recently, several components involved in the regulation of cell differentiation in Dictyostelium have been elucidated. Numerous studies on Dictyostelium cell differentiation as a function of varying in vitro culture conditions have shown that cAMP and cell contact are required for differentiation (see 4,5,6). In addition, ammonia and DIF have been implicated in the regulation of cell differentiation (7,8). Recently, cDNA and genomic clones of genes which are preferentially expressed in prespore and prestalk cells have been used to further analyze the timing and regulation of cell differentiation (9,10). The prestalk specific genes show coordinate regulation while the prespore specific genes show a different pattern of coordinate regulation. In this paper, we show that induction of both classes of genes requires cAMP and a substance(s) secreted by developing cells. However, the two classes differ in their 1) timing of expression during development; 2) requirement for a cell-surface interaction; and 3) cAMP regulation later in development.

RESULTS

Developmental Expression of Cloned Genes

A library of cDNA clones was constructed from poly $(A)^+$ RNA from the culmination stage of development (20 hours, 9).

Twelve clones have been identified whose complementary RNAs were present late in development but not in vegetative cells. RNA and genomic DNA blots have shown that the cloned genes are complementary to different and single-copy genes. To determine whether the genes were differentially expressed in prespore and prestalk cells, the two cell types were separated on Percoll gradients (11). Northern blot analysis of RNAs from the purified cell fractions demonstrated that 7 genes are primarily expressed in prespore cells, 4 genes are primarily expressed in prestalk cells, and mRNA complementary to one gene (10-C3) is present equally in both cell types.

The expression of the genes during development was examined. Cells were harvested throughout development and their RNA extracted, size fractionated, blotted, and probed. Figure 1 shows the expression of representative prespore and prestalk genes and a cell-type-nonspecific gene during the development of a wild type strain, NC-4. The mRNAs complementary to all three genes were undetectable in vegetative cells. The mRNA complementary to the cell-type-nonspecific gene, 10-C3, was induced during early aggregation and peaked prior to culmination (~15 hours). The prestalk mRNA, 2-H6, accumulated during late aggregation and also peaked at 15 hours. All of the prestalk mRNAs we have examined appear coordinately with the same kinetics as 2-H6. In contrast, the prespore mRNA, 2-H3, accumulated only late in development, during culmination. The rapid accumulation and loss of this RNA over a short time interval was especially striking. All except one of the prespore mRNAs we have examined appear coordinately with 2-H3. One prespore mRNA (20-B6) becomes detectable two hours earlier than the other prespore mRNAs. The developmental expression of the same genes in strain FR17 is also shown in Figure 1. FR17 is a temporal mutant which completes development in ~16 hours instead of the ~24 hours required by its wild type parent, NC-4 (12). The rapid developing strain exhibited accelerated appearance and disappearance of prestalk and prespore mRNAs while the expression of 10-C3 was less affected. These results indicate that the expression of the cell-type-specific genes is tightly linked to morphological changes during development.

Induction of Prespore and Prestalk Gene Expression in Suspension

The expression of the cell-type-specific genes has been used as a marker of prespore or prestalk cell differentiation under varying *in vitro* culture conditions (9,10). The effects of cAMP and cell-cell contact on prespore and

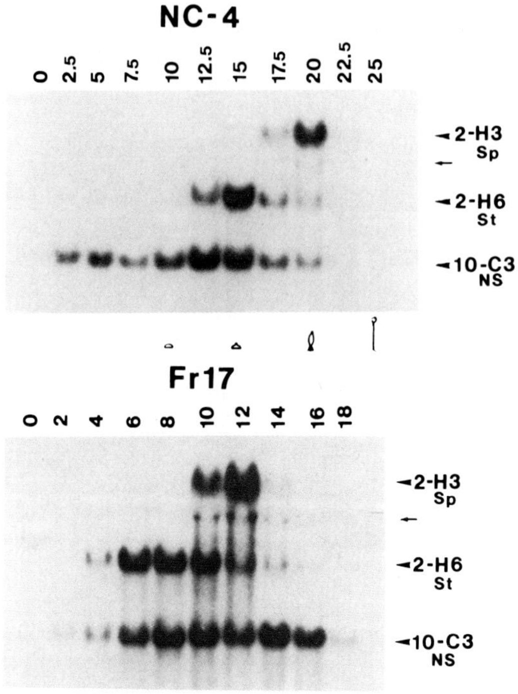

FIGURE 1. Expression of prespore, prestalk and cell-type-nonspecific genes during the development of NC-4 and FR17 strains. Cells were developed on filter pads and the resulting total cell RNAs analyzed by blot hybridization as previously described (9). Hours in development are indicated. 2-H3, a prespore mRNA; 2-H6, a prestalk mRNA; 10-C3, a cell-type-nonspecific RNA.

prestalk gene expression were individually assessed in suspension culture (9). Vegetative AX-3 cells (an axenic strain derived from the wild type strain, NC-4, 13) were starved in a fast shaking suspension for 6 hours. The culture was then divided into continued fast shaking cells (230 rpm) or slowly shaking cells (70 rpm) with or without the addition of cAMP. The fast-shaken cells remained as primarily single cells while the slowly shaken cells formed stable cell-cell contacts and formed large clumps. The top panel of Figure 2 shows that the prestalk gene, 2-H6, was induced in the fast-shaking cells only when cAMP was present. It was also expressed in slowly shaken cells with

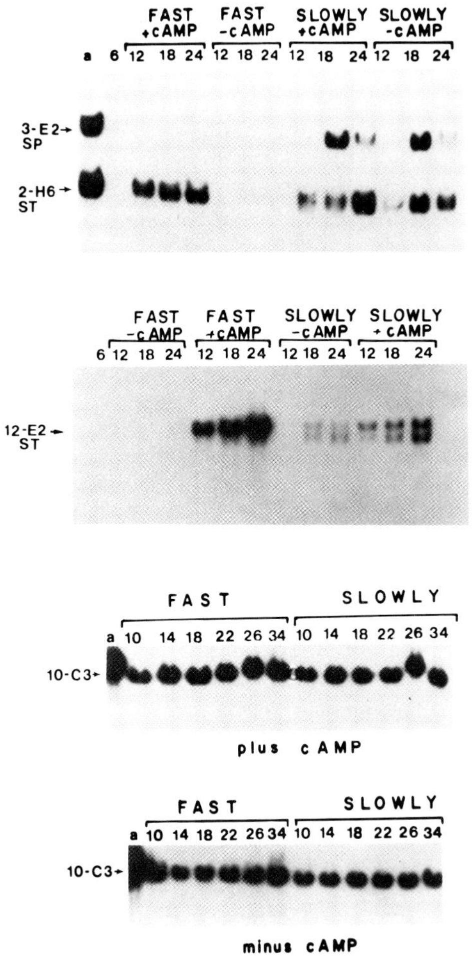

FIGURE 2. The effects of cAMP and cell contact on the induction of late gene expression. AX-3 cells were suspended in buffer and shaken at 230 rpm for 6 hours (9). The suspension was divided into fast and slowly shaking cultures, with and without 200 μM cAMP. cAMP was added to 200 μM every 2 hours to maintain its level. Blot hybridization was performed as Fig. 1. Hours after the cells were suspended in buffer are indicated. Lane a is RNA from culminating aggregates as a reference. 3-E2, a prespore mRNA; 2-H6 and 12-E2, prestalk mRNAs; 10-C3, a cell-type nonspecific mRNA.

or without added cAMP. In contrast, the prespore gene, 2-H3, was expressed only in the slowly shaking cultures. The cell-type-nonspecific gene, 10-C3, showed a third type of regulation; it was induced regardless of shaking speed or the presence of cAMP (Fig. 2, bottom panel). The middle panel shows the expression of another prestalk gene, 12-E2. 12-E2 exhibits regulation similar to that of the other prestalk gene, 2-H6. It was induced in fast-shaking cells only in the presence of exogenous cAMP and it was also expressed in slowly shaken cells. Genomic mapping and cloning experiments have shown that the two differently sized mRNAs are derived from a single gene (Reymond and Firtel, unpubl. observations). It is interesting that both mRNAs were approximately equally prevalent during normal development (data not shown) and in slowly shaking cells while only the higher molecular weight transcript was present in the fast shaken, cAMP treated cells. Thus, the levels of the two transcripts appear to be differently regulated.

These results demonstrate substantial differences in the requirements for induction of prespore and prestalk genes. The prestalk genes can be induced in single cells by exogenous cAMP while prespore gene induction appears to require cell surface interactions. The induction of the prestalk genes in the slowly shaking culture not treated with cAMP was probably attributable to endogenous cAMP production by the large agglomerates. Experiments described below will demonstrate that prespore induction also depends on the presence of cAMP.

Gene Expression in Submerged Cultures

The role of cell-cell contact in prespore and prestalk gene expression was further investigated for two reasons. First, transient cell-cell contact could not be eliminated in shake culture. Second, it has been recently shown that several mutant strains can differentiate into prespore or spore cells as single cells, independent of cell contact (14). To clarify these issues for axenic and wild-type strains, cells were plated on petri dishes, submerged under a layer of buffer. By varying the cell density, the extent of cell contact was precisely controlled. The range of cell densities shown in Table 1 was utilized. After 4 hours, cAMP was added to half of the cultures to 300 µM; we have found that cAMP addition at 0 hour prevents the accumulation of the prespore but not prestalk mRNAs. Cells were harvested after 20 hours. At the lower densities (1X and 5X), the cells existed as widely spaced single cells (Table 1). In the 20X culture, ~90% of the cells were single, while the

DEVELOPMENT IN SUBMERGED CULTURE

CELL DENSITY		CELL DISTRIBUTION
	CELLS/cm^2	
1x	900	SINGLE CELLS
5x	4,500	SINGLE CELLS
20x	18,000	~90% SINGLE CELLS, ~10% SMALL AGGREGATES
400x	360,000	LARGE AGGREGATES

remainder were in 2-3 cell aggregates. In contrast, cells in the 400X culture formed large agglomerates. Figure 3 shows the expression of cell-type-specific genes in these cells.

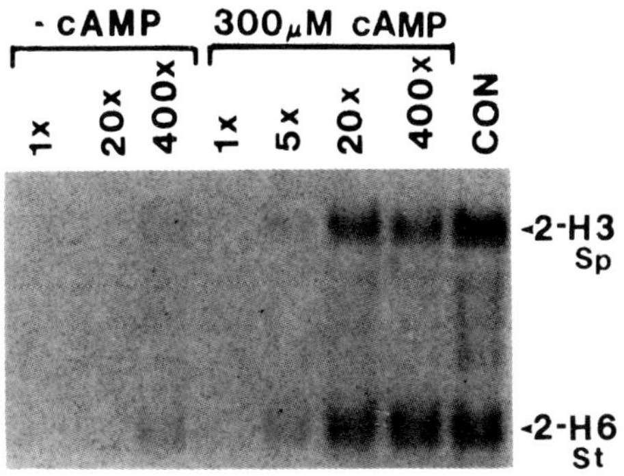

FIGURE 3. The effects of cAMP and cell density on cell-type-specific gene expression. AX-3 cells were suspended in buffer such that 15 mls plated on 15 cm diameter petri dishes gave the indicated cell densities. cAMP was added to 300 μM to half of the cultures after 4 hours. Cells were harvested after 20 hours. Blot hybridization was performed as Fig. 1. Lane labeled CON is RNA from culminating aggregates as a reference. 2-H3, a prespore mRNA; 2-H6, a prestalk mRNA.

In the cultures without added cAMP, both cell-type-specific mRNAs were expressed at low levels only in the 400X culture. Presumably, sufficient endogenous cAMP in the aggregates allowed some prestalk and prespore gene expression. In the cultures treated with cAMP, there was no detectable accumulation of either mRNA in the 1X culture. The 5X cells contained low levels while the 20X and 400X cells contained comparable high levels of both mRNAs. All other cell-type-specific genes examined showed comparable results. Preliminary experiments showed that the level of the cell-type-nonspecific mRNA, 10-C3 increased with increasing cell density and was higher in the presence of cAMP (data not shown). The reasons for cAMP stimulation of 10-C3 mRNA in this culture system and not in suspension culture are unknown. The expression of the prespore and prestalk genes in these cultures was very surprising based on our expectations from the suspension cultures. In suspension (see Fig. 2), the prestalk genes were expressed by single cells in the presence of cAMP, yet there were no detectable expression in the single cells at a 1X density and little expression in the single cells at a 5X density in the presence of abundant cAMP. In addition, the prespore genes were only expressed in agglomerates in suspension (see Fig. 2). However, Figure 3 shows comparable high levels of the prespore mRNA in 20X and 400X cultures despite very disparate levels of cell-cell contact (see Table 1).

The dependence of cell-type-specific gene expression on cell density which then levels off at high cell densities suggested that the cells produce a factor(s) which is required at some threshold level for the expression of these genes. Therefore, the effect of conditioned medium, in which cells at a high density had been incubated, on gene expression in low density cells was tested. AX-3 cells were plated at a low (2X) density in either fresh medium or conditioned medium, both with and without cAMP added after 4 hours. Cells were harvested after 20 hours and RNAs extracted. The results are shown in Figure 4. Neither prespore nor prestalk mRNA was detectable in cells cultured in fresh medium alone (FM-), while low levels of both mRNAs were present in cells cultured in fresh medium supplemented with cAMP (FM+). Conditioned medium alone (CM--) was ineffective in inducing cell-type-specific gene expression, but the combination of cAMP and conditioned medium (CM-+) caused the cells to accumulate high levels of both cell-type-specific mRNAs. Conditioned medium from high density cells treated with cAMP (CM+-, CM++), as expected, was also effective in inducing prespore and prestalk mRNAs. The cell-type-nonspecific mRNA showed regulation similar to that with cell-type-specific genes: low levels in cAMP or conditioned medium alone and high levels in the presence of both. Similar results were obtained with wild-type NC-4 cells (data not shown).

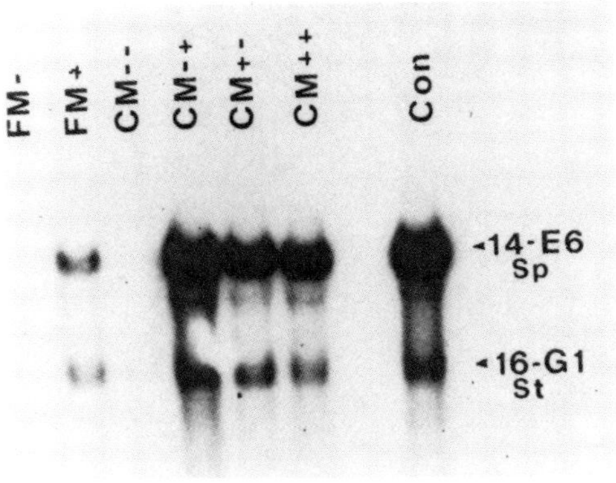

FIGURE 4. The effects of cAMP and fresh and conditioned media on cell-type-specific gene expression. 100X density AX-3 cells were incubated as described in Fig. 3, with or without cAMP added to 300 μM at 4 hours. After 20 hours, the conditioned medium was collected and freed of cells. AX-3 cells were resuspended in the following and plated at a 2X density: a. fresh medium alone (FM-) b. fresh medium with cAMP added to 300 μM at 4 hours (FM+) c. conditioned medium alone from 100X cells not treated with cAMP (CM--) d. same as c plus cAMP added to 300 μM at 4 hours (CM-+) e. conditioned medium from 100X cells treated with cAMP (CM+-) f. same as e plus cAMP added to 300 μM at 4 hours (CM++). Cells were harvested after 20 hours. Blot hybridization was performed as Fig. 1. Lane a is RNA from culminating aggregates as a reference. 14-E6, a prespore mRNA; 16-G1, a prestalk mRNA.

The timing of production of active conditioned medium was assessed by collecting medium which had incubated in the presence of high density cells for varying time intervals. Vegetative cells were plated at a low density (2X) in fresh medium or the conditioned media and all cultures received 300 μM cAMP after 4 hours. Cells were harvested after 20 hours of incubation. Figure 5 shows the accumulation of prespore and prestalk mRNAs in these cells. The levels of both mRNAs were proportional to the duration of high density cell incubation in conditioned medium, up to 12 hours. Approximately equal RNA levels were found in cells treated with either 12 hour or 20 hour (data not shown) conditioned media. We suggest that these results reflect the level of active factor(s) in the conditioned media. The factor(s)

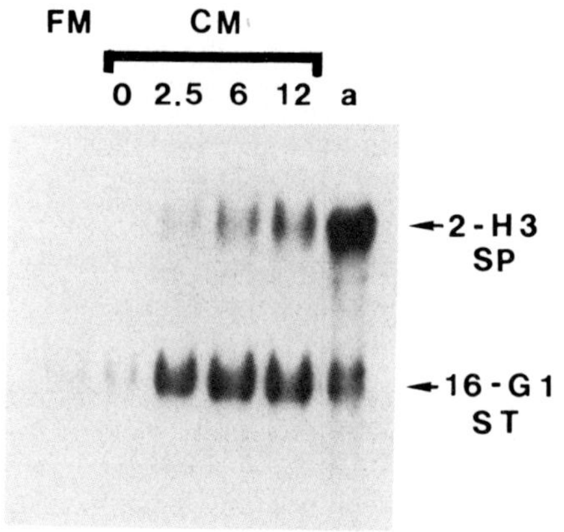

FIGURE 5. Time course of production of active conditioned media. AX-3 cells were plated at 100X density and medium was collected at the indicated hours and freed of cells. Fresh AX-3 cells were plated at 2X density after resuspension in fresh medium (FM) or the conditioned media (CM). All cultures were treated with 300 µM cAMP at 4 hours. Cells were harvested after 20 hours. Blot hybridization was performed as Fig. 1. Lane a is RNA from culminating aggregates as a reference. 2-H3, a prespore mRNA; 16-G1, a prestalk mRNA.

was produced rapidly after the onset of starvation. It was detectable at 2.5 hours and its level increased up to 12 hours. After 12 hours, its level became nonlimiting.

The production of the factor during early development prompted the examination of its effects on early gene activity. We chose to assess actin mRNA levels since actin mRNA is present in vegetative cells, remains relatively high during the first 12 hours of development, then sharply decreases (4). We also examined M4-1, a gene whose mRNA is present in vegetative cells and not late in development (Kimmel, in preparation). Conditioned medium was prepared from high density cell cultures which had incubated for 10 hours. As before, fresh vegetative cells were resuspended in either fresh or 10-hour conditioned media and plated at 2X density. No cAMP was added to the cultures since micromolar cAMP inhibits the expression of numerous genes expressed during early development (15; Mann, Brandis, and Firtel, unpublished results). Figure 6 shows the expression of actin and M4-1 genes in the 2X cultures at the indicated times.

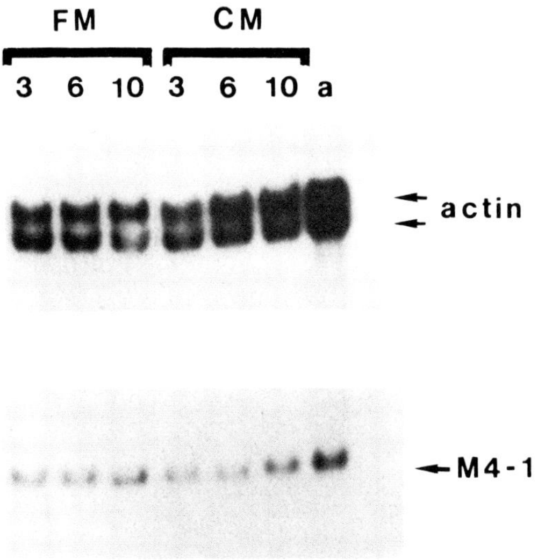

FIGURE 6. The effects of fresh and conditioned media on actin and M4-1 mRNAs during early development. Conditioned medium was prepared from 100X density cells which had incubated for 10 hours. Fresh AX-3 cells were resuspended in fresh (FM) or conditioned (CM) media and incubated for the indicated times (hrs). Blot hybridization was performed as Fig. 1. Lane a is RNA from culminating aggregates as a reference. Actin exists as 2 differently sized mRNAs.

In contrast to the cell-type-specific genes expressed during late development, the level of actin mRNA was constant in fresh and conditioned media treated cells. In addition, the level of actin mRNAs was equal in fresh and conditioned media in the presence of exogenous 300 μM cAMP (data not shown). The standard prespore and prestalk mRNA assay showed that the 10-hour conditioned medium was equal in activity to 20-hour conditioned medium (data not shown). These data indicate that conditioned medium promotes the expression of the cell-type-specific genes but not other genes we tested whose developmental kinetics and regulation are entirely different. Preliminary data indicate that conditioned medium also positively affects another group of developmentally regulated genes. These genes are induced very early in development, then repressed after tight aggregation (10 hours, Mann, Brandis, and Firtel, unpublished results).

The Effect of Surface Interaction and cAMP on Prespore and Prestalk Gene Expression Later in Development

The roles of cAMP and cell surface interaction in regulating cell-type-specific gene expression late in development were examined. AX-3 cells were developed on filter pads to the 15 hour aggregate stage when both cell-type-specific RNAs are present. Dissociation of the aggregates into a suspension of single cells resulted in the rapid loss of both sets of mRNAs in the absence of exogenous cAMP (Figure 7).

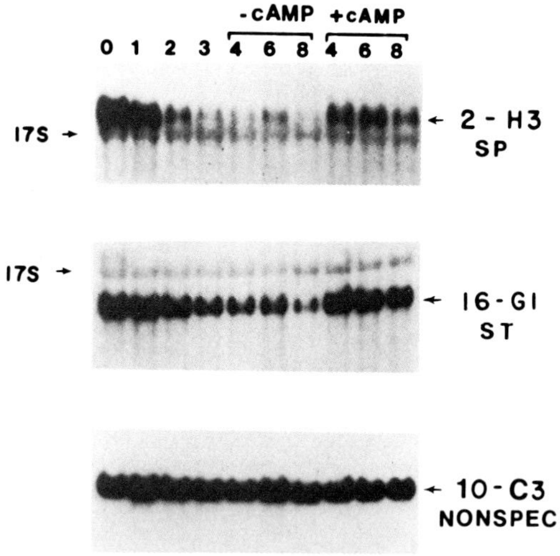

FIGURE 7. The effects of dissociation and cAMP on late gene expression. At the finger stage in development (~15 hours), AX-3 aggregates were shaken off the filters, dissociated, and fast shaken in buffer for 3 hours (9). The suspension was divided, one half received 1 mM cAMP, and the cultures continued to shake an additional 5 hours. cAMP was added to 100 μM every hour. RNA blots were hybridized as Fig. 1. Hours after dissociation are indicated. 2-H3, a prespore mRNA; 16-G1, a prestalk mRNA; 10-C3, a cell-type-nonspecific mRNA. The rRNAs which sometimes cross-hybridize to a low degree are also indicated.

The addition of cAMP after 3 hours caused rapid reaccumulation of the cell-type-specific mRNAs (9,10). The addition of cAMP immediately after disaggregation maintained the cell-type-specific RNAs at nearly the original levels (data not shown). Thus, later in development, both groups of cell-type-specific genes require only cAMP for their continued expression; the prespore genes no longer require cell surface interaction for their expression in suspension. However, the mechanisms of cAMP regulation are different in that cAMP regulation of prespore expression, but not prestalk expression, is inhibited by cycloheximide and, thus, is probably dependent on protein synthesis (9). The level of the cell-type-nonspecific mRNA is unaffected by disaggregation and cAMP, in further support of its independent regulation.

DISCUSSION

In the development of Dictyostelium, a uniform population of vegetative amoeboe differentiate into prespore or prestalk cells and eventually become mature spore or stalk cells respectively. To understand the mechanisms controlling this choice, we and others (9,10) have utilized clones of genes that are preferentially expressed in prespore or prestalk cells. The effects of various culture conditions on cell differentiation has been evaluated by monitoring the expression of the cell-type-specific genes. We have found that the expression of prespore and prestalk genes involves common and cell-type-specific regulatory mechanisms.

The prespore and prestalk mRNAs we have examined differ in their kinetics during development (9). In wild-type cells, the prestalk mRNAs accumulate during late aggregation and decrease after the finger stage. In contrast, the prespore mRNAs are detectable only during culmination, several hours after the completion of aggregation. A cell-type-nonspecific mRNA shows kinetics similar though not identical to that of the prestalk mRNAs. Similar observations have been reported by Barklis and Lodish (10). The rapidly developing mutant, FR17, shows accelerated appearance and disappearance of the cell-type-specific mRNAs. These results complement previous studies showing precocious accumulation of developmentally regulated enzymes in this mutant (1,12). Similar to enzyme markers, the expression of the cell-type-specific genes closely parallels morphological changes during development.

The prespore and prestalk genes differ in the requirements for their expression in suspension culture. The prestalk genes can be induced in single cells in the presence of cAMP while prespore genes are induced only in cells permitted to agglomerate. The cell-type-nonspecific gene is induced independently of cAMP and cell agglomeration. The latter pattern of prespore gene expression indicates that these genes require cell surface interactions or a short-range, diffusible molecule(s) for induction. Expression of prestalk genes in large aggregates not treated with cAMP is likely due to endogenous cAMP production. Therefore, a cAMP requirement for prespore gene induction could not be distinguished in suspension culture.

The cell contact and potential cAMP requirements for prespore gene expression were further investigated in a plated cell culture system. It was shown that cells plated at a very low density in the presence of cAMP do not accumulate either prespore or prestalk mRNAs. However, low density, single cells in the presence of cAMP and conditioned medium accumulated high levels of both cell-type-specific mRNAs, comparable to the levels obtained during normal development. These results show that both micromolar cAMP and a critical concentration of a factor(s) secreted by developing cells are required for prespore and prestalk specific gene expression. Conditioned medium and cAMP were also required for the expression of the cell-type-nonspecific gene in plated cell cultures. Preliminary data indicates that conditioned medium has different effects on gene expression than the previously identified regulatory molecules, DIF (8) and ammonia (7,8).

The active factor(s) is found in high cell density conditioned medium as well as media produced by fast and slow shaking suspension cultures. A substantial level of the factor(s) is found during early development when actin and M4-1 mRNA levels are maximal (4, Kimmel, in preparation). However, in contrast to the cell-type-specific mRNAs, comparable levels of actin and M4-1 mRNA are present in low density cells incubated in either fresh or conditioned medium. Preliminary data (Mehdy and Firtel, unpublished observations) indicate that conditioned medium positively regulates the expression of another set of genes induced during early development then repressed.

The analysis of prespore and prestalk specific gene expression in plated cell cultures has also shown that specific cell-cell contact is not required for prespore gene induction. These results extend previous, similar findings for mutant strains (14) to a wild-type strain, NC-4. Our results, in suspension and plated cell cultures, suggest a requirement for a nonspecific cell surface interaction since prespore genes are not expressed as single cells in suspension but are expressed in cell clumps or as single cells adhered to a plastic substrate.

We have also compared the requirements for prespore and prestalk gene expression later in development. When developing aggregates are dissociated into single cells, both sets of cell-type-specific mRNAs are rapidly reduced. Addition of cAMP restores the mRNA levels substantially. The results show that contiued prespore and prestalk gene expression is dependent on high external cAMP; the prespore genes no longer require cell surface interactions for expression. We suggest that the effect of dissociation is the lowering of the intercellular concentration of cAMP. The mechanism of cAMP regulation differs for the two groups of genes. cAMP regulation of prespore gene expression is inhibited by cycloheximide while this protein synthesis inhibitor has no effect on cAMP dependent prestalk gene expression. In contrast to the cell-type-specific mRNAs, the level of the cell-type-nonspecific mRNA is unaffected by disaggreation or cAMP addition.

We have begun to dissect the common and cell-type-specific regulatory factors controlling gene expression. Present work in the laboratory is directed towards understanding: 1) the spatial pattern of cell differentiation during normal development; 2) the structural features of prespore and prestalk genes involved in their differential expression; and 3) the biochemical nature of the active factor(s) in the conditioned medium.

ACKNOWLEDGEMENTS

M.C.M. was a recipient of a NSF Predoctoral Fellowship and has been supported by a USPHS Training Grant. RAF was the recipient of an ACS Faculty Research Award. This work was supported by Grants from NIGMS to RAF.

REFERENCES

1. Loomis, WF (1975) "Dictyostelium discoideum: a Developmental System." New York: Academic Press.
2. Muller, W, Hohl HR (1973). Pattern formation in Dictyostelium discoideum: temporal and spatial distribution of prespore vesicles. differentiation 1:267.
3. Hayashi M, Takeuchi I (1976). Quantitative studies on cell differentiation during morphogenesis of the cellular slime mold Dictyostelium discoideum. Dev Biol 50:302.

4. Kimmel AR, Firtel RA (1982). The organization and expression of the Dictyostelium genome. In Loomis WF (d): "The Development of Dictyostelium discoideum," New York: Academic Press, p 233.
5. Lodish HF, Blumberg DD, Chisholm R, Chung S, Coloma A, Landfear S, Barklis E, Lefebvre P, Zuker C, Mangiarotti G (1982). Control of gene expression. In Loomis WF (ed): "The Development of Dictyostelium discoideum." New York: Academic Press, p 325.
6. Sussman, M (1982). Morphogenetic signaling, cytodifferentiation, and gene expression. In Loomis WF (ed): "The Development of Dictyostelium discoideum," New York: Academic Press, p 353.
7. Sussman M, Schindler J (1978). A possible mechanism of morphogenetic regulation in Dictyostelium discoideum. Differentiation 10:1.
8. Gross JD, Bradbury J, Kay RR, Peacey MJ (1983). Intracellular pH and the control of cell differentiation in Dictyostelium discoideum. Nature 303:244.
9. Mehdy MC, Ratner D, Firtel RA (1983). Induction and modulation of cell-type-specific gene expression in Dictyostelium. Cell 32:763.
10. Barklis E, Lodish HF (1983). Regulation of Dictyostelium discoideum mRNAs specific for prespore or prestalk cell. Cell 32: 1139.
11. Ratner D, Borth W (1983). Comparison of differentiating Dictyostelium discoideum cell types separated by an improved method of density gradient centrifugation. Exp. Cell Res. 143: 1.
12. Sonneborn DR, White GJ, Sussman M (1963). A mutation affecting both rate and pattern of morphogenesis in Dictyostelium discoideum. Dev Biol 7:79.
13. Loomis, WF (1971). Sensitivity of Dictyostelium discoideum to nucleic acid analogues. Exp Cell Res 64:484.
14. Kay, RR (1982). cAMP and spore differentiation in Dictyostelium discoideum. Proc Nat Acad Sci 79:3228.
15. Williams JG, Tsang AS, Mahbubani H (1980). A change in the rate of transcription of a eucaryotic gene in response to cAMP. Proc Nat Acad Sci 77:7171.

Molecular Biology of Development, pages 309–329

CYTOCHROME P-450 GENES AND THEIR REGULATION

Daniel W. Nebert, Shioko Kimura, and Frank J. Gonzalez

Laboratory of Developmental Pharmacology
National Institute of Child Health and Human Development
National Institutes of Health, Bethesda, Maryland 20205

INTRODUCTION

Estimates of the number of foreign chemicals on this planet range into the millions. Industries synthesize new chemicals (for the first time on earth) at the rate of two to three thousand each year. It is comforting to know that organisms are able to cope with this kind of chemical adversity by (i) metabolizing these compounds to excretable innocuous products (detoxification) and/or (ii) mobilizing greater amounts of certain enzymes to carry out this metabolism (enzyme induction).

The cytochrome P-450 gene family encodes proteins that metabolize most of these foreign chemicals (reviewed in Ref. 1). These same proteins metabolize endogenous substrates such as steroids, fatty acids, prostaglandins, leukotrienes, biogenic amines, and pheromones. The P-450 proteins are principally found in the endoplasmic reticulum (except in Pseudomonads and certain fungi where they are in the cytosol), range in molecular weight between 50,000 and 60,000, and exhibit oxygenative, oxidative and peroxidative catalytic activities. Reductases and sometimes hemoproteins and nonheme iron-proteins comprise the membrane-bound multicomponent system, of which P-450 is the terminal step wherein lies the enzyme active-site. Recommended nomenclature for these activities is "multisubstrate monooxygenases" (1).

The P-450 proteins are ubiquitous; at least some forms are known to exist in certain types of bacteria and all eukaryotes examined. Evidence has been presented for the existence of at least one form of inducible P-450 in the pre-implantation mouse embryo (reviewed in Ref. 2).

Since P-450 proteins function in the biosynthesis and degradation of steroids and fatty acids essential for membrane integrity of all eukaryotic cells, it should be accepted that at least some P-450 genes will be expressed in all cells of all eukaryotes--with the possible exception of early embryogenesis during which time translation of maternal P-450 mRNA may be important.

For the past 15 years this laboratory has been involved in studies of polycyclic hydrocarbon-metabolizing enzymes and their induction by foreign chemicals. Early efforts characterized the enzyme machinery, genetic differences among cell lines in culture and among inbred mouse strains, tissue distribution, developmental aspects, and pharmacokinetic and metabolite studies during carcinogenesis, mutagenesis, teratogenesis, and drug toxicity (3, 4). Using the mouse as a model system, we have more recently purified several forms of P-450, developed polyclonal antibodies, and proceeded to isolate and characterize these genes and the mechanism of P-450 induction (5-9).

The chronologic scheme of our research is outlined in Fig. 1. First, we have chosen a catalytic activity (e.g. benzo[a]pyrene metabolism) preferentially associated with a particular form of P-450 (mouse P_1-450). Following purification of this membrane-bound protein, reconstitution of catalytic activity in vitro (5) and inhibition of activity with the antibody (5) were carried out as necessary prerequisites to make conclusions about the specificity of the purified antigen and the antibody. N-terminal sequence analysis is important for later correlations with the 5' cDNA coding region. The antibody can be used to quantitate P-450 concentrations in the endoplasmic reticulum(10) and to size the particular messenger (11) or purify the particular mRNA with polysome immunoadsorption (12). Our cDNA clones have been used for gene linkage studies (13), nucleotide sequencing (14, 15), and obtaining mouse (8, 12) and human (16) genomic clones. To reproduce catalytic activity via transfection or microinjection of cDNA or genomic P-450 clones has not yet been successfully achieved.

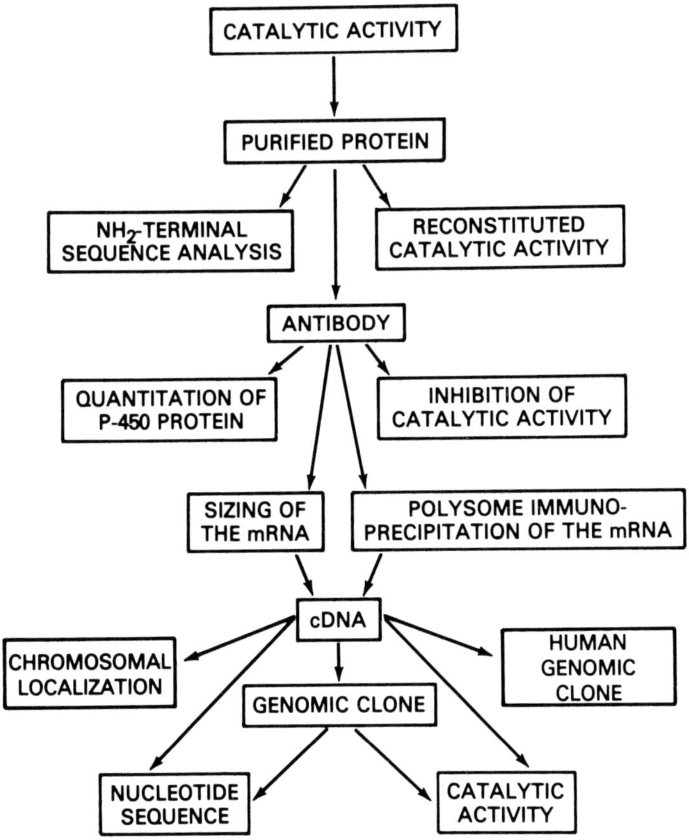

FIGURE 1. Summary of the experimental scheme in this laboratory.

Murine Ah Locus

The P-450 subset on which this laboratory has concentrated most of its efforts involves the Ah locus (Fig. 2). This gene system controls the induction of at least two forms of P-450 by polycyclic aromatic compounds such as 2,3,7,8-tetrachlorodibenzo-p-dioxin (TCDD) and benzo[a]pyrene. TCDD binds avidly to the Ah receptor (apparent $K_d \sim 1$ nM), and the inducer-receptor complex undergoes a temperature-dependent step before gaining high

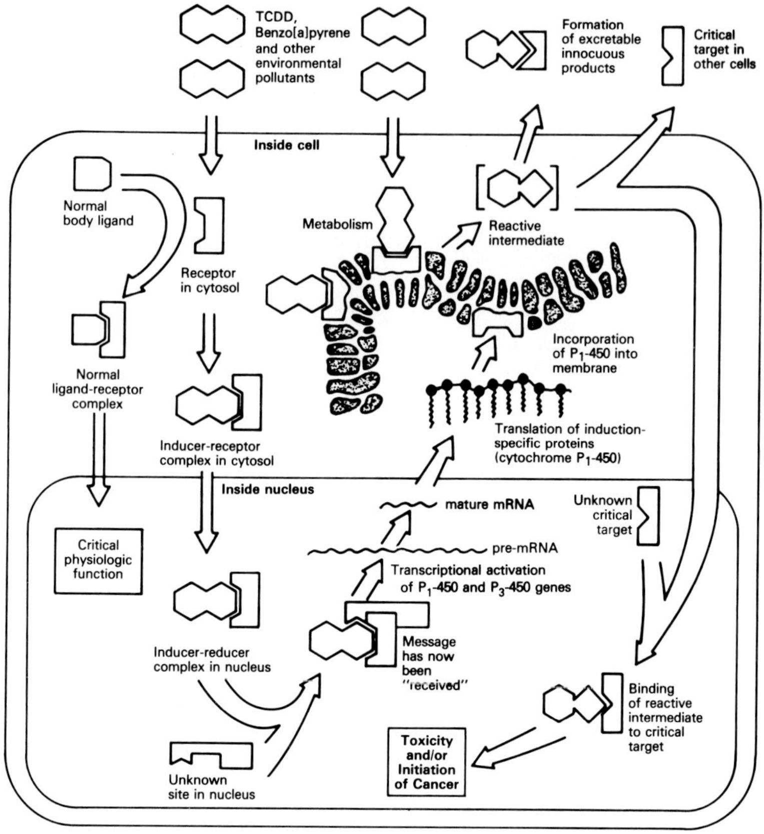

FIGURE 2. Subcellular diagram of the Ah locus. It is possible that the Ah receptor normally has an endogenous ligand and carries out a function critical to life; this could be the reason for dramatic TCDD-induced toxicity (17). If this is the case, the endogenous ligand remains unknown and the receptor would appear to have been appropriated by the P_1-450 induction process (3). The resultant induced P-450 proteins lead to a combination of detoxification and increased levels of carcinogenic and/or toxic reactive intermediates [Modified and reproduced with permission from Dr. W. Junk Publishers (18)].

affinity for DNA (3). C57BL/6N (B6) mice have a high-affinity receptor and therefore the induction process proceeds with ease. DBA/2N (D2) mice have a poor-affinity receptor; the D2 induction process proceeds negligibly with low doses of TCDD or any dose of 3-methylcholanthrene or benzo[a]pyrene. A large TCDD dose can overcome the D2 receptor defect, however, and the induction process proceeds at a rate similar to that found in B6 mice (3). The association of the inducer-receptor complex with nuclear chromatin components (19) is highly correlated with rapid transcriptional activation of at least two P-450 genes (20) and production of the P_1-450 and P_3-450 proteins (10). These two forms of polycyclic hydrocarbon-induced P-450 are most closely associated with induced benzo[a]pyrene and acetanilide metabolism, respectively (5). These forms of induced P-450 controlled by the Ah locus, especially P_1-450 (Fig. 2), are particularly relevant to certain types of polycyclic hydrocarbon-induced toxicity and tumorigenesis (4). A major regulatory gene governing increases in benzo[a]pyrene metabolism has been localized to the distal portion of mouse chromosome 17 (21), although evidence exists for a minimum of two regulatory loci (22). The P_1-450 and P_3-450 genes have been mapped to mouse chromosome 9 (13).

RESULTS AND DISCUSSION

Transcriptional Activation

Accumulation of P_1-450 mRNA and an intranuclear large-molecular-weight pre-mRNA is associated with TCDD- or 3-methylcholanthrene-induced benzo[a]pyrene metabolism (7). These data suggest that the induction process is regulated at the transcriptional level, although mRNA stabilization is not ruled out. Quantitative nuclear transcription assays, similar to those described by McKnight and Palmiter (23), were therefore performed (Fig. 3). Newly synthesized pre-mRNA transcripts were quantitated by filter hybridization to P_1-450 and P_3-450 specific cDNA probes. This assay measures elongation of the RNA chains initiated *in vivo* (25). Levels of specific transcripts detected therefore reflect the number of RNA polymerase molecules in the process of transcribing the

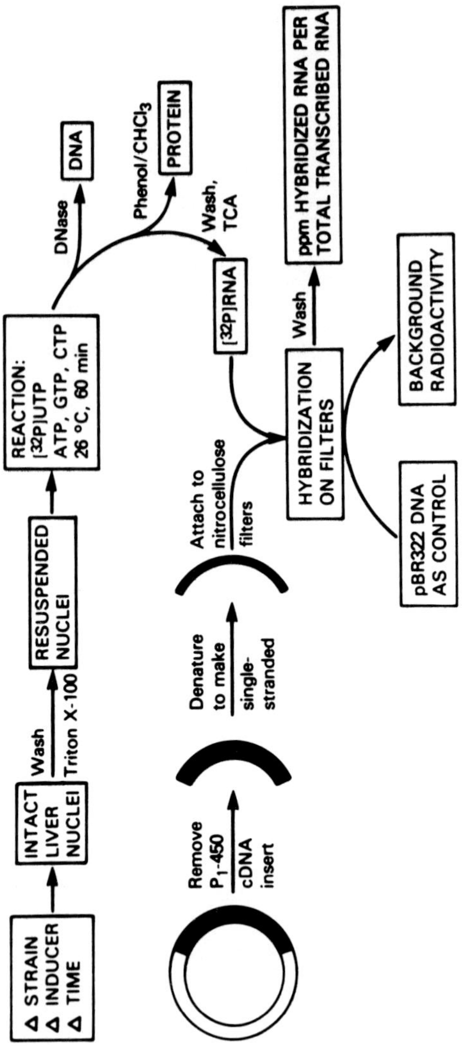

FIGURE 3. Modification (24) of the McKnight and Palmiter transcription assay (23). Following isolation of intact liver nuclei from various inbred mouse strains treated with a particular inducer for differing lengths of time, the activity of RNA polymerase molecules actively transcribing the P_1-450 and P_3-450 genes is measured. Background radioactivity of a pBR322 DNA-bound filter is subtracted from radioactivity of P_1-450 or P_3-450 cDNA-bound filters (20).

P_1-450 and P_3-450 genes. Using this method, we found direct evidence for transcriptional activation of both genes (20), strictly correlated with the genetic differences in intranuclear binding of the inducer-receptor complex in 3-methylcholanthrene- and TCDD-treated B6 and D2 mice.

Quantitation of mRNA Levels

Specific mRNA sequences were quantitated by a filter hybridization assay modified (26) from the original technique described by Spradling and coworkers (27). Poly(A)-containing RNA was partially base-hydrolyzed to lengths of 100-150 nt and labeled with polynucleotide kinase (Fig. 4). These shorter lengths of RNA guarantee

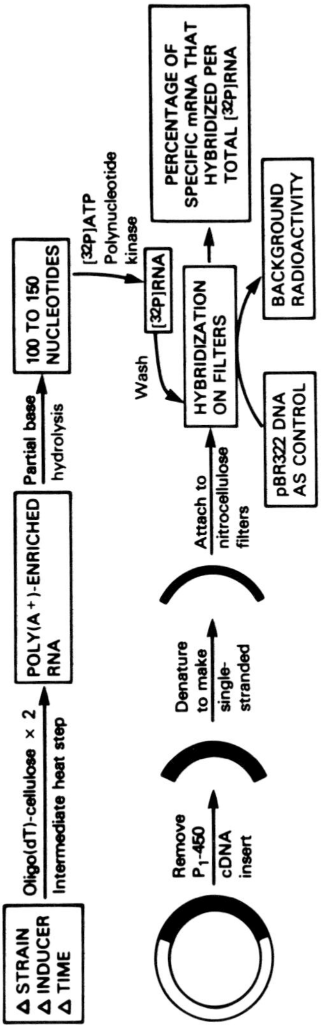

FIGURE 4. Quantition of mRNA levels. Following isolation of liver poly(A)-enriched RNA from various inbred mouse strains treated with a particular inducer for differing lengths of time, small pieces of RNA are end-labeled. Background radioactivity of a pBR322 DNA-bound filter is subtracted from radioactivity of P_1-450 or P_3-450 cDNA-bound filters (20).

a greater amount of radioactivity per unit of RNA. After separation of the RNA from free [^{32}P]ATP, the end-labeled RNA pieces were hybridized to filters containing the P_1-450 and P_3-450 cDNA probes.

Using this method, we found a rise in 3-methylcholanthrene-induced P_1-450 and P_3-450 mRNA levels lagging 3 to 6 h behind the increase in transcriptional rates (20). This lag time is similar to that observed for rat epoxide hydrolase, NADPH P-450 oxidoreductase, and cytochrome P-450b mRNAs following phenobarbital treatment (24). The basal levels of control P_3-450 mRNA were approximately five times greater than those of P_1-450 control mRNA, and the maximal level of 3-methylcholanthrene-induced P_3-450 mRNA was about five times greater than that of induced P_1-450 mRNA in B6 mice (20). Very similar transcriptional rates and quite different induced mRNA levels suggest that some form of posttranscriptional regulation (e.g. differences in pre-mRNA processing or mRNA stabilization) might occur with at least one of these two genes.

Increases in D2 P_1-450 or P_3-450 mRNA do not occur following 3-methylcholanthrene treatment but do occur after high doses of TCDD (20). Again, this result strongly correlates the transcriptional induction process to the availability of inducer-receptor complex having DNA-binding affinity.

Isolation of Full-Length cDNA Clones

A combination of polysome immunoadsorption (26) and use of the Okayama-Berg vector (28) was carried out for isolating full-length cDNA clones for both P_1-450 and P_3-450. From 3-methylcholanthrene-treated B6 mouse liver microsomes, a polyclonal antibody (against P_1-450 and P_3-450, known to have immuno-crossreactivity) was prepared in goat and purified on an antigen column (Fig. 5). The affinity-purified IgG, in combination with S.aureus ghosts, was then used to immunoadsorb P_1-450 and P_3-450 polysomes from 3-methylcholanthrene-treated B6 mouse liver (12). Following removal of protein and two oligo(dT) chromatography steps (Fig. 5), the poly(A) RNA was used to make full-length cDNA clones.

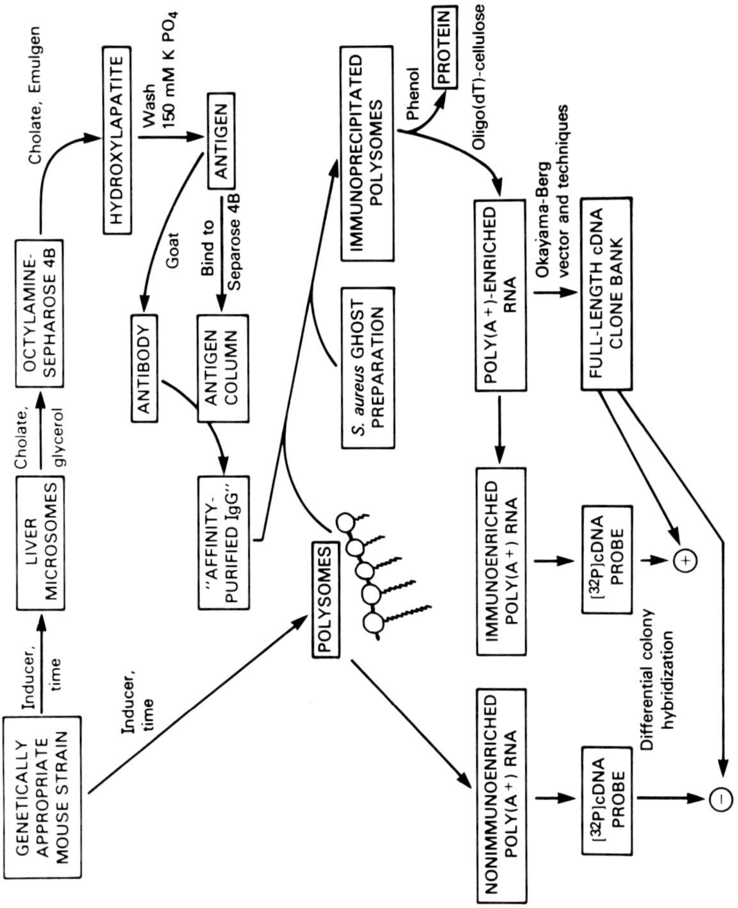

FIGURE 5. Summary of our experimental schemes for (i) development of a polyclonal antibody that crossreacts with P_1-450 and P_3-450, (ii) polysome immunoadsorption with the affinity-purified IgG, and (iii) differential colony hybridization of the full-length cDNA clone bank produced via the Okayama-Berg vector.

The Okayama-Berg vector (Fig. 6) is unique in that (a) it is self-priming and therefore has a low background of noninserted material and (b) S1 nuclease treatment is not used and thus full-length cDNA inserts are more likely to be achieved (28). The length of the first cDNA strand depends on how intact the mRNA is and on the fidelity of the reverse transcriptase, but lengths of 2000 or 3000 nt are easy to achieve by this technique. One shortcoming of this method (28) is that normally there is no size selection step for choosing the appropriate size of insert. To generate the second cDNA strand, RNase H

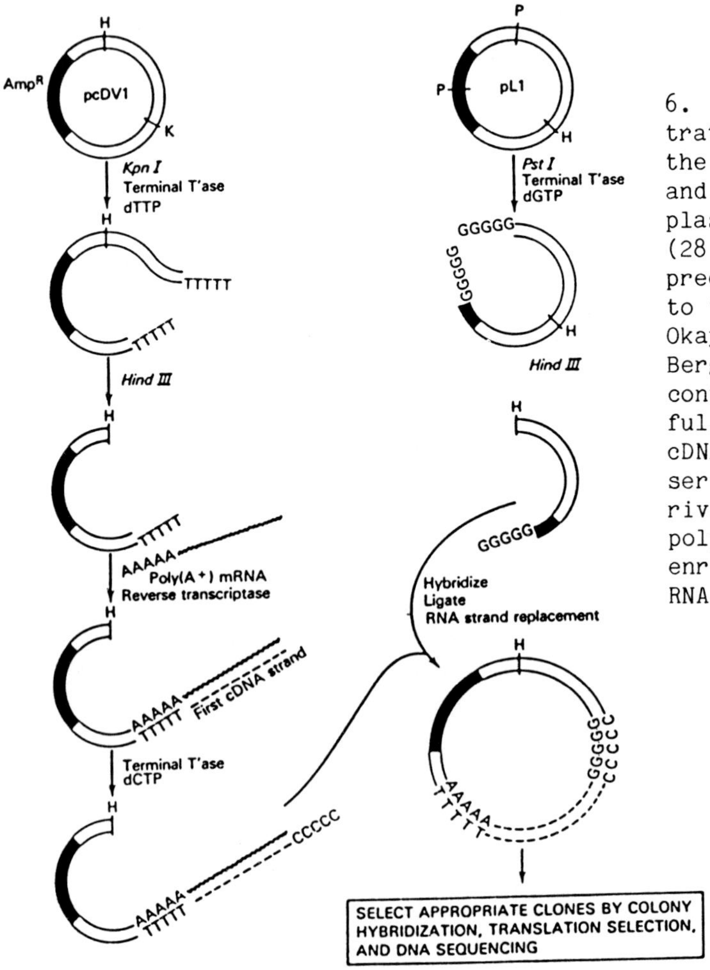

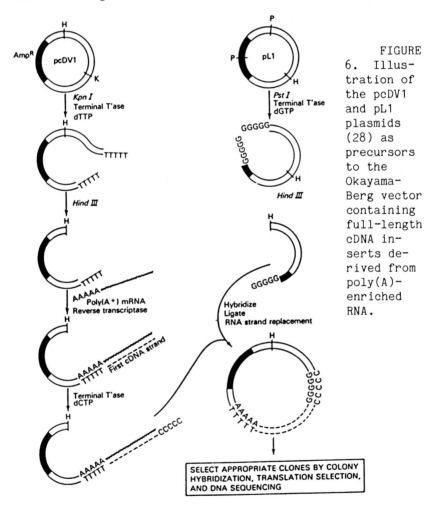

FIGURE 6. Illustration of the pcDV1 and pL1 plasmids (28) as precursors to the Okayama-Berg vector containing full-length cDNA inserts derived from poly(A)-enriched RNA.

is added (to nick RNA in an RNA-DNA hybrid) and then DNA polymerase and ligase are used to complete the strand and close it up (28).

A cDNA clone bank of about 1,000 colonies was screened with [^{32}P]cDNA probes synthesized from the immunoenriched and control poly(A) RNA (Fig. 5). Colonies that reacted with the immunoenriched probe, but not appreciably with the control probe, were selected for restriction enzyme analysis. By restriction mapping, plus Southern blot analyses with known P_1-450 (6) and P_3-450 (29) partial cDNA probes, five P_1-450 and 40 P_3-450 inserts were identified. All of the P_3-450 inserts were about 1900 bp, whereas three of the five P_1-450 inserts were about 2600 bp (12).

Isolation of Genomic Clones

B6 is the inbred mouse best characterized with regard to the _Ah_ locus and P_1-450 and P_3-450 inducibility (1, 3). B6 mice were therefore used for isolating these two genes. B6 liver DNA was ligated to λ Charon 30 arms and the phage was packaged _in vitro_ (12). Screening of this phage library with full-length P_1-450 and P_3-450 cDNA probes yielded 25 clones. A phage containing the entire P_1-450 gene and a phage containing the entire P_3-450 gene were confirmed by heteroduplex analyses with the corresponding full-length cDNA clones (12). Moreover, digestion of either phage with _BamHI_, _EcoRI_ or _HindIII_ produced Southern blot patterns identical to those from B6 liver genomic DNA digested with each of these three enzymes (12). Digestions with another dozen enzymes also have shown no differences between the phage DNA and B6 genomic DNA probed with the full-length cDNA clones (12), indicating that neither insert contains rearranged segments. In contrast, we have found that the P_1-450 gene from a mouse MOPC 41 plasmacytoma genomic library (8) is rearranged at the 5' end (manuscript in preparation).

Evolution of the P-450 Gene Family

Rat P-450e-like genes have nine exons (30-32) spanning about 14 kb (Fig. 7) and are probably located in a cluster (33) at or near the _Coh_ locus on mouse chromosome 7. At least eight genes (and/or pseudogenes) comprise this subfamily (31). Though most of these P-450e-like genes are phenobarbital-inducible, it is unclear if a receptor-mediated mechanism is involved in this induction process.

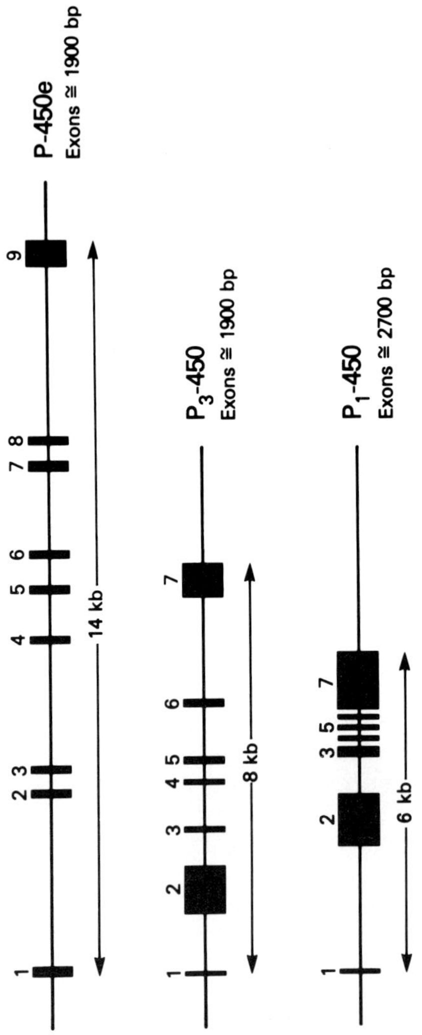

FIGURE 7. Exon-intron patterns of rat P-450e (30), a member of the phenobarbital-inducible subfamily, and mouse P_1-450 and P_3-450 (12), members of the Ah locus-associated P-450 gene subfamily.

Mouse P_1-450 and P_3-450 genes have strikingly similar exon-intron patterns with seven exons spanning 6 and 8 kb, respectively (Fig. 7). Both genes are controlled by the Ah receptor, map to mouse chromosome 9 (13), and possess 68% nucleotide and 73% protein homology (14, 15). Our sequencing data (15) are consistent with divergence of the Ah locus-associated P-450 subfamily from the P-450e-like subfamily more than 170 million years ago (MYA), followed by divergence of the two homologous P_1-450 and P_3-450 genes from each other about 60 MYA (Fig. 8).

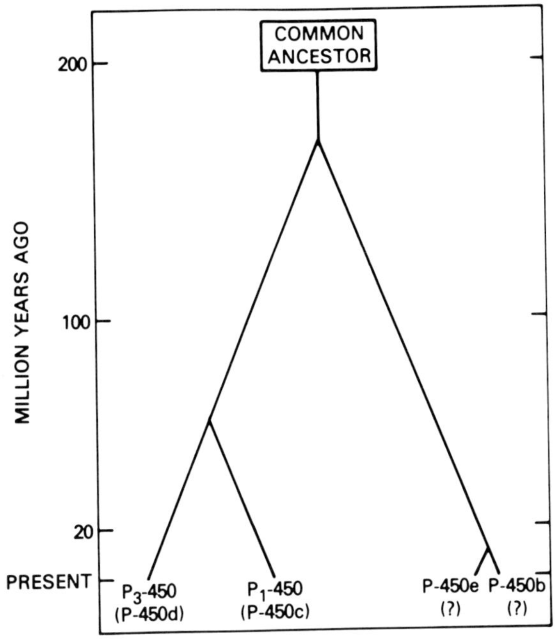

FIGURE 8. Our present knowledge of the P-450 gene family and its divergence from a common ancestral gene. Mouse P_3-450 and P_1-450 correspond to rat P-450d and P-450c, respectively, and rabbit form 6 and form 4, respectively (15). Rat P-450e and P-450b exhibit more than 95% homology (30, 32, 34), but it is unclear whether two different genes are being compared or whether the same gene in dissimilar rat strains in three different countries is being compared. A divergence of 1% every 2.4 million years has been estimated (15), based on rat and mouse P-450 protein comparisons. It should be emphasized that all of these evolution estimates are based on the assumption of linear divergence, and this possibility becomes increasingly unlikely as one proceeds more than 100 MYA back in time.

The general consensus for the rabbit-rodent and the rat-mouse species separations is about 60 and 17 MYA, respectively (36-38). A divergence of at least 170 MYA for the Ah-locus-associated P-450 gene subfamily from the phenobarbital-inducible P-450 gene subfamily would explain why rabbit, rat and mouse exhibit both subfamilies. A division of at least 60 MYA for the two major Ah-locus-associated P-450 genes would explain why rabbit, rat and mouse exhibit these two major homologous 3-methylcholanthrene-inducible genes. Any divergence of genes between 17 and 60 MYA would result in a fundamental difference between rabbits and rodents in genes of either subfamily. Any separation of genes less than 17 MYA would result in basic differences in either subfamily among rabbits, rats and mice. The rat P-450b and P-450e genes are claimed (30, 32, 34) to be between 95% and greater than 99% homologous. If the rat P-450b and P-450e genes have diverged less than 12 MYA, therefore, it is likely that neither rabbit nor mouse will exhibit an equivalent pattern for these recently diverged "P-450e-like genes" in the rat.

Developmental and Tissue-Specific Differences

In cell culture (Table 1), mouse hepatoma Hepa-1 cells exhibit P_1-450 but negligible P_3-450 mRNA inducibility by TCDD, whereas both P_1-450 and P_3-450 mRNA are equally inducible by TCDD in human breast carcinoma MCF-7 cells (39). In B6 mouse liver, both P_1-450 and P_3-450 mRNA are inducible by 3-methylcholanthrene, whereas in B6 kidney P_1-450 but not P_3-450 mRNA is induced by 3-methylcholanthrene (9). P_1-450 mRNA inducibility by polycyclic aromatic compounds occurs developmentally several weeks earlier than P_3-450 inducibility (9); this finding via Northern blot analysis confirms earlier developmental studies with catalytic activities (40, 41) and sister chromatid exchange in cultured mouse embryo (42).

Because the P_1-450 and P_3-450 genes map to the same region of mouse chromosome 9 and are controlled by the Ah receptor, we expect to find these two genes in tandem. The molecular basis of these interesting tissue-specific and developmental differences in gene expression (Table 1) await further experimentation.

TABLE 1

DIFFERENCES IN P_1-450 AND P_3-450 mRNA INDUCIBILITY AS A FUNCTION OF DIFFERENT CELL OR TISSUE TYPES AND DEVELOPMENT

	mRNA Inducibility	
	P_1-450	P_3-450
Mouse Hepa-1 (liver)	+	0
Human MCF-7 (breast)	+	+
Mouse liver	+	+
Mouse kidney	+	0
Developmental expression in mouse liver	early	late

How Many P-450 Subfamilies?

At the level of Northern blot analysis, the 3-methylcholanthrene-induced P_1-450 and P_3-450 mRNAs appear homologous to each other (29) but not to the phenobarbital-induced (7) or pregnenolone 16α-carbonitrile-induced P-450 mRNAs (25). Likewise, the pregnenolone 16α-carbonitrile-induced P-450 mRNAs are not homologous to the phenobarbital-induced P-450 mRNAs (24). These data suggest at least three subfamilies of P-450 genes. Clofibrate (a peroxisome proliferator used in the treatment of hyperlipidemia) dramatically induces at least one form of P-450; this may represent a fourth subfamily (43). We believe that ethanol-induced P-450 may comprise yet another group. If each of these subfamilies contains between two and ten genes (Table 2), the total number of P-450 genes would reside in the 15-to-50 range. Whether the constitutive noninducible P-450 genes will be members of several of these subfamilies, or will comprise separate distinct subfamilies in some instances, is not yet known.

P-450 Diversity

In the face of chemical adversity, it is possible that there exist both acute and chronic mechanisms for diversification of P-450 genes. We define "acute changes" as those occurring in a matter of minutes or hours following chemical exposure. Such a mechanism would allow

TABLE 2

SIZE OF P-450 GENE FAMILY

P-450 Subfamily	Exons	Mouse chromosomal localization	Number of genes and/or pseudogenes
Polycyclic aromatic compounds	7	9	≧2
Phenobarbital	9	7	≧8
*Pregnenolone 16α-carbonitrile	?	?	≧2
Troleandomycin			
Erythromycin			
Clofibrate	?	?	?
Ethanol	?	?	?
Others?	?	?	?

*Troleandomycin and erythromycin appear to induce in rat liver the same P-450 mRNAs as pregnenolone 16α-carbonitrile does [Phil Guzelian, personal communication].

for several (or many) diverse proteins derived from a small amount of DNA. Although this idea would seem attractive (18), there presently is no experimental evidence to support this hypothesis.

We define "chronic changes" as those that have evolved over millions of years. P_1-450 and P_3-450 are most likely products of gene duplication (Fig. 7), though highly conserved regions interspersed with short non-homologous segments provide the proper "target" for gene conversion and/or unequal crossing-over (15). Gene conversion has also been postulated (32, 34) for the rat P-450e-like family, but these cDNA and protein comparisons have involved several outbred strains of rats in Tokyo, New York, Quebec, and New Jersey. The strength of our P_1-450 and P_3-450 data resides in the fact that we have compared two complete cDNA sequences derived from the same 3-methylcholanthrene-treated B6 inbred mouse cDNA clone bank (12). As the sequencing data grow, from numerous laboratories around the world and from comparisons among several species of laboratory animals, our knowledge about the mechanisms of diversification of the P-450 gene family will become more clear.

ACKNOWLEDGMENT

The expert secretarial assistance of Ingrid E. Jordan is greatly appreciated.

REFERENCES

1. Nebert DW, Tukey RH, Eisen HJ, Negishi M (1983). Cloned cytochrome P-450 genes regulated by the Ah receptor. In Hamer D, Rosenberg M (eds): "Gene Expression: UCLA Symposia on Molecular and Cellular Biology, New Series Volume 8," New York: Alan R. Liss, Inc., p 187.
2. Nebert DW, Chen Y-T, Negishi M, Tukey RH (1983). Cloning genes that encode drug-metabolizing enzymes. Developmental pharmacology and teratology. In MacLeod SM, Okey AB, Spielberg SP (eds): "Developmental Pharmacology," New York: Alan R. Liss, Inc., p 51.
3. Eisen HJ, Hannah RR, Legraverend C, Okey AB, Nebert DW (1983). The Ah receptor: Controlling factor in the induction of drug-metabolizing enzymes by certain chemical carcinogens and other environmental pollutants. In Litwack G (ed): "Biochemical Actions of Hormones," New York: Academic Press, Vol X, p 227.
4. Nebert DW (1981). Genetic differences in susceptibility to chemically induced myelotoxicity and leukemia. Environ Health Perspect 39:11.
5. Negishi M, Nebert DW (1979). Structural gene products of the Ah locus. Genetic and immunochemical evidence for two forms of mouse liver cytochrome P-450 induced by 3-methylcholanthrene. J Biol Chem 254:11015.

6. Negishi M, Swan DC, Enquist LW, Nebert DW (1981). Isolation and characterization of a cloned DNA sequence associated with the murine Ah locus and a 3-methylcholanthrene-induced form of cytochrome P-450. Proc Natl Acad Sci USA 78:800.
7. Tukey RH, Nebert DW, Negishi M (1981). Structural gene product of the Ah complex. Evidence for transcriptional control of cytochrome P_1-450 induction by use of a cloned DNA sequence. J Biol Chem 256:6969.
8. Nakamura M, Negishi M, Altieri M, Chen Y-T, Ikeda T, Tukey RH, Nebert DW (1983). Structure of the mouse cytochrome P_1-450 genomic gene. Eur J Biochem 134:19.
9. Ikeda T, Altieri M, Chen Y-T, Nakamura M, Tukey RH, Nebert DW, Negishi M (1983). Characterization of P_2-450 (20 S) mRNA. Association with the P_1-450 genomic gene and differential response to the inducers 3-methylcholanthrene and isosafrole. Eur J Biochem 134:13.
10. Negishi M, Jensen NM, Garcia GS, Nebert DW (1981). Structural gene products of the murine Ah locus. Differences in ontogenesis, membrane location, and glucosamine incorporation between liver microsomal cytochromes P_1-450 and P-448 induced by polycyclic aromatic compounds. Eur J Biochem 115:585.
11. Negishi M, Nebert DW (1981). Structural gene products of the Ah complex. Increases in large mRNA from mouse liver associated with cytochrome P_1-450 induction by 3-methylcholanthrene. J Biol Chem 256:3085.
12. Gonzalez FJ, Mackenzie PI, Kimura S, Nebert DW (1984). Isolation and characterization of full-length cDNA clones and genomic clones of mouse 3-methylcholanthrene-inducible cytochrome P_1-450 and P_3-450. Gene, in press.
13. Tukey RH, Lalley PA, Nebert DW (1984). Localization of cytochrome P_1-450 and P_3-450 genes to mouse chromosome 9. Proc Natl Acad Sci USA 81:3163.
14. Kimura S, Gonzalez FJ, Nebert DW (1984). Mouse cytochrome P_3-450: Complete cDNA and amino acid sequence. Nucl Acids Res 12:2917.
15. Kimura S, Gonzalez FJ, Nebert DW (1984). The murine Ah locus. Comparison of the complete cytochrome P_1-450 and P_3-450 cDNA nucleotide and amino acid sequences. J Biol Chem, in press.

16. Chen Y-T, Tukey RH, Swan DC, Negishi M, Nebert DW (1983). Characterization of the human P_1-450 genomic gene. Pediat Res 17:208A [Abstract].
17. Poland A, Knutson JC (1982). 2,3,7,8-Tetrachloro-dibenzo-p-dioxin and related halogenated aromatic hydrocarbons: Examination of the mechanism of toxicity. Annu Rev Pharmacol Toxicol 22:517.
18. Nebert DW (1979). Multiple forms of inducible drug-metabolizing enzymes. A reasonable mechanism by which any organism can cope with adversity. Mol Cell Biochem 27:27.
19. Tukey RH, Hannah RR, Negishi M, Nebert DW, Eisen HJ (1982). The Ah locus. Correlation of intranuclear appearance of inducer-receptor complex with induction of cytochrome P_1-450 mRNA. Cell 31:275.
20. Gonzalez FJ, Tukey RH, Nebert DW (1984). Structural gene products of the Ah locus. Transcriptional regulation of cytochrome P_1-450 and P_3-450 mRNA levels by 3-methylcholanthrene. Mol Pharmacol, in press.
21. Legraverend C, Kärenlampi SO, Bigelow SW, Lalley PA, Kozak CA, Womack JE, Nebert DW (1984). Aryl hydrocarbon hydroxylase induction by benzo[a]-anthracene: Regulatory gene localized to the distal portion of mouse chromosome 17. Genetics, in press.
22. Robinson JR, Considine N, Nebert DW (1974). Genetic expression of aryl hydrocarbon hydroxylase induction. Evidence for the involvement of other genetic loci. J Biol Chem 249:5851.
23. McKnight GS, Palmiter RD (1979). Transcriptional regulation of the ovalbumin and conalbumin genes by steroid hormones in chick oviduct. J Biol Chem 254:9050.
24. Hardwick J, Gonzalez FJ, Kasper CB (1983). Transcriptional regulation of rat liver epoxide hydratase, NADPH-cytochrome P-450 oxidoreductase, and cytochrome P-450b genes by phenobarbital. J Biol Chem 258:8081.
25. Evans RM, Ziff EB (1978). Coincidence of the promoter and capped 5' terminus of RNA from the adenovirus 2 major late transcription unit. Cell 15:1463.

26. Gonzalez FJ, Kasper CB (1982). Cloning of DNA complementary to rat liver NADPH-cytochrome c (P-450). Oxidoreductase and cytochrome P-450b mRNAs: Evidence that phenobarbital augments transcription of specific genes. J Biol Chem 257:5962.
27. Spradling AC, Digan ME, Mahowald AP, Scott M, Craig EA (1980). Two clusters of genes for major chorion proteins of Drosophila melanogaster. Cell 19:905.
28. Okayama H, Berg P (1983). A cDNA cloning vector that permits expression of cDNA inserts in mammalian cells. Mol Cell Biol 3:280.
29. Tukey RH, Nebert DW (1984). Regulation of mouse cytochrome P_3-450 by the Ah receptor. Studies with a P_3-450 cDNA clone. Biochemistry, in press.
30. Mizukami Y, Sogawa K, Suwa Y, Muramatsu M, Fujii-Kuriyama Y (1983). Gene structure of a phenobarbital-inducible cytochrome P-450 in rat liver. Proc Natl Acad Sci USA 80:3958.
31. Kumar A, Raphael C, Adesnik M (1983). Cloned cytochrome P-450 cDNA. Nucleotide sequence and homology to multiple phenobarbital-induced mRNA species. J Biol Chem 258:11280.
32. Atchison M, Adesnik M (1983). A cytochrome P-450 multigene family. Characterization of a gene activated by phenobarbital administration. J Biol Chem 258:11285.
33. Simmons DL, Kasper CB (1983). Genetic polymorphisms for a phenobarbital-inducible cytochrome P-450 map to the Coh locus in mice. J Biol Chem 258:9585.
34. Affolter M, Anderson A (1984). Segmental homologies in the coding and 3' non-coding sequence of rat liver cytochrome P-450e and P-450b cDNAs and cytochrome P-450e-like genes. Biochem Biophys Res Commun 118:655.
35. Leighton JK, DeBrunner-Vossbrinck BA, Kemper B (1984). Isolation and sequence analysis of three cloned cDNAs for rabbit liver proteins that are related to rabbit cytochrome P-450 (form 2), the major phenobarbital-inducible form. Biochemistry 23:204.
36. Fitch WM, Langley CH (1976). Protein evolution and the molecular clock. Fed Proc 35:2092.

37. Wilson AC, Carlson SS, White TJ (1977). Biochemical evolution. Annu Rev Biochem 46:573.
38. Miyata T, Hayashida H, Kikuno R, Hasegawa M, Kobayashi M, Koike K (1982). Molecular clock of silent substitution: At least six-fold preponderance of silent changes in mitochondrial genes over those in nuclear genes. J Mol Evol 19:28.
39. Jaiswal AK, Eisen HJ, Chen Y-T, Towne DW, Nebert DW (1984). Induction of polycyclic aromatic compounds of cytochrome P-450-mediated monooxygenase activities in human breast carcinoma cell lines. Mol Pharmacol, in press.
40. Guenthner TM, Nebert DW (1978). Evidence in rat and mouse liver for temporal control of two forms of cytochrome P-450 inducible by 2,3,7,8-tetrachloro dibenzo-p-dioxin. Eur J Biochem 91:449.
41. Filler R, Lew KJ (1981). Developmental onset of mixed-function oxidase activity in preimplantation mouse embryos. Proc Natl Acad Sci USA 78:6991.
42. Galloway SM, Perry PE, Meneses J, Nebert DW, Pedersen RA (1980). Cultured mouse embryos metabolize benzo[a]pyrene during early gestation: Genetic differences detectable by sister chromatid exchange. Proc Natl Acad Sci USA 77:3524.
43. Stupans I, Ikeda T, Kessler DJ, Nebert DW (1984). Characterization of a cDNA clone for mouse phenobarbital-inducible cytochrome P-450b. DNA 3:129.

Molecular Biology of Development, pages 331–349

EXPRESSION OF CRYSTALLIN GENE FAMILIES IN THE DIFFERENTIATING EYE LENS

J. Piatigorsky, A. B. Chepelinsky, J. F. Hejtmancik,[1] T. Borrás, G. C. Das, J. W. Hawkins, P. S. Zelenka, C. R. King,[2] D. C. Beebe and J. M. Nickerson

Laboratory of Molecular and Developmental Biology
National Eye Institute, NIH
Bethesda, Maryland 20205

Department of Anatomy, Uniformed
Services University for the Health Sciences
Bethesda, Maryland 20814

ABSTRACT Developmental features of the vertebrate eye lens are described. Lens differentiation involves extensive cell elongation and crystallin gene expression; both can be studied in culture. There are four crystallin gene families (α, β, γ and δ). Each has a characteristic organization of introns and displays different temporal and spatial patterns of expression during lens development. The members within a crystallin gene family (especially β-crystallin) are also differentially regulated in the embryonic lens. Structural and functional studies suggest an unequal promoter strength for the two linked chicken δ-crystallin genes. An expression system using primary explants of embryonic chicken lens epithelia has been developed for studying crystallin gene promoters. Initial experiments indicate that the upstream flanking sequences of the murine αA-crystallin gene can drive the bacterial

1. Present address: Howard Hughes Medical Institute
Houston, Texas 77030
2. Present address: Laboratory of Cellular and Molecular Biology, National Cancer Institute, NIH,
Bethesda, Maryland 20205

CAT gene in the cultured lens cells. Deletion mutants suggest that sequences between 85 and 400 base pairs upstream from the putative cap site are important for expression of the αA-crystallin gene.

INTRODUCTION

The transparent ocular lens of vertebrates is surrounded by a collagenous capsule and is composed of cuboidal, anterior epithelial cells and elongated, posterior fiber cells. Lens induction occurs early in development by the optic vesicle, which promotes invagination of the surface ectoderm to form the lens vesicle. The cells at the posterior of the lens vesicle cease dividing and elongate. The embryonic lens grows by deposition of secondary fiber cells at the equator; the new fiber cells are derived from the anterior epithelial cells. The cell nuclei in the central region of the lens fibers disintegrate, leaving these cells anucleated throughout life. Numerous gap junctions couple the fiber cells. The morphology of the developing lens is shown diagrammatically in Fig. 1A,B. Reviews can be found elsewhere (1,2).

Chicken lens epithelial cells can differentiate in culture under a variety of conditions, providing opportunities for investigating lens cell development (Fig. 1C). Explanted lens epithelia from 6 day-old embryos will form fiber cells when cultured in medium supplemented with fetal calf serum (3,4), insulin (5) or lentropin (6). Lentropin is an extract containing one or more proteins found in the vitreous humor of the eye (behind the lens) which may normally stimulate fiber cell differentiation in vivo. Prolonged cultivation (2-4 weeks) of these explants in fetal calf serum leads to extensive fiber cell differentiation (7). Cells in the explanted lens epithelia divide rather than differentiate if taken from 19 day-old chicken embryos (8) or if confronted with fetal calf serum after preculture for one or more days in serum-free medium (9,10). Finally, groups of cells from confluent cultures of dissociated lens epithelial cells can form lentoid bodies which contain

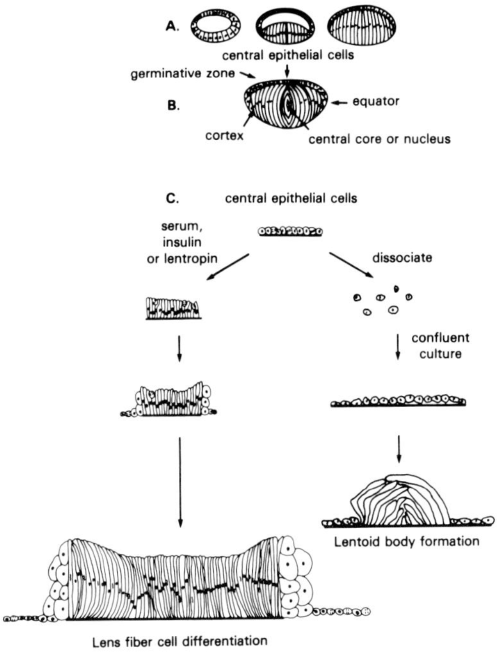

Fig. 1. Schematic illustration of lens development. (A) Early in development, the lens appears as a vesicle with anterior epithelial cells and posterior primary fiber cells. The cells at the posterior region elongate further into the hollow of the vesicle and differentiate into fiber cells. The anterior cells of the lens vesicle remain epithelial and mitotically active. They contribute new cells at the equator, which differentiate into secondary fiber cells. (B) Later in development, mitosis is restricted to the germinative zone that encircles the central epithelial cells. This process of fiber cell formation continues throughout life.(C) Central lens epithelia can differentiate into lens fibers when explanted intact (left) or after dissociation and growth in culture (right). See text.

many features of lens fibers (11). Interestingly, non-lens cells (i.e., embryonic retina, limb or brain cells) may also transdifferentiate into lens-like cells.

Approximately 90% of the soluble protein of the lens are crystallins (13,14). The accumulation of these structural proteins are a characteristic feature of lens differentiation. There are four major antigenically different classes of crystallins (α, β, γ and δ). δ-Crystallin replaces γ-crystallin in lenses of birds and reptiles and is confined to these two vertebrate classes (15). The crystallins are highly conserved, slowly evolving proteins (16). Sequence (17, 18) and structural (19-21) analyses have shown that the β- and γ-crystallins form an evolutionarily related superfamily of proteins. By contrast, the α-and δ-crystallins do not appear to be related to each other or to the β/γ group of crystallins.

Each crystallin class comprises a number of related polypeptides which, except for the γ-crystallins, associate to form polymeric native proteins. Crystallin synthesis is regulated temporally and spatially within the developing lens, making these proteins useful markers for studying differential gene expression during cellular differentiation (2, 22, 23). It is interesting that the pattern of crystallin synthesis differs among species of different vertebrate classes (2, 24). The reasons for the complex developmental pattern of crystallin synthesis are not known but are presumably important for optimizing lens transparency.

CRYSTALLIN GENES

The structures and organization of the crystallin genes are becoming known rapidly (25). Each crystallin gene family has a characteristic organization of introns and exons (Fig. 2). The β and γ-crystallin genes are interesting in that their organization has been directly related to the tertiary structure of their proteins. Each of the four predicted structural motifs of the β-crystallin polypeptide is encoded in a separate exon (21). The γ-crystallin genes retain the central intron which separates the two domains of the protein (19, 20), but have lost the introns which separate the exons encoding the two structural motifs (folding units) within each domain (26, 27). A small intron has been inserted within the first exon of the γ-crystallin gene. All the β- and γ-crystallin genes examined have similar structures. The γ-crystallins form a linked gene family with some six to seven members while the β-crystallins appear to be a dispersed gene family with at least six members (28, 29, 30).

The structure of the αA-crystallin gene is interestingly related to its expression. In four rodents (mouse, rat, squirrel and hamster) there are two αA-crystallin polypeptides (31); one of them (αA^{ins}) has a 22 (rat) or 23 (mouse) amino acid peptide inserted between residue 63 and 64 of the other (αA_2). The αA_2 and αA^{ins} polypeptides are encoded in the same gene; the difference between their mRNAs may be accounted for by the alternative splicing of a small exon located within the first intron (32). S1 protection experiments using an αA^{ins} cDNA probe showed that the insert exon is present in only 10-20% of the mRNAs and its inclusion in the mRNA is not under developmental control (33).

The δ-crystallin genes are a linked family with two members, approximately 4.2 Kb apart, which are arranged with a similar transcriptional polarity (34, 35). In contrast to the other crystallin genes, the δ-crystallin genes are highly interrupted (36, 37), each containing at least 17 introns (35). The two δ-crystallin genes have a similar structure and cross-hybridize extensively (38).

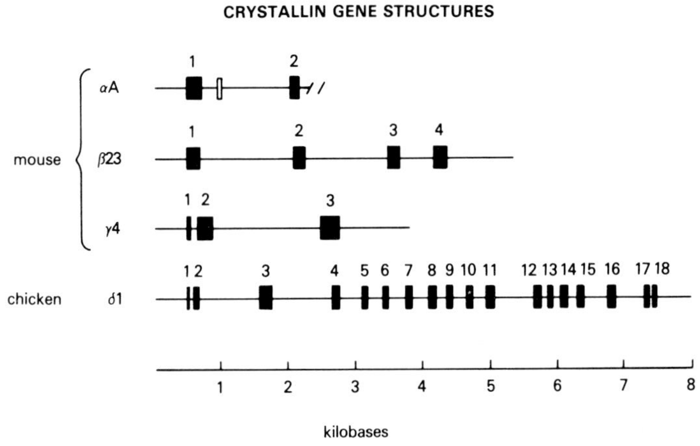

Fig. 2. Schematic representation of one member from each crystallin gene family. αA from (32); β23 from (21); γ4 from (27); δ from (35). Exons are numbered; introns are straight lines. The open box in αA is the alternatively spliced exon encoding the insert peptide in αA^{ins}.

DEVELOPMENTAL EXPRESSION OF THE CRYSTALLIN GENES

Our laboratory has been studying crystallin synthesis in the chicken lens in view of the developmental advantages of using that organism. δ-Crystallin is the first crystallin to appear during development; both the protein (22) and its mRNA (39) are present before the presumptive lens has formed a vesicle and accumulate during development (2, 40). δ-Crystallin synthesis ceases and δ-crystallin mRNA disappears from the lens between three and five months after hatching (41). By contrast, β-crystallins become the predominent protein of the post-hatched chicken lens (see 2). Except for differential synthesis of the β35 polypeptide in elongating lens cells of the embryo (42), there is virtually no information concerning the regulated expression of the individual β-crystallin polypeptides during development.

We have examined the levels δ- and β-crystallin mRNAs in the embryonic and post-hatched chicken lens by dot- and Northern-blot hybridization using cloned cDNA probes (30). Except for one, the cloned β-crystallin cDNAs hybridized specifically to individual mRNAs. The β19/26 cDNA, however, hybridized equally to mRNAs which synthesized polypeptides with molecular weights near 19,000 and 26,000 when translated in vitro. The amounts of δ, β19/26, β23, β25, and β35 mRNA were determined in different regions (central epithelial, equatorial epithelial and fiber cells; see Fib 1B) of the embryonic lens and at different stages of development and maturation. In brief, the δ- and each β-crystallin mRNA displayed a characteristic temporal and spatial pattern and reached a different level in the developing lens. δ-Crystallin mRNA was present in the highest relative frequency in the embryonic lens and the β-crystallin mRNAs were present in the highest relative amounts in the 3 month-old lens. The relative increase in frequency between the 17 day-old embryonic and 3 month-old post-hatched lens was 2-, 4-, 10-and 20-fold for the β23, β25, β35 and β19/26 mRNAs, respectively. The developmental program for expression of the δ-and different β-crystallin mRNAs appears to be preserved, at least in part, during fiber cell differentiation in cultured embryonic lens epithelia. These results indicate that the quantity of each of these mRNAs is controlled independently during development. We do not know whether the amounts of the crystallin mRNAs are regulated at the transcriptional or post-transcriptional level, or both.

PROMOTER REGION OF THE CHICKEN δ1 AND δ2-CRYSTALLIN GENES

The δ-crystallin locus contains two tandemly arranged genes (34, 35) which are so similar (38) that our hybridization studies with cDNA have not been able to distinguish between the mRNAs derived from them. Hybridization studies with gene-specific intron probes have suggested that both δ-crystallin genes are transcribed (43). Considerable evidence, however, suggests that δ1 (located 5' to δ2) is more active than δ2. First, all δ-crystallin cDNAs identified so far are derived from δ1 (44). Moreover, the sequence of this cDNA encodes a putative δ-crystallin polypeptide with a molecular weight near 48,000 (44) which is the size of the principal δ-crystallin polypeptide in the lens (see 15). Microinjection into cultured murine lens cells also indicates that the chicken δ-crystallin gene encodes the predominant δ-crystallin polypeptide (45). The protein product of the δ2-crystallin gene has not been established (see note in ref. 15). It is also not known whether the δ1 and δ2 genes are co-expressed in development.

We have compared the 5' regions of the two δ-crystallin genes. Both genes contain a similar Goldberg-Hogness box structure (TAAAAG) bounded by GC-rich regions. The Goldberg-Hogness box is present at position -28 in the δ1 gene, as judged by S1 mapping (44), and at position -27 in the δ2 gene, as derived from the homology of the first exon of this gene with that of the δ1 gene. Two interesting features were observed which might be factors in the putative differences in expression of δ1 and δ2. The δ1 gene has a consensus CCAAT sequence at position -70 which has not been found in the δ2 gene. In addition, sequences similar to the reverse complement of the viral consensus core enhancer sequence (TGG$^{AAA}_{TTT}$G) (46) are present at positions -298 and -256 in the 5' flanking region of the δ1 gene. These sequences were not detected in the 5'flanking region of the δ2 gene. A viral consensus core-enhancer-like sequence is also present in the second intron at position +352 of the δ1 gene; no counterpart has been detected within this region of the δ2 gene. The CCAAT box (47,48) and core enhancer-like sequences (49) may have significant roles for promoting gene activity in eukaryotic cells. It should be noted, however, that alterations in the canonical CCAAT box sequence did not reduce transcriptional efficiency of the herpes simplex virus tk gene injected into Xenopus oocytes (50), and numerous eukaryotic genes do not contain a CCAAT sequence (i.e., vitellogenin, metallothionein, insulin, ovalbumin, heat shock protein and silk fibroin).

A particularly unexpected finding was that the δ-crystallin genes contain a marked homology with the pentameric repeats (GAGCT, GGGGT, GGGCT) of the immunoglobulin heavy-chain switch region. This is shown for the δ1 gene in Fig. 3. The homology begins in the middle of exon 1 at position +17 and includes most of intron 1. In the δ2 gene the homology includes most of intron 1 (positions +53 to +137) but not the first exon. The homology is greatest with the murine I_gG switch region (51) in the δ1 gene and I_gA (52) switch region in the δ2 gene. This homology was found in a search of the entire Genbank TM database which, at the time of the search, contained about 1.4 million bases in 1561 loci.

```
           +17                      Exon | Intron
δ1    ACCAGCCAGGGCTGAGCTGCGGAGACGGTGAGCAGGGCTGTGCGGGCACTGGGGAGGCTC
                :::::::::       ::::::: : :::: :        :::::: :
IgG   GGTGAGCTGAGCTGAGCTGG------GGTGAGCTGAGCTGAG------CTGGGGTGA---
        50        60             70        80               90

      TGTGCTGCTGTGGGG--------------CTGGGCAGAGCTGAGCTGAGCCAAACAGAGC
        ::::   :::::               :::::  ::::::::::::::      ::::
      ---GCTGAGCTGGGGTGAGCTGAGCTGAGCTGGGGTGAGCTGAGCTGAGCTGGGGTGAGC
              100       110       120       130       140

      TGAACTGAGCCTCGCCTCGCTGTTCCGCAGGT
      ::: ::::::    :     ::::
      TGAGCTGAGCTGGGGTGAGCTGAGCTGAGCTG
      150       160       170       180
```

Fig. 3. Homology between the chicken δ1 gene and the murine I_gG heavy-chain switch region.

Another interesting finding was that the first exons (34 base pairs) of δ1 and δ2 are much more similar to each other than are the second exons (63 base pairs) of these genes. Exon 1 does not encode protein, the coding sequence begins in exon 2. The promoter regions of the δ1 and δ2 genes were subcloned in pBR322 and tested in vitro by analysis of "run-off" transcripts using a Hela cell extract (53). The flanking sequences from both genes initiated transcription. There was approximately 0.7 Kb of 5' flanking DNA in the δ1 gene fragment and about 2.0 Kb of 5' flanking DNA in the δ2 gene fragment being tested. Preliminary experiments examining the size of the transcripts with truncated DNA templates suggest that the expected initiation sites are being utilized. Of particular interest, the experiments indicate that the δ1 DNA is transcribed in vitro approximately four times more efficiently than the δ2 DNA.

PROMOTER REGION OF THE MURINE αA-CRYSTALLIN GENE

Like δ-crystallin in the chicken, α-crystallin is the first crystallin expressed during lens development in the mouse (24). δ-Crystallin is not present in the murine lens (15). We have begun to investigate the murine αA-crystallin gene promoter region, since the 5' region of this gene has been isolated and sequenced (32).

The nucleotide sequence of the 5' region of the murine αA-crystallin gene shows differences from that of the δ-crystallin genes. The putative Goldberg-Hogness box is TATATAG rather than TAAAAG. At position -73 of the αA-crystallin gene there is a CAT sequence instead of a CCAAT box as in the comparable region of the δ1-crystallin gene. In contrast to the chicken δ1 gene, the murine αA gene does not have obvious viral core enhancer-like sequences in the 5' flanking region. A consensus core enhancer-like sequence is present within the αA-crystallin gene in exon 1 and in the insert exon, however.

We have initiated functional studies on the 5' flanking sequences of the murine αA-crystallin gene. The two questions of immediate interest concern the identification of sequences controlling transcription and the delineation of sequences responsible for tissue-specific expression of this gene. We have developed a culture system consisting of primary explants of 14 day-old embryonic chicken lens epithelia for studying the expression of cloned crystallin gene promoters (Fig. 4). The primary lens explants were cultured in collagen-coated tissue culture dishes containing a growth medium supplemented with six growth factors (54). These cells formed epithelial-like outgrowths and maintained the ability to synthesize crystallins.

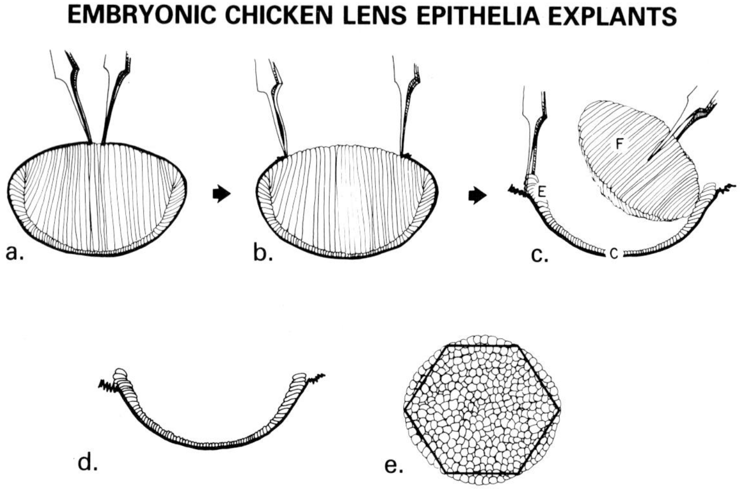

Fig. 4. Preparation of primary embryonic chicken lens epithelia for transfection. Fourteen day-old embryos were used (a-c). Tearing posterior side of lens capsule and removing fiber cells (C, central epithelial cells; E, equatorial epithelial cells; F, fiber cells). (d) One-dimensional view of lens epithelia. (e) Attachment of lens epithelia to dish by cutting with scalpel (dark lines).

Our initial experiments have focused on the 5' flanking region of the murine αA-crystallin gene. We have used the pSVO CAT expression vector of Gorman et al. (55). This contains the bacterial chloramphenicol acetyltransferase (CAT) gene as well as splicing and polyadenylation signals from SV40; pSVO CAT lacks a eukaryotic promoter upstream from the CAT gene. Four constructs were made. These consisted of two sets of

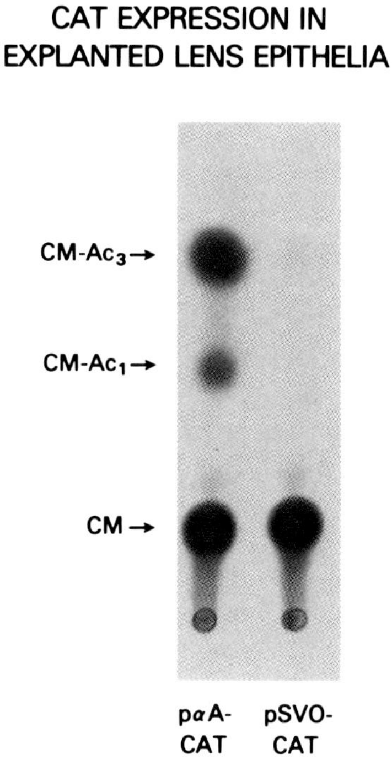

Fig. 5. Autoradiogram of thin-layer chromatography of CAT products in 14 day-old embryonic chicken lens epithelia transfected with pαA-CAT (contains approximately 400 base pairs upstream and 40 base pairs downstream from the cap site of the αA-crystallin gene. pSVO-CAT (no eukaryotic promoter inserted) is the control. Enzyme reaction used C^{14}-chloramphenicol and 1 hour incubation. CM, C^{14}-chloramphenicol; CM-Ac_1 and CM-Ac_3, acetylated products.

sequences (-400 to +40 and -85 to +40) inserted in both orientations upstream of the bacterial CAT gene in the plasmid. The lens cells were transfected by the calcium phosphate method one day after explantation; CAT assays were conducted three days later. Positive controls were performed using the LTR sequence of Rous sarcoma virus (56). Negative controls were conducted using the pSVO CAT plasmid.

The transfected lens cells synthesized CAT when the -400 to +40 sequence from the αA-crystallin gene was inserted in the vector in the same orientation as it exists in the murine gene (Fig. 5). Insertion of the -85 to +40 αA-crystallin gene sequence resulted in CAT activity, but much less than the -400 to +40 sequence. Upstream 5' flanking sequences are known to affect transcription in other eukaryotic genes (50, 57, 58, 59). The Rous sarcoma virus LTR sequence promoted the greatest CAT expression. The pSVO CAT plasmid did not give detectable CAT activity in the cultured explants. The expression of CAT increased when the concentration of vector was raised from 0.8μg/ml to 4.0μg/ml of DNA. Thus the explanted chicken lens epithelia are capable of taking up the recombinant plasmid and of utilizing the murine αA-crystallin promoter region to transcribe the bacterial CAT gene. As in the case of the cloned chicken δ-crystallin promoter (45), the cloned murine αA-crystallin promoter will function in lens cells of a foreign species.

We have also begun to study the tissue specificity of the putative promoter region of the murine αA-crystallin gene. In contrast to the results with the explanted lens epithelia, the -400 to +40 sequence of the αA-crystallin gene did not express detectable levels of CAT activity in NIH 3T3 cells, L cells or BSC-1 cells under conditions that the eukaryotic (which are promoter present in pBR322 sequences that function as a pSVO CAT) are not expressed (see 61). However, cultures containing a mixture of embryonic chicken muscle cells and fibroblasts did show low levels of CAT expression. It is noteworthy that crystallin gene expression has been observed recently in numerous non-lens embryonic tissues (59, 60). We are in the process of quantifying the relative promoter strengths in the different cell cultures by normalizing CAT activity to copy-number of plasmid in the cells.

SUMMARY AND COMMENTS

Structural studies on the crystallin genes have revealed the existence of four major families (α, β, γ and δ) with different organizations within the genome. The present information reveals rather precise relationships between the exon structures of the crystallin genes and the structures of their encoded proteins. Both linked (δ and γ) and apparently dispersed (β) gene families are present. The crystallin gene families may contain as few as two (δ) or at least six to seven (β, γ) members. There is one αA-crystallin gene and presumably one αB-crystallin gene, although the latter has not been isolated.

The different crystallin gene families are differentially expressed in the developing lens, with δ being the first in the chicken and α in the mouse. In addition to differential expression between the families, there is differential expression within the families of the crystallin genes. Each member of the β-crystallin gene family was shown to be expressed with a characteristic temporal and spatial pattern within the embryonic chicken lens.

Experiments are beginning to reveal potentially important DNA sequences for the expression of the αA- and the δ-crystallin genes. Sequences between -400 and -85 of the murine αA-crystallin gene appear necessary for optimal expression of the CAT gene in primary explants of chicken lens cells. The two δ-crystallin genes have different promoter strengths <u>in vitro</u>. There are two reverse complements of consensus viral core enhancer-like sequences and a consensus CCAAT sequence upstream of the δ1 gene which may contribute to its greater expression. Viral consensus core enhancer-like sequences are also present in the second intron of the chicken δ1 gene and in the first exon and the insert exon of the murine αA gene. Comparison of δ1 and δ2 crystallin gene sequences showed, interestingly, that the untranslated, first exon was highly conserved in the two genes. Perhaps this conservation signifies a functional role for these sequences. An unexpected homology was also observed between the first intron in the two chicken δ-crystallin genes and the immunoglobulin heavy-chain switch region of mice. We cannot attribute a function for this sequence in δ-crystallin gene at the present time. An intriguing possibility is that δ-crystallin gene rearrangement is or has been of significance in this specialized avian and reptilian lens protein. At present these structural features of the crystallin genes have not been shown to have functional significance, but they provide useful avenues for further experimentation.

Current experiments suggest tissue-specific expression of the murine αA-crystallin promoter region in chicken lens explants. However, CAT gene expression was also promoted by the upstream sequences of the αA-crystallin gene in primary cultures containing embryonic chicken muscle cells and fibroblasts, suggesting that non-lens cells from embryos may functionally recognize crystallin gene promoters more readily than those from adults.

Clearly, much work remains to be done. The excitement lies in the fact that expression of the numerous crystallin genes can now be approached at several levels simultaneously, i.e., in vivo, in tissue culture, in tranfected lens cells and in cell-free transcription systems, making prospects for advancing our understanding of this remarkable developing tissue very promising.

ACKNOWLEDGMENT

We thank Ms. Dawn Sickles for expert secretarial help.

REFERENCES

1. Papaconstantinou J (1967). Molecular aspects of lens cell differentiation. Science 156:338.
2. Piatigorsky J (1981). Lens differentiation in vertebrates: a review of cellular and molecular features. Differentiation 19:134.
3. Philpott GW, Coulombre AJ (1965). Lens development II. The differentiation of embryonic chick lens epithelial cells in vitro and in vivo. Exp Cell Res 38:635.
4. Piatigorsky J, Webster H deF, Craig SP (1972). Protein synthesis and ultrastructure during the formation of embryonic chick lens fibers in vivo and in vitro. Develop Biol 27:176.
5. Piatigorsky J (1973). Insulin initiation of lens fiber differentiation in culture: elongation of embryonic lens epithelial cells. Develop Biol 30:214.

6. Beebe DC, Feagans DE, Jebens HAH (1980). Lentropin: a factor in vitreous humor which promotes lens fiber cell differentiation. Proc Natl Acad Sci USA 77:490.
7. Piatigorsky J, Rothschild SS, Milstone LM (1973). Differentiation of lens fibers in explanted embryonic chick lens epithelia. Develop Biol 34:334.
8. Piatigorsky J, Rothschild SS (1972). Loss during development of the ability of chick embryonic lens cells to elongate in culture: inverse relationship between cell division and elongation. Develop Biol 28: 382.
9. Philpott GW (1970). Growth and cytodifferentiation of embryonic chick lens epithelial cells in vitro. Exp Cell Res 59:57.
10. Piatigorsky J (1975). Lens cell elogation in vitro and microtubules. Ann NY Acad Sci 253:333.
11. Okada TS, Eguchi G, Takeichi M (1973). The retention of differentiated properties by lens epithelial cells in clonal cell culture. Develop Biol 34:321.
12. Okada TS (1983). Recent progress in studies of the transdifferentiation of eye tissue in vitro. Cell Differentiation 13:6.
13. Harding JJ, Dilley KJ (1976) Structural proteins of the mammalian lens: a review with emphasis on changes in development, aging and cataract. Exp Eye Res 22:1.
14. Bloemendal H (1982). Lens proteins. CRC Critical Rev Biochem 12:1.
15. Piatigorsky J (1984). Delta crystallins and their nucleic acids. Mol Cell Biochem 59:33.
16. de Jong WW (1981) Evolution of lens and crystallins. In Bloemendal H (ed): "Molecular and Cellular Biology of the Eye Lens," New York: John Wiley and Sons, p. 221.
17. Driessen HPC, Herbrink P, Bloemendal H, de Jong WW (1981). Primary structure of the bovine β-crystallin Bp chain. Internal duplication and homology with γ-crystallin. Eur J Biochem 121:83.
18. Inana G, Shinohara T, Maizel JV Jr, Piatigorsky J (1982). Evolution and diversity of the crystallins. J Biol Chem 257:9064.
19. Blundell T, Lindley P, Miller L, Moss D, Slingsby C, Tickle I, Turnell B, Wistow G (1981). The molecular structure and stability of the eye lens: X-ray analysis of γ-crystallin II. Nature 289:771.

20. Wistow G, Turnell B, Summers L, Slingsby C, Moss D, Miller L, Lindley P, Blundell T (1983). X-ray analysis of the eye lens protein γ-II crystallin at 1.9 A resolution. J Mol Biol 170:175.
21. Inana G, Piatigorsky J, Norman B, Slingsby C, Blundell T (1983). Gene and protein structure of a β-crystallin polypeptide in murine lens: relationship of exons and structural motifs. Nature 302:310.
22. Zwaan J, Ikeda A (1968). Macromolecular events during differentiation of the chicken lens. Exp Eye Res 7:301.
23. McAvoy JW (1980). Induction of the eye lens. Differentiation 17:137.
24. Zwaan J (1983). The appearance of α-crystallin in relation to cell cycle phase in the embryonic mouse lens. Develop Biol 96:173.
25. Piatigorsky J, Nickerson JM, King CR, Inana G, Hejtmancik JF, Hawkins JW, Borras T, Shinohara T, Wistow G, Norman B. (in press). Crystallin genes: templates for lens transparency. In Nugent J, Whelan J (eds): "Human Cataract Formation," Ciba Foundation Symp No 106, London: Pitman Publishing.
26. Moormann RJM, den Dunnen JT, Mulleners L, Andreoli P, Bloemendal H, Schoenmakers JGG (1983). Strict co-linearity of genetic and protein folding domains in an intragenically duplicated rat lens γ-crystallin gene. J Mol Biol 171:353.
27. Lok S, Tsui L-C, Shinohara T, Piatigorsky J, Gold R, Breitman M (submitted for publication). Analysis of the mouse γ-crystallin gene family: assignment of multiple cDNAs to discrete genomic sequences and characterization of a representative gene.
28. Schoenmakers JGG, den Dunnen JT, Moormann RJM, Jongbloed R, van Leen RW, Lubsen NH (in press). The crystallin gene families. In Nugent J, Whelan J (eds): "Human Cataract Formation," Ciba Foundation Symp No 106, London Pitman Publishing.
29. Shinohara T, Robinson EA, Appela E, Piatigorsky J (1982). Proc Natl Acad Sci USA 79:2783.
30. Hejtmancik JF, Piatigorsky J (1983). Diversity of β-crystallin mRNAs of the chicken lens. J Biol Chem 258: 3382.
31. Cohen LH, Westerhuis LW, Smits DP, Bloemendal H (1978). Two structurally closely related polypeptides encoded by 14-S mRNA isolated from rat lens. Eur J Biochem 89:251.

32. King CR, Piatigorsky J (1983). Alternative RNA splicing of the murine αA-crystallin gene: protein-coding information within an intron. Cell 32:707.

33. King CR, Piatigorsky J (1984). Alternative splicing of αA-crystallin RNA. J Biol Chem 259:1822.

34. Yasuda K, Kondoh H, Okazaki Y, Shimura Y, Okada TS (1982). Organization and structure of α-and δ-crystallin genes and their expression in cells by microinjection. Fifth International Congress of Eye Research, Eindhoven, The Netherlands p. 38.

35. Hawkins JW, Nickerson JM, Sullivan MA, Piatigorsky, J (in press). The chicken δ-crystallin gene family. Two genes of similar structure in close chromosomal approximation. J Biol Chem.

36. Bhat, SP, Jones RE, Sullivan MA, Piatigorsky J (1980). Chicken lens crystallin DNA sequences show at least two δ-crystallin genes. Nature 284:234.

37. Yasuda K, Kondoh H, Okada TS, Nakajima N, Shimura Y (1982). Organization of δ-crystallin genes in the chicken. Nucl Acids Res 10:2879.

38. Jones RE, Bhat SP, Sullivan MA, Piatigorsky J (1980). Comparison of two δ-crystallin genes in the chicken. Proc Natl Acad. Sci USA 77:5879.

39. Shinohara T, Piatigorsky J (1976). Quantitation of δ-crystallin messenger RNA during lens induction in chick embryos. Proc Natl Acad Sci USA 73:2808.

40. Sun S-T, Tanaka T, Nishio I, Peetermans J, Maizel JV Jr, Piatigorsky J (1984). Direct observation of δ-crystallin accumulation by laser light-scattering spectroscopy in the chicken embryo lens. Proc Natl Acad Sci USA 81:785.

41. Tréton, JA, Shinohara T, Piatigorsky J (1982). Degradation of δ-crystallin mRNA in the lens fiber cells of the chicken. Develop Biol 92:60.

42. Ostrer H, Beebe DC, Piatigorsky, J (1981). β-Crystallin mRNAs: differential distribution in the developing chicken lens. Develop Biol 86:403.

43. Jones, RE, DeFeo D, Piatigorsky J (1981). Transcription and site-specific hypomethylation of the δ-crystallin genes in the embryonic chicken lens. J Biol Chem 256:8172.

44. Nickerson JM, Piatigorsky J (in press). The sequence for a complete chicken δ-crystallin cDNA. Proc Natl Acad Sci USA.

45. Kondoh H, Yasuda K, Okada TS (1983). Tissue-specific expression of a cloned chick δ-crystallin gene in mouse cells. Nature 301:440.
46. Weiher H, Konig M, Gruss P (1983). Multiple point mutations affecting the Simian virus 40 enhancer. Science 219:626.
47. Benoist C, O'Hare K, Breathnach R, Chambon P (1980). The ovalbumin gene-sequence of putative control regions. Nucl Acids Res 8:127.
48. Efstratiadis A, Posakony JW, Maniatis T, Lawn RM, O'Connell C, Spritz RA, DeRiel JK, Forget BG, Weissman SM, Slightom JL, Blechl AE, Smithies O, Baralle FE, Shoulders CC, Proudfoot NJ (1980). The structure and evolution of the human β-globin gene family. Cell 21:653.
49. Gruss P (1984). Magic Enhancers? DNA 3:1.
50. McKnight SL, Kingsbury R (1982). Transcriptional control signals of eukaryotic protein-coding gene. Science 217:316.
51. Dunnick W, Rabbitts TH, Milstein C (1980). An immunoglobulin deletion mutant with implications for the heavy-chain switch and RNA splicing. Nature 286:669.
52. Davis MM, Kim SK, Hood LE (1980). DNA sequences mediating class switching in α-immunoglobulins. Science 209:1360.
53. Manley JL, Fire A, Cano A, Sharp PA, Gefter ML (1980). DNA-dependent transcription of adenovirus genes in a soluble whole-cell extract. Proc Natl Acad Sci USA 77:3855.
54. Ambesi-Impiombato FS, Parks LAM, Coon HG (1980). Culture of hormone-dependent functional epithelial cells from rat thyroids. Proc Natl Acad Sci USA 77:3455.
55. Gorman CM, Moffat LF, Howard BH (1982). Recombinant genomes which express chloramphenicol acetyltransferase in mammalian cells. Mol Cell Biol 2:1044.
56. Gorman CM, Merlino GT, Willingham MC, Pastan I, Howard, BH (1982). The Rous sarcoma virus long terminal repeat is a strong promoter when introduced into a variety of eukaryotic cells by DNA-mediated transfection. Proc Natl Acad Sci USA 79: 6777.
57. Tsuda M, Suzuki Y (1981). Faithful transcription initiation of fibroin gene in a homologous cell-free

system reveals an enhancing effect of 5' flanking sequence far upstream. Cell 27:175.
58. Tsuda M, Suzuki Y (1983). Transcription modulation in vitro of the fibroin gene exerted by a 200-base-pair region upstream from the "TATA" box. Proc Natl Acad Sci USA 80:7442.
59. Agata K, Yasuda K, Okada TS (1983). Gene coding for a lens-specific protein, δ-crystallin, is transcribed in nonlens tissues of chicken embryos. Develop Biol 100:222.
60. Bower DJ, Errington LH, Cooper DN, Morris S, Clayton RM (1983). Chicken lens δ-crystallin gene expression and methylation in several non-lens tissues. Nucl Acids Res 11:2513.
61. Sassone-Corsi P, Corden J, Kedinger C, Chambon P (1981). Promotion of specific in vitro transcription by excised "TATA" box sequences in a foreign nucleotide environment. Nucl Acids Res 9:394.

Molecular Biology of Development, pages 351–360

ACTIVATION OF THE ADENOVIRUS LATE PROMOTER BY CIS- AND TRANS- ACTING ELEMENTS

E. Diann Lewis, Xin-Yuan Fu and James L. Manley

Department of Biological Sciences
Columbia University, New York, New York 10027

ABSTRACT To study the mechanisms regulating transcription initiation from a prototype strong promoter, we fused the adenovirus type 2 (Ad2) late promoter (-403 to +33bp relative to the late mRNA start site) to the SV40 T antigen encoding sequences, and measured T antigen protein and mRNA production following transfection of the recombinant DNAs into different human cell lines. The results obtained establish that there are two fundamentally different mechanisms, with different nucleotide sequence requirements, by which this promoter can be activated.

INTRODUCTION

One of the primary goals of molecular biology today is to obtain an understanding of the mechanisms that regulate the expression of genes in animal cells. As in prokaryotes, it is likely that the primary (although by no means only) step at which regulation occurs is at the level of transcription initiation. The most well studied example of this type of regulatory event is the autoregulation of transcription mediated by the SV40 large T antigen (1-5). This protein appears to function in a manner analogous to bacterial repressors, ie, the protein binds to specific DNA sequences near the start of transcription, thereby presumably preventing RNA polymerase II from initiating transcription.

In contrast, the mechanisms that bring about positive regulation (activation) are largely obscure. Two important examples have been the focus of much study in recent years. Protein products encoded by the E1A region of adenovirus were shown originally to be required for the activation of other adenovirus early genes (6,7). More recently, it has been found that products of the E1A gene can, under some conditions, bring about the expression of otherwise silent cellular genes (8-11). Unfortunately, the mechanism by which this protein functions is completely unclear. This is due in large part to the fact that conditions have not been found that allow E1A function to be detected *in vitro*. However, the fact that a large number of apparently unrelated genes are subject to E1A control suggests that the protein may function by a means other than specific DNA binding.

Another important type of regulatory element is a class of nucleotide sequences termed enhancers (12-14). The prototype for this class of sequence is a 72 base pair sequence that is tandemly repeated in the SV40 origin. In contrast to *trans*-acting factors such as the E1A proteins, these elements function exclusively in *cis* to activate the expression of linked genes. Interestingly, the position of such an enhancer relative to the gene that is under its control does not appear to be important. The elements appear to interact with factors present in limiting amounts in nuclei (15), and some of these elements may be responsible for the tissue specific expression of linked genes (16-17). The mechanism by which these elements function is again largely obscure, due to our inability to reproduce this phenomenon consistently *in vitro* (although see ref.18).

Here we show that the adenovirus late promoter can be activated either by a *trans*-acting E1A protein or by a *cis*-acting enhancer element. Furthermore, the presence of two enhancer elements duplicated head-to-head relative to each other results in at least a two-fold increase in transcription relative to a construct containing only one copy of the enhancer element. In cells constitutively expressing the E1A genes, however, not only is an enhancer element not required to obtain expression from this promoter, but it has no detectable effect on transcription, even when present in duplicate.

MATERIALS AND METHODS

Plasmids. The structure of the plasmids used in these experiments is shown in Figure 1. The enhancer element used in these experiments was obtained from a Bal31 nuclease-generated deletion mutant which had been sealed using Bam HI linkers. The DNA fragment utilized was 176 bp, containing approximately five bp of SV40 DNA on the early side of the 72 bp repeats, and 22 bp on the late side.

The plasmid expressing the E1A proteins was constructed from a plasmid (pHEB4, obtained from S.L.-Hu) that contains all sequences required for expression of the E1A proteins, except that it is terminated at a HpaI site in the 3' untranslated region; ie, it lacks a poly(A) addition signal. To increase expression, a 135bp Bgl II-Bam HI fragment containing the SV40 early polyadenylation signal (also obtained from a Bal 31 deletion mutant), was inserted into pHEB4 to generate pE1Aa (see Figure 1).

Transient Expression Assays. Calcium phosphate-mediated DNA transfections were carried out according to the method of Wigler et al (19), with some modifications (20). The DNA precipitate remained on the cells for 4 hrs for 293 cells and 18 hrs for HeLa cells, after which the cells were shocked with glycerol and fresh medium was added. Approximately 60 hrs after the DNA was added, cells were collected and assayed either for the production of T-antigen by indirect immunofluorescence (21) or else for the production of T-antigen-specific RNA by S-1 analysis (22).

Indirect Immunofluorescence. For testing T-specific immunofluorescence, the cells were grown on coverslips and fixed at 60 hours with 10% formaldehyde. The cells were then permeabilized with nonidet P-40 and the 1st antibody, monoclonal Pab416 (23), applied. The 2nd antibody was goat-anti-mouse conjugated with flourescein. Cells were scanned with a fluorescent microscope.

Nuclease S1 Analysis. For S-1 analysis, total cytoplasmic RNA was prepared and probed with 5' end-labeled DNAs. Reactions containing 10-20μg RNA and 10ng probe were hybridized at 43°C for 4 hrs. S-1 digestion products were analyzed on 8% acrylamide -8M urea sequencing type gels.

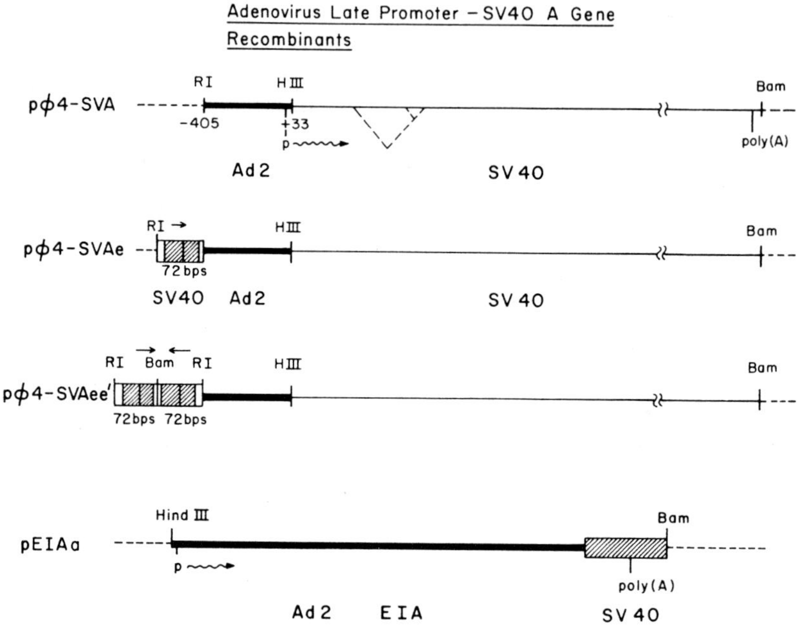

Figure 1. Plasmid constructs used in experiments. All constructs are in pBR322. See Materials and Methods for details.

RESULTS AND DISCUSSION

To begin to study the mechanisms by which RNA polymerase II promoters can be activated in higher cells, we have chosen to analyze the strong adenovirus late promoter. Since the adenovirus late transcription unit is complex (24), we fused the promoter region to sequences encoding a readily assayable product, the SV40 tumor antigen. One such construct, pϕ4-SVA, is shown in Figure 1. As shown previously by others (14,25) the activity of a promoter driving the expression of T antigen in a transient expression assay can be simply and accurately determined by indirect immunofluorescence, in which the percentage of fixed cells that are expressing levels of T antigen detectable by fluorescence is determined.

Table 1. Transient Expression of T-antigen[a]

	% of T-positive cells	
plasmid	293	HeLa
pϕ4-SVA	10%	<0.1%[b]
pϕ4-SVAe	11.2%	18.3%
pϕ4-SVAee'	14%	36%
pHEB4 + pϕ4-SVA	N.D.[c]	0.45%
pE1aA + pϕ4-SVA	N.D.	1.2%

[a]Each experiment was repeated 2-7 times.
[b]The value, <0.1%, means that expression has never been detected; ie, no T antigen-positive cells have ever been observed.
[c]N.D. = not determined.

Table 1 shows that when pϕ4-SVA was transfected into HeLa cells, no T antigen positive cells were detected, consistent with results obtained previously by others (26). However, when the same plasmid was introduced into 293 cells (a human embryonic kidney cell line that had been transformed by the left-hand end of adenovirus DNA, and which constitutively expresses E1A and E1B proteins (27)), significant levels of T antigen were detected. Hen et al (26) showed previously that when the SV40 enhancer element was inserted into a plasmid similar to pϕ4-SVA, expression from the late promoter could be detected in HeLa cells. The data in Table 1 is consistent with this observation: pϕ4-SVe expressed efficiently in HeLa cells. Furthermore, pϕ4-SVAee' gave rise to two-fold more T antigen positive cells than did pϕ4-SVAe. Sassone-Corsi et al observed a similar phenomenon when two enhancer elements were joined head-to-tail (25). Our results thus suggest that two enhancer elements juxtaposed in a head-to-head manner also results in significantly more efficient promoter utilization than does one enhancer.

Interestingly, when these plasmids were transfected into 293 cells, no difference in the level of expression could be detected (Table 1). Thus, in this cell line, not only is an enhancer element not required to obtain efficient expression, but the presence of one (or even two) of these sequences has no effect on expression (see also below).

These results suggested to us that an adenovirus E1A gene product might be able to override the requirement for an enhancer. However, there are obviously many other differences between HeLa and 293 cells than the presence or absence of the E1A proteins. Therefore, to test more directly whether an E1A gene product was responsible for the "enhancerless" expression we observed in 293 cells, HeLa cells were cotransfected with pϕ4-SVA and plasmids capable of expressing the E1A gene products. The results of indirect immunofluorescence assays (Table 1) show that the Ad2 late promoter was indeed activated when cotransfected with an E1A expressing plasmid. The percentage of T antigen positive cells obtained was significantly less than when HeLa cells were transfected with pϕ4-SVAe. This is most likely due to the fact that, in the cotransfection, sufficient quantities of E1A gene products must accumulate before the late promoter can be activated. Note that the inclusion of the SV40 poly(A) addition signal to pHEB4 resulted in a several fold increase in T antigen expression, suggesting that such a signal greatly increases the synthesis of functional E1A mRNA.

The immunofluorescence assay utilized in the experiments described above is a very indirect measure of promoter function. Therefore, we also studied the RNA synthesized in the transfected cells utilizing the S1-nuclease mapping technique (22). Total cytoplasmic RNA was extracted from transfected cells and analyzed as described in Materials and Methods. The results obtained (Figure 2) are completely

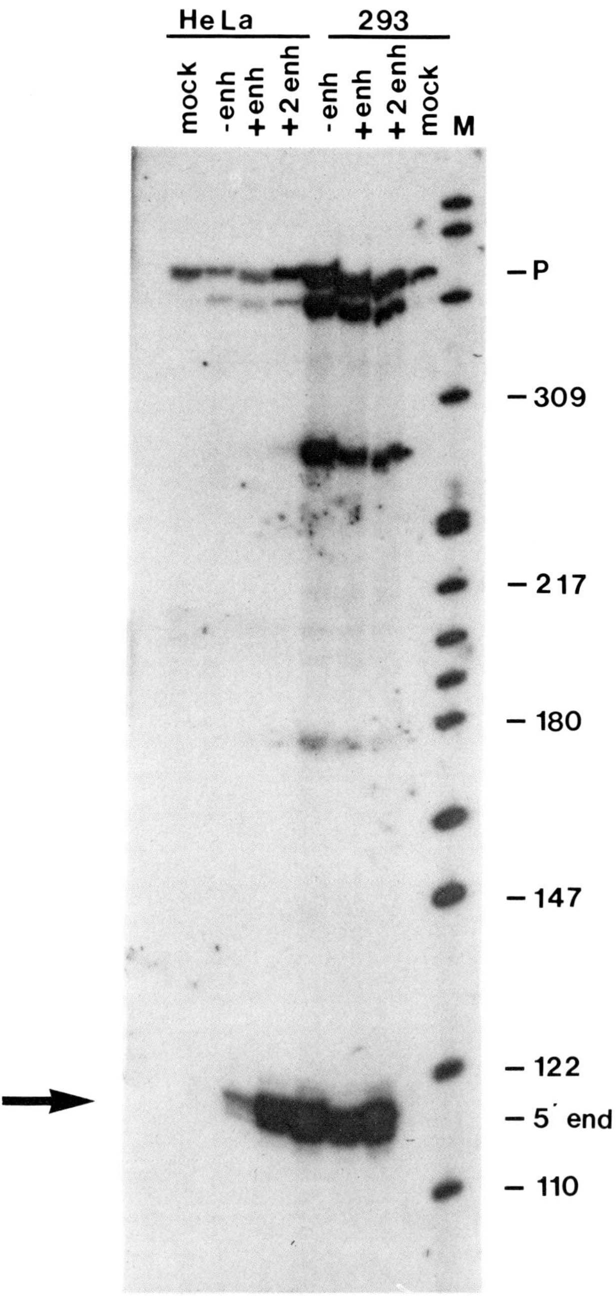
HeLa
293
mock
-enh
+enh
+2 enh
-enh
+enh
+2 enh
mock
M
P
309
217
180
147
122
5' end
110

Figure 2.

Figure 2. Quantitative Nuclease S1 analysis of RNA from HeLa and 293 cells transfected with plasmids plus and minus enhancers. The DNA probe was a 5' end labeled Bst NI fragment from pϕ4-SVA that extended from -126 to +118, relative to the mRNA start site.

consistent with those from immunofluorescence: RNA complementary to the probe could not be detected in HeLa cells transfected with pϕ4-SVA. However, pϕ4-SVAe gave rise to readily detectable levels of RNA and pϕ4-SVAee' produced at least twice this amount. In contrast, all three plasmids led to the synthesis of high, and essentially equal, levels of RNA in 293 cells.

The results described above suggest that there are two distinct mechanisms by which the adenovirus late promoter can be activated during transient expression assays in human cell cultures. One is by the presence (in cis) of an enhancer element. How such activation occurs is an important unanswered question. Another unresolved question is whether enhancer-mediated activation plays a role in activating the late promoter during adenovirus infection. To date, enhancer elements have only been found near the left end of the adenovirus chromosome (28,29), approximately 5,000 bp upstream of the late promoter. Whether or not an enhancer from this region, or perhaps one situated elsewhere in the genome, plays a role in regulating expression from the late promoter is not known.

The mechanism by which an adenovirus E1A protein is able not only to activate the late promoter, but also to obviate the requirement for an enhancer, is likewise completely unclear. Given the fact that a number of cellular and viral genes can be activated by E1A (see Introduction), it is perhaps unlikely that the protein interacts with a specific nucleotide sequence in the promoter region. However, the possibility that a particular structure in the DNA (ie, a stem-loop) is recognized cannot be excluded. In fact, analysis of deletion mutants around the cap site of the late promoter suggests that this region is required for E1A-mediated transcription, but not for enhancer-mediated expression (20). Interestingly, these cap site mutations destroy a region of DNA, first noted by Ziff and Evans (30), that displays a potential for forming a hairpin structure.

An alternative model to explain E1A-mediated gene activation is that an E1 protein modified a subunit of RNA polymerase II, or a factor required for initiation. Such a modification could entail enzymatic modification of a particular polypeptide, or else a direct interaction between the viral protein and the RNA polymerase initiation complex. In any event, the modified polymerase now interacts with DNA in a fundamentally different way: An enhancer element is no longer necessary (and, in fact, very likely of no consequence at all in activating transcription by the modified polymerase), but additional sequences in the promoter region now appear to be required (see above).

To understand the mechanisms responsible for both cis- and trans- promoter activation will require the definition of conditions that allow these phenomena to be reproduced in in vitro transcription systems. As mentioned in the Introduction, this goal has to date proven to be an elusive one. However, given the importance of understanding this mode of gene regulation, it is likely that this problem will be solved in the near future.

ACKNOWLEDGMENT

This work was supported by PHS grant GM 28983.

REFERENCES

1. Tegtmeyer P, Schwartz M, Collins JK and Rundell K (1975). J Virol 16:168.
2. Reed SI, Stark GR, and Alwine JC (1976). PNAS USA 73: 3083.
3. Rio D, Robbins A, Myers R, and Tjian R (1980). PNAS USA 77:5706-5710.
4. Hansen U, Tenen DG, Livingston DM, and Sharp PA (1981). Cell 27:603-612.
5. Rio DC, and Tjian R (1983). Cell 32: 1227-1240.
6. Jones N and Shenk T (1979). PNAS USA 76:3665.
7. Berk AJ, Lee F, Harrison T, Williams J, and Sharp PA (1979). Cell 17:935-944.
8. Kao H-T, and Nevins JR (1983). Molec and Cell Biol 3: 2058-2065.
9. Schrier PI, Bernards R, Vaessen RJ, Houweling A, and van der Eb AJ (1983). Nature 305: 771-775.
10. Green M, Treisman R, and Maniatis T (1983). Cell 35: 137-148.
11. Gaynor RB, Hillman D, Berk AJ (1984). PNAS USA 81:1193-

1197.
12. Moreau P, Hen R, Wasylyk B, Everett R, Gaub MP, and Chambon P (1981). Nucleic Acids Res 9:6047.
13. Gruss P, Dhar R, and Khoury G (1981). PNAS USA 78:943.
14. Banerji J, Rusconi S and Schaffner W (1981). Cell 27:299-308.
15. Scholer HR and Gruss P (1984). Cell 36: 403-411.
16. Gillies SD, Morrison SL, Oi VT, and Tonegawa S (1983). Cell 33: 717-728.
17. Queen C, and Baltimore D (1983). Cell 33: 741-748.
18. Sassone-Corsi P, Dougherty JP, Wasylyk B, and Chambon P (1984). PNAS USA 81: 308-312.
19. Wigler M, Pellicer A, Silverstein S, Axel R, Urlaub G, and Chasin L (1979). PNAS USA 76: 1373-1376.
20. Lewis ED, and Manley JL (1984). Manuscript submitted.
21. Verderame M, Alcorta D, Egnor M, Smith K, and Pollack R (1980). PNAS USA 77: 6624-6628.
22. Berk AJ, and Sharp PA (1978). PNAS USA 75: 1274-1278.
23. Harlow E, Crawford L, Pim DC, and Williamson NM (1981). J Virol 39: 861-869.
24. Chow LT and Brokev TR (1978). Cell 15: 497.
25. Sassone-Corsi P, Corden J, Kedinger C, and Chambon P (1981). Nuc Acids Res 9: 3941-3958.
26. Hen R, Sassone-Corsi P, Corden J, Gaub MP, and Chambon P (1982) PNAS USA 79: 7132-7136.
27. Aiello L, Guilfoyle R, Huebner K, Weinmann R (1979). Virol 94: 460-469.
28. Hearing P and Shenk T (1983). Cell 33: 695-703.
29. Weeks DL and Jones NC (1983). Molec and Cell Biol 3: 1222-1234.
30. Ziff E and Evans R (1978). Cell 15: 1463-1475.

Molecular Biology of Development, pages 361–371

CHROMATIN STRUCTURE IN *DICTYOSTELIUM DISCOIDEUM* RIBOSOMAL DNA[1]

Cynthia A. Edwards[2] and Richard A. Firtel

Department of Biology
University of California, San Diego
La Jolla, California 92093

ABSTRACT We have examined the chromatin structure of both coding and noncoding regions of the extrachromosomal, palindromic rDNA in *Dictyostelium discoideum*. The coding region appears to be arranged in nucleosomes, but the nucleosomes are not phased with respect to DNA sequence. In marked contrast, the spacer regions are phased. We present evidence that the sequences or structures in naked DNA that are most readily attacked by nucleases may also be responsible for directing nucleosome placement in spacer regions. Additional data suggest that the central region of the molecule is organized in a supernucleosomal structure that is disrupted when the nuclei are washed with 0.35 M KCl.

INTRODUCTION

Chromatin is generally thought to be composed of a repeating array of nucleosomes. However, it is apparent that chromatin structure is neither uniform nor static. Many studies have shown that DNA in the chromatin of coding regions is sensitive to nucleases and often shows "hypersensitivity" when actively transcribed (1). However, there

[1]This work was supported by an NIGMS grant to R.A.F.
[2]Present address: The Laboratory of Cellular and Developmental Biology, Building 6, Room B1-10, National Institute of Arthritis, Diabetes, Digestive, and Kidney Diseases, National Institutes of Health, Bethesda, Maryland 20205.

is little information available concerning the chromatin structure of noncoding DNA. One can imagine that phasing in noncoding regions may be crucial for the structural maintenence of regulatory regions, origins of replication and other noncoding but functional domains.

We have chosen to study the rDNA in Dictyostelium since there are several functionally different regions in this repeated (~90 times/haploid genome), palindromic, extra-chromosomal molecule (for review, see ref. 2). There are two transcriptional units on each half of the 88 kb palindrome, one codes for a 36S transcript that is processed to yield the 26S, 5.8S, and 17S rRNAs and the other for a single 5S rRNA. The rest of the molecule (70-75 kb) is non-coding. The map of the rDNA molecules and the probes used in our experiments are shown in Figure 1.

We have used micrococcal nuclease and DNAase I to investigate the structure of chromatin in different regions

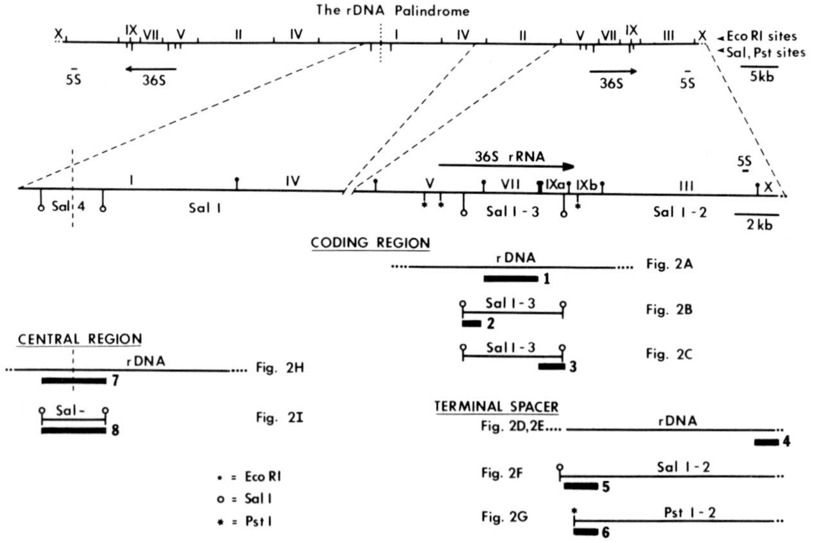

FIGURE 1. The rDNA Palindrome. The experiments we have done are outlined beneath the enlarged map. The upper lines represent the DNAs to which the $[^{32}P]$-labeled probes, indicated here as numbered, solid bars, were hybridized. Dots at the ends of the lines indicate that the DNA has not been cleaved with a restriction enzyme. The figures that show the results of each experiment are indicated.

of the rDNA molecule. In short, our results show that a) the coding region is nucleosomal but not phased with respect to DNA sequence, and b) the noncoding regions we have examined appear to be nucleosomal but, in stark contrast to the coding region, the nucleosomes are distinctly phased. At least half of the ~90 kb molecule is arranged in this stringently phased pattern. Our data suggest that certain sequences or structures in DNA may direct nucleosome placement.

Data obtained by examining changes in chromatin structure when washing the nuclei with varying concentrations of salt suggest that the chromatin is indeed arranged in nucleosomes. In addition, we detect an organization in the central region of the rDNA that is disrupted by a low concentration of salt (0.35 M KCl), suggestive of a supernucleosomal structure; when the structure is disrupted, a regularly repeating array of nucleosomes appears.

RESULTS

For the following experiments, nuclei were isolated then treated with either micrococcal nuclease (MNase) or deoxyribonuclease I (DNase I) (Edwards and Firtel, submitted). The partially digested DNA was purified, cut with restriction endonucleases (if appropriate), separated electrophoretically on agarose gels, and transferred to nitrocellulose. Naked DNA was used as a control in all of the experiments. Hybidization to different types of [^{32}P]-labeled probes yields different kinds of information of chromatin structure. "Full-length" probes, hybridized to Southern blots of MNase-digested nuclear DNA yield information on whether or not the chromatin is arranged in nucleosomes but not whether the nucleosomes are phased. Indirect end-label probes can detect phasing but not the presence of nonphased nucleosomes. Using both approaches we have analyzed coding and noncoding regions in the rDNA chromatin. The probes and the sequences to which they were hybridized are outlined in Figure 1.

The Coding Region

We hybridized probe 1 (see Fig. 1) to MNase-digested rDNA to determine whether or not the coding region is arranged in nucleosomes. The results (Fig. 2A) show a 180

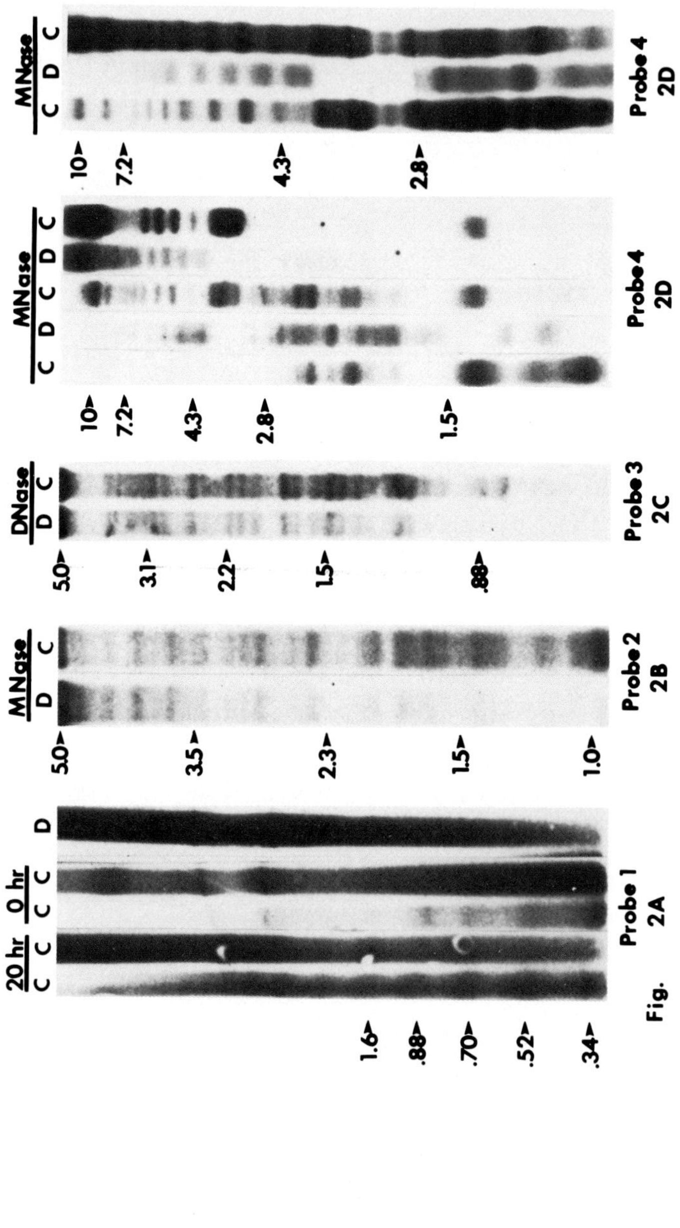

FIGURE 2. The results of the experiments outlined in Fig. 1 are shown here. C=chromatin, the DNA isolated from nuclease-treated nuclei. D=naked DNA control. The arrows indicate size (kb). The nuclease used in each experiment is indicated at the top. Refer to Fig. 1 and the text for descriptions of the experiments. This figure is continued on the next page.

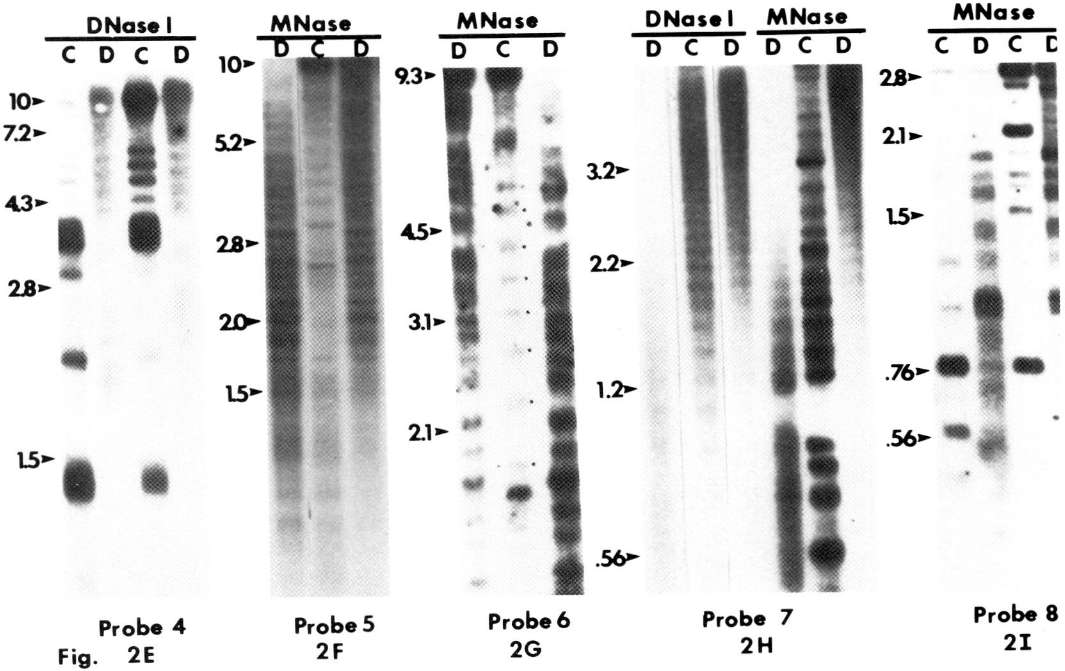

FIGURE 2. Continued. Refer to the text and Fig. 1 for descriptions of the experiments.

bp nucleosomal ladder in both vegetative cells and cells that have completed 20 hours of the developmental cycle. The repeat length is typical of bulk MNase-digested chromatin. It is of interest to note that "longer" ladders are observed for chromatin isolated from 20 hour developed cells (3). At this time in development, the rate of transcription is $<15\%$ the level of vegetative cells (4). Coding region chromatin from vegetative cells is more sensitive to nucleases than bulk chromatin or chromatin from the non-coding regions of the rDNA (data not shown).

To examine the question of phasing in the coding region, probes from each end of rDNA band Sal I-3 (see Fig. 1, probes 2 and 3) were isolated and hybridized to Southern blots of DNA, from MNase- or DNase I-treated nuclei, that had been digested to completion with Sal I. The results are shown in Figs. 2B and 2C. In each case, the naked DNA and chromatin hybridization patterns were identical, suggesting that, in vegetative cells, nucleosomes in the coding region are not phased with respect to DNA sequence. (Note that the short ladder one would expect to find because of the probe length is not detectable on this gel.)

The Terminal Spacer

Much of our work has been focused on the structure of chromatin in the ~10 kb terminal spacer region, the sequences immediately distal to the 3' end of the 36S rRNA transcriptional unit. This region is primarily noncoding, with the exception of a single 5S rRNA gene (2). (Preliminary mapping data suggest that this gene is located near the terminus of the rDNA molecule, ~8 kb from the 3' end of the 36S rRNA gene.) We have used end-labeling probes from both sides of the terminal spacer to examine phasing in this region. Probe 4 (from A. Weiner, ref. 5; see Fig. 1) was used as a terminal indirect endlabel for MNase- or DNase I-digested rDNA; the results are shown in Figs. 2D and 2E. Probes 5 and 6 were used as end-labels for partially-digested rDNA bands Sal I-2 and Pst I-2, respectively (see Fig. 1); these results are shown in Figs. 2F and 2G, respectively.

The results of the terminal spacer experiments can be summarized as follows: a) MNase and DNase I exhibit qualitatively similar patterns of digestion throughout the terminal spacer region (compare Figs. 2D and 2E), b) both nucleases exhibit preferences for certain sequences or structures in naked DNA, but c) the sites in naked DNA at which the nucleases cleave most readily appear to be the same sites that are most protected in chromatin. This striking, highly phased pattern is clearly seen by comparing any naked DNA lane with the adjacent chromatin lane in Figs. 2D, 2E, 2F, or 2G. The sites attacked most frequently in naked DNA and the sites cleaved most often in chromatin are almost mutually exclusive, resulting in a striking "positive/negative" pattern.

It is of additional interest to note that in the center of this terminal spacer, the chromatin sites are spaced roughly 400 bp apart, the length expected for 2 nucleosomes. In between each of the chromatin sites there are 2 naked DNA sites (see Figs. 2F and 2G). This is not seen in Figs. 2D or 2E since the resolution of bands is disrupted by the slightly variable end length of the rDNA molecule (5).

The Central Spacer

When probe 7 (from A. Weiner) is hybridized to Southern blots of partially MNase- or DNase I-digested rDNA (see Fig. 1), one observes a very long, distinct, 180 bp ladder in both naked DNA and chromatin digested with either enzyme, although in chromatin digested with MNase, some of the "rungs" of the ladder are cleaved much more frequently while others are barely visible (Fig. 2H). In this experiment, identical patterns of naked DNA and chromatin could have been generated in 2 ways: the naked DNA and chromatin sites could be a) the same, or b) exactly alternate. Since end-labels from this region are unavailable, we tested these hypotheses by hybridizing probe 8 to MNase-digested, Sal I-digested DNA (Fig. 2I). If hypothesis a is correct, then the naked DNA

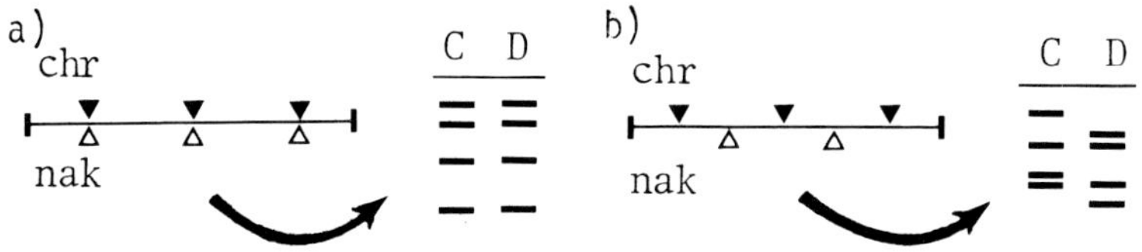

and chromatin fragments should still be the same size. If hypothesis b is correct, then some of the naked DNA fragments will differ in size from chromatin fragments (those in which one end is the Sal I site), while other naked DNA and chromatin fragments will be the same size (those in which both ends are internal to the Sal I sites). Figure 2I shows that hypothesis b is correct. Thus, by extrapolation, at least 20 kb of the central spacer region also appears to be nucleosomal and phased in such a way that the sites attacked most readily by nucleases in naked DNA are the same sites that are most protected in chromatin.

Salt Extractions of Nuclei

When nuclei are washed with different concentrations of salt, chromosomal proteins are dissociated according to their affinity for the DNA. By observing the changes in chromatin structure that occur after salt washes, one can gain some insight into the nature of the DNA-binding proteins (6). We washed nuclei with increasing concentrations of KCl prior to nuclease treatment, then hybridized the resulting DNA to probe 4 (Fig. 3A). Our results show that most of the bands in the standard hybridization pattern (0.06 M KCl) remain unchanged at 0.35 M and 0.55 M KCl (as does the bulk chromatin MNase ladder), but these bands revert to the naked DNA pattern when washed with 0.75 M KCl, coincident with the loss of the MNase ladder in bulk chromatin, presumably because the core histones have been dissociated. These data suggest that most of the phased pattern seen after nuclease digestion of the terminal spacer is due to the presence of nucleosomes.

When the same DNA filter is hybridized to probe 7 (Fig. 3B), one observes a marked change in the normal hybridization pattern when the nuclei are washed in 0.35 M KCl. Under standard conditions (0.06 M KCl), some of the "rungs" of the MNase ladder are cleaved at a much higher frequency than others (see Fig. 2H). When the nuclei have been washed with 0.35 M KCl however, each band appears to be cleaved with roughly equal frequency. These data suggest the presence of a supernucleosomal structure that is disrupted because a binding component of the structure has been dissociated. The resulting pattern observed after washing the nuclei with 0.35 M KCl is a 180 bp nucleosomal ladder.

DISCUSSION

Our results show that the 36S coding region of the rDNA palindrome is composed, at least in part, of nucleosomes

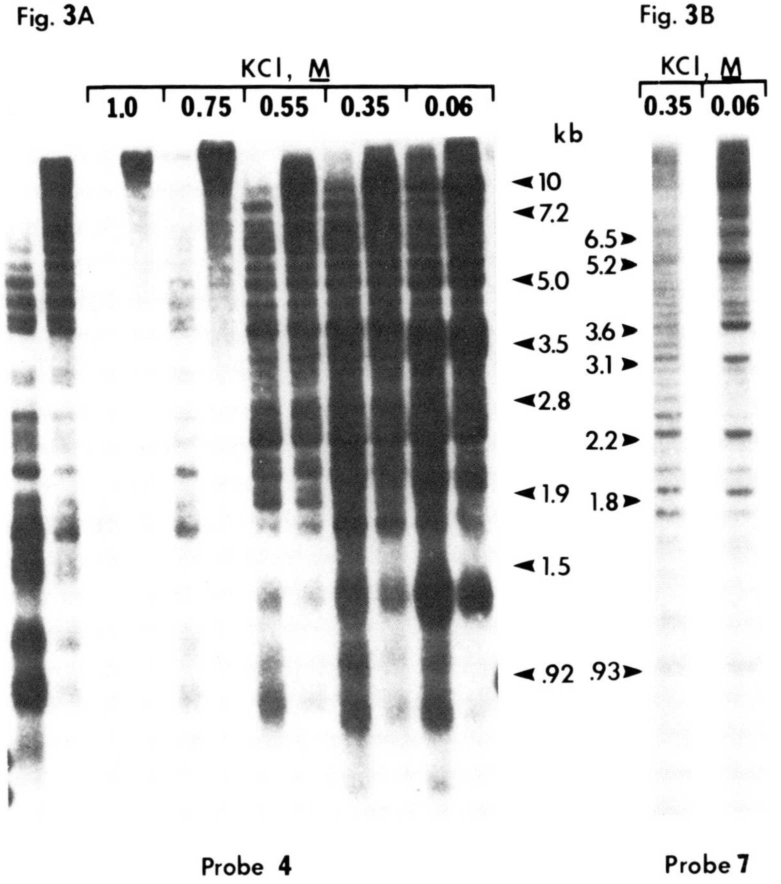

FIGURE 3. Nuclei were washed in a standard nuclease digestion buffer containing 0.06, 0.35, 0.55, 0.75, or 1.0 M KCl, resuspended in buffer containing 0.06 M KCl, then partially digested with MNase. Naked DNA (nak) was used as a control. Southern blots were prepared and hybridized to probe 4 (Fig. 3A) or probe 7 (Fig. 3B).

that exhibit the same linker length as bulk Dictyostelium chromatin. This region is relatively more sensitive to nucleolytic attack than either bulk chromatin or the non-coding regions of the rDNA; similar results have been obtained for a large number of other transcribed genes (1). The structure of chromatin in the 10 kb terminal spacer is strikingly different from that of the coding region. Here, the chromosomal proteins are distinctly phased; furthermore, salt extraction data supports the idea that the chromatin is composed of histone-containing nucleosomes. Many past arguments in favor of phasing have been complicated by the sequence preferences of the "nonspecific" endonucleases, MNase and DNase I (see ref. 7). Our data are not effected by these problems since the preferred sites in naked DNA are actually the sites most protected in chromatin. This observation has led us to postulate that the sequences or structures that are most nuclease sensitive are perhaps the same sequences or structures that direct nucleosome placement. In support of this theory we note that, in the middle of the terminal spacer, the chromatin sites are 400 bp apart, approximately the length expected for a dinucleosome. If there are indeed "phasing sites" for each nucleosome, then one would expect to find 2 naked DNA sites in-between each 2 chromatin sites. This is, in fact, the observed result (see Figs. 2E and 2F). We do not mean to imply that all nucleosomes are phased (certainly those in the coding region are not), nor that nucleosomes in other noncoding regions will necessarily exhibit such a high degree of phasing. However, if "phasing sites" exist, the rDNA terminal spacer appears to be an ideal system in which to look for such sequences.

Finally, we have presented evidence for a supernucleosomal structure in the central region of the rDNA palindrome. Our data shows that relatively low salt concentrations disrupt this structure, which is observed when nuclei are isolated under standard salt conditions. The 0.35 M salt wash is not sufficient to dissociate core histones since the MNase ladders observed for bulk chromatin are unchanged with salt washes of up to 0.55 M KCl. This suggests that there is a loosely bound component that can be disturbed under conditions where nucleosomes remain intact.

ACKNOWLEDGMENTS

We are very grateful to Dr. Robert Simpson for allowing C.A.E. to complete work in his laboratory at the NIH and to Dr. Alan Kimmel for helpful comments on the manuscript.

R.A.F. was the recipient of an ACS Faculty Research Award. C.A.E. was the recipient of an NSF Predoctoral Fellowship and has been supported by a USPHS Training Grant.

REFERENCES

1. Weisbrod S (1982). Active Chromatin. Nature 297:289.
2. Kimmel AR, Firtel RA (1982). The organization and expression of the Dictyostelium genome. In Loomis WF (ed): "The Development of Dictyostelium discoideum," New York: Academis Press, p 233.
3. Ness PJ, Labhart P, Banz E, Koller T, Parish RW (1983) Chromatin structure along the ribosomal DNA of Dictyostelium: regional differences and changes accompanying cell differentiation. J Mol Biol 166:361.
4. Mangiorotti G, Altruda F, Lodish HF (1981). Rates of synthesis and degredation of ribosomal ribonucleic acid during differentiation of Dictyostelium discoideum. Mol Cell Biol 1:35.
5. Emery HS, Weiner AM (1981). An irregular satellite sequence is found at the termini of the linear extrachromosomal rDNA in Dictyostelium discoideum. Cell 26:411.
6. Worcel A, Giuseppe G, Jessee B, Udvardy A, Louis C, Shedl, P (1983). Chromatin fine structure of the histone gene complex of Drosophila melanogaster. Nucl Acids Res 11:421.
7. McGhee JD, Felsenfeld G (1983). Another potential artifact in the study of nucleosome phasing by chromatin digestion with micrococcal nuclease. Cell 32:1205.

Molecular Biology of Development, pages 373–382

TIGHTLY LINKED GENES WITH DIFFERENT MODES OF DEVELOPMENTAL REGULATION IN *DICTYOSTELIUM*

Alan R. Kimmel

Laboratory of Cellular and Developmental Biology
NIADDK (6/B1-12)
National Institutes of Health
Bethesda, Maryland 20205

ABSTRACT The M4 region of the *Dictyostelium* genome contains two divergent transcription units; the 5'-ends of these genes are separated by ~1.5 kb. These genes exhibit completely opposite modes of developmental regulation. M4-1 is expressed in vegetative cells but not in cells late in development. Its expression is repressed by developmental pulses of cAMP. In contrast, the expression of M4-4 increases during development. M4-4 contains a short, developmentally regulated repeat. The increase in expression of the repeated sequence during development parallels that of the unique portion of the M4-4 gene, suggesting coordinate regulation of the M4-4 repeat gene set.

INTRODUCTION

Dictyostelium grow vegetatively as ameboid single cells. When cells are starved by either removing amino acids from the media or plating the cells on filter pads in a low salt buffer, synchronous development is initiated. Approximately, 4 hr after starvation, gradients of cAMP are established; ameba move toward regions with high cAMP concentrations and secrete cAMP to propagate the concentration waves. After relaying the cAMP signal, cells become transiently refractory to further stimulation. During this period, extracellular cAMP is destroyed by membrane bound and secreted forms of phosphodiesterase (PDE) and then PDE is inactivated by PDE-inhibitor. This cycle repeats with cAMP pulses initiating from aggregation centers with periodicities of ~6 min.

The result is the assembly of multicellular aggregates, each containing ∿10^5 cells. Each aggregate then develops into a mature fruiting body consisting of spore and stalk cells.

We have previously isolated a Dictyostelium genomic recombinant plasmid denoted M4 which contains two divergent transcription units (Figure 1). We report here studies on the independent regulation of these genes during the developmental cycle. The single-copy M4-1 gene encodes a 0.9 kb mRNA present at 0.1% of total poly(A)+ RNA in vegetative cells but absent from cells late in development. Its repression is coincident with the developmental cAMP pulse; M4-1 is rapidly reexpressed during dedifferentiation. The second transcription unit M4-4 is composed of repeat and single copy DNA. M4-4 mRNA represents 0.01% of total mRNA in vegetative cells but its relative level of expression increases ∿5 fold during early development.

The repeated sequence associated with the M4-4 gene is found in ∿100 copies in the Dictyostelium genome, and is present on ∿1% of poly(A)+ RNA in vegetative cells. It is not transcribed however, in the M4-4 mRNA. The M4-4 RNA population is heterogeneous in size and ∿90% of its mass is complementary to single-copy sequences. From comparative DNA sequence analyses of various genomic and cDNA sequences complementary to the M4 repeat, we have shown that the common repetitive element is $(AAC/GTT)_n$. This sequence is asymmetrically represented in RNA and hybridization studies suggest that there is a coordinate increase in expression of many members of this repeat gene family during development.

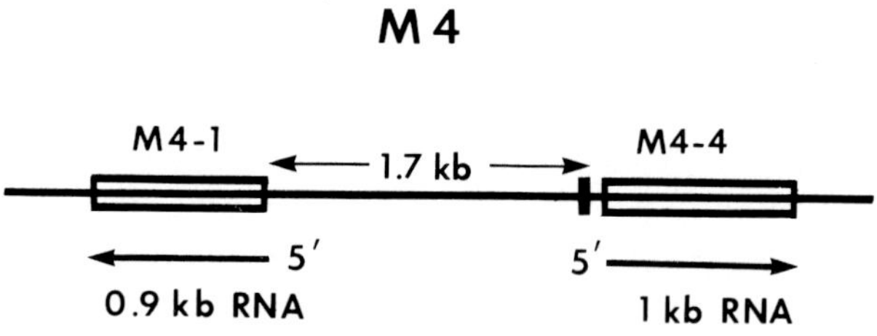

FIGURE 1. The M4 Region. M4-1 encodes a 0.9 kb mRNA and M4-4 encodes a 1 kb mRNA. Directions of transcription are indicated, as is the distance between 5'-ends. The black box in M4-4 indicates the location of the M4-4 repeat.

RESULTS

Developmental Expression of the M4-4 Gene and its Associated Repeat Element.

Poly(A)+ RNA was purified from vegetative cells and cells at various stages in the developmental cycle. Corresponding RNA blots were hybridized to an M4-4 gene probe to monitor its developmental pattern of expression (Figure 2A). M4-4 mRNA is present during vegetative growth, but its level of expression increases 5-10 fold as development procedes. The repeat hybridizes with a heterogeneous pattern at all stages with an increase in developmental expression similar to that observed for the unique M4-4 gene (Figure 2B).

The M4-4 Repeat is $(AAC)_n$ and Lies 5' to the Gene

We have previously identified a structure unique to the 5'-ends of protein coding genes in *Dictyostelium* (1,2) and have suggested that, in addition to the TATA box present in most eukaryotic genes, an oligo(dT) stretch, which lies between the TATA box and the translational start site, is an essential component of the *Dictyostelium* promoter. Using the Berk-Sharp S_1 nuclease technique we have mapped the 5'-end of the M4-4 hnRNA and hence its transcription initiation site. The M4 gene possesses the TATA box and the oligo (dT) stretch both within 40 bp of the 5' end of the mRNA. 5' to this conserved region, we have identified another sequence of particular interest, AACAACAACACAACAACAACAACAACAACAAC, $(AAC)_n$. This sequence would lie within 70 bp of the transcription initiation site of M4-4.

In order to define the limits of the M4 repeat, we constructed and analyzed a series of deletions. The resulting constructions were screened by hybridizing them to *Dictyostelium* genomic DNA blots to determine if they retained a *Dictyostelium* repetitive component; their sequence analyses suggested that the $(AAC)_n$ sequence is the repetitive element. We have hybridized a *Dictyostelium* Southern blot with *in vitro* labelled $(dAAC/dGTT)_n$ synthetic homopolymer (Figure 3). The results show a pattern identical to that of the M4-4 repeat. Approximately 100 $(AAC)_n$ blocks are present in the *Dictyostelium* genome and on a large number of poly(A)+ cytoplasmic RNAs. The sequence is asymmetrically expressed; in most genes associated with the $(AAC)_n$ repeat, the $(AAC)_n$ se-

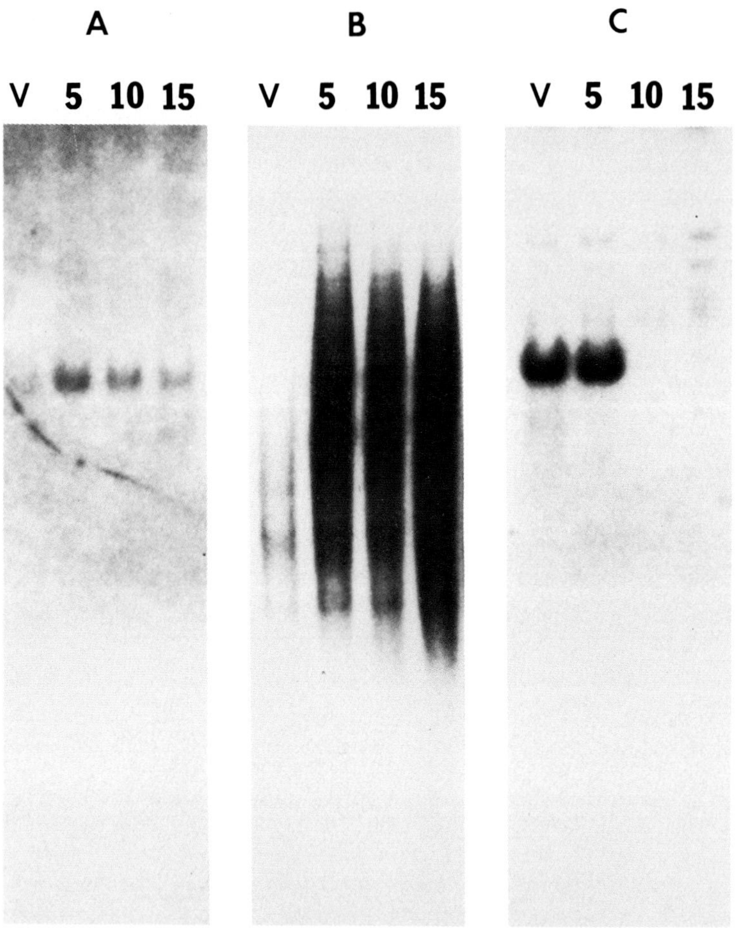

FIGURE 2. Developmental Expression of the M4 Genes. Equal amounts of Poly(A)+ RNA from vegetative (V) cells and cells at 5, 10 and 15 hr in development were hybridized on RNA blots to the M4-4 gene (A), the M4-4 repeat (B) and the M4-1 gene (C).

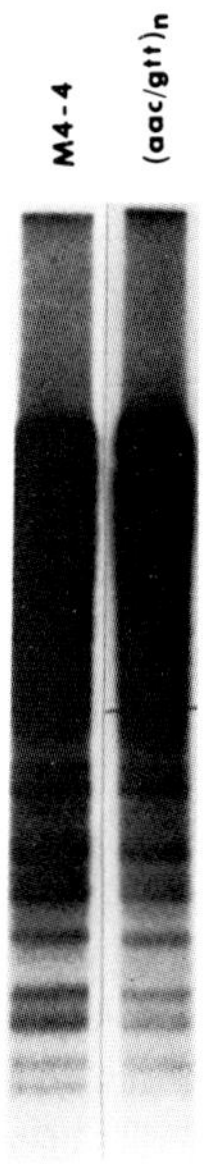

FIGURE 3. DNA blot hybridization of Eco RI, Hae III and Hap II digested Dictyostelium DNA to the M4-4 repeat and an $(ACC/GTT)_n$ homopolymeric probe.

quences are on the mRNA. The M4 mRNA does not have the repeat, although the $(AAC)_n$ sequences lie upstream from the 5' terminus in the same polarity as transcription.

M4-1 is Repressed by Pulses of cAMP during Early Development

As a comparison with the M4-4 gene, the developmental expression of the M4-1 gene was also monitored (Figure 2C). M4-1 mRNA is present at similar levels in bacterially grown vegetative cells and cells early in development. By aggregation (10 hr) the level of M4-1 RNA has decreased to undetectable levels (<1 mRNA/cell). The M4-1 does not reappear during development.

Since the levels of M4-1 RNA are similar in vegetative and 5 hr developed cells it seemed unlikely that the initiation of development *per se* would yield the change in expression observed later in development. Rather, the timing of M4-1 repression appeared to correlate with the kinetics of cAMP signalling during early development. Thus, we examined the expression of M4-1 in cells in shaking culture with and without exposure to cAMP.

Log phase cells were collected from growth media, resuspended in low salt buffer (PDF) at 2 x 10^6 cells/ml and shaken rapidly. Under these conditions the cells neither generate an endogenous cAMP signal as observed in denser cultures shaken at lower speeds nor form agglomerates observable by light microscopy.

One culture received pulses of cAMP to 25 nM at 6 min intervals. Samples were taken for assay to confirm that the cAMP concentration was not cumulative and that indeed these conditions were mimicking the cAMP pulses observed in normal development. A second culture was adjusted to 500 μM cAMP and received cAMP to an additional 100 μM every 60 min. As expected, the cAMP concentration was cumulative. These conditions are sufficient to completely inhibit development. A third culture was not treated with cAMP. Poly(A)+ RNA was isolated from vegetative cells and from the three suspension cultures at incubation times of 2.5, 5, 7.5 and 10 hr. and hybridized to an M4-1 probe (Figure 4).

Cells incubated with high concentrations of cAMP as well as those incubated without cAMP continue to express the M4-1 gene at a level similar to that of vegetative cells. Little or no change in expression is seen in cells pulsed with cAMP for 5 hrs or less. By 7.5 hr, the level of the M4-1 RNA has declined and by 10 hrs, it has decreased more than ten-fold relative to that of vegetative cells. The timing of M4-1 mRNA decay parallels that observed in normal development and the kinetics of the endogenous developmental cAMP pulse. It should be noted that although the cAMP pulse in suspension culture was initiated at the onset of starvation, cells in suspension signal and relay with a timing coincident with that of normally developing cells. Thus, pulses of cAMP appear sufficient to elicit the repression of the M4-1 gene.

M4-1 is Induced During Dedifferentiation

Developing cells will rapidly dedifferentiate if re-exposed to fresh growth media. Complete dedifferentiation should yield vegetative cells that express M4-1 at normal levels. Vegetative cells were pulsed with cAMP in PDF to depress M4-1 expression and then resuspended in fresh media. Within 2 hr, the M4-1 mRNA begins to re-accumulate and reaches vegetative levels within 4 hr (Figure 5). Re-exposing the cells to growth media for 45 min results in a reappearance of

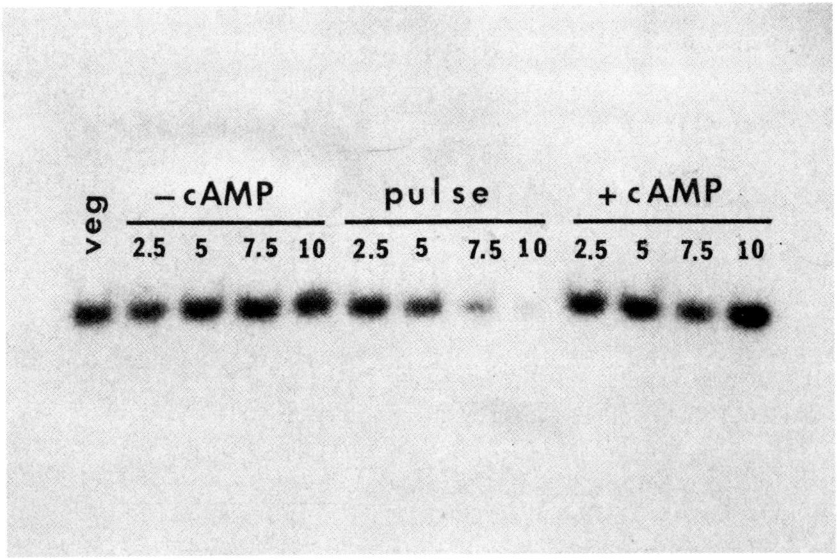

FIGURE 4. Expression of M4-1 in cultures with and without exogenous cAMP. Poly(A)+ RNA was isolated from vegetative cultures (veg), cultures without cAMP (-), with exogenously added pulses of cAMP (pulse), and with continuous high levels of cAMP (+). RNA blots were prepared and hybridized to M4-1 probe.

M4-1 RNA in both the nucleus and the cytoplasm. It is interesting to note that the initial relative accumulation of M4-1 sequences may be greater in nuclei than in the cytoplasm. These results suggest strongly that the M4-1 gene is actively transcribed in dedifferentiated cells and that developmental cAMP pulses effect an actual repression of M4-1 gene activity.

DISCUSSION

We have described the organization and expression of an interesting gene region in Dictyostelium. M4 contains two developmentally regulated genes which are divergently transcribed. In addition, these genes exhibit completely different patterns of regulation; their organizations are suggestive of individual gene regulation rather than control of large

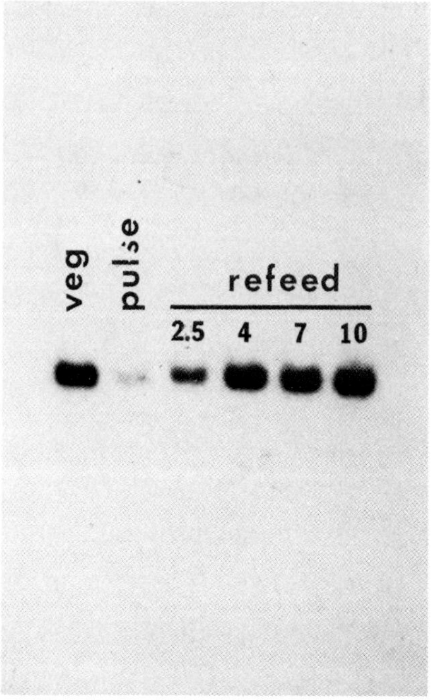

FIGURE 5. Expression of M4-1 during Dedifferentiation. Vegetative (veg) cells were pulsed (pulse) with cAMP and subsequently refed with fresh media for 2.5, 4, 7 and 10 hr. Poly(A)+ RNA was isolated and hybridized on blots to M4-1 probe.

chromosomal regions. If 5' upstream regions were regulatory elements associated with these different modes of developmental expression they would lie within ∿2 kb of each other in the genome. It is likely that all the regulatory regions of both genes are present in the 4.5 kb genomic clone.

M4-1 is the first vegetative gene to be identified whose transcription is negatively modulated by cAMP-pulses early in Dictyostelium development. Previous studies of proteins synthesized in vegetative and developing cells indicate that ∿10 moderately expressed genes would show the same developmental kinetics of expression (3). Hybridization studies also indicate that there is a very limited class of vegetative genes that are repressed at this development stage (1, 3). An identification and comparison of regulatory sequences within such a limited gene set may lead to a clearer understanding of the mechanisms involved in repressing genes during early development as well as reinducing them during dedifferentiation.

The M4-4 gene is part of a large family of single-copy sequences associated with a developmentally regulated, short, interspersed, repeat element. 1% of the poly(A)+ RNA from vegetative cells contains sequences that hybridize to this repeat. The complementary RNA is heterogeneous in size and 90% of its mass hybridizes to single-copy DNA. A series of genomic DNAs and cDNAs derived from poly(A)+ RNA have been isolated which are complementary to the repeat. Comparisons of the various genomic and cDNA sequences indicate that $(AAC/GTT)_n$ is the common sequence element. The repeat does not occur in long arrays (as satellite DNA) but in ∿100 short (∿35 to 150 bp) tandem blocks per haploid genome interspersed with single-copy DNA. Probes from regions adjacent to this element or probes specifically deleted of $(AAC/GTT)_n$ sequence hybridize to unique restriction fragments on DNA blots and unique poly(A)+ RNA species on RNA blots. The $(AAC/GTT)_n$ is asymmetrically transcribed with only $(AAC)_n$ sequences represented in RNA. The repeat has been localized 70 bp 5' to the transcription initiation site of one gene, while the M4 repeat lies toward the 5'-end of other cDNA clones. The $(AAC)_n$ gene family is expressed with a specific developmental pattern. Individual $(AAC)_n$-associated genes have a developmental pattern of expression suggestive of the coordinate expression of many $(AAC)_n$-gene family members.

Very short repeats have been implicated in the control of heat shock genes, metallothionein genes, amino acid metabolism genes, and steroid-inducible genes, among others (see 4). Unlike the M4 repeat however, these repeats are not themselves transcribed. Immunoglobulin genes have recently been shown to contain sequences within their transcribed introns which are required to promote their own expression (see 5). These "enhancers" are cell-type specific and position independent in their effect on gene expression. Enhancers have also been described in viral systems and an H2A sea urchin histone gene located 5' to transcription initiation.

The M4-4 repeat possesses some enhancer characteristics. The M4-4 repeat, located 5' to transcription initiation as well as at the 5'-ends of cDNAs, appears to be position independent and temporally regulated in expression. Unlike most known enhancers, however, the M4-4 sequence is not polarity independent. The asymmetry of transcription reinforces the suggestion that M4-4 is regulatory in nature. It would appear likely that the specific orientation of the repeat to the transcription units is related to function. Its simple sequence would be compatible with a unique protein-sequence interaction.

REFERENCES

1. Kimmel AR, Firtel RA (1982) The organization and expression of the Dictyostelium genome. In Loomis WF (ed): "The Development of Dictyostelium discoideum", New York: Academic Press, p. 233.
2. Kimmel AR, Firtel RA (1983) Sequence organization in Dictyostelium: Unique structure at the 5'-ends of protein coding genes. Nucl. Acids Res. 11: 541.
3. Lodish HF, Blumberg DD, Chisholm R, Chung S, Coloma A, Landfear S, Barklis E, Lefebvre P, Zuker C, Mangiarotti G (1982) Control of gene expression. In Loomis WF (ed): "The Development of Dictyostelium discoideum". New York: Academic Press, p. 325.
4. Davidson EH, Jacobs HT, Britten RJ (1983) Very short repeats and coordinate induction of genes. Nature 301: 468.
5. Marx JL (1983) Immunoglobulin genes have enhancers. Science 221: 735.

Molecular Biology of Development, pages 383–394

EXPRESSION AND REGULATION OF CHICKEN ACTIN GENES IN AVIAN AND MURINE MYOGENIC CELLS

Bruce M. Paterson, Anne Seiler-Tuyns and Juanita D. Eldridge

Laboratory of Biochemistry, National Cancer Institute
National Institutes of Health, Bethesda, Maryland 20205

ABSTRACT Using a primer extension assay that is diagnostic for the mRNA levels from the β cytoplasmic, α cardiac, and α skeletal actin genes of the chicken, we have determined the level of isoform expression in chicken breast muscle during myogenesis. The same assay, in conjunction with 3' S1 analyses, was used to measure the level of expression and regulation of these chicken genes introduced into mouse myogenic and nonmyogenic cells. The α cardiac actin gene is the major skeletal actin isoform expressed in embryonic chicken breast muscle. All the chicken actin genes are expressed in a murine cell background; however, the level of expression is tissue dependent. The chicken α actin genes are not regulated during myogenesis in a murine myogenic cell line even though the endogenous mouse α genes regulate. The chicken β actin gene regulates in parallel with the mouse β gene providing a model system for the down regulation of gene expression during development.

INTRODUCTION

One approach to the study of cellular differentiation involves the introduction of cloned genes into cells which will differentiate _in vitro_ and possibly regulate the expression of the genes of interest. We have isolated the β actin, α cardiac actin, and the α skeletal actin genes from the chicken and have introduced them into mouse L-cells and into mouse myogenic cells (C2C12).

The differentiation process in avian and murine muscle tissue culture systems has been well documented (1-4).

During this process the pattern of actin gene expression changes. In the dividing myoblast β-actin is the predominant cytoplasmic isoform and one cannot detect either of the sarcomeric α actins. Once cell fusion has taken place the synthesis of β actin is greatly reduced and the synthesis of α actin begins. We have determined the levels of actin isoform expression during myogenesis in avian breast muscle and have analyzed the expression and regulation of the chicken actin genes in two different mouse cell backgrounds. The results indicate cell background can modulate the level of expression of the sarcomeric genes. Furthermore, the regulated expression of the chicken β actin gene in murine myogenic cells implies the sequences responsible for this regulation are not species specific. The lack of regulation of the sarcomeric actins suggests a more complicated picture for gene activation compared to gene repression during development and cellular differentiation.

MATERIALS AND METHODS

Cell culture and DNA transfection methods are as previously described (5-10). The chicken actin genes were isolated as reported (11). S1 analyses and primer extension assays were performed as described (PNAS, in press). RNA isolation and Northern analysis were described earlier (7, 12, 13).

RESULTS

Actin Isoform Expression During Avian Myogenesis

There are at least six actin isoforms in adult mammalian tissues (14-15). These include two sarcomeric forms (α cardiac and α skeletal), two smooth muscle forms (α and γ smooth) and two cytoplasmic forms (β and γ cytoplasmic). The conservation in amino acid sequence is striking in that the skeletal muscle isoform differs in 4/375 amino acids from cardiac actin, 6/375 amino acids from the γ smooth isoform, 8/375 amino acids from the α smooth isoform and 24-25/375 from the β and γ cytoplasmic isoforms. To avoid homology problems in the isoform assay a primer extension protocol was employed rather than RNA dot blots or Northern blots to determine the level of expression of a given actin isoform. The method assumes each actin gene initiates transcription at a nucleotide that would yield a primer extended fragment unique for each actin gene. The validity of the approach is demonstrated in Figure 1.

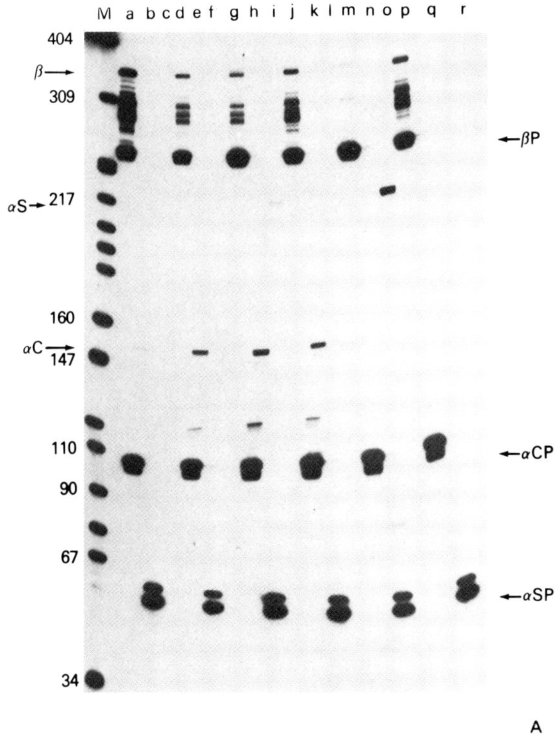

Figure 1. Primer extension analysis of the actin isoform mRNAs during avian myogenesis in vitro and in vivo. α C, α S, and β indicate the positions of the primer extended fragments specific for α cardiac actin (149 base pairs), α skeletal actin (218 base pairs), and β cytoplasmic actin (347 base pairs), respectively. P designates the primer location. M is the HpaII cut pBR322 markers. Source of the RNA: (a-c) 36 hour prefusion myoblast cultures, (d-f) 92 hour fused cultures, (g-i) one day post

hatch cardiac muscle, (j-1) 15 day embryonic breast muscle, (m-o) 5 week post hatch breast muscle, (p-r) 15 day embryonic brain tissue.

Greater than 90% of the sarcomeric actin mRNA transcripts represent α cardiac actin in 15 day embryonic chick breast muscle, in vitro and in vivo. Five weeks after hatching there is no detectable α cardiac mRNA in breast muscle and expression has switched entirely to the α skeletal actin isoform. No sarcomeric actin mRNA was detectable in embryonic chick brain although β actin is abundantly expressed. This latter observation is in contradiction to a previously published report by Ordahl and coworkers (16). Our results are in agreement with the results of Minty et al (4) who found a similar situation in embryonic mouse muscle. Thus the predominant actin isoform expressed in embryonic avian and murine skeletal muscle is α cardiac actin, the adult cardiac actin sarcomeric isoform.

Introduction of the Chicken Actin Genes into Murine Myogenic and Nonmyogenic cells

The PSV2-Gpt constructs containing the chicken actin genes are illustrated in figure 2. The arrow above the restriction map of each gene gives the orientation of the gene with respect to the SV-40 early promotor and the Eco GPT gene.

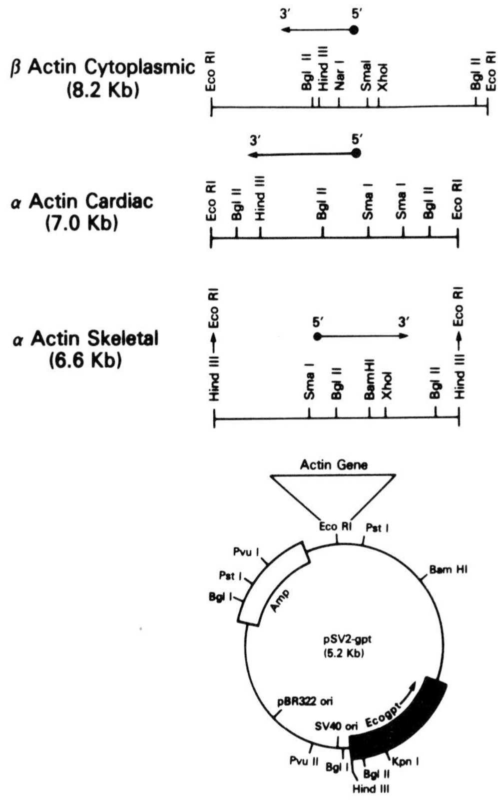

Figure 2. PSV2-gpt constructs containing the chicken actin genes. The actin genes were inserted into the Eco-Rl site in the vector in the 5' to 3' orientation indicated by the arrow. The PSV2-gpt vector is 5.2Kb in length. The β actin gene is 8.2Kb, the α cardiac gene is 7.0Kb, the α skeletal gene is 6.6Kb in length. The direction of transcription of Eco-gpt is shown by the small arrow.

The actin genes introduced into the different mouse cell backgrounds were initially tested for expression and regulation utilizing an Sl nuclease assay with 3' specific noncoding probes. This was compared to the GPT expression from the vector. As shown in figure 3, no discrete fragments were observed with mRNA prepared from the murine cells that did not contain one of the chicken actin genes.

Figure 3A

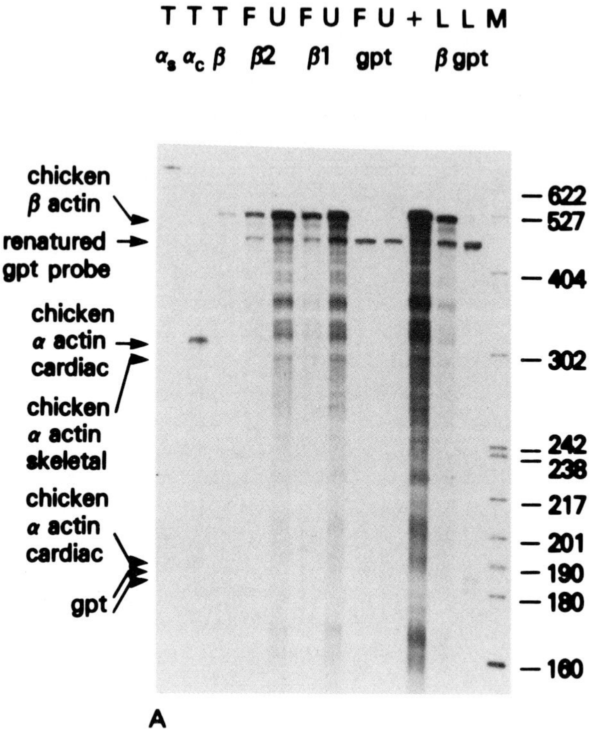

Figure 3. 3' S1 analysis of the RNA from murine cells containing the PSV2-gpt actin constructs. A) the β actin transformants. B) the α cardiac actin transformants. C) the α skeletal actin transformants. The symbols used are: (M), HpaII digest of pBR322; (L-GPT), L-cells transformed with PSV2-gpt; (Lb,LaC,LaS) L-cells transformed with the β actin, α cardiac actin, and α skeletal actin PSV2-gpt constructs, respectively; (+), the positive control with chicken muscle RNA; (U) unfused myogenic cell RNA; (F), fused myogenic cell RNA; (T), transient expression of the indicated gene 60 hours after fusion.

Figure 3B

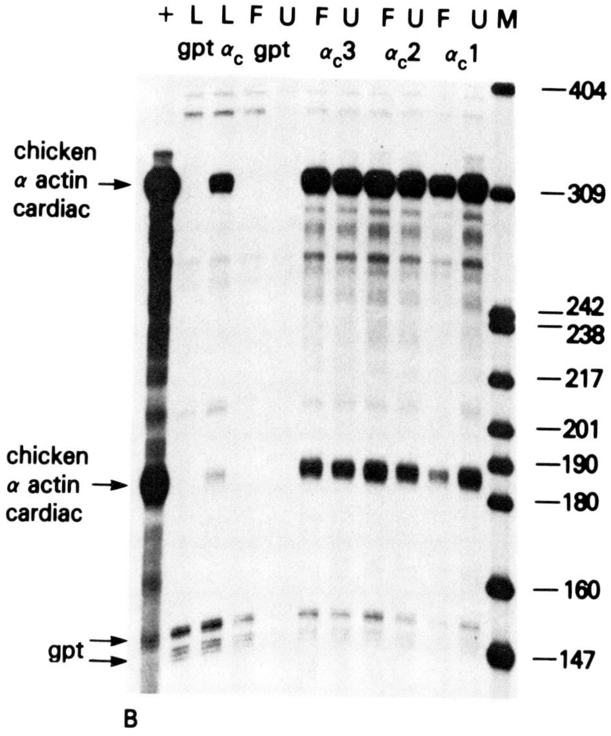

The intensity of the band is proportional to the actin mRNA transcript level expressed in the different mouse cell backgrounds. As shown, all of the chicken actin genes are functional in both myogenic and nonmyogenic murine cells. However, the level of expression is dependent upon the particular mouse cell background. The α sarcomeric actins are expressed poorly in L-cells whereas the β cytoplasmic actin transcripts are clearly more abundant. In the myogenic C2 cells the α cardiac actin and the β actin genes are efficiently transcribed whereas the α skeletal actin gene is expressed at a much lower level. Following cell fusion the

Figure 3C

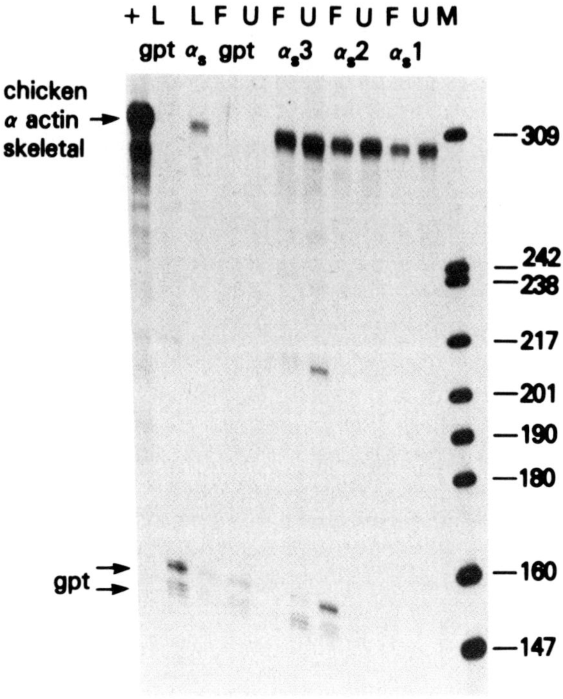

C

level of RNA from the α actin genes remains unchanged; however, there is a substantial decrease in the β actin transcripts when compared to Eco-gpt transcripts: densitometric scans reveal β actin transcripts decrease 7-8 fold whereas Eco-gpt varies no more than 2-fold. Thus by S1 analysis all of the chicken actin genes are functional in a mouse cell background. Only the β actin transcript level decreases in parallel with the endogenous mouse gene. An induction of either of the α actins is not seen even though the endogenous mouse α actin gene expression is induced with cell fusion (4,6).

The Chicken Actin Transcripts Initiate Correctly in the Murine Cell Background

The primer extension assay presented in figure 1 was employed to determine if the various actin gene transcripts were initiated and processed correctly in the murine cell background. As shown in figure 4, the RNA transcribed from the transfected β and α cardiac actin genes produce extended fragments identical in length to those synthesized with control RNA. Furthermore, since all three genes contain large introns in the 5' noncoding leader region, the results suggest the correct processing has occurred. The low level of expression of the α skeletal actin gene precluded routine primer extension analysis yet results with long term exposures show the correct 5' end for these transcripts.

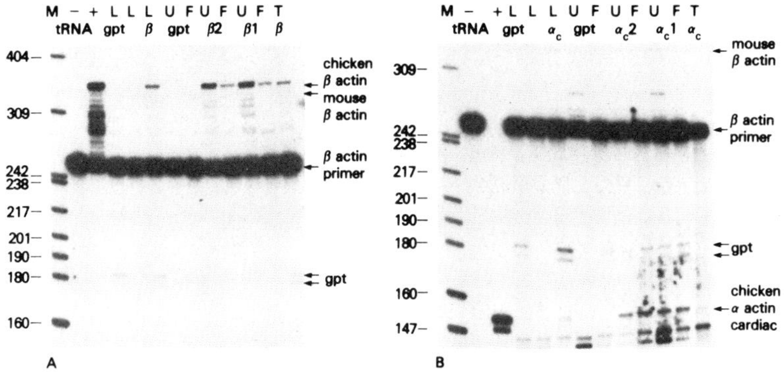

Figure 4.

Figure 4. Primer extension analysis of the 5' ends of the chicken actin transcripts in murine cells. A) The β actin transformants. B) the α cardiac transformants. (-tRNA), extension with 10 μg of tRNA, (+) positive control with chicken muscle RNA, (L-GPT) L-cells transformed with PSV2-gpt, (L) L-cells alone, (Lb,La) L-cells transformed with the β actin and α cardiac constructs, respectively, (U) unfused myogenic cells, (F) fused myogenic cells, (T) transient expression of the indicated actin gene in myogenic cells.

DISCUSSION

The preliminary results reported here clearly demonstrate the chicken actin genes are transcriptionally active in two types of mouse cells and produce RNA templates with the correct 5' and 3' termini. Whether these templates are translationally active remains open.

The level of expression from the different chicken actin genes in mouse cells is a reflection of the cell background and mirrors, to a degree, the situation in vivo. Both the sarcomeric actins are poorly expressed in L-cells whereas the β actin gene, active in most dividing cells, is expressed at relatively high levels. By comparison, all the chicken sarcomeric actin gene transcripts are more abundantly expressed in myogenic cells. This again suggests the quantitative differences in the levels of expression are tissue related.

The striking result is the reduction in the level of the β actin transcripts in parallel with the endogenous mouse β actin mRNA during myogenic cell differentiation. All of the transformants to date have given the same result with a 7-8 fold reduction in β actin transcripts relative to a 2-fold variation in GPT expression. Previously reported results (17, 18) suggest the quantitative differences in β actin gene expression during myogenesis are not explained by differential mRNA stability.

Transcription from both the sarcomeric genes is unchanged with myogenic cell differentiation, even though the endogenous mouse α actin transcripts increase (4, 6). A variety of explanations can be considered. Species differences may play a role but heterokaryon studies (6) and examples of regulation in heterologous systems (19-22, 23, 24) do not support this interpretation. Chromosomal location and/or structure (25, 26), the extent of flanking DNA sequence (27), and methylation (28) are likely to play substantial roles in the regulation of tissue specific gene expression with differentiation. The down regulation of the β actin gene with myogenesis provides a model system for the analysis of gene repression in eukaryotes during development.

REFERENCES

1. Paterson BM, Strohman RC (1972). Dev Biol 29:113-138.
2. Yaffe D, Dym H (1972). Cold Spring Harbor Symposium Quant Biol 37:543-547.
3. Emerson CP, Bechner SK (1975). J Mol Biol 93:431-447.
4. Buckingham ME, Minty AJ In "Eukaryotic Genes: Their Structure, Activity and Regulation", Butterworths (in press).
5. Yaffe D, Saxel O (1977). Nature 270:725-727.
6. Blau HM, Chiu CP, Webster C (1983). Cell 32:1171-1180.
7. Seiler-Tuyns A, Birnstiel ML (1981). J Mol Biol 151: 607-625.
8. Chu G, Sharp P (1981).Gene 13:197-202.
9. Tsui L-C, Breitman M, Siminovitch L, Buchwald M (1982). Cell 30:499-508.
10. Gorman CM, Moffat LF, Howard BH (1982). Mol Cell Biol 2:1044-1051.
11. Zehner Z, Paterson BM (1983). Proc Nat Acad Sci USA 80:911-915.
12. Paterson BM, Roberts BE (1981). In "Gene Amplification and Analysis" 2:417-437.
13. Alwine JC, Kemp DJ, Stark GR (1977). Proc Nat Acad Sci USA 74:5350-5353.
14. VandeKerckhove J, Weber K (1979). Differentiation 14: 123-133.
15. VandeKerckhove J, Weber K (1978). J Mol Biol 126:783-802.
16. Ordahl CP, Tilghman SM, Ovitt C, Fornwald J, Longen MT (1980). Nucleic Acids Res 8:4989-5005.
17. Singer RH, Kessler-Icekson G (1978). Eur J Biochem 88: 395-402.
18. Kessler-Icekson G, Singer RH, Yaffe D (1978). Eur J Biochem 88:403-410.

19. Kurtz DT (1981). Nature 291:629-631.
20. Robins DM, Pack J, Seeburg P, Axel R (1982). Cell 29:623-631.
21. Mantei N, Weissmann C (1982). 297:128-132.
22. Banerji J, Rusconi S, Schaffner W (1981). Cell 27:299-308.
23. Pavlakis GN, Hamer DH (1983). Recent Prog in Hormone Res 39:353-385.
24. DiMaio D, Treisman R, Maniatis T (1982). Proc Nat Acad Sci USA 79:4030-4034.
25. Spradling AC, Rubin G (1983). Cell 34:47-57.
26. Lacy E, Roberts S, Evans EP, Burtenshaw MD, Costantini FD (1983). Cell 34:343-358.
27. McKnight SL, Kingsbury R (1982). Science 217:316-324.
28. Busslinger M, Flavell RA (1983). Cell 34:197-206.

Molecular Biology of Development, pages 395–405

A SINGLE GENE LOCUS ENCODES THE FAST SKELETAL MYOSIN LIGHT CHAIN 1 AND 3 ISOFORMS.

Muthu Periasamy, Emanuel E. Strehler, and Bernardo Nadal-Ginard

Laboratory of Molecular and Cellular Cardiology
Department of Cardiology, Children's Hospital
Department of Pediatrics, Harvard Medical School,
Boston, Massachusetts 02115

ABSTRACT To elucidate the structural relationships between fast myosin light chain 1 and 3 isoforms, we have isolated cDNA clones specific for the corresponding mRNAs and determined their primary nucleotide sequence. $MLC1_f$ and $MLC3_f$ mRNAs display complete sequence identity in the 3' untranslated sequences and in the codons specifying the 141 carboxyl terminal amino acids (aa). In contrast, sequences encoding the amino termini of the $MLC1_f$ (49 aa) and $MLC3_f$ (8 aa) and their 5' untranslated region are unrelated. Using the cloned cDNAs as probes we have isolated a single gene locus encoding both $MLC1_f$ and $MLC3_f$ mRNA in four genomic clones spanning over ∿25kb DNA. In this report we present evidence that $MLC1_f$ and $MLC3_f$ mRNAs are produced from the same gene through a novel process of alternative RNA splicing and possibly differential transcription from two separate promotors.

INTRODUCTION

During the course of muscle development, the fusion of mononucleate myoblasts to form multinucleate myotubes is accompanied by the cytoplasmic accumulation of large amounts of contractile proteins some of which are tissue-specific and developmentally regulated. These components of the contractile apparatus include myosin heavy chain, actin, three myosin light chains, three troponins and two tropomyosins.

In fast muscle the alkali myosin light chains exist in two isoforms of different size: $MLC1_f$ (190 aa long) and $MLC3_f$ (149 aa long). These two isoforms have complete sequence homology for the first 141 amino acids from their carboxyl termini and differ in length and amino acid sequence of their amino termini ($MLC1_f$ = 49 aa, $MLC3_f$ = 8 aa) (1). Based on the sequence conservation of the two carboxyl termini of $MLC1_f$ and 3_f in chicken and rabbit, it has been suggested that these two proteins originate from a single gene (2,3). Recently a number of eucaryotic genes that produce multiple mRNAs (which in some cases code for different proteins by a process of alternative splicing and/or differential transcription at the 5' or 3' end of the gene (4-8) has been documented. In light of such additional mechanisms of gene regulation and the peculiar structure of $MLC1_f$ and $MLC3_f$ proteins, we set out to investigate the structural organization and expression of $MLC1_f$ and $MLC3_f$ gene(s).

RESULTS

$MLC1_f$ and $MLC3_f$ mRNAS have Identical 3' Coding and Untranslated (UT) Regions but Differ at Their 5' Coding and UT Regions.

In order to determine the structure of the $MLC1_f$ and $MLC3_f$ mRNAs it was necessary to isolate full length cDNA clones that contain the sequences specific for each of these two mRNA's. Using a cDNA probe pMLC-84 (9) that has sequences homologous to $MLC1_f$ and $MLC3_f$ mRNAs, three independent clones pMLC-5, pMLC-91 and pMLC-35 were identified from a rat skeletal muscle cDNA library constructed by the RNA-DNA hybrid cloning procedure using poly A+ mRNA (10,11). The three newly isolated clones were sequenced in their entirety. The primary nucleotide sequence of $MLC1_f$ and $MLC3_f$ mRNAs deduced from these clones is presented in Fig. 1.

Comparison of DNA and derived amino acid sequences of the $MLC1_f$ (pMLC-5, pMLC-91) and $MLC3_f$ (pMLC-35) cDNA clones reveals an interesting structural relationship between them. The region coding for the common carboxyl terminal 141 amino acids of $MLC1_f$ and $MLC3_f$ and the entire 3' UT region is 100% identical among the four cDNA clones. DNA sequences located 5' from the codon specifying the first common amino acid (aa 50) are specific for either $MLC1_f$ or

```
MLC1f: ...GTGGGTCCACCCTCCAGACGCTGGGCTGCGGGTCAAGCGAGCCACCACTGCTCTTCCAAGTGTCATTTTGACAGGTTCTTCTGGAGGAGATCCTTCTTTTGTTAACCC

MLC1f:   ATCGTTTTAATCAAA ATG GCA CCA AAG AAA GAC GTG AAG AAG CCC GCT GCT GCT GCT CCT GCT CCT GCC CCA GCC CCG GCC CCG
                              A   P   K   K   D   V   K   K   P   A   A   A   A   P   A   P   A   P   A   P   A   P  aa 22
MLC3f:                                                                                                 ...GAGACTCCACT

MLC1f:   GCC CCT GCC CCA GCC AAA CCT AAA GAA GAA AAA ATC GAT CTG TCT GCT ATT AAG ATC GAG TTC TCT AAG GAA CAA CAG GAG
          A   P   A   P   A   K   P   K   E   E   K   I   D   L   S   A   I   K   I   E   F   S   K   E   Q   Q   E  aa 49

MLC3f:   CAGGGACCCACTGGAGGTGTCTTCAGTTCATTCTCCAGTCCCGCTGCTGTTTTGCCTCCCAGAACTCCATC ATG TCC TTC AGT GCT GAC CAG ATT GCC
                                                                                      S   F   S   A   D   Q   I   A


         GAA TTC AAG GAG GCA TTT CTC CTC TTC GAC AGA ACC GGT GAA TGC AAG ATC ACC TTA AGT CAG GTG GGC GAC GTC CTC CGG
          E   F   K   E   A   F   L   L   F   D   R   T   G   E   C   K   I   T   L   S   Q   V   G   D   V   L   R  aa 76

COMMON   GCT CTG GGC ACC AAT CCC ACA AAT GCA GAA GTC AAG AAG GTT CTC GGG AAC CCT AGC AAT GAA GAG ATG AAT GCT AAG AAA
          A   L   G   T   N   P   T   N   A   E   V   K   K   V   L   G   N   P   S   N   E   E   M   N   A   K   K  aa 103

BODY     ATC GAG TTT GAA CAG TTT CTG CCC ATG ATG CAA GCC ATC TCC AAC AAC AAG GAC CAG GGA GGC TAT GAA GAT TTC GTT GAG
          I   E   F   E   Q   F   L   P   M   M   Q   A   I   S   N   N   K   D   Q   G   G   Y   E   D   F   V   E  aa 130

MLC1f/   GGT CTG CGT GTC TTC GAC AAG GAG GGC AAT GGC ACC GTC ATG GGT GCT GAG CTC CGC CAT GTC CTC GCC ACT CTG GGA GAG
          G   L   R   V   F   D   K   E   G   N   G   T   V   M   G   A   E   L   R   H   V   L   A   T   L   G   E  aa 157

MLC3f    AAG ATG AAG GAG GAG GAG GTA GAA GCA TTG CTG GCG GGC CAG GAG GAC TCC AAT GGC TGC ATC AAC TAT GAA GCT TTT GTG
          K   M   K   E   E   E   V   E   A   L   L   A   G   Q   E   D   S   N   G   C   I   N   Y   E   A   F   V  aa 184

         AAA CAC ATC ATG TCT GTC TAA AACGAGAATTCAAGAACACAGTGTTGGGGAAGACTGGCCAGAAAGTCAGTTCAAGAACACCTATGGCTAACTGTCAACA
          K   H   I   M   S   V

         CCAGCTTGCCACCACCCGGAAGAACAAACAATCTAGACCATTCTAGATTGAAGGATTCCCTGAAGTTTTATCAACCTTGAGTTTTTCATGGCACACACATCAGGTT

         GATAATCTCAAGTCCATGTACCACCTTTATGATCTACTTGGGTCCACAACATTCATAAGAGCACTACATCAATAAACTTGGTCAGTCTGGAAGACTC (A)n
```

Figure 1. Nucleotide and derived amino acid sequences of $MLC1_f$ and $MLC3_f$ mRNAs as deduced from cDNA clones.

$MLC3_f$. The $MLC1_f$ and $MLC3_f$ specific domains code for different amino terminal regions ($MLC1_f$ = 49aa, $MLC3_f$ = 8aa) and also have unrelated 5' UT sequences. From the above structural analysis it is clear that $MLC1_f$ and $MLC3_f$ mRNAs encode two different protein isoforms differing at their amino termini. The fact that $MLC1_f$ and $MLC3_f$ mRNAs are completely identical with not a single base substitution even in the third codon position, in the common coding sequence and the 3' UT region, argues strongly in favor of their common origin from a single gene.

$MLC1_f$ and $MLC3_f$ mRNA's are Encoded by a Single Gene.

In order to conclusively ascertain whether in fact $MLC1_f$ and $MLC3_f$ mRNAs are encoded by a single gene, the corresponding genomic sequences were isolated and characterized. The complete $MLC1_f$ and $MLC3_f$ gene locus was obtained in four overlapping genomic clones by successive screening of two different charon 4A rat genomic (12) libraries. The structural organization of $MLC1_f$ and $MLC3_f$ common and isoform specific sequences of the genomic clones was determined by Southern blotting, DNA sequence analysis and S1-nuclease mapping. As shown in Fig. 2 the $MLC1_f$ and $MLC3_f$ mRNA common and specific sequences are in physical continuity and part of a single gene locus.

The Distribution of $MLC1_f$ and $MLC3_f$ Exons in the Gene Reveals a Novel Type of Gene Organization.

As illustrated in Fig. 2 the sequences coding for the 141 amino acids common to $MLC1_f$ and $MLC3_f$ proteins and the common 3' UT region are split into at least four exons which are grouped in a region of 5kb of DNA. Interestingly, the 5' specific domains for $MLC1_f$ and $MLC3_f$ are separated from the common body exons by $\sim$15kb and $\sim$5kb of genomic DNA, respectively (Fig. 2). The 5' UT region and aa 1-40 of $MLC1_f$ are encoded by the 5'-most exon of the gene. $\sim$9kb downstream lies the next exon that contains only the 5' UT region of $MLC3_f$, while the $MLC3_f$ specific 8 amino terminal aa are located in a separate "mini-exon" 741bp further downstream. 244bp downstream from this $MLC3_f$ mini-exon lies a $MLC1_f$-specific mini-exon coding for aa 41-49. This $MLC1_f$ mini-exon is separated from the upstream $MLC1_f$ exon by $\sim$12kb of genomic DNA that includes two $MLC3_f$-specific exons.

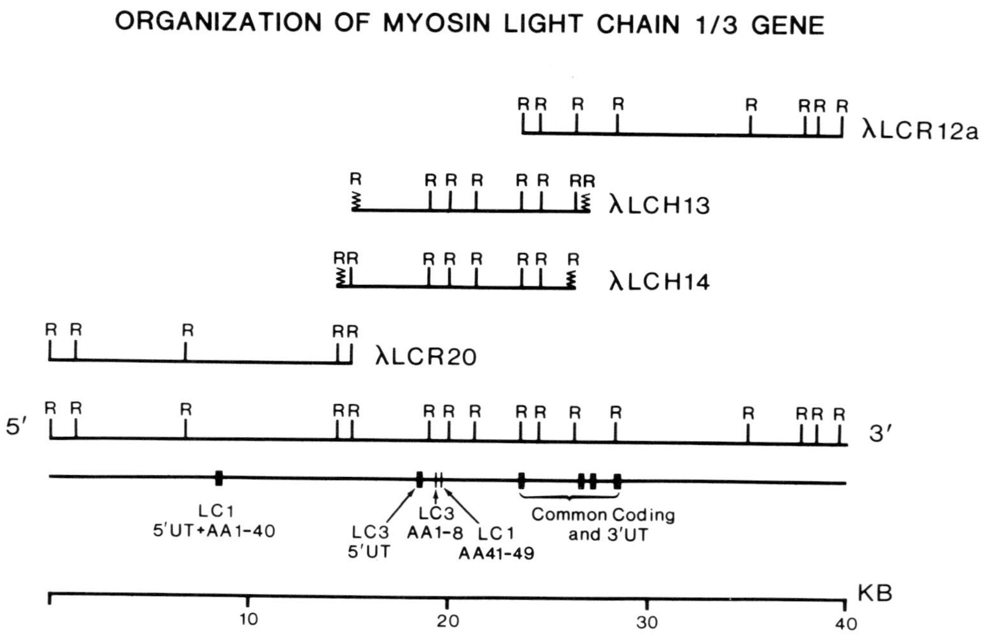

Figure 2. Structure of $MLC1_f/3_f$ gene locus as defined by four overlapping genomic clones (R = EcoRI). At the bottom, the organization of $MLC1_f$ and $MLC3_f$ specific and common exons are depicted.

Given this bizarre organization differential splicing must be involved in the processing of the primary $MLC1_f/3_f$ gene transcript(s) regardless of whether one or two promotors are present in the gene.

$MLC1_f$ and $MLC3_f$ mRNAs are Transcribed From Different Promotors.

The sequence of the 5' UT region of the $MLC1_f$ and $MLC3_f$ mRNAs does not reveal the existence of a common 5' exon that is shared by the two mRNAs as it should be the case if both were transcribed from a common promotor and suggest the existence of a distinct promotor for each one of the two mRNAs. In order to further explore this possibility, the flanking sequences 5' to the exons containing the UT regions of $MLC1_f$ and $MLC3_f$ were determined. As expected, TATAA-like and CAT sequences were found (27 and 75 nt for $MLC1_f$; 20 and 45 nt for $MLC3_f$) upstream from the first nucleotide present in the cDNA clones. The initiation of transcription site for both mRNAs was determined by S1-nuclease mapping using end labelled genomic fragments (data not shown). The putative cap site of $MLC1_f$ and $MLC3_f$ mRNAs is 4 and 2 bases upstream, respectively, from the first nucleotide shown in Fig. 1. These results, taken together, indicate that the two mRNAs generated from the $MLC1_f/3_f$ gene are transcribed from two different promotors that are located $\sim$10 kb apart.

$MLC1_f$ and $MLC3_f$ mRNAs are Differentially Accumulated during Development.

The fact that $MLC1_f$ and $MLC3_f$ mRNAs are the products of the same gene and generated by a process that necessarily involves differential splicing, raises interesting questions about the developmental and tissue specific expression of these two proteins. The accumulation of $MLC1_f$ and $MLC3_f$ mRNAs during muscle development follows an interesting pattern, as determined by Northern blot analysis. $MLC1_f$ accumulation precedes that of $MLC3_f$, and in 20 day old rat fetuses the ratio of $MLC1_f$ to $MLC3_f$ is about 4:1. This ratio decreases to 2:1 by one day after birth and reaches about 1:1 by adulthood. This pattern of accumulation of $MLC3_f$ relative to $MLC1_f$ mRNA during development parallels that of the corresponding proteins and demonstrates that mRNA abundance is the primary determinant of $MLC1_f$ and $MLC3_f$ protein accumulation during muscle development.

DISCUSSION

A number of genes producing more than one mRNA have been recently described. In most cases, the multiple mRNAs originate from different transcriptional units utilizing alternative promotors (4,5) and/or poly(A) addition sites (6,7,14,-17). Although most of these multiple mRNAs produced from a single gene translate into identical proteins, in a few instances the utilization of alternative 3' splice acceptor sites results in the production of proteins with different primary sequences, either internally (8) or at the carboxyl terminal end (6,7). The $MLC1_f/3_f$ gene is the first documented example of a gene coding for two proteins with different amino terminal ends and identical carboxyl terminal sequences whose differential expression is developmentally regulated.

The organization of the $MLC1_f/3_f$ gene reveals an unusual structure and a novel pattern of RNA splicing. The fact that in the $MLC1_f/3_f$ gene, the exon containing the 5' UT sequences of $MLC3_f$ mRNA is located 10kb downstream from the first $MLC1_f$ specific exon that contains the 5' UT region and sequences coding for aa 1-40, is a unique feature of this gene. Yet the most interesting aspect of the $MLC1_f/3_f$ gene is the existence of the $MLC1_f$-specific exon coding for aa 41-49 next to and downstream from an exon coding for aa 1-8 of $MLC3_f$. The existence of these two mini-exons in this particular arrangement indicates that a novel type of differential splicing is involved in generating both $MLC1_f$ and $MLC3_f$ mRNAs.

The $MLC3_f$-specific exons need to be spliced out from the $MLC1_f$ primary transcript in order to generate $MLC1_f$ mRNA, while the mini-exon coding for $MLC1_f$ aa 41 to 49 must be removed to produce $MLC3_f$ mRNA (Fig. 3). The splicing out of the 5' terminal $MLC3_f$-specific exon is easily explained since the 5' end of this exon lacks the consensus splice acceptor sequence. More interestingly, however, no known mechanism of RNA splicing can account for the alternative use of the $MLC1_f$ and $MLC3_f$-specific mini-exons to form $MLC1_f$ or $MLC3_f$ mRNAs.

These two mini-exons are flanked by functional acceptor and donor sites, as shown by the DNA sequence (Strehler et al., in preparation) and by the fact that both of them are incorporated into different mature mRNAs. Surprisingly, however, they are not spliced together into the same RNA molecule, nor are they both spliced out or interchanged in

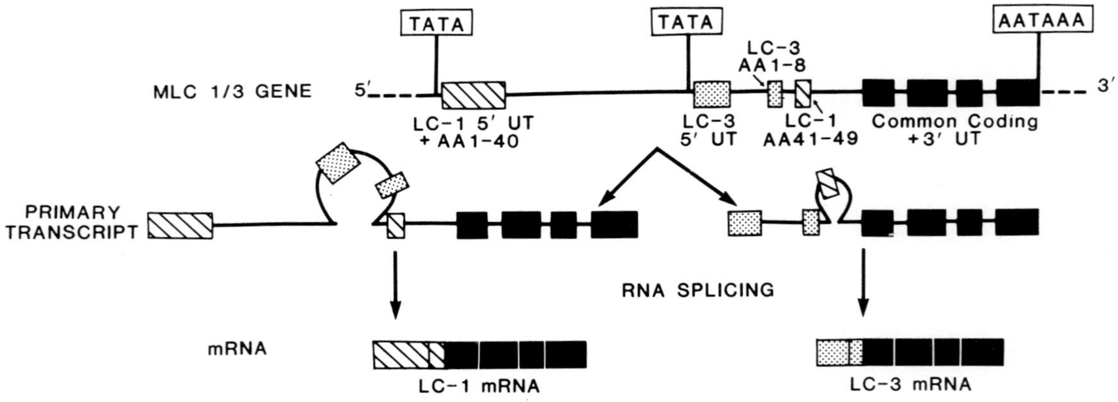

Figure 3. Diagramtic representation of the splicing pattern of the $MLC1_f/3_f$ primary transcripts originated from two different promoters

the production of $MLC1_f$ and $MLC3_f$ mRNAs. Yet, the 3' end of both is spliced to the same 5' end of the first common exon. This type of alternative splicing is fundamentally different from the process found in a mutant β -globin gene (13), growth hormone (14), IgD (15), IgM (16), ADH (5), γ-fibrinogen (17) and fibronectin (8) in which one or more donor (or acceptor) sites compete for a single acceptor (or donor) site, but where no splicing combinations between an adjacent pair of sites is forbidden. Moreover, the generation of the mature $MLC1_f$ and $MLC3_f$ mRNAs can not be explained by a 5' to 3' or a 3' to 5' linear scanning model of RNA splice selection (18,19), nor by the rules outlined by Kuhne et al (1983) (20) for duplicated splice sites in the β -globin gene.

The organization and behavior of $MLC1_f$ and $MLC3_f$ specific mini-exons suggests three main possibilities for the production of mature $MLC1_f$ and $MLC3_f$ mRNAs: a) the existence of sequence specificity in the splice junctions of the $MLC1_f$ and $MLC3_f$ specific exon b) the existence of unknown trans acting factors (sn RNAs? proteins?) that specifically activate and/or inactivate donor and/or acceptor sites c) the existence of conformational differences between the $MLC1_f$ and $MLC3_f$ mRNA precursors that are responsible for generating preferential splicing pathways. Although not yet proven for the production of any mature mRNA at the present time, we favor this third possibility because it is readily explained by the existence of two different promotor sites that would give rise to two mRNA precursors of significantly different size (∿ 10 and ∿ 20kb, respectively) and, potentially different conformation. In addition, this possibility is conceptually more appealing than the other two because it only requires a promotor selection choice in order to produce each of the two mRNAs. Experiments to test among these possibilities are now in progress.

REFERENCES

1. Frank G, Weeds AG (1974). The amino acid sequence of the alkali light chains of rabbit skeletal muscle myosin. Eur J Biochem 44:317.
2. Matsuda G, Maita T, Umegane T (1981). The primary structure of L-1 light chain of chicken fast skeletal muscle myosin and its genetic implication. FEBS Lett 126:111.

3. Nabeshima Y, Fujii-Kuriyama Y, Muramatsu M, Ogata K (1982). Molecular cloning and nucleotide sequences of the complementary DNAs to chicken skeletal muscle myosin two alkali light chain mRNAs. Nucl Acids Res 10:6099.
4. Hagenbuchle O, Tosi M, Schibler V, Bovey R, Wellauer PK, Young RA (1981). Mouse liver and salivary gland α - amylase mRNAs differ only in 5' non-translated sequences. Nature 289:1053.
5. Benyajati C, Spoerel N, Haymerle H, Ashburner M (1983). The messenger RNA for alcohol dehydrogenase in Drosophila melanogaster differs in its 5' end in different developmental stages. Cell 33:125.
6. Alt FW, Bothwell ALM, Knapp M, Siden E, Mather E, Koshland M, Baltimore D (1980). Synthesis of secreted and membrane-bound immunoglobulin Mu heavy chains is directed by mRNAs that differ at their 3' ends. Cell 20:293.
7. Amara SG, Jonas V, Rosenfeld MG, Ong ES, Evans RM (1982). Alternative RNA processing in calcitonin gene expression generates mRNAs encoding different polypeptide products. Nature (London) 298:240.
8. Schwarzbauer JE, Tamkum JW, Lemischka IR, Hynes RO (1983). Three different fibronectin mRNAs arise by alternative splicing within the coding region. Cell 35:421.
9. Garfinkel LI, Periasamy M, Nadal-Ginard B (1982). Cloning, identification, and characterization of α - actin, myosin light chains 1, 2, and 3, α-tropomyosin and troponin C and T. J Biol Chem 257:11078.
10. Wood KO, Lee JC (1976) Integration of synthetic globin genes into an E. coli plasmid. Nucl Acids Res 3:1961.
11. Zain S, Sambrook J, Roberts RJ, Keller W, Fried M, Dunn AR (1979). Nucleotide sequence analysis of the leader segments in a cloned copy of adenovirus-2 fiber mRNA. Cell 16:851.
12. Blattner FR, Blechl AE, Denniston-Thompson K, Faber HE, Richards JE, Slighton JL, Tucker PW, Smith D (1978). Cloning human fetal γglobin and mouse α -type globin DNA: preparation and screening of shotgun collections. Science 202:1279.
13. Busslinger M, Maschonas N, Flavell RA (1981). β+ Thalassemia: Aberrant splicing results from a single point mutation in an intron. Cell 27:259.
14. deNoto FM, Moore DD, Goodman HM (1981). Human growth hormone - DNA sequence and mRNA structure: possible alternative splicing. Nucl Acids Res 9:3719.

15. Maki R, Roeder W, Traunecker A, Sidman C, Wabl M, Raschle W, Tonegawa S (1981). The role of DNA rearrangement and alternative processing in the expression of immunoglobulin delta genes. Cell 24:353.
16. Rogers J, Early P, Carter C, Calame K, Bond M, Hood L, Wall R (1980). Two mRNAs with different 3' ends encode membrane-bound and secreted forms of immunoglobulin. Cell 20:303.
17. Crabtre GR, Kant JA (1982). Organization of the rat γ -fibrinogen gene: alternative mRNA splice patterns produce the A and B (γ) chains of fibrinogen. Cell 31:159.
18. Sharp PA (1981). Speculations on RNA splicing. Cell 33:643.
19. Lang KM, Spritz RA (1983). RNA splice site selection: evidence for a 5 ' 3' scanning model. Science 220:1351.
20. Kuhne T, Wieringa B, Reiser J, Weissman C (1983). Evidence against a scanning model of RNA splicing. EMBO J 5:727.

Molecular Biology of Development, pages 407–416

EXPRESSION OF DROSOPHILA MUSCLE TROPOMYOSIN I GENE: A SINGLE GENE ENCODES DIFFERENT MUSCLE TROPOMYOSIN ISOFORMS

Guriqbal S. Basi, Mark Boardman and Robert V. Storti

Department of Biological Chemistry
University of Illinois Health Sciences Center
Chicago, IL 60612

ABSTRACT *Drosophila melanogaster* contains three tropomyosin (TM) genes clustered within an 18 kbp segment of DNA. Two of these genes, mTM I and mTM II, encode muscle forms of Tm, whereas the third gene, (cTm), encodes a cytoplasmic form of the protein. The mTM I gene expresses a major transcript of 1.3 kb and lesser amounts of a 1.6 kb transcript during muscle development in embryos late in development. In the adult thorax, however, the mTM I gene expresses two new transcripts of 1.7 and 1.9 kb. The change in mRNA expression of the mTM I gene is accompanied by a change in the expression of muscle tropomyosin isoforms. In late embryos the 1.3 and 1.6 kb mRNAs synthesize an embryonic form of tropomyosin. In adult thorax the 1.7 and 1.9 kb mRNAs express a thorax form of tropomyosin. Thus, a single gene expresses different tropomyosin isoforms in different muscle tissue. The structure of the mTM I gene is reported, and the mechanism for generating the different mTM I isoforms is discussed.

INTRODUCTION

Vertebrate myogenesis is characterized by the expression of different developmentally regulated and muscle-specific isoforms of contractile proteins (1). The contractile protein isoforms that have been investigated have been shown to be encoded by different genes (1). We have been studying contractile protein genes in *Drosophila melanogaster* and have identified three closely linked *tropomyosin* genes (2).

The tropomyosin genes map to the 88 F region of chromosome 3R, the same chromosomal map region as one of the muscle actin genes and a region where several flight muscle mutations have been mapped (3-5). The tropomyosin genes, mTM I and mTM II, encode muscle tropomyosin proteins of 34,000 and 35,000 daltons, respectively, in embryos and muscle tissue culture cells. Both muscle genes are developmentally and coordinately expressed during embryonic and larval myogenesis and in muscle cell cultures. The third tropomyosin gene (cTM) is expressed in both muscle and non-muscle tissue. In embryos and tissue culture cells it encodes a 31,000 dalton cytoplasmic form of tropomyosin (6). Each of the different tropomyosin genes is single copy in the genome. In this preliminary communication we analyze in more detail the expression of the mTM I gene. We show that the same mTm I gene synthesizes a new set of transcripts that are abundant in the indirect flight muscle of adult thorax and that these new transcripts direct the synthesis of a unique isoform of tropomyosin. The structure of the mTM I gene is described and the mechanism for generating the different mTM I isoforms is discussed.

RESULTS

A schematic diagram of the arrangement of the tropomyosin genes is shown in Figure 1A. The expression of the mTM I gene was determined by hybridization of a nick-translated plasmid DNA, pAS85-1, containing the entire gene and flanking DNA regions, to total cellular RNA extracted from early and late embryos and adult thorax. The RNA was extracted from frozen tissue, separated by electrophoresis under denaturing conditions and bound to nitrocellulose filter paper (7-9) (Fig. 1B). The two mTM genes and their mRNAs do not cross-hybridize under the stringent hybridization conditions used in this assay (2). The results show, as reported previously, that the mTM I gene transcripts are not expressed at detectable levels in early embryos and only become abundant in late embryos during the time of muscle development (Fig. 1B). In late embryonic development the mTM I gene synthesizes an abundant transcript of 1.3 kb and a less abundant 1.6 kb transcript (2).

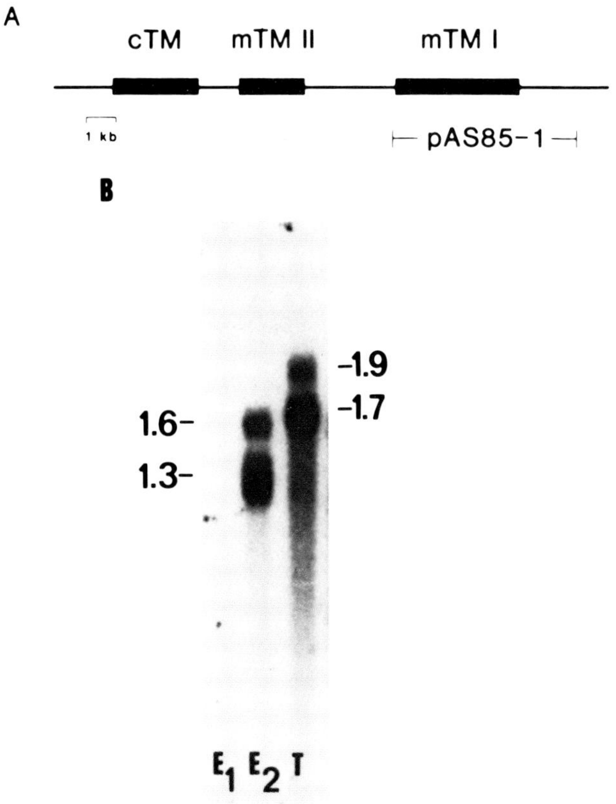

Figure 1. A. A schematic map of the cloned Drosophila genomic DNA (λDm85) that contains the tropomyosin genes (2). The location of the tropomyosin genes is indicated by solid bars. The 5.5 kb Bam HI DNA fragment, containing the entire mTM I gene and flanking regions, was subcloned into pACYC184 and designated pAS85-1. B. Muscle tropomyosin I (mTM I) expresses different transcripts in embryos and adult thorax. Equal amounts of total RNA (5 μg) extracted from staged 2-4 hr embryos (E_1); 17-19 hr embryos (E_2); or thoraces (T) was denatured, separated in formaldehyde agarose gels and transferred to nitrocellulose filters. The filters were hybridized under stringent conditions with DNA nick-translated ^{32}P-labelled pAS85-1 DNA.

The size estimates of RNA reported here were determined in formaldehyde-agarose gels and differ slightly from those previously reported using glyoxol-agarose gel electrophoresis. When the mTM I gene plasmid is hybridized to thoracic RNA, the 1.3 and 1.6 kb transcripts are now present at much lower levels; instead, two new transcripts of 1.7 and 1.9 kb are detected (Fig. 1B). The major 1.7 kb transcript comprises about 80% of the total amount of RNA homologous to the mTM I gene. The thoracic transcripts have also been found in reduced amounts in pupae, the stage at which thoracic muscle development begins, and in leg tissue.

To determine the tropomyosin(s) encoded by the thoracic transcripts, RNA preparations from embryos and thorax were translated in a rabbit reticulocyte lysate cell-free protein-synthesizing system. Tropomyosin proteins among the ^{35}S-methionine products of the in vitro translation were identified by immunoprecipitation with rabbit anti-chicken gizzard tropomyosin antibody. This antibody cross-reacts with the Drosophila muscle and cytoplasmic tropomyosins (6). The total translations and the immunoprecipitated products were separated by electrophoresis in two-dimensional gels (Fig. 2).

The total translation products and the immunoprecipitates from late embryo RNA show the previously characterized mTM I and mTM II proteins (9) (Fig. $2E_2$). The translation products of total thoracic RNA, however, show a new abundant protein spot that has a slightly slower mobility in the gel and is slightly more basic than mTM I (Figure 2T). This abundant protein is specifically immunoprecipitated with anti-chicken tropomyosin antibody along with mTM II and trace amounts of mTM I present in this preparation. We conclude that thoracic mRNA encodes a new isoform of tropomyosin. Since this protein is synthesized in vitro by thorax RNA and not by embryo RNA, it is unlikely that it is the result of in vitro protein modification of the mTM I or mTM II protein.

The new form of tropomyosin was shown to be encoded by the mTM I gene by a hybrid-selection translation assay in which the filter bound mTM I gene plasmid was used to purify the homologous 1.7 and 1.9 kb thoracic RNAs (10). The hybrid-selected RNA was translated in vitro, and the ^{35}S-methionine labelled product(s) were electrophoresed in a two-dimensional gel (Fig. 3). The result shows that the thorax-specific tropomyosin isoform is the product of

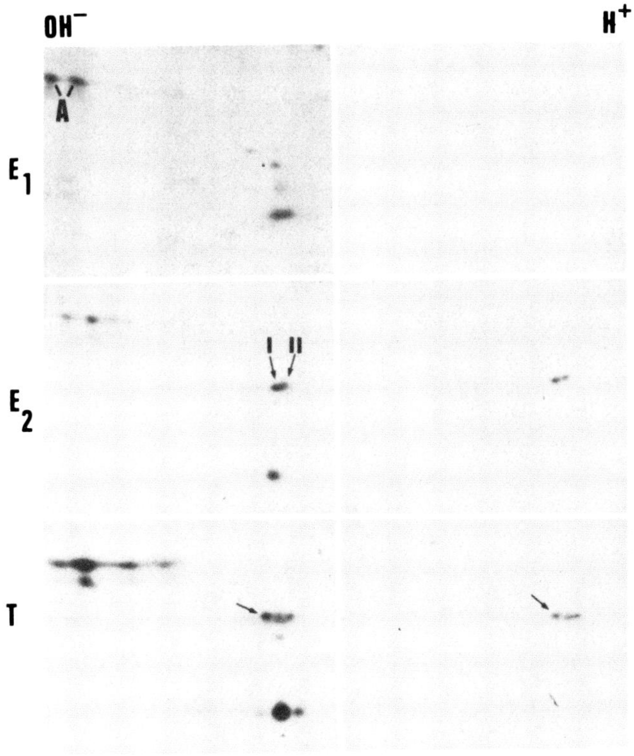

Figure 2. Fluorogram of ^{35}S-methionine labelled cell-free translation products of total RNA of 2-4 hr embryos (E_1), 17-19 hr embryos (E_2) and thoraces (T) separated by two-dimensional polyacrylamide gel electrophoresis. The translation, immunoprecipitation and two-dimensional gel electrophoresis have been described (13,19). The regions depicted show only the electrophoretic pattern of spots in the vicinity of the muscle tropomyosins. The total translation products are on the left, and the translation products after immunoprecipitation with rabbit anti-chicken gizzard tropomyosin antibody are shown on the right. A indicates actin proteins; I and II indicate mTM I and mTM II proteins, respectively. The arrow in the thoracic samples indicates the thorax-specific isoform of tropomyosin.

thorax RNA hybrid-selected by the mTM I gene plasmid and is, therefore, encoded by the mTM I gene. The synthesis of a small amount of the embryonic form of mTM I protein seen in Fig. 3 is due to the small amounts of the 1.3 kb transcript seen in long exposures of the RNA blot autoradiograms. The 1.3 kb transcripts in thorax RNA are probably due to embryonic or larval-like muscle in adult thorax or possibly due to other adult tissue contaminating the thorax preparation. The mTM I gene, therefore, synthesizes two thorax transcripts that code for a new isoform of tropomyosin.

Figure 3. Fluorogram of ^{35}S-methionine labelled translation products of RNA hybrid-selected from total thorax RNA by pAS85-1 DNA (mTM I gene) and electrophoresed in a two-dimensional polyacrylamide gel, as in Figure 2. The arrow at the right indicates the position of mTM I, and the arrow at the left indicates the thorax-specific form of tropomyosin.

The structure of the mTM I embryonic and thorax gene has been determined by DNA sequence, S1 nuclease mapping and primer-extension analysis. The gene spans approximately 2.4 kbp of DNA. Figure 4 shows the structure of the gene encoding the major 1.3 kb mRNA. It is split into three exons and two introns. Preliminary results indicate that the minor 1.6 kb mRNA differs from the 1.3 kb transcript by additional sequences in the untranslated regions. Both transcripts function in vitro to translate mTM I.

The major embryonic and thoracic transcripts share identity in transcriptional initiation and termination

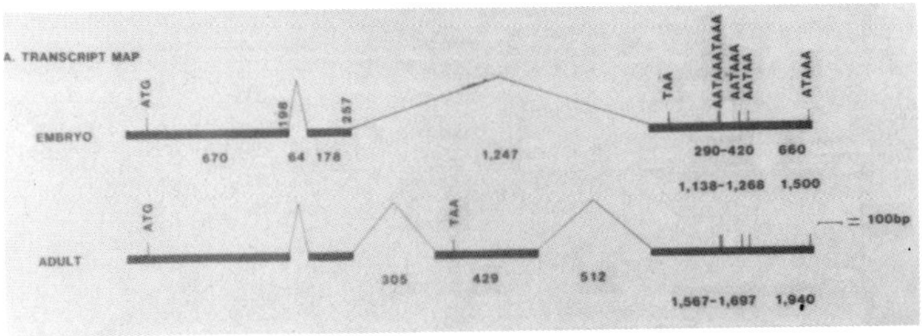

Figure 4. Schematic diagram of mTm I gene. The thick lines represent exons and thin lines represent introns. The ATG is the site of translational initiation and amino acid 1. Numbers 198 and 257 above the line refer to amino acid positions in the tropomyosin sequence. TAA is the site of translational termination and follows amino acid 284 in the tropomyosin sequence. The numbers below the line refer to exon and intron sizes in base pairs. Five consensus adenylation signals are indicated and are utilized in vivo as determined by S1 analysis.

sites and in the first two exons. The thoracic transcript, however, is generated by alternative splicing of an additional exon located within the 1.2 kb second intron of the embryonic gene. The alternative splicing results in tropomyosin isoforms with different C-terminal amino acids.

DISCUSSION

We have shown previously that the mTM I gene is developmentally regulated and expresses a 1.3 and 1.6 kb set of transcripts in late embryos and muscle cell cultures. We have also determined, though not shown here, that both transcripts synthesize mTM I protein when translated in vitro, indicating that both transcripts are probably functional transcripts *in vivo*. These results indicate that the difference between the two mRNAs probably resides in a non-coding region of the mRNAs, since both mRNAs encode the same protein. Preliminary S1 nuclease mapping studies indicate that the 1.6 kb transcript may differ from the 1.3 kb mRNA by an additional 3' extension of the mRNA. The 1.6 kp mRNA may thus differ from the 1.3 kb mRNA by transcriptional termination at a different site. Since both thoracic transcripts also encode the same protein, and they both differ in size by an amount similar to that of the two embryonic transcript, it is possible that both thoracic transcripts will differ in a way similar to the embryonic transcripts.

S1 mapping and primer-extension studies indicates that the 1.3 kb embryonic and 1.7 kb thoracic mRNAs have the same transcriptional initiation site. The mTM I gene that encodes the embryonic mRNA consists of three exons and two introns. The thoracic transcript, on the other hand, contains a fourth exon that is generated by alternative splicing and results in a difference in the 27 amino acids at the C-terminus of the different isoforms.

The thorax mTM I transcripts therefore, arise by a process similar to that shown recently for multiple *Drosophila* myosin heavy chain transcripts (11). In that study a single myosin heavy chain gene was shown to encode different size transcripts in embryos and pupae. The different transcripts were shown to result from differential splicing at the 3' end of the gene. Whether the different myosin transcripts encode different myosin isoforms has not been determined.

The adult thorax in *Drosophila* consists primarily of indirect flight muscle, and we presume that this tissue is the origin of the thoracic tropomyosin transcripts. Indirect flight muscle is morphologically and physiologically different from embryo and larval muscle (12). This muscle produces asynchronous contractions to control wing movement, and it is reasonable to suggest that the synthesis of a new tropomyosin isoform is in some way necessary to meet the specialized needs of this muscle. The mTM II gene transcripts do not change qualitatively in thoracic muscle (data not shown) and do not express a new tropomyosin isoform (Fig. 2).

In vertebrates different contractile-protein isoforms are encoded by multiple genes rather than by a single gene. The synthesis of multiple tropomyosin isoforms by a single Drosophila gene, rather than different genes, may represent a fundamental difference in the way insects and vertebrates regulate the expression of different contractile-protein isoforms during myogenesis to meet the specific functional requirements of different muscles. This difference may be the result of evolutionary constraints on the total number of genes in Drosophila due to its small genome size, or it may be the result of other more fundamental differences in the way some muscle genes are regulated in the two groups of organisms.

ACKNOWLEDGMENTS

We thank Dr. Jim Lin for the tropomyosin antibody. This work was supported by grants from the N.I.H. to R.V.S.. R.V.S. is the recipient of a Research Career Development Award from the N.I.H.

REFERENCES

1. Buckingham ME, Minty AJ (1983). Contractile protein genes. In Maclean M, Gregory S, Flavell R (eds): "Eukaryotic genes: their structure, activity and regulation. London: Butterworths.
2. Bautch VL, Storti RV, Mischke D, Pardue ML (1982). Organization and expression of Drosophila tropomyosin genes. J. Mol. Biol. 162:231.
3. Mogami K, Hotta Y (1981). Isolation of Drosophila flightless mutants which affect myofibrillar protein of indirect flight muscle. Mol. Gen. Genet. 183:409.
4. Tobin SS, Zulauf E, Sanchez F, Craig EA, McCarthy BJ (1980). Multiple actin related sequences in the Drosophila melanogaster genome. Cell 19:121.
5. Fyrberg EA, Kindle KL, Davidson N, Sodja A (1980). The actin genes of Drosophila: a dispersed multigene family. Cell 19:365.
6. Bautch VL, Storti RV (1983). Identification of a cytoplasmic tropomyosin gene linked to two muscle tropomyosin genes in Drosophila. Proc. Natl. Acad. Sci. USA. 80:7123.

7. Chirawin JM, Przybyla AE, MacDonald RJ, Rutter WJ (1979). Isolation of biologically active ribonucleic acids from sources enriched in ribonuclease. Biochem. 18:5294.
8. Maniatis T, Fritsch EF, Sambrook J (1982). Molecular Cloning: A Laboratory Manual. New York: Cold Spring Harbor Laboratories.
9. Rave N, Crkuenjakov R, Boedtker H (1979). Identification of procollagen mRNAs transferred to diazobenzyloxymethyl paper from formaldehyde agarose gels. Nuc. Acids Res. 6:3559.
10. Storti RV, Szwast AE (1982). Molecular cloning and characterization of Drosophila genes and their expression during embryonic development and in primary muscle cell cultures. Devel. Biol. 90:272.
11. Rozek CE, Davidson N (1983). Drosophila has one myosin heavy-chain gene with three developmentally regulated transcripts. Cell 32:23.
12. Crossley C (1978). The Genetics and Biology of Drosophila Vol. 2b. Ashburner M, Wright TRF (eds): London: Academic Press. p 499.

Molecular Biology of Development, pages 417–426

VARIANT FIBRONECTIN SUBUNITS ARE ENCODED BY DIFFERENT mRNAS ARISING FROM A SINGLE GENE[1]

John W. Tamkun, Jean E. Schwarzbauer,
Jeremy I. Paul and Richard O. Hynes

Center for Cancer Research and Department of Biology,
Massachusetts Institute of Technology
Cambridge, Massachusetts 02139

ABSTRACT Three fibronectin mRNAs in rat liver differ by the presence of 0, 285 or 360 bases at an identical position in the coding region. This results in the production of fibronectin subunits varying by the insertion of 0, 95 or 120 amino acid residues between the cell and fibrin binding regions and within the C-terminal heparin binding region. These mRNAs arise from a single gene by a pattern of alternative splicing of a common transcript in which one 5' splice site can use any one of three different 3' splice sites.

Fibronectins mediate interactions between many cell types and a variety of extracellular molecules including collagen types I-V, fibrin, and proteoglycans(1,2,3). These interactions are important in cellular adhesion, microfilament organization, hemostasis and thrombosis, malignant transformation, and cell migration. Fibronectins are widely distributed in vivo. Fibronectin is present at considerable levels in plasma and is synthesized and secreted by hepatocytes(4) There is also an insoluble form in fibrillar matrices surrounding fibroblasts, as well as numerous other cell types.

[1]This work was supported by a grant from the USPHS, National Cancer institute(POI CA 26712). JES is a Charles A. King Trust Fellow.

Fibronectins Are Composed of Multiple, Complex Subunits

Plasma and cellular fibronectins are both disulfide-bonded dimers. The two forms, as well as their 230,000 D subunits, are very similar but it is clear that they are not identical. The differences between plasma and cellular fibronectins and among their subunits cannot be accounted for by posttranslation modifications alone(5), suggesting primary sequence differences between subunits. This possibility raises obvious questions concerning the number of mRNAs encoding fibronectin subunits as well as the number of fibronectin genes. The precise identification of the nature of the structural differences between fibronectin subunits would be of great use in the determination of the functional differences between fibronectins, should they exist. About one-half of the amino acid sequence of bovine plasma fibronectin has been determined(6). While this study provided no evidence for primary sequence differences between the subunits of plasma fibronectin, a striking pattern of homologous repeating sequences was observed. These repeating sequences fall into three categories. The type I repeats each contain approximately 50 amino acids and are from 18-60% homologous. The 60 residue type II repeats and 90 residue type III repeats are approximately 50% and 30% homologous respectively. Might additional repeats of these three types, or others, exist in the unsequenced portion of the fibronectin molecule? How is the fibronectin gene(or genes) organized to encode this complicated pattern of repeating units? How many fibronectin mRNAs exist, and what types of subunits do they encode?

To address these questions we, and others, have recently isolated cDNA and genomic clones encoding fibronectins. Since a considerable fraction of plasma fibronectin had been sequenced, and we had shown previously that hepatocytes synthesize plasma fibronectin(4), we decided to isolate cDNA clones from rat liver. We recently reported the isolation of such clones from a rat liver cDNA library constructed in the phage expression vector λgt11(7). These clones together span more than 3 kb of the 3' end of the approximately 8 kb fibronectin mRNA. Analysis of these clones has revealed that a single fibronectin gene encodes multiple mRNAs, some of which arise by a pattern of alternative RNA splicing not previously seen for other cellular genes.

Multiple Fibronectin mRNAs Exist In Rat Liver

Sequencing of these clones and S1 nuclease analysis demonstrated the existence of at least three different fibronectin mRNAs in rat liver. The mRNA sequences predicted from the cDNA sequence are identical except for a region 827 bases 5' of the termination codon. One type of mRNA contains 360 nucleotides at this position, while another contains this sequence minus its 5' 75 nucleotides, resulting in the addition of a total of 285 bases at this position. Another mRNA lacks the entire 360 nucleotide segment. The reading frames in the regions flanking this segment are the same in all three types of mRNA.

The portion of the fibronectin subunits encoded by the 3' one- third of the fibronectin mRNAs is depicted in figure one. The smallest mRNA encodes the C-terminal one third of fibronectin, spanning the cell, heparin, and fibrin binding sites mapped to this region of the molecule. This sequence includes seven type III and three type I repeats and is greater than 90% identical with previously reported amino acid sequences of bovine plasma fibronectin(6) and the 96 C-terminal residues of human fibronectin(8). The next largest mRNA encodes a subunit differing by the insertion of 95 amino acids between the last two type III repeats in the subunit. The largest mRNA encodes an additional 120 amino acids at this position. The 95 and 120 amino acid segments do not fit the consensus sequence for any of the three previously identified types of homology repeats.

Given the existence of multiple mRNAs which encode fibronectin subunits varying by the presence of discrete blocks of amino acids, several questions immediately follow. 1) How are the different mRNAs generated? 2) What forms of fibronectin correspond with each of the predicted subunits? 3) What function might the additional amino acid segments serve?

Origin of the Multiple Fibronectin mRNAs

The absolute identity of nucleotide sequence on either side of the region of difference suggests strongly that the three fibronectin mRNAs found in rat liver are encoded by a single gene, and therefore must arise by alternative processing of a common transcript. Although several patterns of RNA splicing could give rise to the three mRNAs, the presence of 3' splice sites within the coding region led

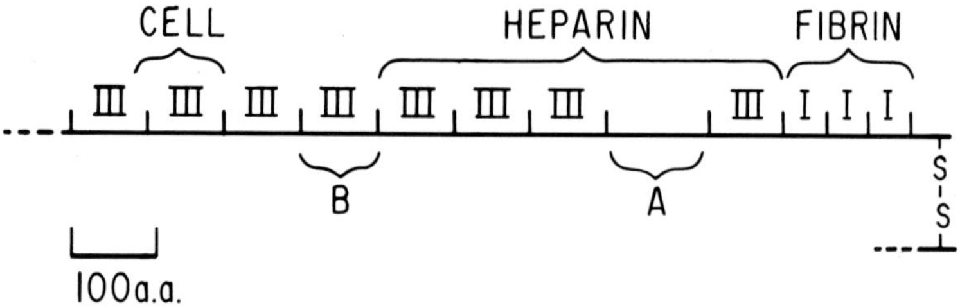

FIGURE 1. Structure of the C-terminal one-third of the fibronectin subunits, showing the type I and III homology repeats and the cell, heparin and fibrin binding regions. a) The 120 amino acid segment encoded by one class of rat liver fibronectin mRNA(7). Two other types of mRNA encode subunits lacking either the entire segment or only its N-terminal 25 residues. b) The additional type III homology repeat encoded by one class of human fibronectin mRNA found in a human tumor cell line(13). This segment is not encoded by mRNAs found in liver tissue, and presumably is never present in plasma fibronectin subunits.

us to propose a model in which a single 5' splice site can be paired with any one of three 3' splice sites(7).

In order to test this hypothesis, we determined the number of fibronectin genes in the rat genome, and isolated genomic clones containing the sequences encoding the three mRNAs(9). Southern blot analysis of rat DNA digested with several restriction enzymes was carried out under conditions of low stringency using as probe a restriction fragment from a cDNA clone which covers the 3' boundary of the region of variability observed in the three mRNAs. If the three different mRNAs are products of three different genes, it is likely that this probe would hybridize to more than one band. However, single bands were observed with each restriction enzyme used, suggesting the existence of a single fibronectin gene per haploid rat genome(figure 2). Similar results were derived using probes spanning the 3' one third of the fibronectin mRNA. While highly diverged genes might have escaped notice in these analyses, it is clear that the three mRNAs found in rat liver are encoded by a single gene, as suggested by the identity of nucleotide sequence in regions flanking the position of difference.

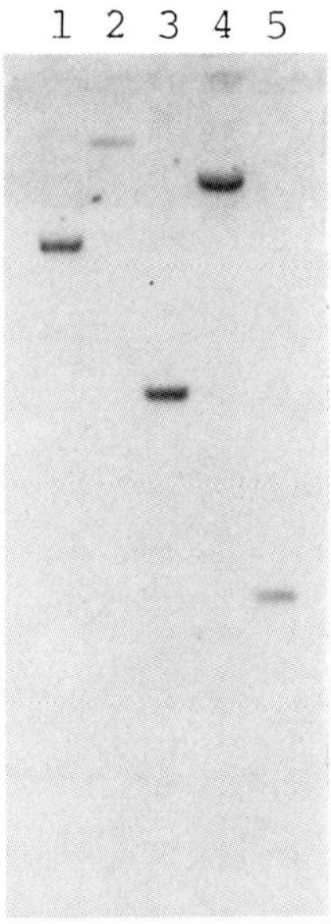

FIGURE 2. Determination of fibronectin gene copy number by blot hybridization analysis. Fisher rat genomic DNA was digested with Bgl II(1), Hind III(2), Bam H1 and Eco R1(3), Eco R1(4) or Xba I(5). Following electrophoresis on a 1.0% agarose gel and transfer to Zetabind, filters were probed with the 401 base pair Bgl II/Mbo II fragment of the fibronectin cDNA clone λrlf-2(9). Conditions of low stringency were used for both hybridization(6X SSC, 60°C) and washing(2X SSC,50°C).

We have recently isolated a number of fibronectin genomic clones from a rat genomic library constructed in the vector λEMBL 3B(9). Two of these clones span the 3' 29 kb of the fibronectin gene, which has been reported to be approximately 48 kb in length(10). The sequencing of a part of one of these clones clearly shows the pattern of alternative splicing which gives rise to the three fibronectin mRNAs(figure 3). The 90 bases of coding sequence immediately 5' of the region of difference are encoded by a single exon. This exon is followed by a 613 nucleotide intron containing multiple termination codons in all three reading frames. The next exon is 467 nucleotides in length and is quite complex. The 5' 360 bases of this

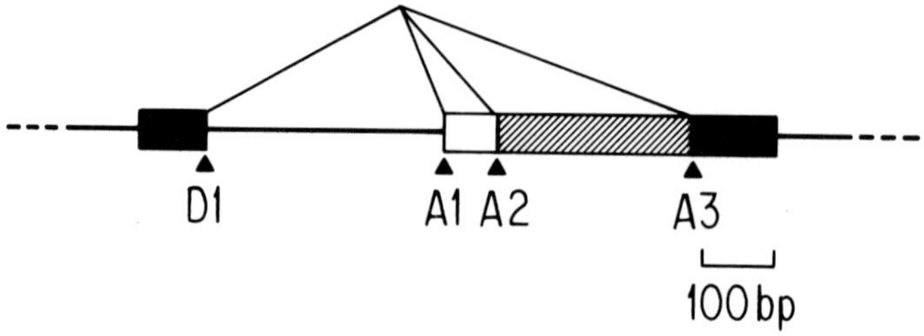

FIGURE 3. Pattern of alternative splicing giving rise to the three forms of rat liver fibronectin mRNA. Regions encoding sequence common to all three forms of mRNA are shown as black segments. The 285 and 75 nucleotide inserted segments are shown as crosshatched and open segments, respectively. Introns are represented by thin lines. Splice donor(D) and acceptor(A) sites are marked by arrows.

exon correspond exactly to the 360 base sequence present in the largest of the mRNAs. This segment is then immediately followed by the 107 nucleotides of sequence common to all three mRNAs and located 3' of the region of difference.

It is therefore clear that pairing a single 5' splice site(D1) with one of three different 3' splice sites(A1-3) can result in the production of three different fibronectin mRNAs from a common transcript. When A1 is chosen, only the 613 base intron is deleted. When A2 is chosen, this intron as well as the first 75 nucleotides of the 476 base exon are deleted. A3 can be used to generate the smallest of the three mRNAs by deletion of the first 360 nucleotides of the exon as well as the intron. This novel pattern of alternative splicing allows the production of three forms of mRNA by using different 3' splice sites, one at the beginning of, and two within, a single exon.

What else does the sequence of this region of the fibronectin gene tell us about its organization? The additional amino acid segments encoded by two of the mRNAs are found between two type III repeats(see figure 1). Thus the 90 nucleotide exon encodes only the last 30 amino acids of the 90 amino acid repeat preceding the additional blocks of sequence. The first third of the next type III repeat is encoded by the same exon as the inserted segments. It is therefore apparent that type III repeats are not necessarily encoded by single exons. Further analysis of this gene has revealed that the type III repeats are organized in a striking pattern(11). The details of this work will be reported elsewhere.

Fibronectin Subunits Encoded by the Different mRNAs

Is one of the predicted subunits unique to either plasma or cellular fibronectins? Several lines of evidence address this problem. S1 mapping of the region of mRNA variability has been carried out using mRNAs from a variety of cell types(7). This showed that all three forms of mRNA are found in both rat hepatocytes and fibroblasts, which produce plasma and cellular fibronectins, respectively. Thus none of the three forms of mRNA encodes a plasma or cellular fibronectin specific subunit, but the relative levels of the subunits comprising the two forms of fibronectin appear to vary(assuming the mRNAs are translated with equal efficiency). Obviously, the existence of at least three distinct fibronectin subunit types could allow the production of up to six different dimeric fibronectins. The division of fibronectins into plasma and cellular categories is thus overly simplistic.

The difference in size between a subunit containing the 95 or 120 amino acid segment and one lacking these segments could result in the molecular weight difference between the subunits of plasma or cellular fibronectins. Is this indeed the case? An immunological approach is currently being used to analyze further the subunits encoded by the different mRNAs. The phage expression vector λgt 11 contains a unique Eco R1 site near the 3' end of the gene for B-galactosidase. We have subcloned portions of the rat liver cDNA clones into this site in order to produce B-galactosidase/fibronectin fusion proteins to use as immunogens. This approach has allowed us to produce antisera directed against defined regions of the fibronectin molecule including the inserted amino acid segments. An antiserum directed against a portion of the 95 amino acid segment has been used to show that this segment is unique to the larger subunits of rat plasma and cellular fibronectins(12). This suggests that the smaller subunit is encoded by the mRNA lacking any additional nucleotides at the position of difference, and that the larger subunit observed on polyacrylamide gels is actually a mixture of 95 and 120 amino acid segment containing subunits. The use of such antisera should prove very useful in the continued analysis of fibronectin structure and function.

The inability of the three fibronectin mRNAs found in rat liver to account for all observed fibronectin subunits(5) raises the possibility that additional forms of fibronectin mRNA are also encoded by the single fibronectin gene. It has, in fact, recently been shown by Kornblihtt et al(13) that fibronectin mRNAs produced by a human cultured cell line vary by the presence or absence of a 270 base segment encoding a type III homology repeat(see figure one). This segment is located between two type III homology repeats between the cell and heparin binding sites. S1 mapping experiments were used to show that the mRNA encoding the subunit containing the extra type III repeat is not present in liver tissue. It was therefore concluded that this repeat is unique to cellular fibronectin subunits, but that not all cellular fibronectin subunits contain this repeat. This sequence is not present in the mRNAs predicted from the rat liver cDNA clones and the nucleotide sequence across the region of difference seen in these clones was not reported for the human cDNA clones. However, if both forms of variation occur in both species, at least six different fibronectin mRNAs could be produced from a single gene. These findings are particularly striking when one considers the fact that cDNA sequence representing less than half the length of a fibronectin subunit has been isolated. We are currently attempting to identify any additional positions of variation among fibronectin mRNAs, should they exist.

Functional Significance of the Additional Segments

The inclusion or omission of specific blocks of coding sequence might result in functional differences among fibronectin subunits. Several possible functions may be provided by the 95 and/or 120 amino acid segments. A fibronectin-fibronectin self association site has been mapped to this region of plasma fibronectin(14). This portion of the molecule has also been implicated in heparin binding(15, see figure 1). The presence of clusters of basic amino acid residues in the 95 residue segment might indicate a role of this segment in such an interaction, or in the interaction with hyaluronic acid, with which cellular fibronectin interacts more strongly than does plasma fibronectin(16). We are currently testing these hypotheses through the use of fusion proteins and the antisera directed against them.

The results reviewed here provide answers to several of the questions raised and offer the means to answer the others. The fact that a single gene encodes all forms of fibronectin raises interesting questions about its versatility of expression and about the regulation and significance ot the various forms of RNA processing which generate multiple mRNAs for related but different proteins.

ACKNOWLEDGEMENTS

We would like to thank Jennifer Lee and Minka Van Beuzekom for their excellent technical assistance.

REFERENCES

1. Hynes RO and Yamada KM (1982). Fibronectins: multifunctional modular glycoproteins. J Cell Biol 95:369-377.
2. Yamada KM (1983). Cell surface interactions with extracellular materials. Ann Rev Biochem 52:761-799.
3. Furcht LT (1983). Structure and function of the adhesive glycoprotein fibronectin. Mod Cell Biol 1:53-117.
4. Tamkun JW and Hynes RO (1983). Plasma fibronectin is synthesized and secreted by hepatocytes. J Biol Chem 258:4641-4647.
5. Paul JI and Hynes RO, submitted for publication.
6. Petersen TE, Thogersen HC, Skorstengaard K, Vibe-Pedersen K, Sahl P, Sottrup-Jensen L, and Magnusson S (1983). Partial primary structure of bovine plasma fibronectin: three types of internal homology. Proc Natl Acad Sci USA 80:137-141.
7. Schwarzbauer JE, Tamkun JW, Lemischka IR and Hynes RO (1983). Three different fibronectin mRNAs arise by alternative splicing within the coding region. Cell 35:421-431.
8. Kornblihtt AR, Vibe-Pedersen K and Baralle FE (1983). Isolation and charecterization of cDNA clones for human and bovine fibronectins. Proc Natl Acad Sci USA 80:3218-3222.

9. Tamkun JW, Schwarzbauer JE, and Hynes RO, submitted for publication.
10. Hirano H, Yamada Y, Sullivan M, DeCrombruggee B, Pastan I and Yamada KM (1983). Isolation of genomic DNA clones spanning the entire fibronectin gene. Proc Natl Acad Sci USA 80:46-50.
11. Odermatt E, Tamkun JW, Schwarzbauer JE and Hynes RO, submitted for publication.
12. Schwarzbauer JE, Paul JI and Hynes RO, submitted for publication
13. Kornblihtt AR, Vibe-Pedersen K, and Baralle FE (1984). Human fibronectin: molecular cloning evidence for two mRNA species differing by an internal segment coding for a structural domain. EMBO J 3:221-226.
14. Ehrismann R, Roth DE, Eppenberger HM and Turner DC (1982). Arrangement of attachment-promoting, self-association and heparin-binding sites in horse serum fibronectin J Biol Chem 257:7381-7387.
15. Hayashi M and Yamada KM (1983). Domain structure of the carboxyl-terminal half of human plasma fibronectin. J Biol Chem 259:3332-3340.
16. Laterra J and Culp LA (1982) Differences in hyaluronate binding to plasma and cell surface fibronectins. Requirement for aggregation. J Biol Chem 257:719-726.

Molecular Biology of Development, pages 427–435

GENE REGULATION DURING EARLY DEVELOPMENT OF THE CELLULAR SLIME MOLD *DICTYOSTELIUM DISCOIDEUM*[1]

James A. Cardelli, George P. Livi[2], Robert C. Mierendorf, David A. Knecht[3] and Randall L. Dimond

Department of Bacteriology, University of Wisconsin, Madison, WI 53706

ABSTRACT We have analyzed the changes occurring in the spectrum of proteins synthesized during early development of the cellular slime mold *Dictyostelium discoideum* using two dimensional gel electrophoresis. The relative rate of synthesis of 56 proteins changes during the first 90 minutes of development in cells previously grown in axenic culture. The majority of these changes (40) represent an increase in the relative rate of synthesis of proteins while the remainder (16) represent proteins decreasing in relative rate of synthesis. Seven additional proteins increase in rate of synthesis in cells previously grown on bacteria. Two dimensional gel electrophoresis of proteins synthesized *in vitro* directed by RNA extracted from developing cells indicates four control mechanisms operating at pretranslational and translational levels to regulate the synthesis of these proteins during early development. Genetic studies reveal that at least 28 of these regulated proteins are integrated into the developmental program and that only a few developmentally essential genes may be necessary for their expression.

[1]This work supported by NIH grant GM26632 and NSF grant PCM 81-10987
[2]Present address: Cold Spring Harbor Laboratory, NY
[3]Present address: Dept. of Biology B-022, UCSD, LaJolla, CA.

INTRODUCTION

The cellular slime mold, Dictyostelium discoideum, has proven to be an excellent system in which to study regulation of gene expression during differentiation (1). When cells growing on bacteria or in axenic (broth) culture are removed from their food source they undergo a well defined synchronous developmental cycle which culminates in the formation of a fruiting body (sorus) containing dormant spore cells situated on top of a column of supporting stalk cells. Throughout development changes occur in the specific activities of many enzymes (2), in the relative rate of synthesis of many proteins (3,4), and in the cellular population of mRNAs (5,6,7). Both transcriptional and translational control mechanisms have been proposed to regulate these changes during development (1,8-12). We have been studying the regulation of gene expression during early development using a combined biochemical and genetic approach and have found that a number of control mechanisms operate during the first few hours of development to control the synthesis of at least 63 proteins.

RESULTS

Identification by Two Dimensional Gel Electrophoresis of Proteins Regulated during Early Development.

Cells growing in axenic culture and cells developing on filters for various periods of time were pulse labeled with ^{35}S-methionine and the labeled proteins resolved by two dimensional gel electrophoresis (4). Our 2D gel system was optimized to detect a larger range of cellular proteins than previously reported (3). Examples of fluorograms of 2 of our gels are shown in Fig. 1 and represent proteins labeled during vegetative growth and at 4 hours of development. By this approach we have determined that during the first few hours of development 40 proteins increase in rate of synthesis with 19 of these proteins apparently being synthesized for the first time (see spots labeled 4,5,30,56,88 in Fig. 1 as examples). Decreases occur in the relative rate of synthesis of 16 proteins with most of the decreases occurring during the first 30 min of development (eg. spots 1,7,8,41,43,66, Fig. 1). Seven additional proteins increase in rate of synthesis in cells previously grown on bacteria (results not shown). All of these proteins are already being synthesized in cells growing in axenic cultures, consistent with the model proposing that cells in liquid culture have initiated the earliest portion of the developmental program (13).

Control Mechanisms Regulating Early Developmental Gene Expression.

To investigate the molecular mechanisms regulating early developmental gene expression, polysomal RNA was isolated from growing and developing cells, translated *in vitro*, and the resulting products separated on 2D gels. Greater than 50% of the proteins labeled *in vivo* are found in the same gel position as products synthesized *in vitro* (compare Fig. 1 with Fig. 2). This correspondence in spot location enabled us to determine changes in the concentration of functional mRNA coding for developmentally regulated proteins. Many of the proteins increasing in relative rate of synthesis are accompanied by increases in the concentration of polysomal mRNAs encoding them (spots labeled 4,5,10,30,88 in Fig. 1 and 2). The increase in the concentration of polysomal associated mRNA reflects an increase relative to total cytoplasmic mRNA (results not shown) and suggests regulation at a pretranslational level. One protein, however, increases in rate of synthesis during the first 30 min of development but is not accompanied by an increase in the relative concentration of cellular mRNA (9). Instead the efficiency of translation of this mRNA dramatically increases early in development (marked with arrow in Fig. 1 and 2).

The reduction in the relative rate of synthesis of at least 16 proteins is paralleled by a drastic decrease in the relative concentration of polysomal mRNA coding for them (eg. spots labeled 1,8,18,41,43,66, in Figs 1 and 2). However, most of these mRNAs are not degraded but instead shift off polysomes and are stored in an inactive state (8-10). Some of the changes in the distribution of specific mRNAs between functional and nonfunctional classes are quite dramatic. For instance,

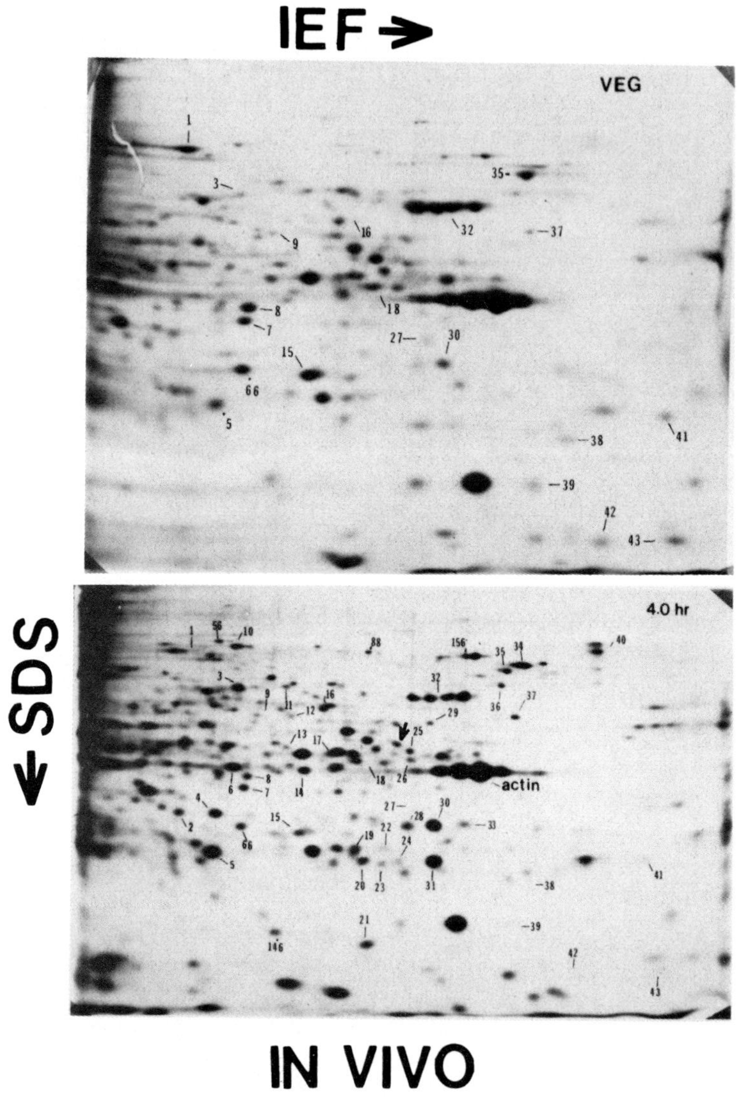

Figure 1. Two dimensional gel electrophoresis of proteins pulse labeled for 1 hour with ^{35}S-met in growing cells and cells developing for 4 hours.

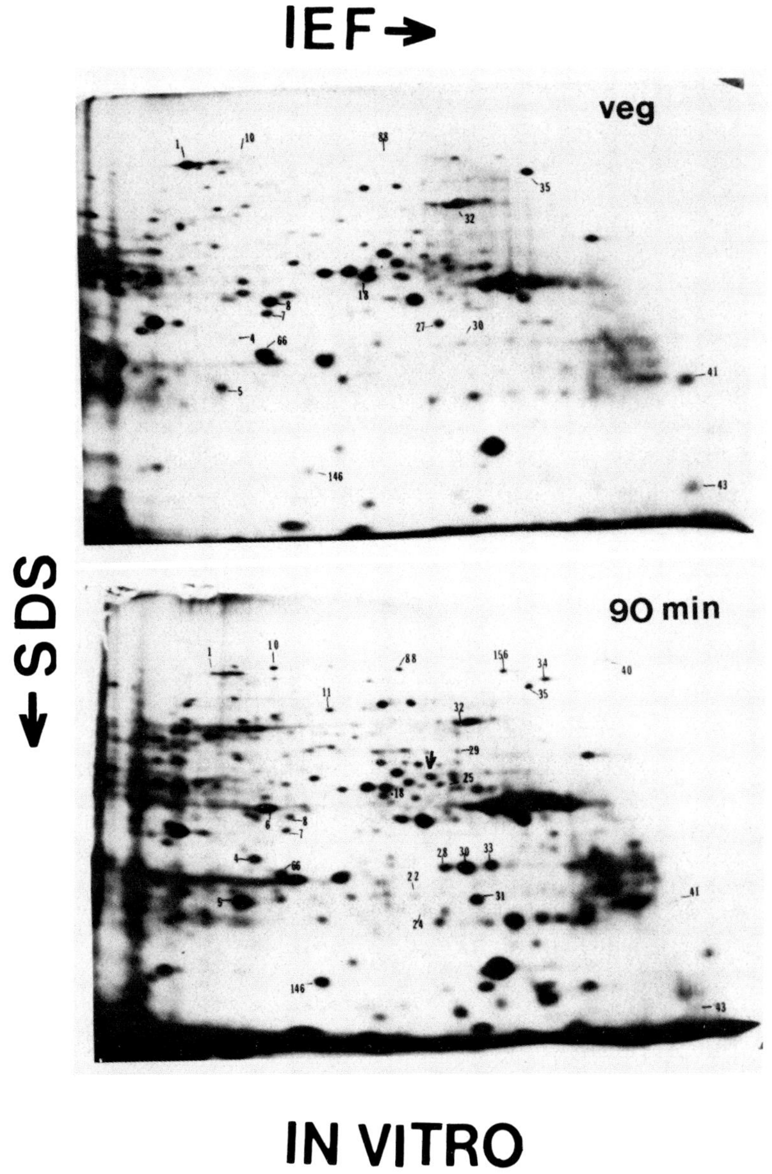

Figure 2. Two dimensional gel electrophoresis of proteins synthesized _in vitro_ directed by polysomal RNA isolated from growing cells and cells developing for 90 minutes.

60-80% of the cellular mRNAs coding for proteins 41 and 43 are polysomal associated in growing cells while in developing cells only 0-10% are being translated (9). This developmentally induced change in the intracellular distribution of these 2 mRNAs far exceeds that observed for the total population of cytoplasmic mRNA which changes from 85% polysomal associated in growing cells to 45% in developing cells (8,10). These nonpolysomal mRNAs are not irreversibly inactivated and can under certain conditions (eg. refeeding) rapidly reassociate with polysomes (8-10). These results indicate the existence of a reversible translational control mechanism regulating the synthesis of proteins early in development.

At least 2 proteins reduced in synthesis early in development are accompanied by a drastic reduction (9) in the relative cellular concentration of functional mRNA (eg. spot labeled 66 in Figs 1 and 2). This suggests an additional regulatory mechanism in which the developing cell selectively degrades or inactivates certain mRNAs while preserving others intact in an untranslated state.

Developmental Regulation of Proteins in Cells Previously Grown on Bacteria.

At least 7 proteins have been identified on 2D gels that increase in rate of synthesis during development in cells previously grown on bacteria. All of these proteins are constitutively synthesized in cells grown in axenic culture. Three of the 7 regulated proteins have been identified as belonging to the discoidin I family (results not shown). Discoidin I is a group of 3 developmentally regulated carbohydrate binding proteins encoded by a small multi-gene family (1). Consistent with previous reports we find that in wild type cells but not in a particular developmental mutant (see next section) all 3 proteins show similar temporal regulation (results not shown).

Another well studied example of this type of protein is the lysosomal enzyme α-mannosidase (α-man), which increases in specific activity during development in cells previously grown on bacteria (14) but is fully induced in cells growing in axenic culture (13). The increase during development in the relative amount of

Table 1

SUMMARY OF CHANGES IN PROTEIN SYNTHESIS PATTERNS DURING EARLY DEVELOPMENT

Growth conditions	Change in rate of synthesis[a]	Number of proteins in class[b]	Regulatory mechanism[c]
Axenic	increase	39 (16)	1
"	increase	1 (0)	2
"	decrease	14 (6)	3
"	decrease	2 (1)	4
Bacteria	increase	7 (5)	1

[a]This column represents proteins changing in rate of synthesis during early development as revealed by 2D gel electrophoresis

[b]Numbers in parentheses represent proteins misregulated in the early developmental mutant HMW 404.

[c]Changes in the concentration of mRNAs coding for proteins in each class was determined as described in the text. Regulatory mechanisms are as follows: 1) increase in functional mRNA, 2) no change in mRNA concentration but more efficient translation, 3) decrease in efficiency of translation of mRNA but no change in concentration 4) degradation or inactivation of mRNA.

α-man and discoidin proteins is accompanied by increases in both their relative rate of synthesis and in the cellular concentrations of functional mRNAs suggesting this class of protein is regulated at a pretranslational level (15,16).

Genetic Approaches to Dissecting Early Development.

As described above we have identified 63 proteins on 2D gels which change in rate of synthesis during the first few hours of development. Its essential to determine if the synthesis of these proteins is regulated by the developmental program or simply in response to starvation. We consider a protein to be developmentally controlled if it can be shown that its expression is determined by a gene essential for development. As one approach we have isolated a number of mutants which are blocked in development prior to aggregation and also fail to accumulate α-mannosidase, one of the earliest enzymes to be induced. These mutants define developmentally essential early acting pleiotropic genes. One of the mutants, HMW 404, misregulates 28 of the 63 regulated proteins defining the minimum number of proteins under developmental control. Another mutant, HMW 456, misregulates the expression of many proteins but more interestingly regulates the discoidin I gene family noncoordinately. In this mutant the developmental increase in the rate of synthesis of discoidin Ib and Ic lags behind that of discoidin Ia by a few hours (results not shown). These studies will eventually indicate how many of the remaining 63 proteins are developmentally controlled, how many early pleiotropically acting genes regulate their expression, and how many of these proteins are coordinately regulated.

DISCUSSION

We have utilized 2D gel electrophoresis to analyze the changes which occur in the patterns of proteins synthesized during early development. A change in the rate of synthesis of 56 proteins occurs during the first few hours of development in cells grown in axenic culture. These proteins can be divided into 4 classes based on the regulatory mechanisms governing their synthesis: 1) those that increase in rate of synthesis due to an increase in functional mRNA, 2) those that increase in rate of synthesis due to a more efficient translation of preexisting mRNA, 3) those that decrease in rate of synthesis due to a decrease in the initiation rate of translation, and 4) those that decrease in rate of synthesis due to an inactivation or degradation of mRNA. A fifth class represented by 8 proteins including the lysosomal enzyme α-man increase in rate of synthesis during early development in cells previously grown on bacteria (summarized in Table 1). Genetic studies have indicated that almost half of these proteins are under strict developmental control. Future studies will be directed at understanding the genetic and molecular mechanisms regulating the expression of this large class of genes during early development.

REFERENCES

1. Loomis WF (1982). "The Development of Dictyostelium discoideum", Academic Press, New York
2. Loomis WF (1975). "Dictyostelium discoideum: A Developmental System" Academic Press, New York
3. Alton TA and Lodish HF (1977). Dev Biol 60: 180.
4. Cardelli JA, Knecht DA and Dimond RL (1984). submitted
5. Margolskee JP and Lodish HF (1980). Dev Biol 79: 285.
6. Jacquet M, Part D and Felenbok B (1981). Dev Biol 81: 155.
7. Firtel RA (1972). J Mol Biol 66: 363.
8. Cardelli JA and Dimond RL (1981). Bioch 20: 7391.
9. Cardelli JA and Dimond RL (1984). submitted
10. Alton TA and Lodish HF (1977). Cell 12: 301.
11. Landfear SM, Lefebvre P, Chung S and Lodish HF (1982). Molec Cell Biol 2: 1417.
12. Williams JG, Lloyd MM and Devine JM (1979). Cell 17: 903.
13. Burns RA, Livi GP and Dimond RL (1981). Dev Biol 84: 407.
14. Loomis WF (1970). J Bacter 103: 375.
15. Livi GP, Cardelli JA, Mierendorf RC and Dimond RL (1984). this volume
16. Livi GP, Cardelli JA and Dimond RL (1984). submitted

Molecular Biology of Development, pages 437–446

MULTIPLE TROPONIN T PROTEINS ENCODED BY A SINGLE GENE. DEVELOPMENT AND TISSUE-SPECIFIC REGULATION BY DIFFERENTIAL RNA SPLICING[1]

Hanh T. Nguyen,[2] Russell M. Medford,[3] Antonia T. Destree, Eric Summers,[4] B. Nadal-Ginard

Laboratory of Molecular and Cellular Cardiology
Department of Cardiology, Children's Hospital
Department of Pediatrics, Harvard Medical School,
Boston, Massachusetts 02115

ABSTRACT The molecular basis of troponin T heterogeneity was studied at the level of gene organization and RNA expression, by using an adult rat skeletal muscle troponin T cDNA clone. Two structurally distinct forms of troponin T mRNA, whose expression is developmental - and tissue-specific, were found to be derived from a single skeletal troponin T gene. The present study demonstrates that a novel mechanism of alternative RNA splicing, involving the mutually-exclusive interchangeable expression of internal, protein-coding exons is responsible for the troponin T mRNA heterogeneity.

INTRODUCTION

A salient feature of eukaryotic cell differentiation is the developmentally-regulated expression of structurally distinct forms of the same protein. The molecular mechanisms responsible for protein diversity include systems

[1]This work was supported by grants from the Muscular Dystrophy Association and the NIH.
[2]Present address: Department of Cellular and Developmental Biology, Harvard University, Cambridge, MA 02138.
[3]Present address: Department of Medicine, Beth Israel Hospital, Harvard Medical School, Boston, MA 02115.
[4]Present address: Albert Einstein College of Med., Bronx, NY

that select a particular member of a multigene family for expression and systems that generate protein heterogeneity from a single gene. DNA rearrangement (1) and alternative RNA splicing (2) are mechanisms that result in the differential expression of internal gene sequences, thereby contributing to the generation of multiple proteins from a single gene.

Tissue-specific properties of muscle contraction is partly dependent on the composition of the structural (myosin, actin) and regulatory (troponins, tropomyosin) components of the myofibrillar apparatus. The mechanisms responsible for the expression of a particular set of contractile protein genes in a specific muscle type at a defined stage of development must involve the coordinate and differential expression of single gene isotypes from eight distinct genes or gene families (3). In this paper, we present evidence for a novel mechanism of alternative RNA splicing which contributes towards the generation of multiple developmentally-regulated troponin T mRNAs and proteins from a single gene.

RESULTS

The Developmental and Tissue-Specific Expression of Structurally Distinct Troponin T mRNAs.

Biochemical and immunological studies have shown that there exist developmental and tissue-specific troponin T (TnT) isoforms (4,5). An adult rat skeletal muscle TnT cDNA clone (6), was used to identify structurally distinct TnT mRNAs. Results from RNA blotting experiments indicate that skeletal muscle TnT mRNA sequences are highly divergent from cardiac TnT mRNA sequences. The varying degree of hybridization of the cDNA clone to skeletal muscle from different developmental stages suggests that (1) skeletal TnT mRNAs exhibit partial sequence homology, (2) the same TnT mRNA accumulates to different levels at various stages, or (3) a combination of both.

S1 nuclease mapping was done to distinguish among the possibilities presented above. The TnT cDNA clone was 3' end-labelled at amino acid 70 and recut in the 3' untranslated region. As shown in Fig. 1, a fully-protected fragment is detected in abundance with RNA from adult skeletal muscle and much less with RNA from newborn, fetal, and embryonic skeletal muscle. The fully-protected fragment represents TnT mRNA sequences, designated as TnT -α mRNA, that are totally homologous to the TnT cDNA. The differential signals show that TnT -α mRNA accumulates to different levels during skeletal muscle development. A partially-protected fragment, representing an alternate TnT mRNA sequence designated as TnT - β mRNA, is detected with all skeletal muscle RNA. The size of the partially-protected fragment indicates that the TnT -β mRNA sequence is homologous to the TnT -α mRNA sequence from the labelled end at amino acid 70 to ~ amino acid 228. TnT -β mRNA is the predominant TnT mRNA species in embryonic, fetal, and newborn skeletal muscle whereas in the adult skeletal muscle, TnT -α and TnT -β mRNAs are accumulated to comparable levels.

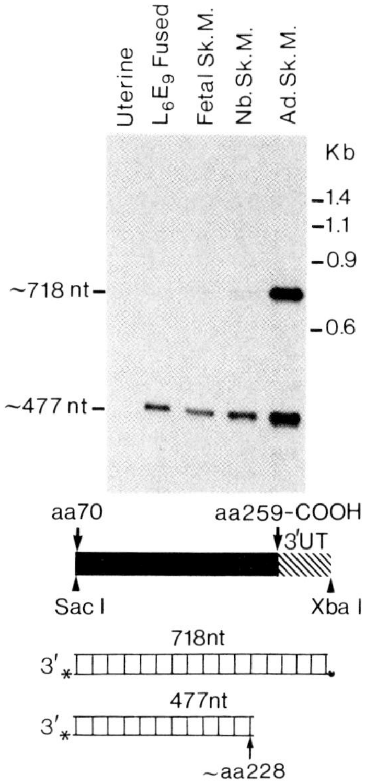

Figure 1. TnT-α and TnT-β are differentially -expressed during skeletal muscle development.

The structure of TnT -β mRNA beyond the point of sequence divergence was also examined. In this S1 nuclease mapping study, the TnT cDNA clone was 5' end-labelled in the 3' untranslated region and recut at amino acid 70. The results of this analysis indicate that TnT -β is homologous to TnT -α from the labelled end in the 3' untranslated region to ~amino acid 243.

From both S1 nuclease mapping studies, each exploring a different side of the point of sequence divergence, it is evident that TnT -α and TnT -β mRNAs are structurally identical from amino acid 70 to the carboxyl terminus plus the 3' untranslated region, except for a region which codes for an internal peptide covering amino acid 229-242. Each TnT mRNA contains either the α- or β- specific internal coding sequence but not both, as indicated by the absence, in either S1 nuclease mapping study, of any additional partially-protected fragment whose size would correspond to a TnT mRNA containing both sequences.

The distribution of TnT -α and TnT -β mRNAs in various skeletal muscle types was also explored by S1 nuclease mapping using RNA from a panel of anatomically-defined adult skeletal muscles. The results suggest that the differential expression of each TnT mRNA isotype is correlated with anatomically-defined muscle group, rather than muscle fiber type as classically defined.

Evidence for a Single Skeletal Troponin T Gene.

The two structurally distinct skeletal TnT mRNAs could arise from a single gene through differential RNA splicing. Alternatively, they could be encoded by two separate genes. Results from Southern blotting analysis are consistent with a single genomic copy of the TnT sequences under examination.

To examine the nature of the TnT gene, an EcoRI rat genomic Charon 4A library was screened. Southern blotting analysis and partial DNA sequencing analysis of one of the genomic clones, using the Sanger (7) or Maxam-Gilbert (8) methods, indicate that 80% of the protein-coding sequences of the TnT gene, extending from amino acid 44 to the carboxyl-terminus plus the entire 3' untranslated region, resides in an 8.5kb EcoRI genomic fragment. Thus, skeletal TnT is encoded by a single gene in the rat genome.

TnT -α and TnT -β mRNAs Share a Common Nuclear Precursor.

DNA sequencing analysis shows that two small exons contain sequences coding for structurally distinct peptides covering the same region of the TnT protein, namely amino acid 229-242 (Fig. 2). The sequences in the β- specific exon are not homologous to the α- specific exon but rather to a new rabbit TnT mRNA sequence (9). These exons correspond to the two forms of TnT mRNA identified by S1 nuclease mapping. These results represent the first evidence for the mutually-exclusive interchangeability of internal exons, contributing to the structural heterogeneity of a protein. Henceforth, the term "exchangeon" will be used to refer to any set of exons whose members are incorporated into the mature mRNA in a mutually-exclusive manner.

At the DNA structural level, cells actively expressing muscle-specific proteins could undergo DNA rearrangement to remove one of the two exchangeons. Southern blotting analyses with genomic DNA from sperm, muscle, and non-muscle

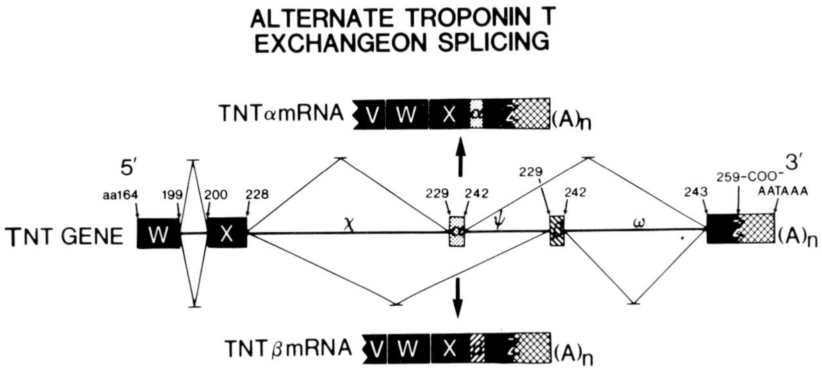

Figure 2. Scheme for the generation of TnT-α and TnT-β mRNAs.

sources show that TnT genomic sequences do not appear to undergo DNA rearrangement. Moreover, results from RNA blotting analyses using, DNA probes specific for the α- or β- exchangeon (see Fig. 2), strongly indicate that both α- and β - exchangeons are present on a common nuclear RNA transcript in cells ultimately producing differential amounts of mature TnT -α or TnT -β mRNA. Whether the RNA transcript synthesized in the α- and β- producing cells uses the same transcriptional promoter is not known.

The mechanism by which the α- or β- exchangeon is incorporated into the mature TnT mRNA must involve an active regulation of splicing choices of a common nuclear RNA transcript with multiple and optional 5' and 3' splicing sites. Preliminary structural analysis shows, however, that both exchangeons are flanked by junction sequences that are consistent with the consensus splicing junction sequence (10). Further analysis of this region, using DNA probes specific for the α- or β- exchangeon (see Fig. 2), reveals that this region contains sequences coding for a small RNA species of ~4S whose direction of transcription is opposite to that of TnT mRNA. The relationship between the small RNA and the mutually-exclusive expression of the exchangeons remains to be clarified. Nonetheless, the results suggest that a unique mechanism of alternative RNA splicing is operating to select, in a mutually-exclusive manner, between two exchangeons sharing a common nuclear transcript.

DISCUSSION

Structurally-Distinct TnT mRNAs are Expressed in a Tissue - and Developmental-Specific Manner.

At least three structurally distinct TnT mRNAs have been identified. Cardiac and skeletal TnT mRNAs do not share significant sequence homology, strongly suggesting a separate TnT gene for each muscle type with each under tissue-specific regulatory control. Conversely, the developmentally - and tissue-regulated TnT -α and TnT -β mRNAs, differing in the region detected by the cDNA clone by an internal 39 nucleotides, are coded by a single skeletal TnT gene. Preliminary studies indicate that additional skeletal TnT mRNAs, differing in the 5' terminal coding sequences, are expressed differentially during development and in skeletal muscle types (11). This high degree heterogeneity at the RNA level is consistent with protein data (4,5). The results argue strongly for the existence of two different modes of regulation for the generation of multiple TnT mRNAs and proteins.

Constitutive versus Alternative RNA Splicing.

During the study of alternate RNA splicing or splicing in general, a major question arises: what makes a particular splice junctive active? It appears that most splice site activations are constitutive and fixed, resulting in the linking of the appropriate exons to form a functional, mature mRNA (12). DNA sequence analyses and reconstruction experiments have identified factors that are not involved in splice site selection (13,14). The results indicate that factors other than the nucleotide sequence near the constitutive splice site, such as proteins or small nuclear RNAs (15) may influence the final splicing pattern by activating the normal splicing site junction.

In contrast to constitutive RNA splicing, alternative RNA splicing is a selective mechanism in which splice site activation is differentially-regulated. Thus far, models of alternative RNA splicing share a common feature characterized by a single-operation decision: which one of several alternative splice site junctions is to be activated? Once a splice site is selected for activation, however, the final structure of the mRNA reflects the constitutive utilization of all the normal splice site junctions distal to the selected junction. In most systems, differences in the 5' or 3' end of the mRNA are associated with differential usage of transcriptional promotors or of polyadenylation sites, resulting in some cases in altered terminal protein-coding sequences (2,16,17). Alternative RNA splicing resulting in internal peptide alteration has been shown in only two systems, α- crystallin (14) and fibronectin (18) but neither has been proven to be specifically-regulated.

Exchangeons: A Novel Class of Alternative RNA Splicing.

The concept of exchangeons, illustrated by the skeletal TnT -α and TnT -β mRNAs, represents a new class of alternative RNA splicing. This model system provides the first conclusive evidence that alterations in internal peptide sequences results from alternative RNA splicing which is developmentally-regulated. The exchangeon model differs from previous models discussed in several aspects. The α- and β- exchangeons are of the same size, code for the same amino acid positions in the TnT protein, and link up to a common downstream 3' exon. Most importantly, the incorporation of the α- or β- exchangeon into the mature TnT mRNA is mutually-exclusive.

The mutually-exclusive expression of the α- and β- exchangeons suggests a distinctive feature of alternative RNA splicing within this region of heterogeneity, hereafter designated as the isotype switch region (ISR). The ISR contains two flanking introns, χ and ω , and two internal exons, α- and β- , separated by an internal intron ψ (see Fig. 2). In contrast to other alternative RNA splicing models, two operational decisions must be made: (1) Selection of either the χ- or ψ-3' splice site to be linked to the common 5' splice in χ and (2) Inactivation of the other non-selected 3' splice site. Although the 5' and 3' splice sites within intron ψ are functional and are utilized by intron ω and χ splicing sites, respectively, the ψ intron 5' and 3' splicing sites are not compatible with each other. Preliminary structural analysis of this region does not reveal any dramatic DNA sequence or tertiary structure that may influence the incompatibility of these two apparently functional splicing sites. Whether an alteration in the structural conformation imparted by upstream sequences or whether transacting factors, such as proteins or small RNAs, are responsible is not known. Nonetheless, the elimination of this splicing option contributes to the mutually-exclusive nature of exchangeon expression. Moreover, it is evident that a scanning model type of RNA splicing is not a plausible mechanism for generating TnT -α and TnT -β mRNAs.

The structural conformation of the ISR could be modulated by alterations in the structure of the TnT primary transcript. The generation of alternate TnT primary transcripts could be derived from differential usage of transcriptional promoters, generating primary transcripts differing in size and nucleotide composition and thus altered secondary conformation. Preliminary studies on the 5' end of the TnT mRNA strongly suggest that there is a correlation between the 5' exon composition of the TnT RNA transcript and the expression of the α- or β- exchangeon (11).

The precisely defined and differentially-regulated α- and β- internal peptides are highly suggestive of functional domains. These peptides represent a region of the TnT protein that is included in a chymotryptic fragment which has been shown to interact with tropomyosin and TnC (19). Alteration of the peptide structure of this region may result in changes in the interactions with the tropomyosin - TnC complex which in turn may alter the physiologic response of the complex to changes in the intracellular environment.

The finding of TnT isotype heterogeneity from a single gene implies that other contractile protein genes may exhibit additional degrees of diversity beyond the differential expression of multigene families (3). Indeed, preliminary studies indicate that protein diversity arising from a single gene may apply to the LC1/LC3 and tropomyosin genes (20,21). The construction of subtly modified proteins from a single gene may therefore represent a general mechanism.

ACKNOWLEDGMENTS

The expert secretarial assistance of Mrs. M. Hager and B. Trudeau is greatfully appreciated.

REFERENCES

1. Leder P, Max EE, Seidman JG, Kwan SP, Nau M, Norman B (1981). Recombination events that activate, diversity, and delete immunoglobulin genes. CSHQB 45:859.
2. Benyajati C, Spoerel N, Haymerle H, Ashburner M (1983). The messenger RNA for alcohol dehydrogenase in Drosophila melanogaster differs in its 5' end in different developmental stages. Cell 33:125.
3. Nadal-Ginard B, Medford RM, Nguyen HT, Periasamy P, Wydro RM, Hornig D, Gubits R, Garfinkel LI, Wieczorek D, Bekesi E, Mahdavi V (1982). Structure and regulation of a mammalian sarcomeric myosin heavy-chain gene. In Pearson ML, Epstein HF (eds): "Muscle Development: Molecular and Cellular Control," New York: Cold Spring Harbor Laboratory, p 143.
4. Dhoot GK, Perry SV (1979). Distribution of polymorphic forms of troponin components and tropomyosin in skeletal muscle. Nature (London) 278:714.
5. Toyota N, Shimada Y (1981). Differentiation of troponin in cardiac and skeletal muscles in chicken embryos as studied by immunofluorescence microscopy. J Cell Biol 91:497.
6. Garfinkel LI, Periasamy M, Nadal-Ginard B (1982).

Cloning, identification, and characterization of α-actin, myosin light chains 1, 2, and 3, α-tropomyosin, and troponin C and T. J Biol Chem 257:11078.
7. Sanger F, Nicklen S, Coulson AR (1977). DNA sequencing with chain-terminating inhibitors. Proc Natl Acad Sci (USA) 74:5463.
8. Maxam AM, Gilbert W (1977). A new method of sequencing DNA. Proc Natl Acad Sci (USA) 74:560.
9. Putney SD, Herliky WC, Schimmel P (1983). A new troponin T and cDNA clones for 13 different protein, found by shotgun sequencing. Nature (London) 302:718.
10. Mount S (1982). A catalogue of splice junction sequences. Nucl Acids Res 10:459.
11. Nguyen HT, Medford RM, Destree AT, Ruiz-Opazo R, Nadal-Ginard B (1984). Manuscript in preparation.
12. Ziff EB (1980). Transcription and RNA processing by the DNA tumor viruses. Nature (London) 287:491.
13. Kuhne T, Wiernga B, Reiser J, Weissman C (1983). Evidence against a scanning model of RNA splicing. The EMBO Journal 2 (5):727.
14. King CR, Piatigorsky J (1983). Alternative RNA splicing of the murine A-crystallin gene: protein-coding information within an intron. Cell 32:707.
15. Busch H, Reddy R, Rothblum L, Choi YC (1982). SnRNAs, snRNPs, and RNA processing. Ann Rev Biochem 51:617.
16. Schibler U, Hagenbuchle O, Wellauer PK, Pittet AC (1983). Two promoters of different strengths control the transcription of the mouse alpha-amylase gene Amy-1^a in the parotid gland and the liver. Cell 33:501.
17. Rosenfeld MG, Mermod JJ, Amara SG, Swanson LW, Sawchenki PE, Rivier J, Vale WW, Evans RM (1983). Production of a novel neuropeptide encoded by the calcitonin gene via tissue-specific RNA processing. Nature (London) 304:129.
18. Schwarzbauer JE, Tamkum JW, Lemischka IR, Hynes RO (1983). Three different fibronectin mRNAs arise by alternative splicing within the coding region. Cell 35:421.
19. Pearlstone JR, Smillie LB (1982). Binding of troponin T fragments to several types of tropomyosin. J Biol Chem 257:10587.
20. Periasamy M, Strehler EE, Garfinkel LI, Gubits RM, Ruiz-Opazo N, Nadal-Ginard B (1984). Manuscript submitted.
21. Ruiz-Opazo N, Weinberger J, Nadal-Ginard B (1984). Manuscript in preparation.

Molecular Biology of Development, pages 447–457

DEVELOPMENTAL CONTROL OF α-MANNOSIDASE-1 SYNTHESIS IN *DICTYOSTELIUM DISCOIDEUM*[1]

George P. Livi[2], James A. Cardelli, Robert C. Mierendorf and Randall L. Dimond

Department of Bacteriology, University of Wisconsin, Madison, WI 53706

ABSTRACT The lysosomal enzyme α-mannosidase-1 is one of the earliest developmentally controlled gene products in *D. discoideum*. Monoclonal antibodies prepared against purified α-mannosidase-1 have been used to identify the different enzyme forms synthesized *in vivo* and *in vitro*, and to study the regulation of enzyme synthesis, localization, modification, processing and secretion during growth and development. α-Mannosidase-1 is first synthesized as a large (140K) precursor protein which is modified and proteolytically processed to mature enzyme with subunits of 58K and 60K. The primary translation product for the 140K enzyme precursor is a 120K protein which is synthesized from membrane-bound polysomal mRNA. The linear increase in cellular α-mannosidase-1 activity observed during the first 20 hr of development is due to a change in the relative rate of precursor biosynthesis, which can be accounted for by an increase in the level of functional α-mannosidase-1 mRNA. Aggregation-deficient mutants prematurely terminate the developmental accumulation of α-mannosidase-1 activity due to an enhanced rate of enzyme inactivation late in development. This phenotype correlates

[1]This work was supported by the College of Agricultural and Life Sciences, University of Wisconsin, Madison, and by NIH grants GM 29156 and GM 31181.

[2]Present address: Cold Spring Harbor Laboratory, Cold Spring Harbor, NY.

with a failure to induce a developmentally controlled change in the post-translational modification system. Mutations in any one of the many aggregation-essential genes exhibit pleiotropic effects on the developmental program that prevent normal changes in lysosomal enzyme modification required for activity. In addition, aggregation-deficient mutants have been isolated which define a small number of genes, expressed very early in development, that are required for the accumulation of α-mannosidase-1 activity above the normal vegetative level. Some of these strains are defective in the post-translational modification and/or processing of the enzyme precursor, whereas others are deficient in the accumulation of α-mannosidase-1 mRNA.

INTRODUCTION

Development in multicellular organisms is controlled by the temporal expression of many genes. The products of these genes may appear at distinct stages of development and are often required for differentiation. The cellular slime mold, Dictyostelium discoideum, is an ideal organism for the study of developmental gene regulation because its simple developmental cycle is amenable to both genetic and biochemical analysis. Upon starvation unicellular vegetative amoebae aggregate to form groups of approximately 10^5 cells that undergo a series of defined morphological changes which result in the formation of fruiting bodies containing two differentiated cell types, spore and stalk cells. Many proteins appear (or disappear) at particular times in development due to a change in their relative rate of synthesis (1-3). Some of these proteins are cell-type-specific and are coded for by genes regulated at the level of transcription (1-9).

Several years ago a model was proposed to explain the pleiotropic effect of morphological mutations in this organism (10). It had been observed that mutants blocked in development fail to express many developmentally controlled genes. The pattern of this pleiotropy was revealed by analyzing the accumulation of several stage-specific enzymes in a number of morphological mutants. The results were consistent with a model describing the Dictyostelium developmental program as a linear dependent sequence of stages, with each stage defined by a unique set of developmentally regulated genes. Each morphological mutant identifies a gene required for orderly progress through the developmental program and the expression of all subsequent developmental events (or proteins). Genetic estimates indicate that at least 250 genes are required for completion of this developmental program (11-13).

We have attempted to understand the complex genetic control of development in this organism by initially studying the expression of a single gene. One of the earliest developmentally controlled gene products in Dictyostelium is the lysosomal enzyme α-mannosidase-1 (14). This very acidic glycoprotein accumulates intracellularly and is secreted during growth and development (15). The enzyme is not required for morphogenesis since structural gene mutants develop normally (16). Although this enzyme has been well-characterized biochemically (see below) and its structural gene has been genetically mapped (16), the molecular and genetic mechanisms involved in controlling its synthesis during development have not been studied. We have used monoclonal antibodies prepared against purified α-mannosidase-1 (17) to analyze the different forms of the enzyme synthesized in vivo and in vitro (18), and to study the regulation of enzyme synthesis, localization, modification, processing and secretion during growth and development. Our results reveal some of the mechanisms whereby the expression a single gene is regulated during development.

RESULTS AND DISCUSSION

1. α-Mannosidase-1 Biosynthesis during Development

Cellular α-mannosidase-1 specific activity is low in vegetative amoebae growing on bacteria, but increases dramatically upon starvation and accumulates with linear kinetics through the first 20 hr of development to a maximum level just prior to fruiting body formation (14). Using monoclonal antibodies prepared against purified α-mannosidase-1 (17,18) we have identified the different forms of the enzyme synthesized in vivo and in vitro and have measured their relative synthetic rates during development. Purified cellular α-mannosidase-1 consists of equimolar amounts of two subunits with molecular weights of

58,000 and 60,000 daltons (18). α-Mannosidase-1 is first synthesized as a large (M_r = 140,000) species which has been identified as a precursor to the mature enzyme on the basis of several criteria including antibody recognition, enzymatic activity, peptide mapping, pulse-chase experiments and its absence in a structural gene mutant (17-19). Both the 140K protein and the mature enzyme are secreted in an enzymatically active form (18) and carry an antigenic determinant shared by many Dictyostelium lysosomal enzymes, which consists of a sulfated N-linked mannose-rich oligosaccharide (20,21).

Cellular accumulation of α-mannosidase-1 activity during development is controlled by the rate of de novo enzyme precursor biosynthesis (22). The relative synthetic rate of the 140K precursor protein increases about 10-fold between 30 min and 8 hr, and remains at a maximum level for at least the next 10 hr of development (Fig. 1). We estimate that the amount of 140K protein synthesized within the period of peak synthesis equals approximately 0.14% of the total ^{35}S-methionine incorporated. At about 20-22 hr of development there is a significant reduction in the relative rate of precursor synthesis. Figure 1 shows that throughout development the relative rate of synthesis closely parallels the rate of accumulation of cellular enzyme activity.

We have compared these data to the amount of functional α-mannosidase-1 mRNA present in cells during development. Immunoprecipitation following cell-free translation of Dictyostelium mRNA reveals an α-mannosidase-1-specific polypeptide of 120K that is synthesized exclusively on membrane-bound polysomes (22-24). This 120K protein appears to be the primary translation product for the 140K precursor found in vivo, based on antibody recognition, peptide mapping, and by its absence from translation products derived from mRNA extracted from an α-mannosidase-1 structural gene mutant (23). In addition, the 120K protein can be processed in vitro by dog pancreas microsomes into a form indistinguishable from the 140K precursor found in vivo on the basis of molecular weight, endoglycosidase H sensitivity and isoelectric point (Fig. 2, ref. 23). Analysis of SDS-polyacrylamide gels containing immunoprecipitated translation products from total RNA extracted at various times of development indicates that the level of functional cellular α-mannosidase-1 mRNA increases in parallel with the change in relative synthetic rate of the 140K precursor protein in vivo (Fig. 1, ref. 22).

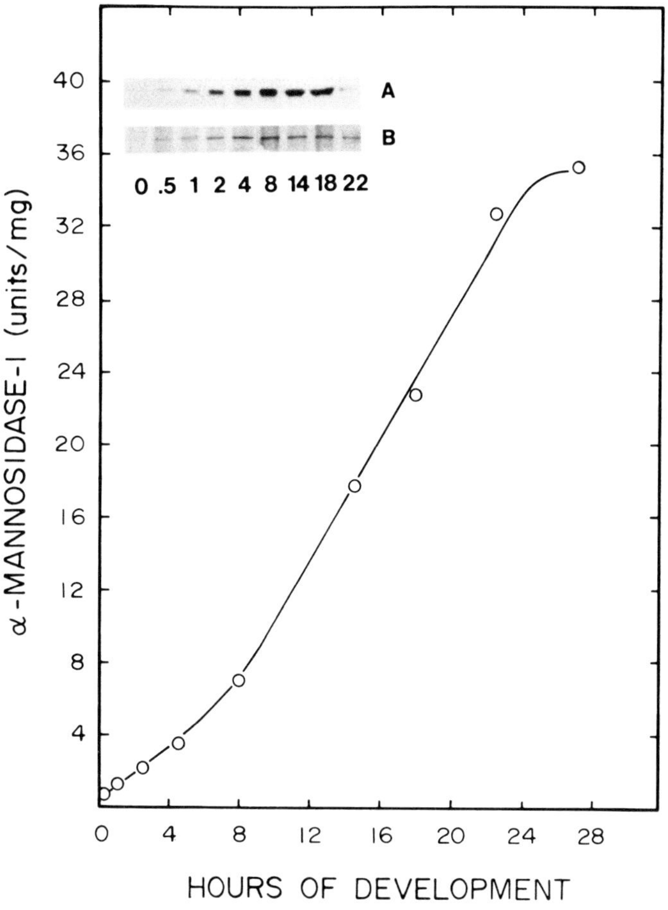

FIGURE 1. Relative rate of α-mannosidase-1 precursor synthesis, accumulation of functional α-mannosidase-1 mRNA, and increase in cellular enzyme activity during development. The increase in cellular α-mannosidase-1 specific activity during synchronous development is shown for the wild-type strain Ax3. (A) Synthesis of 140K protein _in vivo_: Within the same experiment cells were pulse-labeled (30 min) with ^{35}S-methionine at the indicated times (hr) and equal numbers of TCA-precipitable counts of each cell extract were immunoprecipitated with α-mannosidase-1 monoclonal antibody and subjected to SDS-polyacrylamide gel

electrophoresis. (B) Synthesis of 120K polypeptide in vitro: Total cellular RNA was extracted from developing cells, translated in vitro in the presence of ^{35}S-methionine, and an equal number of counts was incubated with rabbit anti-α-mannosidase-1 antiserum, followed by electrophoresis and fluorography.

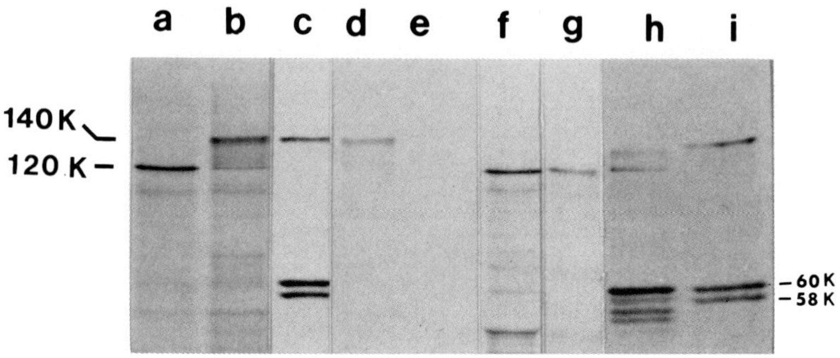

FIGURE 2. Synthesis and processing of α-mannosidase-1 precursor in vitro. In Lanes (a) and (f) total RNA was extracted from growing cells, translated in vitro and immunoprecipitated as described in Figure 1. Lane (b) addition of dog pancreas microsomal membranes. Lane (d) same as (b) plus trypsin. Lane (e) same as (d) plus Triton X-100. Lane (g) same as (b) plus endoglycosidase H. These results are compared to the α-mannosidase-1 species synthesized in vivo. Lanes (c) and (i) a pulse-labeled extract of developing cells immunoprecipitated with monoclonal antibody. Lane (h) same as (c) plus endoglycosidase H.

We have also measured the rate at which the 140K precursor is processed in vivo to the mature enzyme (22). Pulse-chase data indicate that cells in early development efficiently process the α-mannosidase-1 precursor to the 58K and 60K mature enzyme subunits with a half-life of less than 10 min. The rapid conversion of the 140K protein into mature enzyme also supports the conclusion that the increase in cellular enzyme activity is primarily controlled by the rate of precursor synthesis. Since the relative rate of precursor biosynthesis is strictly related to the level of functional mRNA present in the cell, we conclude that the synthesis of α-mannosidase-1 is regulated at a pre-translational level.

2. Enzyme Regulation in Aggregation-deficient Mutants

Aggregation-deficient (AGG⁻) mutants are defective in the cellular accumulation of α-mannosidase-1 activity (25). Developmental α-mannosidase-1 accumulation occurs initially at similar rates in mutant and wild-type cells, but prematurely terminates in AGG⁻ mutants within a particular time period in development. Cellular enzyme activity remains constant following the shutoff of accumulation. This enzymatic defect is not due to (a) an increased rate of enzyme secretion, (b) a change in the rate of total protein synthesis, (c) a premature termination of *de novo* enzyme precursor biosynthesis, (d) a defect in enzyme precursor processing to mature enzyme or (e) preferential enzyme degradation (25,26). Instead, the plateau in cellular α-mannosidase-1 activity results from an increased rate of enzyme inactivation. We have immunologically detected the presence of a significant quantity of inactive α-mannosidase-1 protein in cells of AGG⁻ mutants late in development (26). In addition, these AGG⁻ mutants fail to induce a specific change(s) in the post-translational modification of several lysosomal enzymes, including α-mannosidase-1, which normally occurs as part of the wild-type developmental program. This defect in modification has been revealed by differences in migration of lysosomal enzymes on nonequilibrium isolelectric focusing gels, differences in enzyme thermostability, and differences in enzyme precipitability using rabbit antiserum specific to the common carbohydrate modification (20,26, 27). Mutations in any one of the more than 100 genes required for aggregation prevent normal changes in enzyme modification due to their pleiotropic effects on the developmental program (25,26). These changes in modification may be requried for enzyme activity.

3. Genes Controlling Enzyme Precursor Synthesis, Modification and Processing

We have used a direct screening procedure to assay thousands of mutagenized cell clones for α-mannosidase-1 activity, and have isolated a large number of mutants with reduced enzyme activity (27,28). The majority of these mutants (59/70) are also aggregation-deficient. Most of these AGG^- strains accumulate intermediate levels of α-mannosidase-1 activity and thus define the same large class of developmentally essential genes required for the continued accumulation and stability of enzyme activity. A subset of these strains, however, are completely deficient in developmental accumulation of enzyme activity. Two strains fruit normally and contain mutations in the α-mannosidase-1 structural gene, manA (16,29). In addition, six enzyme-deficient AGG^- mutants exist. The mutations in these strains genetically complement lesions at manA in constructed diploids (29). Moreover, fruiting revertants of these strains accumulate wild-type levels of enzyme activity, indicating that the enzymatic defects are caused by mutations in nonstructural genes essential for early development (28).

We have used our α-mannosidase-1-specific monoclonal antibodies to examine the synthesis and processing of the enzyme precursor *in vivo* in these mutants (28). Several strains are drastically reduced in the relative rate of enzyme precursor synthesis throughout development. One strain synthesizes no detectable precursor and contains no functional α-mannosidase-1 mRNA. Some strains synthesize an altered precursor protein and are deficient in its conversion to the mature enzyme subunits (Fig. 3). These include both α-mannosidase-1 structural gene mutants which may produce enzyme molecules defective in sites important for modification and processing, as well as AGG^- strains which are also defective in other lysosomal enzymes. The AGG^- strains identify a small number of very early dependent sequence genes required for the synthesis of α-mannosidase-1 activity. These genes may act either to control the expression of manA or to regulate the normal post-translational modification or processing of the enzyme precursor. We are presently trying to determine the exact number and location of genes directly involved in regulating the activity of α-mannosidase-1 by genetic complementation and linkage analysis.

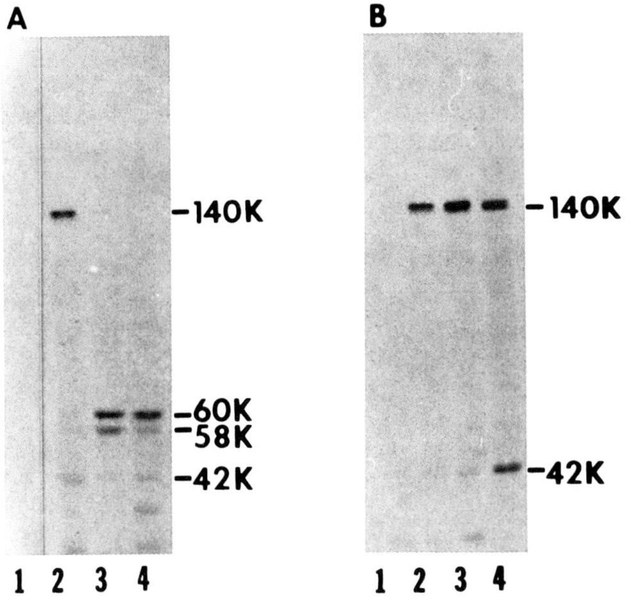

FIGURE 3. A structural gene mutant defective in α-mannosidase-1 precursor processing. Cells of strain Ax3 (A) and HMW-464 (B) were pulse-labeled with ^{35}S-methionine for 30 min at 4 hr of development, then chased in the presence of unlabeled methionine. Equal numbers of TCA-precipitable counts were immunoprecipitated with α-mannosidase-1 monoclonal antibody and subjected to SDS-polyacrylamide gel electrophoresis. In both fluorographs, Lane (1) control precipitation using purified normal mouse IgG, (2) 30 min pulse, (3) 1 hr chase, (4) 5 hr chase.

CONCLUDING REMARKS

α-Mannosidase-1 is an example of an early developmentally controlled protein that accumulates intracellularly due to an increase in its relative rate of synthesis. Developing cells of D. discoideum regulate the level of this one protein not only by controlling its synthesis, but also by controlling the modification and processing of the enzyme precursor, the stablilty of the mature enzyme, and its cellular localization. Our studies have clearly shown that complex cellular and molecular mechanisms are involved in determining the fate of a single macromolecule, and that many genes control these mechanisms during development.

REFERENCES

1. Alton, TA and Lodish, HF (1977a) Dev Biol 60: 180-206.
2. Alton, TA and Lodish, HF (1977b) Cell 12: 301-310.
3. Cardelli, JA, Knecht, DA, Wunderlich, R and Dimond, RL (1984) submitted.
4. Alton, TA and Brenner, M (1979) Dev Biol 71: 1-7.
5. Barklis, E and Lodish, HF (1983) Cell 32: 1139-1148.
6. Coloma, A and Lodish, HF (1981) Dev Biol 81: 238-244.
7. Margolskee, JP and Lodish, HF (1980) Dev Biol 74: 50-64.
8. Morrissey, J, Farnsworth, P and Loomis, WF (1981) Dev Biol 83: 1-8.
9. Mehdy, MC, Ratner, D and Firtel, RA (1983) Cell 32: 763-771.
10. Loomis, WF, White, S and Dimond, RL (1976) Dev Biol 53: 171-177.
11. Coukell, MB (1975) Molec Gen Genet 142: 119-135.
12. Williams, KL and Newell, PC (1976) Genetics 82: 287-307.
13. Coukell, MB and Roxby, NM (1977) Molec Gen Genet 151: 275-288.
14. Loomis, WF (1970) J Bacteriol 103: 375-381.
15. Burns, RA, Livi, GP and Dimond, RL (1981) Dev Biol 84: 407-416.
16. Free, SJ, Schimke, RT and Loomis, WF (1976) Genetics 84: 159-174.
17. Mierendorf, RC and Dimond, RL (1983) Anal Biochem 135: 221-229.
18. Mierendorf, RC, Cardelli, JA, Livi, GP and Dimond, RL (1983) J Biol Chem 258: 5878-5884.
19. Pannell, R, Wood, L and Kaplan, A (1982) J Biol Chem 257: 9861-9865.
20. Knecht, DA and Dimond, RL (1981) J Biol Chem 256: 3564-3575.
21. Freeze, HH, Yeh, R, Miller, AL and Kornfeld, S (1983) J Biol Chem 258: 14874-14879.
22. Livi, GP, Cardelli, JA, Mierendorf, RC and Dimond, RL (1984) submitted.
23. Cardelli, JA, Mierendorf, RC and Dimond, RL (1984) submitted.
24. Mierendorf, RC, Cardelli, JA and Dimond, RL (1984) submitted.
25. Livi, GP and Dimond, RL (1984) Dev Biol 101: 503-511.
26. Livi, GP and Dimond, RL (1984) submitted.

27. Dimond, RL, Knecht, DA, Jordan, KB, Burns, RA and Livi, GP (1983) Meth Enzymol 96: 815-828.
28. Livi, GP, Cardelli, JA and Dimond, RL (1984) submitted.
29. Singleton, C, Livi, GP and Dimond, RL (1984) in preparation.

Molecular Biology of Development, pages 459–480

ACTIVE TRANSCRIPTION OF REPEAT SEQUENCES DURING TERMINAL DIFFERENTIATION OF HL60 CELLS[1]

Chuan-Chu Chou*, Richard C. Davis**, Arlen Thomason[2], Janet P. Slovin[3], Glenn Yasuda[4], Sunil Chada*, Jocyndra Wright**, Michael L. Fuller*, Robert Nelson** and Winston Salser* **

Department of Biology* and Molecular Biology Institute**, University of California at Los Angeles, Los Angeles, California 90024

ABSTRACT. cDNA clones of mRNAs whose expression is regulated during terminal differentiation of DMSO-treated HL60 cells were selected by differential hybridization and then probed using labeled total human DNA to detect the presence of repeat sequences. We found that about 50% of these cDNA clones contain repeat sequences, a level 20-fold higher than in the unselected clone library. This implies that repeat sequences are largely or completely restricted to mRNAs which are regulated during terminal differentiation of HL60 cells. Some repeats are members of the Alu family. Examples of mRNAs containing one copy, two copies, or a partial copy of Alu repeats have been found. The Alu sequences in these cDNAs are sufficiently diverged that some of them cross-hybridized poorly in dot blot tests. One gives undetectable hybridization with the BLUR 2 Alu probe and little hybridization with a human total DNA probe. Other repeats that have no sequence homology to the Alu repeat have also been found. One of these hybridizes as strongly with a bulk human DNA probe as any Alu

[1]This work was supported by grants from the National Institutes of Health (CA 32186 and GM 18586) and the National Science Foundation (PCM82-04182).
[2]Present address: AMGen Inc., Newbury Park, CA 91320.
[3]Present address: Plant Hormone Lab, B050 Rng 4 HH4, USDA Agriculture Research Center, Beltsville, MD 20705.
[4]Present address: Department of Genetics, University of Washington, Seattle, WA 98105.

repeat we have tested (signal corrected for repeat sequence length). We hope that further studies will clarify why the appearance of mRNAs with repeat sequences is correlated with terminal differentiation in these cells.

INTRODUCTION

The HL60 human promyelocytic leukemia cell line (1) can be induced to terminally differentiate and become either macrophage-like with phorbol esters (TPA)(2,3) or granulocyte-like after treatment with dimethyl sulfoxide (DMSO) (4)(Fig. 1). In both differentiation pathways dramatic changes in RNA transcription, protein synthesis, and morphology accompany cessation of cell division. Morphologically, the induced granulocytes show nuclear constriction and pycnosis while the macrophages show some nuclear distortion and become adherent to tissue culture glassware. Both cell types exhibit phagocytosis.

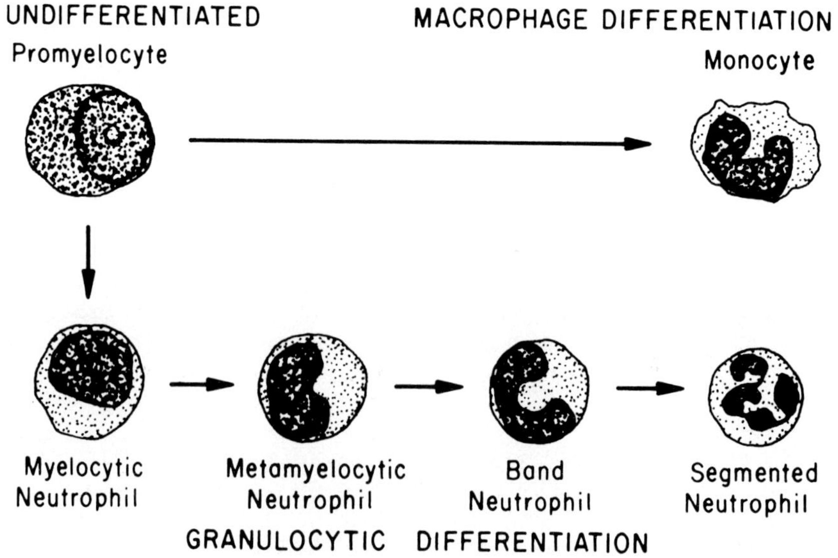

FIGURE 1. Morphological changes in HL60 promyelocytes induced to differentiate by treatment with DMSO or TPA mimic those in normal granulocytes or macrophages (Redrawn from Diggs et al., ref 5).

DMSO-treated cells show characteristic changes in activities of enzymes such as chloroacetate esterase and peroxidase (1). Lysozyme excretion and expression of monocyte-specific cell surface antigens are induced by TPA (6). The expression of many proteins is strongly regulated during these differentiation steps as shown by the representative small regions of 2-dimensional O'Farrell gels for ^{35}S labeled proteins in Fig. 2.

It has been shown that the HL60 granulocytes and macrophages have many of the biological attributes of normal granulocytes and macrophages. For instance, HL60 granulocytes actively phagocytize Candida albicans particles (4), and HL60 macrophages are like normal macrophages in their ability to selectively kill cancer cell lines. Weinberg (7) showed that with normal fibroblasts there is about 77% survival after exposure to HL60 macrophages, while only 3.2% of HeLa cells survive HL60 macrophage "attack".

From these observations it is clear that HL60 cells, when induced with DMSO or TPA, are making important differentiation "decisions" and that the HL60 system provides a model for the study of both commitment and terminal differentiation.

We have constructed cDNA libraries from differentiated and undifferentiated HL60 cells. From these libraries we have selected cDNA clones corresponding to mRNAs whose synthesis is regulated during differentiation. Our long-term goal is to study terminal differentiation using site-directed mutagenesis techniques in which genomic counterparts of these regulated mRNAs are modified and returned to cells to test their response to differentiation signals.

In the course of these studies we have found that several different repeat sequences are present in the cytoplasmic RNAs of HL60 cells and that such repeat-containing transcripts appear to be regulated during differentiation.

METHODS

Cell Growth and Differentiation

HL60 cells were obtained from Dr. Robert Gallo (1) and were passaged at densities from 5×10^4 to 2×10^6 per ml in alpha-MEM (Gibco) supplemented with 20% fetal calf serum and

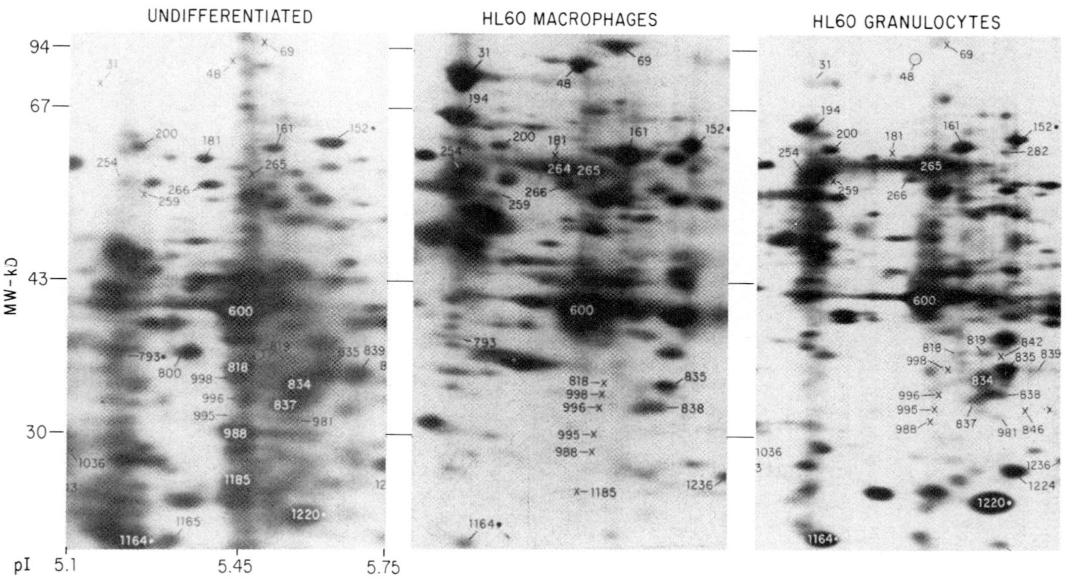

FIGURE 2. ^{35}S Fluorography of O'Farrell gels of HL60 promyelocytes, HL60 macrophages (10^{-7} M TPA for 4 days), and HL60 granulocytes (1.25% DMSO for 4 days with 2 ugm/ml Amiloride to enhance differentiation (8)). The proteins shown are 23-95 kD with pI's of 5.1 to 5.75. Proteins which appear or disappear during differentiation are marked with an "X" on gels where they are absent or at very low intensity. Spots #152 and #600 (actin) are of relatively constant intensity; other spots change dramatically during treatment. Spots #998 and #996 decrease and spot #194 increases dramatically during both differentiations. Spot #48 is present much more strongly in HL60 macrophages than in promyelocytes or granulocytes. Many other examples of regulated synthesis may be observed.

antibiotics. The cells were induced in normal growth medium with either 10^{-7}M TPA for 2-4 days or with 1.25% DMSO with 2 ug/ml Amiloride added to enhance granulocytic differentiation (8).

cDNA Clone Selection

HL60 clones were selected from cDNAs corresponding to messenger RNAs strongly regulated during DMSO-induced differentiation (Davis et al., in preparation). M13 subclones for sequencing or for use as hybridization probes were constructed by insertion into the universal cloning sites of M13 strains mp8, mp9 or mp11 (9).

DNA Sequencing

Sanger's dideoxy method (10) was used for sequence analysis. In some cases duplex restriction fragments of the cloned insert were used as sequencing primers instead of the universal primers. Computerized sequence comparison was carried out using programs developed by Staden (11).

Southern and Northern Blots

Cytoplasmic polyA^{+} RNAs were prepared as described by Maniatis et al. (12). Southern (13) and Northern (14) blots were prepared and probed by standard methods (12) but using positively charged nylon membrane filters which permit multiple reuses of each blot. This permitted the regulation of several different mRNA species to be directly compared by sequential probing of the same blot, thus eliminating possible sources of error. ^{32}P-labeled probes were synthesized by random priming (12) of double-stranded pBR322 clones or by primer extension (9) on single-stranded M13 clones. Northern blots were hybridized at 42 C in 5X SSC, washed twice for 30 minutes at 65 C in 2X SSC, and washed twice for 30 minutes at room temperature in 0.1X SSC.

Dot Blot Analysis

cDNAs were denatured with NaOH and neutralized with HCl and Tris·HCl before being bound to nitrocellulose filters in a BRL dot blot manifold (15). Blots were washed twice for 30 minutes at 60 C in 0.2X SSC, 0.2% SDS. Hybridization signals produced by cDNA clones in response to probes of random-primed human genomic DNA or specific cloned repeat sequences were estimated using densitometric scans of autoradiographs and, in some cases, scintillation counting.

Two Dimensional Gel Electrophoresis

Regulation of the synthesis and accumulation of proteins in differentiated HL60 cells was measured by two dimensional gel electrophoresis (Davis et al., in preparation) carried out by the method of O'Farrell (16). Autoradiography was accomplished using PPO-impregnation of the gels as described by Laskey (17).

RESULTS

Repeat Sequences Present in Regulated HL60 Cytoplasmic Trancripts

cDNA clone libraries have been prepared from the cytoplasmic polyA$^+$ RNA of HL60 promyelocytes and the differentiated HL60 cells. Most of the work reported here deals with cDNA clones derived from HL60 granulocytes. The clones were prepared by reconstruction of Pst I sites in pBR322 (18) although more recently our clones have been prepared in or transferred to the Pst I sites of the mini-plasmid piVX (19) and related plasmids piAN7, piSL13 and piSDL12 (B. Seed, personal communication), or into the M13 sequencing vectors.

Differential hybridization was used to select for cDNA clones corresponding to mRNAs regulated during differentiation of HL60 cells. Colony hybridization filters were prepared as described by Grunstein and Hogness (20). Oligo-dT-primed cDNA from HL60 promyelocytes and from HL60 granulocytes were ^{32}P-labeled and used as differential probes.

Repeat sequences were found in a high proportion of the clones selected by differential hybridization. Since earlier work had shown that repeated sequences were rarely present in cytoplasmic mRNAs (6-10% in HeLa cells (21,22),

8% in Friend cells (23), less than 3% in sea urchin (24,25)), we were surprised to find that slightly more than half of the first group of regulation candidates tested contained repeated sequences as shown by the following criteria:

- When used as radioactive probes they give smears on genomic Southern blots of human DNA. (13 out of 25 clones tested.)
- When total human DNA is labeled and used to probe dot blots of these clones, a positive signal is seen under conditions where no detectable signal is seen with clones of known single or low copy genes (Table 1).

TABLE 1
HYBRIDIZATION TEST OF HL60 cDNA CLONES

Clone	Dot Blot (30 ng/spot)		Southern Blot†
	Total human*	BLUR 2*	
pHL 203	+	+	smear
pHL 507	-	-	smear
pHL 1007	+	-	smear
pHL 1009	-	-	smear
pHL 1078	+	-	multiple bands
pHL 1201	-	-	smear
pHL 1204	+	-	smear
pHL 1206	+	+	smear
pHL 1208	+	-	N/D
pHL 1214	+	+	multiple bands
pHL 1219	+	+	smear
pHL 1222	+	+	N/D
m8HL 1507	+	N/D	N/D

* In dot-blot experiments ^{32}P-labeled total human DNA or Alu BLUR 2 DNA probes were hybridized to the indicated cDNA clone DNA bound to nitrocellulose.

† Southern blots of human genomic DNA were probed with ^{32}P-labeled DNA of individual cDNA clones.

When labeled human genomic DNA is used to probe the total cDNA libraries from either undifferentiated or DMSO-treated cells, positive signals are seen with only about 3% of the clones (data not shown). Since a high proportion of the clones selected by differential screening contain repeats, this suggests that repeat sequences may be largely or completely absent from mRNA species which are not strongly regulated in this differentiation pathway.

Alu and Non-Alu Repeats found in HL60 cDNAs

Several different kinds of repeats were found among the regulated clones. We first divided the repeats found in this regulation study into two groups, the Alu repeats and non-Alu repeats, based upon their cross-hybridization properties as shown in Tables 1 and 2.

In Table 1, nitrocellulose dot blots of each cDNA clone (30 ng/spot) were probed with either ^{32}P-labeled human genomic cDNA or labeled insert from the repeat-bearing clone BLUR 2. Hybridization to human DNA indicates that the cDNA sequence is present as a highly repeated sequence in the genome. Hybridization to BLUR 2 shows that the cDNA has homology to the Alu family of repeats.

TABLE 2:
CROSS-HYBRIDIZATION OF REPEAT SEQUENCES

	PROBE:						
	TOTAL HUMAN DNA	BLUR 2	pHL 203	pHL1009	pHL1201	pHL1204	pHL1218*
CLONE:							
pHL 203	+	+	+	+	-	-	-
pHL 1009	-	-	-	+	-	-	-
pHL 1078	+	-	-	-	-	-	-
pHL 1201	-	-	-	-	+	-	-
pHL 1204	+	-	-	-	-	+	-
pHL 1214	+	+	+	-	-	-	-
pHL 1219	+	+	+	-	-	-	+
M8HL 1507**	+	N.D.	N.D.	N.D.	N.D.	N.D.	N.D.
BLUR 2	+	+	+	-	-	-	+
TOTAL HUMAN DNA	+	+	+	-	-	+	+

* THIS PROBE WAS A SUBCLONE OF pHL 1219 WHICH CONTAINS BOTH COPIES OF ALU SEQUENCES.

** CROSS HYBRIDIZATION WAS NOT TESTED BUT DIRECT SEQUENCE ANALYSIS SHOWS THAT THIS CLONE SHARES NO HOMOLOGY WITH ALU-CONTAINING CLONES (E.G. pHL 203, pHL 1204 OR pHL 1219) OR WITH CLONES pHL 1078 OR pHL 1201.

As shown in Table 2, similar dot blots were used to measure cross-hybridization between the repeat-bearing clones. pHL 203 hybridizes strongly to BLUR 2 and other known Alu repeats (data not shown) and sequence analysis shows that it bears very high homology to the consensus Alu sequence (26). pHL 1204 does not cross-hybridize to the repeats tested or human genomic DNA, but sequence analysis later showed it to contain an Alu repeat with 78% homology to BLUR 2 over one region of 122 bases.

Although hybridization with a known Alu sequence is useful in identifying clones containing an Alu repeat, we find that the degree of cross-hybridization between members of the Alu family varies greatly. Even when the sequence homology is as high as 78% no cross-hybridization will be seen if the 22% mismatches are distributed in a way that leaves no long regions of stable base-pairing, as is the case with clone pHL 1204 (see sequence comparisons in Fig. 6).

Sequence analysis (see below) confirmed that clones pHL 1201, pHL 1078 and m8HL 1507, have no sequence homology to the Alu family or to each other.

Northern Blot Analysis Reveals Distinct mRNA Species Corresponding to Each Repeat

Alu Repeats. Since Alu sequences cross-hybridize it seemed likely that an Alu probe might hybridize to a featureless smear of many mRNA species in Northern blot analysis of cytoplasmic polyA$^+$ mRNA. The actual experiments showed that each of the HL60 Alu sequence probes hybridized to a few major species which were different in each case. For instance, using a probe from one strand of clone pHL 203 (Fig. 3) we observed a strong sharp band at 10.5 kb, a diffuse band at 4.7 kb, and a sharp band at about 300 bases. The latter is within the expected size range for an RNA polymerase III (Pol III) transcript of the repeat (27) (see Fig. 3, Lane A). There is also a background smear in the size range below 5 kb. When the complementary strand of the pHL 203 insert was used as probe we observed that transcription of this sequence is highly asymmetric, since there is much more intense labeling with a band at about 11 kb and an intense smear extending from 11 kb down to a few hundred bases (Lane B).

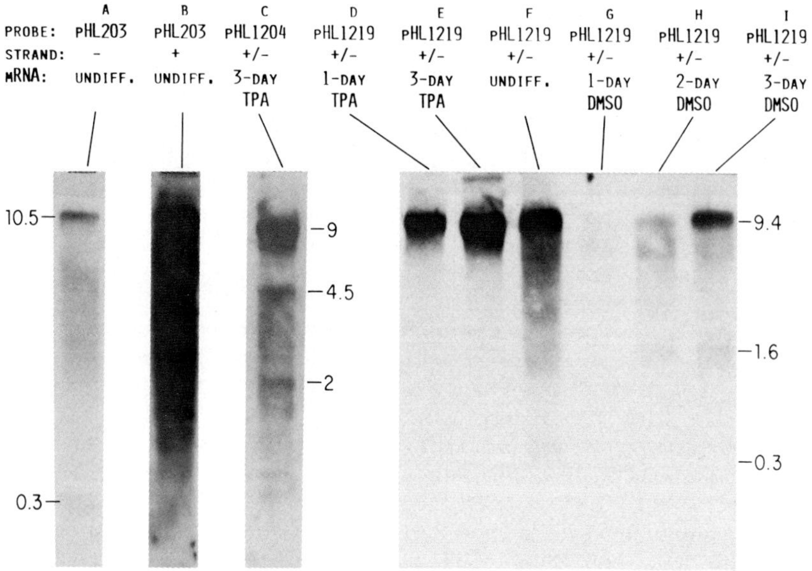

FIGURE 3. Northern blot analysis of the regulation of Alu repeat sequences in HL60 cytoplasmic polyA$^+$ mRNA. Cytoplasmic polyA$^+$ mRNA was prepared from undifferentiated HL60 cells or from the indicated differentiation stages of HL60 cells. "+" or "-" indicates that one of the two possible homologous cDNA-insert strands in the single stranded M13 clones were labeled and used as probe. "+/-" indicates that either both homologous orientations of the M13 cDNA clones (pHL 1204) were labeled or that double stranded pBR322 clones were random-primed and used as probe (pHL 1219).

The hybridization pattern of pHL 1204 is very different (Fig. 3, Lane C). Unlike the results with pHL 203, there is virtually no hybridization when the RNA in the lanes is from uninduced HL60 cells (data not shown). Transcription is strongly induced during TPA treatment and gives species of 9, 4.5 and 2 kb (Lane C) but no induction is observed during granulocytic differentiation (data not shown). The probe used in this experiment, as in all those reported below, is a mixture of labeled + and - strands.

Quite different regulation is seen when we use the Alu repeats carried in clone pHL 1219 as probes (Fig. 3, Lanes

D-I). A strong band seen at about 9.4 kb is present at a high level in uninduced cells and at an increased level in 3-day TPA-induced cells. This band disappears early in DMSO-induction and reappears by 3 days. The smear of smaller species present in uninduced cells is strikingly reduced during TPA induction and is replaced by a discrete band of about 1.6 kb during DMSO induction. A 325-base band that may be a Pol III transcript is reproducibly seen in uninduced cells, usually as a sharp band, and disappears in the differentiating cells.

Non-Alu Repeats. Analogous results are seen with the non-Alu repeats. For instance, the pHL 1201 repeat sequence labels sharp Northern blot bands at very high molecular weights (estimated at approximately 21 and 33 kb) as well as a major band 0.89 kb in size. In this case the high molecular weight species disappear early in DMSO induction, and the largest and smallest bands reappear with different kinetics whereas all three species disappear coordinately in TPA induction (Fig. 4, Lanes J-O).

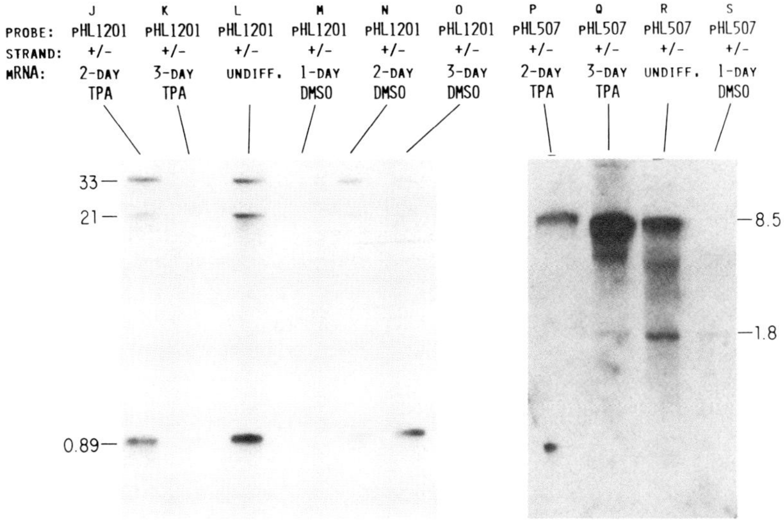

FIGURE 4. Northern blot analysis on the regulation of non-Alu repeat sequences in HL60 cytoplasmic polyA+ mRNA. RNA was prepared in the same way as that in Fig. 3. pHL 1201 is a pBR322-based probe, while the pHL 507 probe is a mix of homologous inserts in M13.

In the case of repeat sequence pHL 507 a prominent 8.5 kb sequence is induced during TPA treatment and is down-regulated early in DMSO induction whereas a 1.8 kb sequence shows almost reciprocal behavior (Fig. 4, Lanes P-S).

The presence of discrete bands in Figures 3 and 4 argues strongly that our cytoplasmic RNA preparations are free from nuclear contamination. Recall that Alu repeats, and possibly the non-Alu repeats studied here as well, are transcribed in the heterogeneous nuclear RNA at high levels and in such a diversity of different molecules that a continuous smear of sizes results.

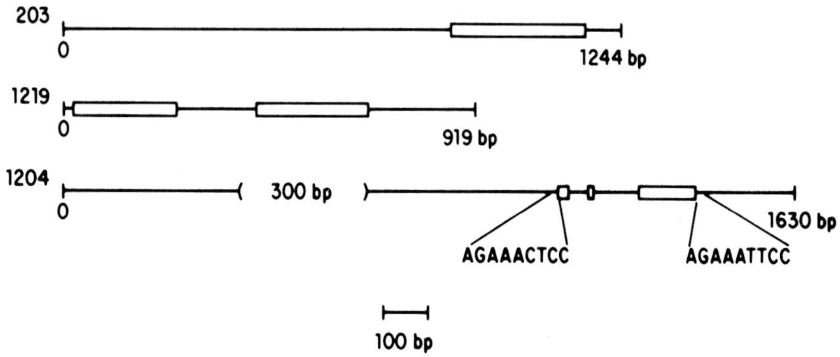

FIGURE 5. Distribution of Alu-sequence homology in the HL60 cDNA clones. The Alu-homologous regions are represented as boxes. In all cases the Alu is regarded from left (5') to right (3') in the "canonical" orientation, which means that the characteristic homopolymer stretches of A's are at the right. The entire sequence of each clone has been determined except for the 300 bp region indicated in the left half of insert 1204. The direct terminal repeats flanking the pHL 1204 Alu sequence are shown by insets.

Sequence Analysis of HL60 Repeats

Alu Family Repeats. The Alu sequences present in HL60 transcripts include examples of tandem direct repeats, interrupted homologies, and cases lacking direct terminal repeats as shown in Fig. 5. The only direct repeats found flanking the Alu sequences were in insert 1204 and are indicated in the figure.

```
                        10         20         30         40         50
Alu cons.       GGCTGGGCGT GGTGGCTCAC ACCTGTAATC CCAGCACTTT GGGAGGCCGA
Blur 2             G          A                    AT      C      T
203 Alu         CTTGGCACCAGGCACAGT   T                             T
1204 Alu        ATGAGGAAGAAGCAAAAGCAGAA   T  (A)   A     T  GATCTCATGAGACTTATT
1219 Alu.1      ---------- ---------- ---------- ---------- ----------
1219 Alu.2      TCTCAAATAGGGAAAGAGGTTTTAAAATCAAATTTGAGGCCAGGTGCAGTGGCTCATG  G

                        60         70         80         90        100
Alu cons.       GGTGGGTGGA TCACCTGAGG TCAGGAGTTC AAGACCAGCC TGGCCAACAT
Blur 2                      A  T   C                     T   A
203 Alu           CA           --            A     C    T      T
1204 Alu        CACTATCATGAGAATAGCACAGGAA ---------- ------  A      CC
1219 Alu.1      --------A    C T    C C       G G   G  A A  -CA    --
1219 Alu.2        A   CA    T  T     CT

                       110        120        130        140        150
Alu cons.       GGTGAAACCC CGTCTCTACT AAAAATACAA AAATTAGCCG GGCGTGGTGG
Blur 2                      A                          G A   GC
203 Alu                  G T                    A A(A)5          C
1204 Alu          ATTCAATTACCTCTCCCTGGGTCCCTCCCACAACACGTGGGAATTCTGGGAGATACAATTCAAG
1219 Alu.1      ------      A T        T   CA        G         A
1219 Alu.2                  A         G                  --  -- A A

                       160        170        180        190        200
Alu cons.       CGCGCGCCTG TAATCCCAGC TACTCGGGAG GCTGAGGCAG GAGAATCGCT
Blur 2           A   A        C            A                         A
203 Alu          AG T        G        C                           G A/T
1204 Alu        CTGAGATTTGGGTGGGACACAGCCAAACCATATCACAA    A TG     G   A
1219 Alu.1       AT T        G          T
1219 Alu.2      T  A       C G           A      T         A

                       210        220        230        240        250
Alu cons.       TGAACCCAGG AGGTGGAGGT TGCAGTGAGC CGAGATCGCG CCACTGCACT
Blur 2                 TA      CA                          AT   T    T -
203 Alu                        CA   C                      T  A   G
1204 Alu           G  TG     T C    AC          A   A    A A       G/C
1219 Alu.1         G     A     A                    A
1219 Alu.2                  A                          T T   TG

                       260        270        280        290        300
Alu cons.       CCAGCCTGGG CAACAGAGCG AGACTCCATC TCAAAAAAAA AAAAAAAAAA
Blur 2                             T     T                             G
203 Alu                     G            T
1204 Alu                           T       TG                       CC
1219 Alu.1          T              T    C TG       T  T   T  T  T  T
1219 Alu.2                  TG A A AC  G GTG     C    G            G

                304
Alu cons.       AAAA
Blur 2
203 Alu             AAAGGATGGATATTCTGATATATGCTACAATGTGGATGAACCTTGAGGATGTTAACTAAAG
1204 Alu         GAAATTCCACTCCCATTATCAGTCACCTCCCCATTCCCCTCCCCCACCCAGCCCTTGGGAACCA
1219 Alu.1       TGATGATACTCTAAGAAAAAAATCTCAACATACTTCATTTAATAGCTCGTTACCAAGTGTGAAT
1219 Alu.2      TCAAATTCAAATATCATCTGGACATGTCACAATGGATCGCGGATCCTTATGAGTGATTTTCCCCA
```

FIGURE 6. Comparison of HL60 cDNA Alu with Consensus Alu. Flanking sequences and internal regions which share no homology with the Alu repeat are in italics. Blanks indicate homology between the HL60 Alu repeats and the Alu consensus sequence. A dash (-) indicates a deletion, and a series of dashes stands for a blank or a gap. The circled nucleotides stand for insertions.

In Fig. 6, the Alu sequences found in HL60 cDNA (clones pHL 203, pHL 1204 and pHL 1219) are compared with BLUR 2 and the consensus Alu sequence deduced by Deininger et al. (26). Interestingly, all the HL60 Alu sequences found in HL60 transcripts are either partial or interrupted copies, as compared to the consensus Alu sequence derived from the genomic DNA. The 9-nucleotide direct terminal repeats flanking the pHL 1204 Alu sequence are indicated in the Figure by underlining. The location of the Pol III promoter is also indicated (nucleotides 75-86) (27,28).

We have further examined the divergence among the Alu sequences found in HL60 cDNA clones as shown in the divergence matrix shown in Table 3. This matrix shows the percent divergence within all regions of homology except in the case of pHL 1204, where the computation was limited to the largest of the three homology regions (122 bp).

Among the four Alu sequences thus far analyzed, three lack the flanking direct repeats which were reported to be characteristic of the mammalian Alu family (29,30).

In one cDNA clone, pHL 1219, we have found the first instance of a cytoplasmic RNA containing tandem Alu direct repeats. It is interesting that these two repeats have highly divergent sequences.

TABLE 3:
THE DIVERGENCE AMONG THE ALU REPEATS FOUND IN HL60 cDNAs*

	ALU CONSENSUS	BLUR 2	203 ALU	1204 ALU	1219 ALU.1	1219 ALU.2
ALU CONSENSUS	0					
BLUR 2	11%(33/314)	0				
203 ALU	12%(35/290)	18%(53/290)	0			
1204 ALU	18%(22/122)	22%(27/122)	23%(28/122)	0		
1219 ALU.1**	21%(50/243)	24%(58/243)	25%(61/243)	20%(25/122)	0	
1219 ALU.2***	15%(39/255)	21%(54/255)	25%(65/255)	29%(35/119)	28%(66/240)	0

* COMPARISON WAS BASED UPON THE LONGEST CONTINUOUS HOMOLOGOUS REGION WHEN THE ALU REPEATS ARE COMPARED BY PAIRS.

** 1219 ALU.1 IS THE FIRST PARTIAL COPY OF ALU IN INSERT 1219.

*** 1219 ALU.2 IS THE SECOND PARTIAL COPY OF ALU IN INSERT 1219.

Non-Alu Repeats. Non-Alu repeats present in regulated HL60 transcripts include those present in clones pHL 1507, pHL 1078, and pHL 1201. m8HL 1507, which was isolated from a cDNA library in M13 mp8, is very highly repeated in the human genome. In cross hybridization with total human DNA the signal seen with clone m8HL 1507 is equal to that seen with those Alu clones which give the highest signals in our hands (BLUR 2 and pHL 203) and twice as high as for pHL 1219 (when all signals are corrected for the length of the repeat). This could mean that there are 500,000 copies of the m8HL 1507 repeat family (if they are as highly diverged as the Alu sequences) or that there are a lesser number of more highly conserved sequences. We have compared this sequence with the entire Los Alamos sequence data bank using a program run on the Cray computer (31). No significant homologies were found.

The repeat-containing clones m8HL 1507, pHL 1078 and pHL 1201 have been completely sequenced and show no homology with Alu, with each other, or with the BLUR 16 Kpn I family repeat sequence (26). These sequences are shown on the following page. Sequences for inserts 1078 and 1201 include the homopolymer tails and terminal Pst I sequences introduced during cloning.

INSERT 1507

```
        10         20         30         40         50
CTGCAGCAAG GCTATGATCC ACAGTCCGGA GCTGCTTTTA AGATTTGTCT
        60
TTTGTGGGCT GCAG
```

INSERT 1078

```
        10         20         30         40         50
CTGCAGGGGG GGGGGGGGGG TAAGGAAATT GTTAAGGAAG TCAGCACTTA
        60         70         80         90        100
CATTAAGAAA ATTGGCTACA ACCCCGACAC AGTAGCATTT GTGCCAATTT
       110        120        130        140        150
CTGGTTGGAA TGGTGACAAC ATCCTGGAGC CAAGTGCTAA CATCCCTTGG
       160        170        180        190        200
TTCAAGGGAT GGAAAGTCAC CCGTAAGATG GCAATGCCAG TGGAACCACC
       210        220        230        240        250
GCTGCTTGAG GCTCTGGACT GCATCCTACC ACCAACTCGT CCAACTGACA
       260        270        280        290        300
AGCCCTTGCG CCTGCCTCTC AGGATGTCTA CAAAATTGGT GGTATTGGTA
       310        320
CTGTTCCTGT CCCCCCCCCC CTGCAG
```

INSERT 1201

```
       10         20         30         40         50
CTGCAGGGGG GGGGGGGGGG GGGGGGGTTT AGACGTCCGG GAATTGCATC
       60         70         80         90        100
TGTTTTTAAG CCTAATGTGG GGACAGCTCA TGAGTGCAAG ACGTCTTGTG
      110        120        130        140        150
ATGTAATTAT TATACGAATG GGGGCTTCAA TCGGGACTAC TACTCGATTG
      160        170        180        190        200
TCAACGTCAA GGAGTCGCAG GTCGCCTGGT CTAGGAATAC CCGGGAAGTA
      210        220
TGTACCCCCC CCCCCCCCCC CCCTGCAG
```

Both clones pHL 1078 and pHL 1201 contain inverted short repeats of 9 nucleotides and 6 nucleotides respectively which are indicated by underlining.

DISCUSSION

We have presented evidence that several repeat sequences are not only transcribed but are transported to the cytoplasm in HL60 cells. Some of these repeats share homology with the Alu family while others have no homology to previously reported repeat sequences. Both the Alu and non-Alu repeats hybridize a variety of discrete mRNA species on Northern blots of HL60 cytoplasmic polyA$^+$ RNA. Each of these discrete species has its own characteristic pattern of regulation during terminal differentiation of HL60 cells.

Because repeat sequences are enriched by 20-fold in regulated cDNAs over unselected cDNAs of HL60 we suggest that such repeats may play a strong role in gene regulation during the differentiation of human macrophages and granulocytes. It has long been suggested that interspersed repeat sequences are involved in gene regulation (32). A major problem with this suggestion has been that there seem to be many more interspersed repeats than regulated genes (or even total genes), so that it is difficult to imagine how more than a small minority of repeats could be actively involved in regulating transcription. The proposal that Alu sequences are mobile genetic elements, as suggested by the presence of the flanking direct repeats typical of retrogenes, suggests that most of the Alu sequences could be inactive pseudogenes with only a minority functional in regulation.

It is, therefore, important to ask which of the many

repeat sequences are actually involved in regulation and whether they have any distinguishing characteristics. Our analysis of the non-Alu repeats is too preliminary to shed light on this point, but examination of the Alu repeats found in HL60 cytoplasmic RNAs reveals interesting features. From earlier sequence analysis of the Alu repeats found in chromosomal DNA clones it has been concluded that most of them are flanked by direct repeats (30). We would expect a similar pattern in the Alu sequences presented here if they were included in their respective RNAs by chance rather than because of some meaningful association with regulation of these genes during HL60 cell differentiation. In fact three of these four Alu sequences lack flanking repeats. We propose that these experiments have successfully identified repeats which are actively involved in gene regulation and that this approach may be generally useful in selecting such repeats from among hundreds of thousands of repeats present as pseudogenes not actually participating in regulation.

It has also been noted in previously published Alu sequences that the 5' "end of the Alu repeat is conserved among different family members and can be precisely defined" (33). By contrast, all four of the regulation-involved Alu sequences presented here lack the 5' end of the consensus sequence. The possibility must also be considered, however, that the 5' terminus was present in the chromosomal DNA but was removed by splicing during the processing of these RNAs. The flanking direct repeats argue against this explanation in the case of the pHL 1204 Alu repeat, but in the other three cases this question cannot be answered until we sequence the corresponding chromosomal clones. The pHL 1204 Alu repeat is also noteworthy in lacking the Pol III promoter (positions 75-86 in Figure 6) although there is a partial homology between the sequence GATTCAATTACC, found 72 nucleotides to the right of the direct repeat which defines the left end of this extensively modified Alu sequence, and the sequence GAGTTCAAGACC which has been proposed as the most likely Pol III control region in the Alu consensus sequence (27).

The regulation-involved repeat sequences presented here share some common features with the tissue-specific "identifier" sequences in mouse brain messengers (34,35,36) and the Set 1 repeat described by Murphy et al. (37). Comparisons of these with our regulation-involved Alu sequences are summarized in Table 4.

TABLE 4
COMPARISON WITH OTHER REPEATS

	Alu repeat in HL60	ID sequence in mouse brain	"Set 1" Repeat (Murphy et al.)
Present in a small % of mRNAs	+	+	+
Regulated during differentiation	+	+	+
Include Pol III promoter	+	+	?
Diversity	high	low	low
Detected in "inappropriately small" molecules	yes (300 bases)	160 bases	?

Sutcliffe et al. and Milner et al. (35,36) have proposed that the smallest RNAs containing identifier sequences are Pol III transcripts and that Pol III transcription serves a pivotal role in making adjacent regions of chromatin accessible to transcription by Pol II. Alu repeats are also viewed as potential Pol III transcription sites. Probes containing regulation-associated repeat sequences consistently hybridize to small RNAs in the size range expected for Pol III transcripts terminating near the end of the Alu repeat (320 nucleotides).

In our experiments these small Pol III-like RNAs are seen only in the undifferentiated HL60 cells. The "Set 1" repeats of the mouse are also at least potentially transcribable by Pol III since they belong to the B-2 repeat family which is generally considered to contain Pol III promoters and to be Alu-like in its behavior (38, 39). We do not yet know whether the non-Alu repeats described here are transcribable by Pol III.

In the case of the mouse identifier sequence (34) and the Set 1 repeat sequence (37) found, respectively, in mouse brain and fibroblasts, single repeat sequences have been correlated with tissue-specific differentiation. In HL60 cells, by contrast, we have identified a number of different repeat sequences which are involved in a greater variety of regulatory events than can be inferred from the behavior of the Set 1 repeat in mouse F9 cells. Each of the regulation-involved repeats in HL60 cells hybridizes to different distinct bands on Northern blots of the HL60 cell cytoplasmic RNA and each of these repeats identifies RNAs with a distinct pattern of regulation during the macrophage and granulocyte differentiations. Some of the HL60 regulation-involved repeats are turned off during differentiation rather than on, as in the case of the Set 1 repeats. We do not know if any of our regulation-involved repeats would resemble the Set 1 repeats in hybridizing to a more complex set of mRNAs from embryonic tissue.

In order to directly test the role of these repeat sequences in the regulation of gene expression during HL60 cell differentiation we have subcloned flanking single-copy regions into members of the pi-plasmid family (R. Nelson and S. Teraoka, in preparation). We are using the pi-plasmid recombination selection technique for rapid isolation of chromosomal clones of single-copy gene sequences and have begun transformations of HL60 cells with tagged genomic clones of regulated genes (Concannon et al., work in progress).

ACKNOWLEDGMENTS

The authors wish to thank Dr. H.P. Koeffler for supplying HL60 strains and for making the original observation that HL60 regulation candidate cDNA clones gave smears when used as probes on genomic Southern blots. We also thank Dr. Carl W. Schmid for the generous gift of BLUR 2 clones. We thank Patrick Concannon and Hun-Chi Lin for valuable discussions, Michele Lau for technical assistance and Susan Salser for typing the manuscript.

REFERENCES

1. Gallagher R, Collins S, Trujillo J, McCredie K, Ahearn M, Tsai S, Metzgar R, Aulakh G, Ting R, Ruscetti F, Gallo R (1979). Characterization of the continuous, differentiating myeloid cell line (HL-60) from a patient with acute promyelocytic leukemia. Blood 54:713.
2. Rovera G, Sautoli D, Damsky C (1979). Human promyelocytic leukemia cells in culture differentiate into macrophage-like cells when treated with a phorbol diester. Proc Natl Acad Sci (USA) 76:2779.
3. Murao S-i, Gemmell MA, Callahan MI, Anderson NL, Huberman E (1983). Control of macrophage cell differentiation in human promyelocytic HL-60 leukemia cells by 1,25-dihydroxyvitamin D_3 and phorbol-12-myristate-13-acetate. Cancer Res 43:4989.

4. Collins SJ, Ruscetti FW, Gallagher RE, Gallo RC (1978). Termininal differentiation of human promyelocytic leukemia cells induced by dimethyl sulfoxide and other polar compounds. Proc Natl Acad Sci (USA) 75:2458.
5. Diggs LW, Sturm D, Bell A (1978). "The Morphology of Human Blood Cells." North Chicago Illinois: Abbott Laboratories, p 14.
6. Rovera G, Ferrero D, Pagliardi GL, Vartikar J, Pessano S, Bottero L, Abraham S, Lebman D (1982). Induction of differentiation of human myeloid leukemias by phorbol diesters: phenotypic changes and mode of action. Ann New York Acad Sci 397:211.
7. Weinberg JB (1981). Tumor cell killing by phorbol ester-differentiated human leukemia cells. Science 213:655.
8. Carlson J, Dorey F, Cragoe Jr E, Koeffler HP (1984). Amiloride potentiates dimethylsulfoxide-induced differentiation of the human promyleocytic cell line, HL-60. J Nat Cancer Inst 72:13.
9. Messing J, Vierira J (1982). A new pair of M13 vectors for selecting either DNA strand of double-digest restriction fragments. Gene 19:269.
10. Sanger F, Coulson AR, Barrell BG, Smith AJH, Roe BA (1980). Cloning in single-stranded bacteriophage as an aid to rapid DNA sequencing. J Mol Biol 143:161.
11. Staden R (1979). A strategy of DNA sequencing employing computer programs. Nucleic Acids Res 6:2601.
12. Maniatis T, Fritsch EF, Sambrook J (1982). "Molecular Cloning: A Laboratory Manual." Cold Spring Harbor New York: Cold Spring Harbor Laboratory, p 187, p 131.
13. Southern E (1975). Detection of specific sequences among DNA fragments separated by gel electrophoresis. J Mol Biol 98:503.
14. Thomas PS (1980). Hybridization of denatured RNA and small DNA fragments transferred to nitrocellulose. Proc Natl Acad Sci (USA) 77:5201.
15. Kafatos FC, Jones CW, Efstratiadis A (1979). Determination of nucleic acid sequence homologies and relative concentrations by a dot hybridization procedure. Nucleic Acids Res 7:1541.
16. O'Farrell PH (1975). High resolution two-dimensional electrophoresis of proteins. J Biol Chem 10:4007.
17. Laskey RA (1980). The use of intensifying screens or organic scintillators for visualizing radioactive molecules resolved by gel electrophoresis. Meth Enzymol 65:363.

18. Goeddel DV, Heyneker HL, Hozumi T, Arentzen R, Itakura K, Yansura DG, Ross MJ, Miazzari G, Crea R, Seeburg P (1979). Direct expression in Escherichia coli of a DNA sequence coding for human growth hormone. Nature 281:544.
19. Seed B (1983). Purification of genomic sequences from bacteriophage libraries by recombination and selection in vivo. Nucleic Acids Res 11:2427.
20. Grunstein M and Hogness D (1975). Colony hybridization: a method for the isolation of cloned DNAs that contain a specific gene. Proc Natl Acad Sci (USA) 72:3961.
21. Klein WH, Murphy W, Attardi G, Britten RJ, Davidson EH (1974). Distribution of repetitive and nonrepetitive sequence transcripts in HeLa mRNA. Proc Natl Acad Sci (USA) 71:1785.
22. Bishop JO, Morton JG, Rosbash M, Richardson M (1975). Three abundance classes in HeLa cell messenger RNA. Nature 250:199.
23. Birnie GD, MacPhail E, Young BD, Getz MJ, Paul J (1974). The diversity of the messenger RNA population in growing Friend cells. Cell Diff 3:221.
24. Goldberg RB, Galau GA, Britten RJ, Davidson EH (1973). Nonrepetitive DNA sequence representation in sea urchin embryo messenger RNA. Proc Natl Acad Sci (USA) 70:3516.
25. Davidson EH, Hough BR, Klein WH, Britten RJ (1975). Structural genes adjacent to interspersed repetitive DNA sequences. Cell 4:217.
26. Deininger PL, Jolly DJ, Rubin CM, Friedmann T, Schmid CW (1981). Base sequence studies of 300 nucleotide renatured repeated human DNA clones. J Mol Biol 152:17.
27. Elder JT, Pan J, Duncan CH, Weissman SM (1981). Transcriptional analysis of interspersed repetitive polymerase III transcription units in human DNA. Nucleic Acids Res 9:1171.
28. Fowlkes DM, Shenk T (1980). Transcriptional control regions of the adenovirus VA I RNA gene. Cell 22:405.
29. Fukamaki Y, Collins F, Kole R, Stoeckert CJ, Jagadeeswaran P, Duncan CH, Weissman SM (1982). Sequences of human repetitive DNA, non-alpha-globin genes, and major histocompatibility locus genes. Cold Spring Harbor Symp Quant Biol 46:1079.
30. Jelinek WR, Haynes SR (1982). The mammalian Alu family of dispersed repeats. Cold Spring Harbor Symp Quant Biol 46:1123.
31. Smith TF, Waterman MS (1981). Identification of common molecular subsequences. J Mol Biol 147:195.

32. Davidson EH, Britten RJ (1979). Regulation of gene expression: possible role of repetitive sequences. Science 204:1052.
33. Jelinek WR, Schmid CW (1982). Repetitive sequences in eukaryotic DNA and their expression. Ann Rev Biochem 51:813.
34. Sutcliffe JG, Milner RJ, Bloom FE, Levner RA (1982). Common 82-nucleotide sequence unique to brain RNA. Proc Natl Acad Sci (USA) 79:4942.
35. Sutcliffe JG, Milner RJ, Gottesfeld JM, Lerner RA (1984). Identifier sequences are transcribed specifically in brain. Nature 308:237.
36. Milner RJ, Bloom FE, Lai C, Lerner RA, Sutcliffe JG (1984). Brain-specific genes have identifier sequences in their introns. Proc Natl Acad Sci (USA) 81:713.
37. Murphy D, Brickell PM, Latchman DS, Willison K, Rigby PW (1983). Transcripts regulated during normal embryonic development and oncogenic transformation share a repetitive element. Cell 35:865.
38. Georgiev GP, Kramerov DA, Ryskov AP, Skryabin KG, Lukanidin EM (1982). Dispersed repetitive sequences in eukaryotic genomes and their possible biological significance. Cold Spring Harbor Symp Quant Biol 46:1109.
39. Ryskov AP, Ivanov PL, Kramerov DA, Georgiev GP (1983). Mouse ubiquitous B2 repeat in polysomal and cytoplasmic poly(A)$^+$ RNAs: unidirectional orientation and 3'-end localization. Nucleic Acids Res 11:6541.

Molecular Biology of Development, pages 481–490

DNA Elements Controlling Cell Specific Expression of Insulin and Chymotrypsin Genes[1]

Michael D. Walker, Thomas Edlund[2], Anne M. Boulet and William J. Rutter

Department of Biochemistry and Biophysics and Hormone Research Laboratory, University of California, San Francisco, San Francisco, California 94143

ABSTRACT By introducing hybrid genes containing putative regulatory sequences into a variety of tissue culture cells, we were able to define DNA sequences located in the 5' flanking region of insulin and chymotrypsin genes which play a role in cell specific gene expression. The 5' flanking DNA of the rat insulin gene contains an element which exhibits many of the properties of viral enhancers: however, it shows activity only in insulin-producing cells.

INTRODUCTION

In common with many other tissues of higher eukaryotic organisms, the pancreas exhibits a striking ability to express at very high levels a subset of genes. These include the amylase, trypsin, chymotrypsin, elastase, and carboxypeptidase genes expressed in pancreatic exocrine cells (1) and the glucagon, insulin and somatostatin genes expressed in the A, B and D cells respectively of the endocrine pancreas (2). With a view towards understanding the molecular basis of this tissue specificity of gene expression, we and our colleagues have over the past few years undertaken to characterize at the DNA sequence level those genes which are expressed at high level in the pan-

[1]This work was supported by NIH Grants GM-28520 and AM-21344 and by March of Dimes Grant 1-745.
[2]Present address: Department of Microbiology, University of Umea, Umea, Sweden.

creas. A comparison of nucleotide sequences in the region surrounding the coding domain, and in particular the 5' flanking DNA sequences, of pancreas-specific genes has failed to reveal an unambiguous pancreas specific sequence which may be involved in selective gene expression.

We have therefore developed a complementary biological approach to the detection of genetic determinants of cell specific expression. We constructed hybrid genes containing putative regulatory regions of pancreas genes and tested their relative expression following introduction into a variety of tissue culture cells including cells which express pancreas specific products. These procedures have enabled us to detect and map sequences located in the 5' flanking DNA that are required for high level of expression in differentiated cells.

RESULTS

Our experimental approach (3) involved linking the 5' flanking DNA sequences of pancreas specific genes to the coding sequence of the bacterial enzyme chloramphenicol acetyltransferase (CAT). The use of this enzyme as a reporter function has ben pioneered by Gorman et al. (4) who have demonstrated that it represents a convenient and sensitive way of monitoring expression of DNA species introduced into mammalian cells. Our initial experiments involved plasmids containing the 5' flanking region of human and rat insulin and rat chymotrypsin (Figure 1). We introduced these plasmids into a variety of tissue culture cells including an insulin-producing line derived from hamster islet cells (HIT; ref. 5), a chymotrypsin-producing line from rat pancreas (AR4-2J; ref. 6) and control cells of non-pancreatic origin (CHO and BHK). 48 hr after induction of DNA to the cells we measured the activity of CAT.

The flanking DNA sequences from the insulin and chymotrypsin genes are substantially more active in directing CAT activity in cells where the endogenous cognate gene is active (Figure 2). In order to quantitate the effects observed it was necessary to control for the variability in efficiency of uptake and expression between different cultured cell lines. For this we used plasmids which contained the promoters from the Herpes simplex virus thymidine kinase (TK) gene and the Rous Sarcoma virus long terminal repeat (RSV LTR). These promoters are active in a

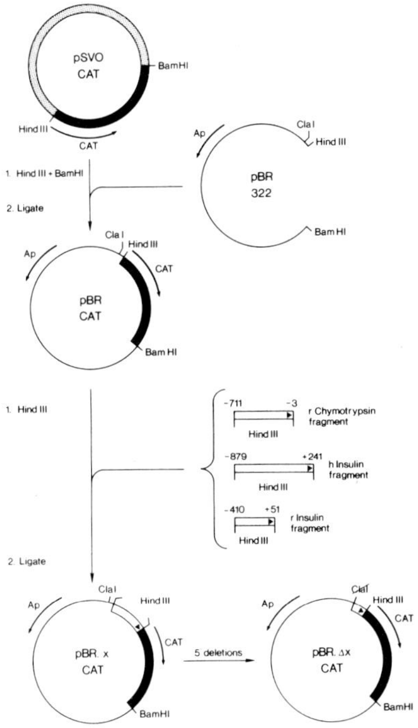

FIGURE 1. Construction of recombinants containing 5' upstream fragments of insulin and chymotrypsin genes linked to the CAT coding sequences (3). The numbers above the fragments of insulin and chymotrypsin genes refer to the location of the fragment in bp relative to the transcription start site.

wide variety of cells. Since they exhibited a similar ratio of activity in the three cell lines tested (Table 1) they represent essentially neutral promoters with regard to their cell specificity of expression in these cells. In marked contrast to the RSV and TK promoters, the DNA sequences containing the 5' flanking regions of the rat and human insulin genes and the rat chymotrypsin gene exhibited very different ratios of activities in the related differentiated cells than in other cells (Table 1): both the rat

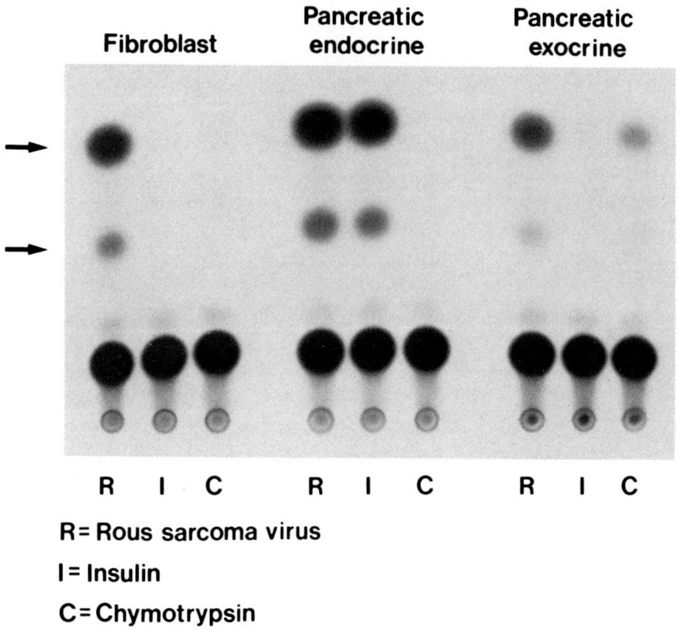

FIGURE 2. CAT activity directed by eucaryotic 5' flanking sequences in different cells. Recombinant plasmids containing 5' flanking sequences (see Figure 1) were applied to tissue culture cells as a calcium phosphate coprecipitate. 48 hr later CAT activity from extract containing 25 μg of protein was determined (3,4). The photograph shows an autoradiogram of the TLC separation of CAT reaction products. The cell types used were CHO fibroblasts, HIT-T15 pancreatic endocrine cells (5) and AR4-2J pancreatic exocrine cells (6). The upper two spots, indicated by arrows, correspond to the two forms of C^{14} chloramphenicol mono-acetate while the lower intense spot corresponds to unreacted C^{14}-chloramphenicol. The reaction times used were 20 min (CHO), 60 min (HIT), and 12.5 hr (AR4-2J).

and human insulin gene fragments were 50-200 fold more active in HIT cells than in either CHO or AR4-2J cells. The activity of the rat insulin gene fragment was also very low in COS 7 cells (SV40 transformed monkey kidney cells) and Alexander cells (human hepatoma cells) (data not shown). Thus, efficient CAT expression directed by the rat insulin gene fragment was only observed in insulin-producing cells. In an analogous fashion, the chymotrypsin gene 5' flanking region showed 50-200 fold higher activity in AR4-2J cells than in either CHO or HIT cells (Table 1) or in the rat fibroblast line, rat 2 (data not shown).

TABLE 1
ACTIVITY OF DNA CONTAINING 5' FLANKING SEQUENCES

Fragment	Cell Type		
	CHO (%)	HIT (%)	AR4-2J (%)
RSV	100	100	100
TK	15	14	17
Rat insulin	1.2	71	<0.4
Human insulin	<0.1	4	ND
Rat chymotrypsin	0.9	0.2	42
Rat growth hormone	<0.1	<0.2	ND

Relative CAT activities directed by DNA containing 5' flanking sequences (3). For each cell type, activity is expressed as a percentage of that directed by the RSV fragment. Results shown represent the mean of two independent determinations. The range of activities around a particular value varied by up to 30% of that value. ND = not determined.

In order to define better the region involved in mediation of the cell specific expression, we deleted portions of the 5' flanking DNA and measured the effect of these deletions in the appropriate differentiated cell types. Interestingly, a dramatic loss of activity in all cases resulted when DNA sequences located between about -168 and -300 were removed (Figure 3). This is considerably upstream from known transcriptional regulatory sequences defined for other mammalian protein coding genes (7,8) and suggest that novel upstream elements are involved.

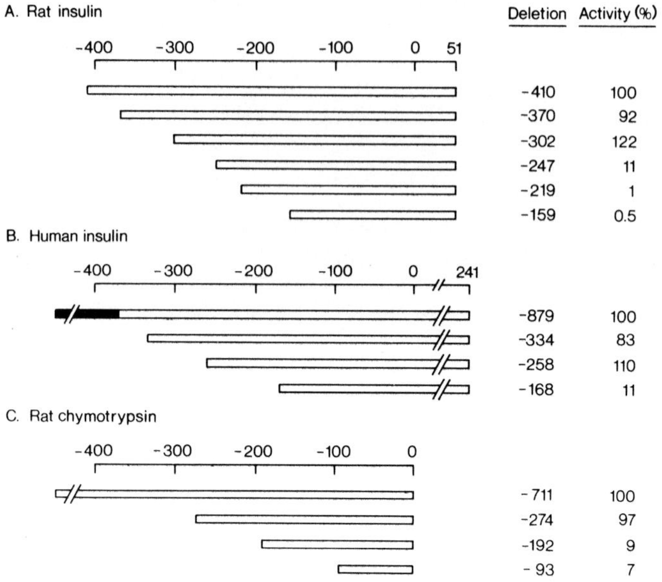

FIGURE 3. Relative activity of insulin and chymotrypsin gene fragments containing 5' deletions. Activities were determined following transfection of HIT cells for CAT plasmids containing insulin gene sequences and AR4-2J cells for CAT plasmids containing chymotrypsin gene sequences (3).

It was not possible for us to map directly the 3' border of the active fragment since deletion of components of the promoter of the insulin and chymotrypsin gene, e.g. TATA box, would most likely lead to loss of activity even if sequences important for cell specific expression were not present here. We therefore adopted an alternate approach: to search for cell specific elements capable of affecting the activity of a heterologous promoter. For this we used a fragment derived from the rat insulin gene containing 410 bases upstream of the transcription start site. This fragment was inserted into a plasmid containing the Herpes simplex thymidine kinase promoter located immediately upstream of CAT coding sequences. The fragment from the rat insulin gene was inserted either immediately upstream of the TK promoter (plasmid 2, Table 2) or 600 bases upstream (plasmid 3). In both cases the orientation was such that the rat insulin promoter could not be used to transcribe the CAT gene. The presence of the rat insulin gene fragment had essentially no effect on transcription in fibroblasts (Table 2). However, the fragment substantially augmented activity in HIT cells. The effect was stronger when the fragment was adjacent to the TK promoter (18-fold) but still substantial when 600 bp distant from the TK promoter (3-fold). This fragment of the rat insulin I gene clearly manifests at least some of the properties of enhancer elements. Enhancers are short (100-200 bp) DNA elements which are able to augment transcription from a heterologous promoter in a manner independent of orientation or precise positioning relative to the transcription

TABLE 2
EFFECT OF RAT INSULIN 5' FLANKING SEQUENCES ON TK PROMOTER ACTIVITY

				Relative activity	
Plasmid	Upstream sequence	Promoter	Coding Sequence	BHK	HIT
1	pBR322	TK	CAT	1	1
2	rat insulin	TK	CAT	0.8	18
3	rat insulin-pBR322	TK	CAT	1.2	3

Relative CAT activities directed the HSV TK promoter in the presence of 5' flanking sequence of the rat insulin gene. Plasmid 1 contains the HSV TK promoter (-109 to +51) located immediately upstream of CAT coding sequences. Plasmid 2 was produced by inserting the rat insulin I 5' flanking sequences (-410 to +51) immediately upstream of the TK promoter in an orientation opposite to that of TK. Plasmid 3 was produced by introducing the same fragment from the rat insulin gene upstream of the TK promoter in the same orientation as before but with 600 bases of pBR322 DNA intervening. The activities of the plasmids in fibroblasts (BHK) and HIT cells are expressed relative to plasmid 1.

unit (reviewed in 9). Our fragment also exhibits a strong cell-type preference. In more recent experiments we have unequivocally demonstrated the orientation-independent nature of the activation and shown that the element can be active when located at the 3' end of the CAT gene (Edlund et al., unpublished results).

DISCUSSION

The introduction of cloned DNA segments into eukaryotic cells in tissue culture is a powerful tool for identifying functionally significant DNA sequences. For example, this approach has enabled substantial progress towards defining promoters for eukaryotic protein coding genes (7,8) and the regulatory domain of steroid responsive genes (10,11). Our studies together with those of other investigators examining expression of immunoglobulin genes (12,13), form a basis for identifying the DNA sequences involved in the mediation of cell specific gene expression.

The presence of cell specific enhancer elements within introns of immunoglobulin genes (12,13) can be elegantly rationalized in terms of the special properties of this unusual gene family. However, it is by no means clear that other mammalian genes, the vast majority of which are not believed to rearrange during development, employ similar kinds of control mechanism. Unlike the immunoglobulin system we have not observed substantial activities associated with intron sequences (data not shown); we do however observe cell-specific expression mediated by the 5' flanking DNA sequences. These control sequences are in close proximity or perhaps overlapping with promoter sequences. However, as with the immunoglobulin system, at least part of the cellular specificity seems to be contributed by the action of enhancer-like elements. We are currently examining the flanking DNA sequences in greater detail with a view to more closely defining the domain of response. Furthermore, our results and those of others (10,12,13) are most easily explained by a model where control elements are activated by the presence of positive, trans-acting regulatory factors which we term differentiators. We are using a number of biochemical and genetic approaches to detect and isolate these factors in pancreas cells.

ACKNOWLEDGMENTS

We would like to thank Dr. C. Gorman for supplying pSVO CAT and pRSV CAT, Dr. R. Santerre for providing HIT cells, Dr. Y. Kim for providing AR4-2J cells, and K. Raneses for preparation of the manuscript.

REFERENCES

1. Sanders TG, Rutter WJ (1974). The developmental regulation of amylolytic and proteolytic enzymes in the embryonic rat pancreas. J Biol Chem 249:3500.
2. Pictet R, Rutter WJ (1972). Development of embryonic endocrine pancreas. In Steiner DF and Freinkel (eds): "Handbook of Physiology", section 7, Endocrinology, Williams and Wilkins, Baltimore.
3. Walker MD, Edlund T, Boulet AM, Rutter WJ (1983). Cell specific expression controlled by the 5' flanking region of insulin and chymotrypsin genes. Nature 306:557.
4. Gorman CM, Moffat LR, Howard BH (1982). Recombinant genomes which express chloramphenicol acetyltransferase in mammalian cells. Mol Cell Biol 2:1044.
5. Santerre RF, Cook RA, Crisel RMD, Sharp JD, Schmidt RJ, Williams DC, Wilson CP (1981). Insulin synthesis in a clonal cell line of simian virus 40-transformed hamster pancreatic beta cells. Proc Natl Acad Sci USA 78:4339.
6. Jessop NW, Hay RJ (1980). Characteristics of two rat pancreatic exocrine cell lines derived from transplantable tumors. In vitro 16:212.
7. McKnight SL, Kingsbury R (1982). Transcriptional control signals of a eukaryotic protein coding gene. Science 217:316.
8. Dierks P, van Ooyen A, Cochram MD, Dodkin C, Reiser J, Weissmann C (1983). Three regions upstream from the cap site are required for efficient and accurate transcription of the rabbit β-globin gene in mouse 3T6 cells. Cell 32:695.
9. Khoury G, Gruss P (1983) Enhancer elements. Cell 33:313.
10. Chandler VL, Maler BA, Yamamoto KR (1983) DNA sequences bound specifically by glucocorticoid receptor in vitro render a heterologous promoter hormone responsive in vivo. Cell 33:489.

11. Karin M, Haslinger A, Holtgreve H, Cathala G, Slater E, Baxter JD (1984). Activation of a heterologous promoter in response to dexamethasone and cadmium by metallothionein gene 5' flanking DNA. Cell 36:371.
12. Gillies SD, Morrison SL, Oi VT, Tonegawa S (1983). A tissue-apecific transcription enhancer element is located in the major intron of a rearranged immunoglobulin heavy chain gene. Cell 33:717.
13. Banerji J, Olson L, Schaffner W (1983). A lymphocyte-specific cellular enhancer is located downstream of the joining region in immunoglobulin heavy chain genes. Cell 33:729.

Molecular Biology of Development, pages 491–508

TRANSCRIPTION OF DIRS-1: AN UNUSUAL *DICTYOSTELIUM* TRANSPOSABLE ELEMENT[1]

Stephen M. Cohen, Joe Cappello, Charles Zuker
and Harvey F. Lodish

Department of Biology
Massachusetts Institute of Technology
Cambridge, MA 02139

Dictyostelium Intermediate Repeat Sequence 1 (DIRS-1) is a transposable element with several unusual features. DIRS-1 consists of 4.1 kb of unique internal sequence flanked by approximately 330 bp inverted terminal repeats (1). There are about 40 copies of the intact DIRS-1 element interspersed throughout the *Dictyostelium* genome, as well as an additional 200 copies of related sequences. DIRS-1 and related sequences are transcribed into a heterogeneously sized population of RNAs (1, 2). Transcription is induced during development and can be induced in vegetative cells by heat shock and other stresses (3). Sequence analysis of the terminal repeats of DIRS-1 has revealed the presence of heat shock promoters that appear to be responsible for directing transcription of DIRS-1 RNAs (4). Sequences related to DIRS-1 were originally identified as a moderately repetitive family of developmentally regulated genes (2). DIRS-1 has been independently identified by Rosen *et al.* (5) who call it Tdd-1.

Transposition and Genomic Organization of DIRS-1

Several lines of evidence suggest that DIRS-1 might be a transposable element. We have isolated a number of genomic clones that contain all or part of the same 4.7 kb DIRS-1 sequence (1). Sequences flanking different intact genomic copies of DIRS-1 have different restriction maps. In addition, hybridization of labeled DIRS-1 DNA to genomic DNA

[1] This work was supported by grants No. GM3275 from the National Institutes of Health and PCM 79-00839 from the National Science Foundation.

digested with restriction enzymes that cut DIRS-1 only once labels a large number of fragments of different sizes (1,2, 5). These observations suggest that the DIRS-1 elements are scattered throughout the genome with different flanking sequences. The most compelling evidence that DIRS-1 is a transposable element comes from a comparison of Southern blots of genomic DNAs from different strains of Dictyostelium. While all strains have approximately 40 copies of the same 4.7 kb DIRS-1 element, the sequences flanking some of the DIRS-1 elements are different in different strains. These results indicate that the element has transposed since these strains were separated (1,5).

A schematic representation of DIRS-1 is presented in Figure 1. The intact element consists of a 4.1 kb internal EcoRI fragment flanked by inverted terminal repeats (1,5). DNA sequence analysis of a number of genomic copies of DIRS-1 indicated that the terminal repeats have some unusual structural features (4,5). Both left and right repeats consist of a similar core sequence 332 bp in length that

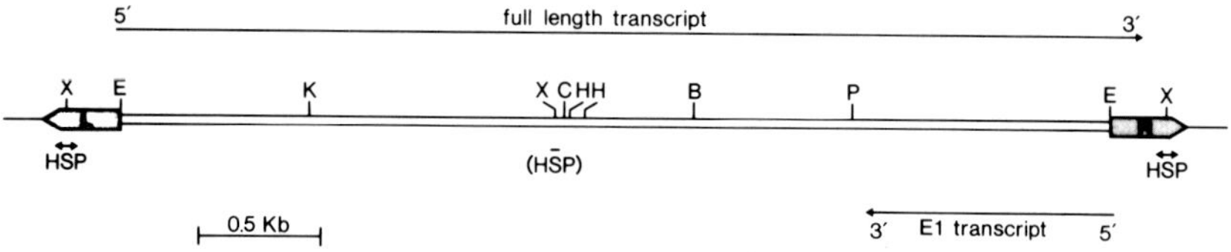

FIGURE 1. Structure of DIRS-1. This restriction map is a consensus of 6 cloned DIRS-1 elements. DIRS-1 is 4.7 kb in length and contains 330 bp inverted terminal repeats (shaded boxes). The unique internal 4.1 kb sequence is indicated by the open box. The terminal repeats contain heat shock promoters (HSP). The heat shock promoter in the left repeat is responsible for directing rightward transcription of the full length 4.5 kb RNA transcript. The HSP in the right terminal repeat directs leftward transcription of the E1 RNA. The extent and direction of these transcripts are indicated by arrows above and below the element. An internal heat shock promoter has been identified by DNA sequence but has not yet been shown to be functional. Restriction enzymes: B, BglII; C, ClaI; E, EcoRI; H, HindIII; K, KpnI; P, PvuII; X, XbaI.

begins at the EcoRI restriction sites of the DIRS-1 element. Certain characteristic structural differences were observed between the 4 left and 5 right repeats that we have sequenced (4). All right repeats contain a 28 bp A + T rich extension of the 332 bp core repeat sequence, adjacent to flanking sequences, that is not found in any left repeat. All left repeats contain a 4 bp insertion near the EcoRI site that is not found in any right repeat. The significance of these differences has not been determined, but the fact that they are highly conserved suggests that they may be of some importance.

Hybridization of DIRS-1 DNA to genomic DNA, digested with EcoRI, indicates that the 4.1 kb internal DIRS-1 fragment is present in about 40 copies per haploid genome. There are also an additional 200 DIRS-1 related EcoRI fragments of widely differing sizes (1,2). These other EcoRI fragments that contain DIRS-1 sequences apparently represent rearranged or deleted copies of the intact element. Some of these rearrangements appear to be the result of transpositional insertion of intact DIRS-1 elements into pre-existing genomic copies of DIRS-1. The original isolate of DIRS-1, genomic clone SB-41, is a good example of such a transpositionally rearranged element (Cappello *et al.* manuscript submitted).

Transcription of DIRS-1

Transcription of DIRS-1 is induced soon after *Dictyostelium* cells are plated for development and results in the production of a population of heterogeneously sized RNA species (1-3). Figure 2 presents an RNA transfer hybridization experiment in which size fractionated *Dictyostelium* polyadenylated RNAs are hybridized to labeled DIRS-1 DNA. DIRS-1 RNAs are present at very low levels in vegetative cells (lane 1), but begin to accumulate within the first hour of development (lane 3) and reach a maximal level by 15 hours (lane 4)(3). These RNAs can also be induced in vegetative cells by a variety of stresses including heat shock (lane 2), high cell density, and plating of cells (3).

Figure 3 depicts the relative rates of DIRS-1 transcription and of DIRS-1 RNA accumulation during development. DIRS-1 transcription rates were determined by hybridization of RNA, labelled *in vitro* by run off transcription of isolated nuclei, to cloned DNAs spotted onto filters (6). DIRS-1 transcripts are synthesized in nuclei isolated from

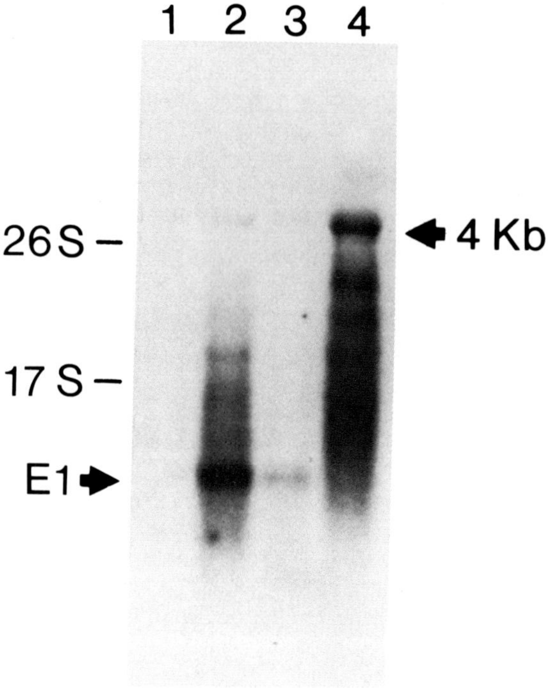

Figure 2. Developmental regulation of DIRS-1 RNAs. Transcription of DIRS-1 RNAs is regulated during Dictyostelium development. Four μg of poly (A^+) RNA isolated from vegetative cells (lane 1), heat shock-treated vegetative cells (lane 2), 1 hour (lane 3) and 15 hour (lane 4) filter developed cells were size fractionated on a formaldehyde--agarose gel. The gel was transferred to "Gene Screen" and hybridized with labeled DIRS-1 DNA. DIRS-1 is transcribed into a heterogenous population of RNA species. The 4.5 kb RNA is presumed to represent a full length transcript of the intact element (1). E1 RNA is predominantly expressed during early development (lane 3) and is induced to high levels by heat shock (lane 2)(3). The heterogeneously sized smear of RNAs are transcribed from the same strand as the 4.5 kb RNA, and are presumed to be derived from the many deleted and rearranged copies of DIRS-1 found in the genome. 26S and 17S are ribosomal RNA size markers.

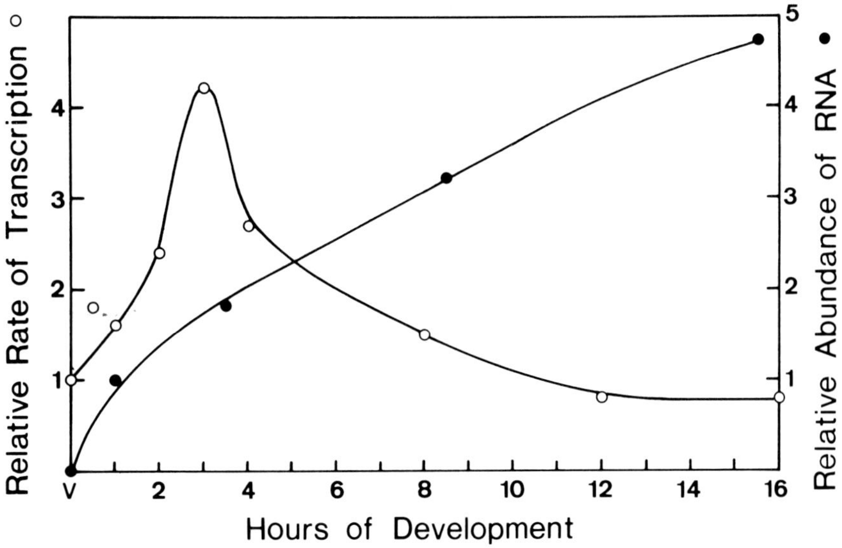

FIGURE 3. Transcription and accumulation of DIRS-1 RNAs. The relative rate of transcription of DIRS-1 RNAs during development was determined by hybridization of total nulear RNA, labeled by run off transcription in isolated nuclei, to cloned DIRS-1 DNAs immobilized on nitrocellulose filters (open circles). The relative accumulation of DIRS-1 RNAs in cytoplasmic polyadenylated RNA was determined by densitometric scanning of gel blots of RNA probed with labeled DIRS-1 DNA (closed circles). The amount of poly (A^+) RNA was normalized to contain equal amounts of a constitutively expressed mRNA (6).

vegetative cells, however very little DIRS-1 RNA can be detected in polyadenylated nuclear or cytoplasmic RNA from vegetative cells. DIRS-1 RNAs transcribed in vegetative cells may be rapidly degraded. Alternatively, this discrepancy may reflect artifactual induction of DIRS-1 transcription caused by stress during the preparation of nuclei from vegetative cells. There is a burst of DIRS-1 transcription at about 3 hours of development followed by a drop in the rate to approximately the level observed in vegetative cells.

Processing and nuclear transit times of DIRS-1 transcripts in developing cells were assayed by continuous labeling with [^{32}P]PO_4 and measuring the accumulation of poly (A^+) and poly (A^-) RNAs in the nucleus and the cytoplasm that hybridize to DIRS-1 DNAs (7). As is the case with all *Dictyostelium* RNAs, polyadenylation of DIRS-1 RNA occurs within 45 min after synthesis. However, DIRS-1 RNAs begin to accumulate in the cytoplasm only after a lag of 135 min. This nuclear transit time is much longer than that of constitutively expressed or other developmentally regulated *Dictyostelium* RNAs, which range from 25 to 60 minutes, and from 50 to 100 minutes, respectively. DIRS-1 RNAs are also unusual in that only 25% of the nuclear transcripts accumulate in the cytoplasm (7). The remaining 75% are apparently degraded within the nucleus, or immediately after export to the cytoplasm. Processing efficiency of most other *Dictyostelium* mRNAs ranges from 60 to 70%, although some mRNAs unrelated to DIRS-1 are processed at about 20% efficiency.

Transcription and processing of DIRS-1 RNAs have many features in common with those of the *Drosophila* copia-like transposable elements. 90% of copia transcripts are also rapidly degraded in the nucleus. The 10% of poly (A^+) copia RNA that is exported to the cytoplasm is stable (8, 9). In both cases the reasons for the low efficiency of processing and transport of nuclear RNAs is not known.

DIRS-1 RNAs transcribed during development are heterogeneous in size, with a predominant band of approximately 4.5 kb. The regions of DIRS-1 which are transcribed were determined by hybridization of a series of subclones of DIRS-1 to replicas of gel separated polyadenylated RNAs (1). Each of 10 fragments of DIRS-1 hybridizes to the 4.5 kb RNA, supporting the notion that it represents a full length transcript of DIRS-1. The subclones also hybridized to a heterogeneous population of smaller RNA species, which may be transcribed from the approximately 200 rearranged or deleted copies of the element present in the genome. The majority of these RNAs are transcribed from the same strand of DIRS-1, from left to right as indicated in Figure 1 for the full length transcript. The direction of transcription was determined by hybridization of strand-specific DNA probes to RNA dot blots (6) and to gel separated RNAs (5) and by hybridization of *in vivo* labeled RNA to strand separated DNAs (6).

The RNA designated E1 (Figure 2) is unusual in several respects. E1 is transcribed exclusively during early development. Unlike the majority of DIRS-1 RNAs E1 reaches its maximal accumulation by 1 hour and is almost undetectable by 2 hours of development (3). E1 is transcribed from the opposite strand than the other DIRS-1 RNAs, beginning near the right terminal repeat and extending leftward into DIRS-1 (Figure 1, also ref. 5). In addition, transcription of E1 is induced to a much greater extent by heat shock than is the case for the other DIRS-1 RNAs (Figure 2, also ref. 5). The reasons for the differences in the regulation of E1 and the other DIRS-1 RNAs are not known.

Transcription of DIRS-1 and related sequences produces at least two discrete RNA species: the "full length" 4.5 kb transcript and E1. The full length transcript must be transcribed from an intact DIRS-1 element. We do not know whether E1 is transcribed from intact copies of DIRS-1 or from some of the large number of partial copies of DIRS-1 found in the genome. As noted above, other heterogeneously sized DIRS-1 related RNAs may also be derived from partial copies of the DIRS-1 element. We cannot exclude the possibility that some of these may be processing products of the full length 4.5 kb transcript.

DIRS-1 RNAs are polyadenylated and are associated with polysomes, suggesting that they may encode protein(s)(6). A subclone of DIRS-1 (pB41-6) that consists of approximately the 3' half of DIRS-1 has been sequenced (3). The strand equivalent to the full length transcript contains three long overlapping open reading frames (621, 705 and 789 bp respectively). The strand equivalent to E1 has only one relatively short open reading frame of 243 bp. By analogy to other transposable elements (reviewed in 10), it seems likely that DIRS-1 may encode proteins that may be required for its transposition.

cDNA Clones Complementary to DIRS-1

In order to establish whether the 4.5 kb RNA represents a full length transcript of an intact DIRS-1 element we have isolated cDNA clones complementary to DIRS-1 RNAs (11). The structure of two of these clones is presented in Figure 4. pCC31 is a 4.1 kb cDNA clone delimited by EcoRI restriction sites. Restriction mapping and DNA sequence analysis of the ends of pCC31 indicate that it represents a complete transcript of the internal 4.1 kb EcoRI segment of a DIRS-1 element. Clone pCC31 therefore presumably reflects the full length transcription of an intact DIRS-1 element predicted by the existence of the 4.5 kb RNA described above. Since the cDNA molecules were truncated by an EcoRI digestion during construction of the cDNA library (see ref. 11 for

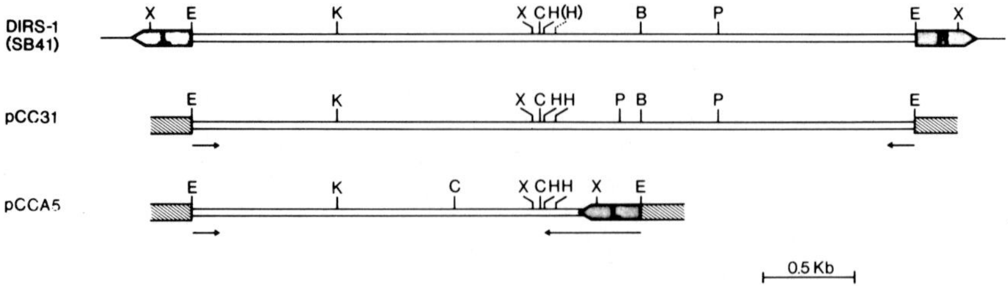

FIGURE 4. cDNA clones complementary to DIRS-1. Restriction maps of DIRS-1 and two DIRS-1 related cDNA clones are presented. cDNA was prepared from poly (A^+) RNA from 8 hour filter developed cells, ligated to EcoRI linkers and cloned into λgt11. DIRS-1 cDNAs were selected by hybridization to a DIRS-1 genomic clone, and were subcloned into pBR322. Restriction enzymes: B, BglII; C, ClaI; E, EcoRI; H, HindIII; K, KpnI; P, PvuII; X, XbaI. The HindIII site indicated by a dotted line is missing in SB-41 due to a single base change. This HindIII site is found in all other DIRS-1 genomic clones. The arrows under the restriction maps of the cDNA clones indicate regions that have been sequenced. The identity of the shaded box in pCCA5 as a left terminal repeat was confirmed by DNA sequencing. Hatched boxes represent vector sequences.

details), we cannot determine from pCC31 whether transcription extends beyond the EcoRI sites on either side.

A previously identified cDNA clone complementary to DIRS-1, pLZ12, contains sequences that are homologous to 230 bp at the 3' end (i.e. right end) and that extend at least 152 bp into a right terminal repeat sequence (6). We have no direct evidence that pLZ12 was derived from a transcript of an intact DIRS-1 element. However, if this is the case, the structure of pLZ12 suggests that transcription of DIRS-1 may proceed through at least half of the right terminal repeat. Transcription termination near the right end of the right terminal repeat, adjacent to flanking sequences, is suggested by the presence of several polyadenylation signals (AATAAA) in the distal 70 nucleotides of the repeat sequence. The suggestion that a full length transcript of

DIRS-1 may contain right terminal repeat sequences is in contrast to the observations of Rosen et al. (5) who did not detect hybridization of repeat sequences to their 4.5 kb DIRS-1 transcript.

The restriction map of part of the cDNA clone pCCA5 is similar to that of the left half of DIRS-1 (Figure 4). We define left and right as the 5' to 3' orientation of the full length DIRS-1 transcript. The 5' end of pCCA5 shares the same nucleotide sequence with pCC31 and DIRS-1. The similarity in the restriction maps extends to the internal HindIII restriction sites. pCCA5 then contains an XbaI and an EcoRI site spaced 240 bp apart, a pattern reminiscent of the DIRS-1 terminal repeat structure. We have sequenced 551 nucleotides of pCCA5 from the 5' HindIII site to the EcoRI site at the 3' end (11). The sequence of the first 223 nucleotides from the 5' HindIII site is virtually identical to the equivalent internal region of genomic DIRS-1 clone SB-41. There are a small number of single base differences, one of which accounts for the absence of the second HindIII restriction site in SB-41 (Figure 4). This HindIII site is present in all other genomic copies of DIRS-1. Beginning at nucleotide 224 from the 5' HindIII restriction site the cDNA sequence is no longer homologous to the internal region of DIRS-1 but is nearly identical to that of the SB-41 left terminal repeat. Transcription of the RNA from which pCCA5 was derived is presumed to have initiated upstream from the 5' EcoRI site of the cDNA in a DIRS-1 left terminal repeat. pCCA5 reflects transcription of a genomic copy of a rearranged DIRS-1 element that would appear as a small EcoRI fragment on a genomic Southern blot. The presumed genomic template of pCCA5 could have arisen from a DIRS-1 into DIRS-1 transposition event similar to the one proposed in the generation of genomic clone SB-41 (Cappello et al., manuscript submitted).

Transcription Initiation of DIRS-1 RNAs

We have determined the initiation site of the E1 RNA transcript by an S1 nuclease protection experiment (11). An RsaI fragment that extends from a point within the E1 transcript to beyond the right terminal repeat was isolated from a plasmid containing an intact DIRS-1 element (Figure 5). The fragment was end labeled and digested with HindIII to generate an asymmetrically labeled probe for S1 mapping. The labeled probe was hybridized to RNA from vegetative and

```
0    ATTAAAAATTTAATTTATTAAATTATATTTTATATTAAATATATATATATATATTATGAATAACATTTATTTATTTGAATTTCCCAAATATTTAAGATAA
          -330      -320      -310      -300      -290      -230      -270      -260      -250      -240

             HSP
     -* ***  *** **
               ** ***  ***  **
-100 TTTTTTAGAATGTTCTAGAACATTCGAAGAATAAAAAATTTTCGAAAGAAAAGTAAAAATTTCGAACCGGCACAATGACGGCGATAATTGCGCAAGGTCG
               XbaI                   -200      -190      -180      -170      -160      -150      -140

-200 AAAAACTGAAAAATTCCGAACCGACGAGACTATGCACAAATTTGTGAAGGGTCGAAAACTCTTATTTTTTTGAGTTTTGCGAAATTTTAAGAAAATAAAA
          -130      -120      -110      -100      -90       -80       -70       -60       -50       -40

     TATA box                              ↓    ↓
-300 ACGTATAAATAGTGGCACTAAAAACTAAAACAATTTGATTTATGAATTCCGACTAAGGGCGGGTGGGATGGGACTAATATATATATATATATTTAATATA
          -30       -20       -10      -1        EcoRI    20        30        40        50        60

-400 AATATAATTTAATAAATTAAATTTTTAATTTATATTATCATATATATATATATATTATGAAATTTGGACGACAGTATTTAAGAAACCACCAGATTTTCAC
          70        80        90        100       110       120       130       140       150       160

-500 CAATGATTATTTTATCGTAGAAGGTATCTACAGTATCATTTGATTTCCAACGACCCATCTTTTTGACAACGTGGAACGAAACTTTATTGGACAACAGCAG
          170       180       190       200       210       220       230       240       250       260

-600 AGAAGCCATAGCGAACGGGTAGAGTGAGATTTGAACTTGACAATATCAATACCTGACTTTGAGAGTGGTTGATAGTAC
          270       280       290       300       310       320       330     RsaI
```

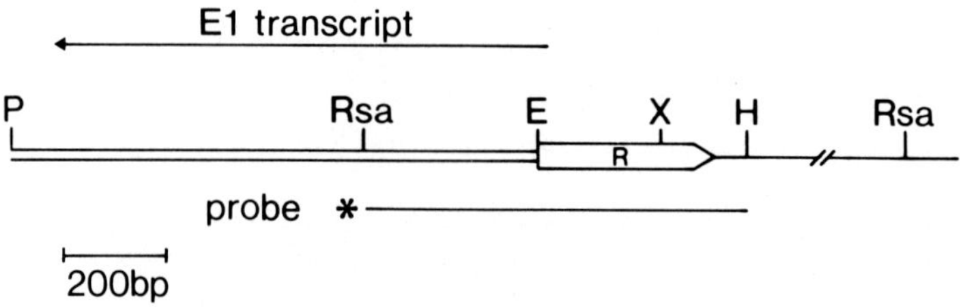

Figure 5.

FIGURE 5. Initiation of transcription of the E1 RNA. The site of initiation of E1 RNA was determined by S1 nuclease protection. An RsaI fragment that begins within the E1 transcribed region and spans the right terminal repeat was isolated from a plasmid containing an intact DIRS-1 element. The fragment was 5' end labeled, and cleaved with HindIII to generate an asymmetrically labeled probe. The asterisk indicates the position of the label in the probe molecule. The structure of the probe and its relationship to the right end of DIRS-1 is indicated in the restriction map. The right terminal repeat of DIRS-1 is indicated by the box labeled R. The direction of E1 transcription is indicated. Restriction enzymes are RsaI, E, EcoRI; H, HindIII; P, PvuII; X, XbaI. This probe was gel purified and hybridized to RNA from vegetative and heat shock treated cells, digested with S1 nuclease, and size fractionated on a sequencing gel.

The sequence of the right terminal repeat and the internal region of DIRS-1 extending to the RsaI site is presented (3,4). This sequence is written from the distal end of the terminal repeat to the RSA site (right to left in the restricton map). The sequence is numbered in both directions from the 5'-most initiation site of transcription. Transcription initiates at 2 closely spaced sites within the right terminal repeat (arrows). A possible TATA box is located within the terminal repeat at position -25 and the heat shock prmoters (HSP) are located at position -207 and -218. Homology to the Drosophila consensus HSP is indicated by asterisks above the sequence.

heat shock-treated cells, digested with S1 nuclease and size fractionated on a DNA sequencing gel. Two predominant fragments were protected by RNA from heat shock treated cells, but not by RNA from vegetative cells. The lengths of the protected fragments indicated that initiation of E1 transcription ocurs at two closely spaced sites within the right terminal repeat, indicated by arrows in Figure 5. A possible TATA box is located at position -25 from the 5' most initiation site. This sequence is a perfect match with the Dictyostelium consensus TATA sequence "TATAAA(T/A)A" (12). The heat shock promoters are located at positions -207 and -218. Transcription initiates at the end of two short oligo (dT) sequences. Oligo (dT) stretches immediately preceeding the transcription initiation site are a unique feature of Dictyostelium genes, and have been observed in actin,

```
                                                                                HSP
0   TTTATATTATCATATATATATATATATATTATGAATAACATTTATTTATTTGAATTTCCCAAATATTTAAGATAATTTTTTTAGAATGTTCTAGAACATTCGAAGAATAAAAAATTTTCGAAA
         -300      -290      -280      -270      -260      -250      -240      -230      XbaI      -210      -200      -190

120 GAAAAGTAAAAATTTCGAACCGGCACAATGACGCGATAATTGCGCAAGGTCGAAAAACTGAAAAATTCCGAACCGAGACTATGCACAAATTTGTGAAGGGTCGAAAACTCTTATTTTTTT
         -180      -170      -160      -150      -140      -130      -120      -110      -100      -90       -80       -70

                                      TATA box                    ↓
240 GAGTTTTGCGAAATTTTAAGAAAATAAAAACGTATAAATAGTGGCACTAAAAACTAAAACACTTTTTATTTGGAATTCCTCCTATATATATTTAAGATACACACACACACAAATTTTATA
         -60       -50       -40       -30       -20       -10       -1   EcoRI         20        30        40        50

360 CAAATAATTATTTCTCTTTATTAATTTTTAATATTATTATTTTTCTCATATTATAATTTTAGATATCAACAAAAAATATGTCTACCACTGTTAATAATAATGATGCCTCTAGTAGCAATA
         60        70        80        90        100       110  EcoRV          130       140       150       160       170

480 CCTCTGCCTCTAATAGCGCTGAATCCTTTGATTTAAGAATAAATCAATCGAGGATC
         180       190       200       210       220  SauIIIa
```

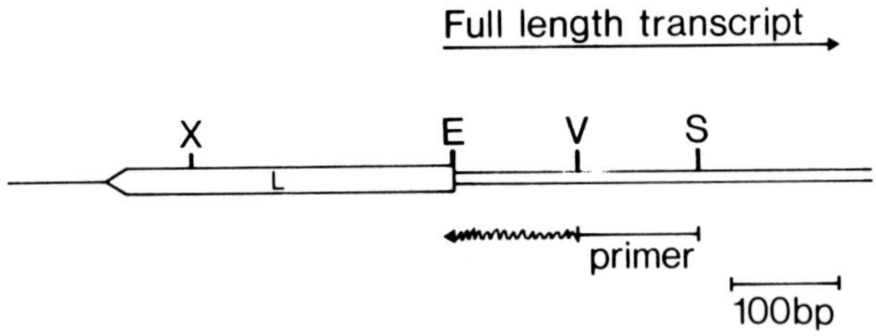

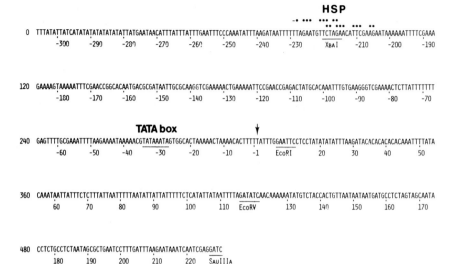

FIGURE 6. Initiation of transcription of the full length DIRS-1 RNAs. The initiation site of transcription was determined by primer extension. The structure of the primer and its relationship to the left end of DIRS-1 is indicated in the restriction map. The left terminal repeat is indicated by the box labeled L. The direction of transcription of the "full length" RNA is indicated. A SauIIIa(S) to EcoRI (E) fragment containing internal sequences from the 5' end of DIRS-1 was cloned into M13mp8, asymmetrically labeled, and the double stranded primer excised by digestion with SauIIIa and EcoRV (V). The primer was isolated from an acrylamide gel and hybridized to RNA from vegetative, heat shock-treated and 15 hour filter

developed cells. The primer was extended by reverse transcription (indicated by the wavy line) and size fractionated on a DNA sequencing gel. In all cases a single predominant band of 230 bp was defining the initiation point.

The DNA sequence of the left terminal repeat and internal region of DIRS-1, extending to the SauIIIa site, is presented (4). The sequence is written from the distal end of the terminal repeat to the SauIIIa site (left to right in the restriction map). The size of the cDNA product places the initiation point of transcription 230 bp to the left of the SauIIIa site (arrow) within the left terminal repeat at the end of a stretch of T nucleotides. The sequence is numbered from the initiation site, in both directions. A TATA box is located in the terminal repeat at position -27 and the heat shock promoters (HSP) are located at positions -205 and -216. Homology to the Drosophila consensus HSP is indicated by asterisks.

discoidin and several other genes (12).

We have also mapped the initiation site for the full length 4.5 kb RNA. These experiments were performed using primer extension in addition to S1 nuclease protection because the nucleotide sequence of left terminal repeats adjacent to the EcoRI restriction site differs in different genomic copies of DIRS-1 (4). A SauIIIa to EcoRI fragment at the 5' end of DIRS-1 was cloned into M13 phage and used to synthesize an asymmetrically labeled primer (Figure 6). Double stranded M13 DNA was cleaved with SauIIIa and EcoRV and a 112 bp fragment was purified from an acrylamide gel for use as the primer. This fragment was hybridized to RNA from vegetative, heat shock-treated and 15 hour filter developed cells, extended by reverse transcription, and size fractionated on a sequencing gel (11). One predominant extension cDNA was generated. The size of the extended primer places the transcription initiation site at the end of a stretch of T nucleotides within the left terminal repeat, indicated by an arrow in Figure 6. A perfect match for the consensus TATA box sequence is located at position -27. The heat shock promoters are located at position -205 and -216. The data agrees with the location of transcription initiation determined by S-1 nuclease protection.

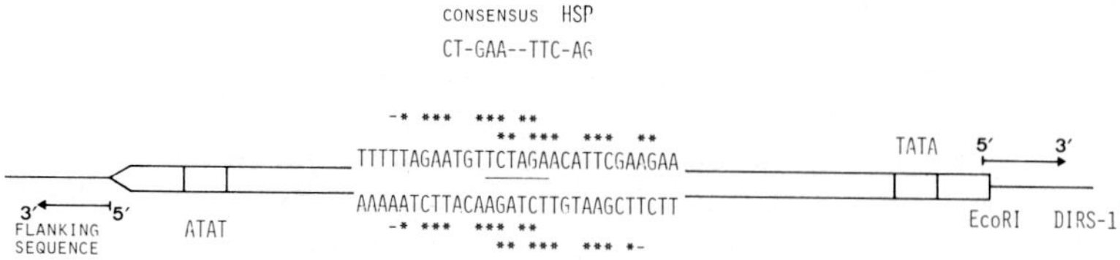

FIGURE 7. Organization of the DIRS-1 terminal repeat. A schematic represenation of the oganization of the heat shock promoter sequences and TATA box sequences contained in the terminal repeats is presented. The sequence of the region of consensus terminal repeat containing the Dictyostelium heat shock promoters is shown. Homology to the Drosophila consensus heat shock promotor (HSP) "CT-GAA--TTC-AG" is indicated by asterisks. Dashes indicate mismatched nucleotides. Two tandem overlapping copies of the HSP are present on each strand of the terminal repeat. The positons of TATA boxes at each end of the repeat are indicated. Transcription into the DIRS-1 sequences from the left and right terminal repeat produces the full length and E1 RNAs respectively. Both in Dictyostelium and in yeast transformants, transcription directed into internal DIRS-1 sequences initiates within the repeat near the junction with internal DIRS-1 sequences. In yeast transformants, transcription outward from the repeat into flanking sequences initiates just outside of the repeat (16).

DIRS-1 Transcription is Directed by the Heat Shock Promoters Contained in the Terminal Repeats

The DIRS-1 terminal repeat contains two tandem, overlapping copies of the Drosophila consensus heat shock promoter, "CT-GAA--TTC-AG", defined by Pelham and Beinz (3, 4,13,14,16). The portion of a terminal repeat containing these sequences is presented in Figure 7; the consensus promoter sequences are indicated by asterisks. Since the consensus heat shock promoter consists of a palindromic sequence, equivalent copies of the two promoters are contained in both strands of the DNA. Drosophila heat shock promoters and synthetic oligonucleotide promoters that confer heat shock inducibility on heterologous genes typically share 8 to 10 bp of homology with the consensus sequence. The DIRS-1 heat shock promoters match the consensus sequence at 9 or 10 of 10 positions. The evolutionary conservation of the sequence of the heat shock promoter is also suggested by numerous observations that heat shock genes and promoters can function in heterologous systems and be regulated according to the host cell type (13-16, see 13 for additional refs).

A consequence of the palindromic structure of the consensus heat shock promoter is the presence of an equivalent copy of the promoter in both strands of the DNA. This observation suggests that heát shock promoter function should be bidirectional. In order to test this prediction, Cappello et al. have transformed yeast with plasmids containing an isolated DIRS-1 terminal repeat. Previous experiments had shown that a larger subclone of DIRS-1, pB41-6, generates a specific heat shock inducible transcript in yeast and that the promoter responsible for the heat shock inducibility of this transcript in contained within the terminal repeat sequence (16). Plasmids containing either the isolated right or left terminal repeat are capable of directing heat shock inducible transcription of flanking vector sequences in both directions. These transcripts initiate within the flanking vector sequences near the junction with the terminal repeat (Cappello et al., in preparation). These results support the prediction that the heat shock promoter initiates transcription bidirectionally.

In summary, both terminal repeats of DIRS-1 contain all of the regulatory elements required to direct transcription of DIRS-1 RNAs. Transcription from the left repeat into the unique part of DIRS-1 produces the full length 4.5 kb RNA transcript. Transcription from the right repeat into internal sequences produces E1 RNA (Figure 1). In both cases transcription initiates within the terminal repeat, near the EcoRI site at the junction with the unique DIRS-1 sequences. When cloned into yeast, DIRS-1 terminal repeats are capable of directing transcription bidirectionally, and all sequences responsible for this transcription in yeast cells are contained within the DIRS-1 terminal repeat. This sequence contains two tandem, overlapping heat shock promoters, either of which is capable of directing transcription in both directions. In addition, the TATA box and oligo (dT) sequences used to position transcriptional initiation are found at each end of the terminal repeats on both sides of the bidirectional promoter. This organization is schematized in Figure 7.

The developmental cycle of Dictyostelium results in the production of spores in response to unfavorable environmental conditions. Unlike yeast, in which sporulation involves meiotic recombination, sporulation in Dictyostelium is a parasexual process. It is possible that the developmental induction of DIRS-1 transcription may result in synthesis of a transposase and in transpositional mobilization of the element. If transcription outward from the DIRS-1 terminal repeats into flanking sequences does occur in Dictyostelium, transposition of DIRS-1 could place novel sequences under developmental regulation. Such a process could be of evolutionary significance. Unfortuantely, due to the large number of copies of DIRS-1 and related sequences in the Dictyostelium genome we have not been able to determine whether or not bidirectional transcription occurs in Dictyostelium. At present the functions of the DIRS-1 RNAs are not known. As noted above, indirect evidence suggests that they may encode protein(s). Determining the functions of these RNAs may help to explain why transcription of this transposable element is developmentally regulated in Dictyostelium.

ACKNOWLEDGEMENTS

We would like to thank Naomi Cohen and Karl Handelsman for excellent technical assistance, as well as Miriam Boucher for expert secretarial assistance, and a willingness to work on Saturdays. SC was supported by a postdoctoral fellowship from the Natural Sciences and Engineering Research Council of Canada, JC was supported by a postdoctoral fellowship from the American Cancer Society.

REFERENCES

1. Chung S, Zuker C, Lodish HF (1983). A repetitive and apparently transposable DNA sequence in Dictyostelium discoideum associated with developmentally regulated RNAs. Nucl Acids Res 11:4853.
2. Zuker C, Lodish HF (1981). Repetitive DNA sequences cotranscribed with developmentally regulated Dictyo-

stelium discoideum mRNAs. Proc Nat Acad sci USA 78:5386.

3. Zuker C, Cappello J, Chisholm R, Lodish HF (1983). A repetitive Dictyostelium gene family that is induced during differentiation and by heat shock. Cell 34:997.
4. Zuker C, Cappello J, Lodish HF, George P, Chung S (1984). A Dictyostelium transposable element DIRS-1 has 350 bp inverted repeats which contain a heat shock promoter. Proc nat Acad Sci USA 81:(in press).
5. Rosen E, Sivertsen A, Firtel RA (1983). An unusual tranposon encoding heat shock inducible and developmentally regulated transcripts in Dictyostelium Cell 35:243.
6. Zuker C (1983) PhD Thesis. M.I.T.
7. Mangiarotti G, Zuker C, Chisholm R, Lodish HF (1983). different mRNAs have different nuclear transit times in Dictyostelium discoideum aggregates. Mol Cell Biol 3:1151.
8. Stanfield SW, Lengyel JA (1980). Small circular deoxyribonucleic acid of Drosophila melanogaster: homologous transcripts in the nucleus and cytoplasm. Biochemistry 19:3873.
9. Falkenthal S, Graham ML, Korn EL, Lengyel JA (1981). Transcription, processing and turnover from the Drosophila mobile genetic element copia. Dev Biol 92:294.
10. Shapiro JA (1982). "Mobile Genetic Elements" New York: Academic Press.
11. Cohen SM, Cappello J, Lodish HF (1984). Transcription of the Dictyostelium transposable element DIRS-1 manuscript submitted.
12. Kimmel AR, Firtel RA (1983). Sequence organization in Dictyostelium: unique structure at the 5' ends of protein coding genes. Nucl Acids Res 11:541.
13. Pelham HRB (1982). A regulatory upstream promoter element in the Drosophila HSP 70 heat hsock gene. Cell 30:517.
14. Pelham HRB, Beinz M (1982). A synthetic heat-shock promoter element confers heat-inducibility on the Herpes simplex virus thymidine kinase gene. EMBO J 1:1473.
15. Lis J, Costlow N, deBanzie J, Knipple D, O'Connor D, Sinclair L (1982). Transcription and chromatin structure of Drosophila heat shock genes in yeast. "Heat Shock From Bacteria To Man" Cold Spring Harbor: Cold Spring Harbor Laboratory, p. 57, Schlesinger MJ,

Ashburner M, Tissiers A (eds.).

16. Cappello J, Zuker C, Lodish HF (1984). Repetitive Dictyostelium heat shock promoter functions in Saccharomyces cerevisciae. Mol Cell Biol 4:591.

IV. GENE EXPRESSION IN HEMATOPOIETIC CELL LINEAGES

Molecular Biology of Development, pages 511–525

TOWARD A MOLECULAR BASIS FOR GROWTH CONTROL IN T LYMPHOCYTE DEVELOPMENT[1]

Ellen Rothenberg, Barry Caplan,
and Rochelle D. Sailor

Division of Biology, California Institute of Technology
Pasadena, California 91125 USA

ABSTRACT Among the most important immunological roles of helper T lymphocytes is the production of interleukin-2, a polypeptide hormone that controls proliferation of all types of T cells. We have examined the helper T-cell precursors in the thymus to determine when competence to secrete this hormone first appears. The results suggest that thymocytes can acquire this capability before complete maturity, while they are still in a proliferating stage. The first stage of analysis of these cells at the nucleic acid level has been the utilization of a oligonucleotide probe complementary to human interleukin-2 mRNA, which cross-hybridizes specifically with the corresponding mouse mRNA.

INTRODUCTION TO THE SYSTEM

Growth Regulators and Lymphocyte Development

Lymphocytes, the effector cells of the immune system, are progeny of the same pool of hematopoietic stem cells that gives rise to erythrocytes, macrophages, and various types of granulocytes (1). Like other hematopoietic cells, lymphocytes differentiate asynchronously and continuously throughout life. More so than the others, however, they retain the capacity to proliferate in their mature "effector" state. In fact, extensive

[1]This work was supported by two grants from the National Institutes of Health, AI 19752 and CA 34181, and by a fellowship to B. C. from the the Alberta Heritage Foundation for Medical Research.

proliferation under particular circumstances is a crucial part of the mature lymphocytic response to foreign antigens. The last few years have brought considerable understanding of the polypeptide hormones that control the proliferation of lymphocytes, both in vitro and in vivo. Lymphocytes respond to a unique class of growth hormones which act only on these cells, and they fail to respond to hormones that act on other hematopoietic or nonhematopoietic cells (2-4). The growth control mechanisms used by lymphocytes must therefore develop exclusively as an outcome of differentiation in a specific lymphoid cell lineage.

We have been interested in the developmental events through which T lymphocytes, the killer and regulatory cells of the immune system, commit themselves to growth regulation by the T-cell growth factor interleukin-2 (IL2). T-cell precursors migrate to one specialized organ, the thymus gland, where they mature before they are exported to peripheral lymphoid organs (5). From late fetal life to puberty, the thymus is a convenient source of highly enriched T-lineage cells in various stages of maturation (5). Animals deprived of a thymus from birth are severely deficient in functional T cells, although they retain precursors that can differentiate normally in a thymus graft (6). Neither the functional properties nor the cell-surface markers of mature T cells are evident in prethymic T-cell precursors (7). Cells being exported by the thymus, on the other hand, are phenotypically and functionally mature as they emerge (8). Thus, the thymus is the site where the ultimate differentiation events in T cell development take place. The growth signals that act on the cells as they undergo these events are still unknown.

Mechanism of T-cell Growth Control

The role of IL2 in controlling proliferation of mature T cells is relatively well understood (2). It acts as a growth hormone for all subclasses of normal T cells but not for any other known cell type (2,3). The only source of IL2 is the specialized "helper" subset of the T cells themselves. Therefore, developing T cells need to mature not only to be able to respond to IL2, but also, in one sublineage, to be able to produce it.

In mature cells, both secretion of IL2 and responsiveness to it are tightly regulated components of the immune response. As shown in Figure 1, no IL2 is made except when a helper cell is triggered by contact with its particular antigen on the surface of an antigen-presenting macrophage (9-11). The IL2 that is secreted can then act on other responsive T cells in the immediate vicinity. Responsiveness is also inducible, rather than constitutive: T cells do not express receptors for IL2 except when stimulated with their own specific target antigens (12, 13). When stimulated, however, they can bind IL2 produced by any helper T cell, whether the antigens triggering the two cells were the same or different. It is presumed that the helper T cell can divide in response to its own IL2 (14, 15).

Recent work has shown that IL2 is a single glycoprotein. A sensitive biosassay for IL2 has been developed, using cloned tissue culture lines of killer T cells that retain absolute dependence upon IL2 and become responsive to it constitutively (Figure 1, bottom line) (16, 17). Using this bioassay, IL2 has been purified to homogeneity (18, 19). Proof that this polypeptide is responsible for the IL2 growth factor activity has come from the cloning of human IL2 and the production of biologically active IL2 in bacteria (20, 21). Thus, a key aspect of helper T cell function is the appropriately regulated expression of a single gene product.

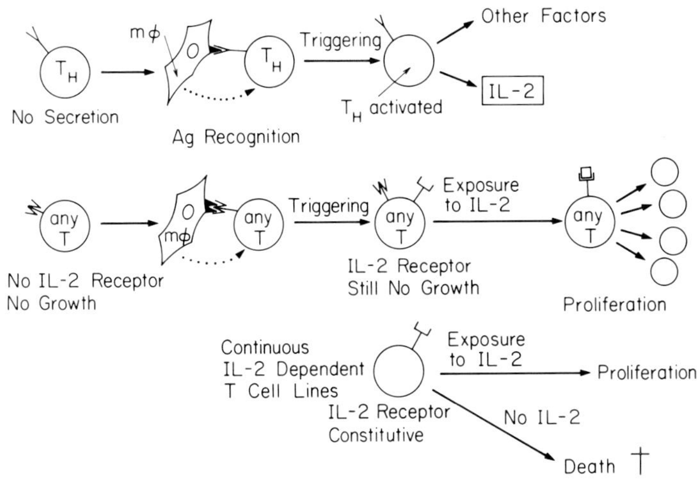

FIGURE 1. IL2-mediated control of T cell growth. Interactions leading to production of IL2 (top line) and responsiveness to IL2 (middle line) are diagrammed. To make IL2, a helper T (T_H) cell may require additional hormonal stimulation from the macrophage (MΦ). For other details, see text.

Developmental Background

In this study, we have focused on the maturation of the helper T cells, i.e., the development of a regulated supply of IL2 for the immune system. We wish to determine at what stage in the thymus pre-helper T cells could be induced to secrete IL2. The results are also germane to several general questions about the way the thymus functions. One is the extent to which cells in the immature precursor population in the thymus have already diverged into separate lineages. Another is the relationship between intrathymic cell division and differentiation.

The thymus is a site of both rapid cell division and massive cell death (Figure 2) (5). The dividing cells include the precursor cells for other thymocytes and for all T lymphocytes. The majority of their descendants, peripheral thymocytes that represent 30-40% of all lymphoid cells in the body, are destined to die within 3-4 days without ever leaving the thymus (22, 23). There

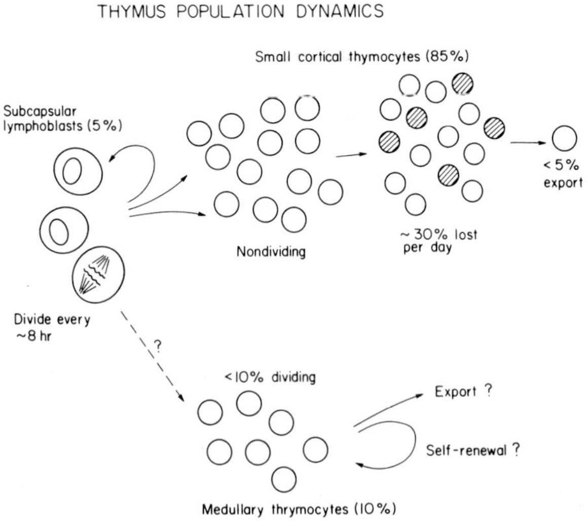

FIGURE 2. Cell lineages in the thymus. Top line, cortical thymocytes. Bottom line, medullary thymocytes. Functional T cells for export may physically originate in either domain, but their phenotypes resemble those of cells that remain in the thymic medulla, not those of most cortical cells.

are two anatomical domains in the thymus, the cortex and the medulla, and the population dynamics are different in each. The cortex contains the majority of the lymphocytes, most of the proliferating cells (lymphoblasts), and most of the nonviable cells, but it may not be the source of competent peripheral T cells (5).

There has been little solid evidence of nascent function in any immature thymocyte population. Until now, helper and killer T cell activity has only been reported in the minority of thymocytes that is indistinguishable from mature T cells (24, 25). In these reports, however, the lymphoblasts were overlooked as a source of truly immature cells. Accordingly, we have reopened the question of thymocyte functional competence, paying close attention to distinctions between lymphoblasts and their postmitotic descendants. As evidence of early helper cell activity, we ask simply how many classes of cells in the thymus have developed as helpers to the point where they are inducible to secrete IL2. By this narrow criterion, we find clear evidence of helper activity in a strictly limited subset of the "immature" dividing cells.

MATERIALS AND METHODS

Animals, cell preparation, and cell culture

C57BL/6 and C57BL/6-$\underline{Tla}^{a}$ mice, 3-6 week old, were used. Thymocyte preparation, fractionation, and cell culture procedures were as described by our laboratory previously (26-28).

IL2 Production Assay

The procedure has been described elsewhere in detail (28). Briefly, thymocyte populations were cultured in the presence of 3-10 μg/ml of Concanavalin A (Con A) and 10-25 ng/ml of 12-0-tetradecanoyl phorbol-13-acetate (TPA) for 16-24 hr. The cell-free supernatants were collected and assayed on MTL.2.8.2 indicator cells (17) in a 24 hr T cell growth factor assay (29).

RNA Extraction and Oligonucleotide Hybridization

Total RNA was extracted from human and mouse T-cell lymphoma lines by a modification of the method of Chirgwin et al. (30). To induce IL2 message expression, human

JURKAT cells (31) were stimulated with 10 ng/ml TPA and 0.2% phytohemagglutinin (GIBCO) and harvested for RNA extraction after 6 hr. Stimulated murine EL4.E1 cells (32) were collected after 10 hr of incubation with TPA. Polyadenylated RNA was isolated by two cycles of oligo(dT)-cellulose chromatography and fractionated by electrophoresis in 1.7% or 2.0% (w/v) agarose gels with 2.2 M formaldehyde (33). The RNA was then blotted on nylon filters (Gene Screen, New England Nuclear) and hybridized under the conditions recommended by the manufacturer, except at 37°C instead of 42°C. Three 30-mer oligonucleotides were synthesized for us by Dr. Susanna Horvath in the Caltech Microchemical Facility (Lee Hood, director). These probes will be described fully in another publication (Caplan, Sailor and Rothenberg, in preparation). They were end-labeled with T4 polynucleotide kinase for use as hybridization probes.

RESULTS

We have used two techniques for preparative separations of different classes of thymocytes. The 10-15% of thymocytes that are dividing (lymphoblasts) are much larger than the majority of the cells, which are nondividing, so that centrifugal elutriation can be used to separate them on the basis of size (26). Cells in different anatomical domains of the thymus tend to bear different cell surface carbohydrates, so that cells from the cortex and cells from the medulla can be separated by their differential binding of lectins. One lectin which binds much better to cortical cells than to medullary cells is peanut agglutinin (PNA) (34). The "PNA-negative" medullary cells, both dividing and nondividing, account for only 10-15% of the cells in the thymus. By combining size fractionation with PNA binding, we can resolve at least four populations of cells, with proliferating cells separated from postmitotic cells and both classes separated into presumptive "cortical" and "medullary" cell types (26, 28). Most important is the fact that each population has a unique biosynthetic phenotype, with synthesis of some polypeptides higher than in any other class and synthesis of others much lower (26, 27). Thus, whether or not each population is homogeneous, the differentiation states of cells that predominate in the different fractions are significantly distinct at the time of isolation.

To look for functional differences between these cell types, we have used a short-term assay for their ability to produce IL2. The assay is diagrammed in Figure 3. Key points include the following. First, the measurement of IL2 itself is sensitive, reproducible and specific. The indicator cells respond to picomolar concentrations of the hormone, and only IL2 enables them to survive--much less proliferate (17). Second, the thymocytes are stimulated by a lectin and a phorbol ester (Con A

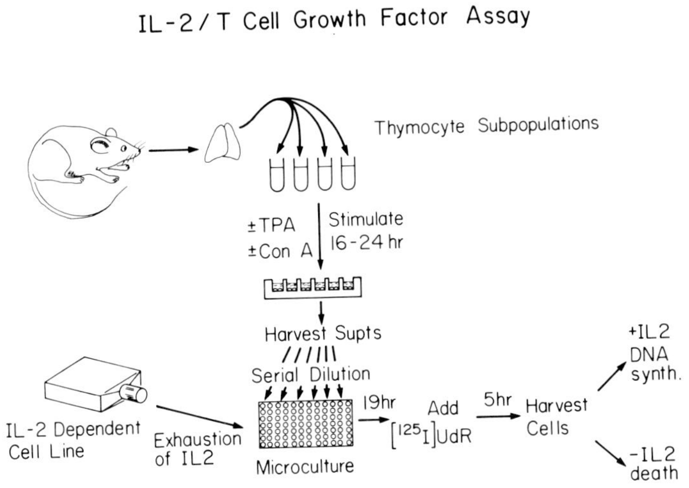

FIGURE 3. IL2 production assay. For details, see text.

and TPA, respectively) that act on all T cells regardless of their antigen-binding specificities. The Con A functions as a proxy for antigen and the phorbol ester both potentiates the response and makes it essentially independent of nonlymphoid accessory cells. Therefore, these activators obviate the need for cell-cell contact, which otherwise introduces a complicating variable (28). In the absence of stimulation, thymocytes do not produce detectable levels of IL2 (Figure 4). Third, we measure IL2 secreted by thymocytes in their first 24 hr in culture or less. This is not enough time for proliferation or extensive cell death, so that the cells scoring in the assay correspond to those that were initially isolated. Altogether, the assay system enables us to test freshly-isolated thymocytes directly for one component of T-cell function that is a simple, autonomous cellular response.

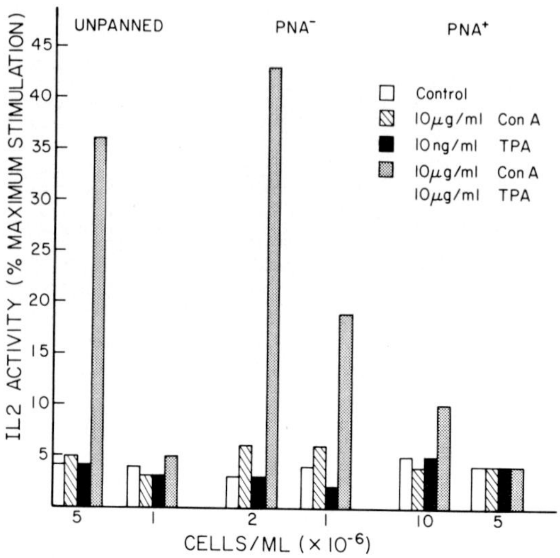

FIGURE 4. Response of thymocytes to Con A and TPA. Thymocytes were separated into PNA^+ and PNA^- fractions by panning. The indicated number of cells from each fraction were then stimulated with medium ± Con A ± TPA. Note that 10 x 10^6 PNA^+ cells make less IL2 than 1 x 10^6 PNA^- cells.

As shown in Figure 4, not all thymocytes are capable of making large amounts of IL2. PNA-binding (PNA^+) cortical cells, the vast majority of cells in the thymus, are virtually inert; any IL2 produced by this fraction is consistent with a low level of PNA^- cell contamination. The PNA^- population is correspondingly enriched for IL2-producing cells. The PNA^+ cells remain inert in mixing experiments with PNA^- cells, neither revealing a latent ability to make IL2 nor interfering with production by the PNA^- cells. Thus far, the results agree with those of other laboratories. The common interpretation would be that

PNA^-, predominantly medullary thymocytes are mature, and that IL2 is only secreted by cells that have finished their intrathymic differentiation (24, 25).

The situation is in fact more complex. When dividing and nondividing cell populations are assayed, by far the most prolific in IL2 production are the populations of dividing cells (Figure 5). The dividing cells in the normal thymus, however, are largely or

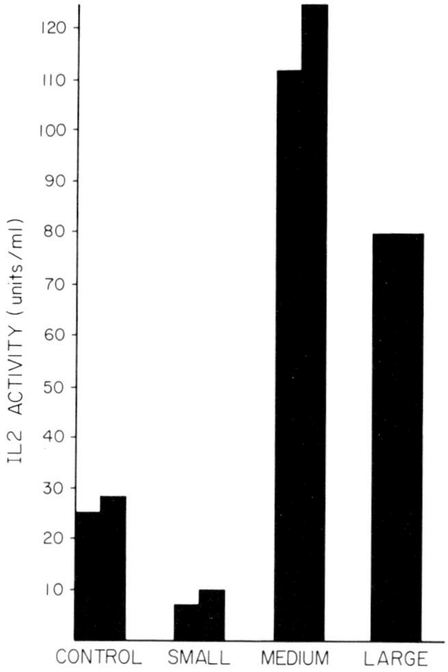

FIGURE 5. IL2 produced by size-fractionated thymocytes. Thymocytes were fractionated by elutriation. The small cells represented 76% of the total, medium cells 22%, and large cells 2%. Equal numbers of cells from each fraction were stimulated with Con A and TPA at 10×10^6 cell/ml, and their supernatants were assayed for IL2. One unit per ml = concentration of IL2 resulting in 30% maximal stimulation.

exclusively a precursor population. Measurements on populations fractionated by size as well as by PNA binding show that even among PNA$^-$ cells, the dividing cell class makes more IL2 per cell than the nondividing cell class (28). Thus, even within the "medullary" population, cells can efficiently carry out the mature T-cell function of IL2 secretion before they have finished their intrathymic proliferation.

To characterize the IL2 producers further, we will need assays that can detect them in situ and assays that can determine the frequency of IL2-producing cells in each population. For this reason, we have begun work on the cloning of mouse IL2 cDNA. This has been a difficult enterprise because of the unexpectedly poor homology between mouse IL2 sequences and the widely-available human IL2 clone. Our work is still in progress, but we have identified a region of the human IL2 gene which is sufficiently homologous to mouse sequences to allow specific cross-hybridization.

Three oligonucleotides of 30 residues each were synthesized for us by Dr. Susanna Horvath of the Caltech Microchemical Facility. They were chosen to be complementary to sequences for the 5' (IL2.1), middle (IL2.2), and 3' (IL2.3) portions of the IL2 gene (20). Their patterns of hybridization to T-cell RNAs are shown in Figure 6. In each case, the probe hybridized well with the homologous RNA in human JURKAT cells stimulated to produce IL2. The probes behaved quite differently with RNAs from the mouse line, EL4.E1. The IL2.1 probe failed to hybridize detectably with up to 20 μg of polyadenylated EL4.E1 RNA whether or not the cells had been stimulated to secrete IL2. The IL2.2 probe, on the other hand, hybridized to unstimulated as well as to stimulated cell RNA. Only IL2.3 showed the desired specificity: it hybridized strongly with RNA from induced EL4.E1 cells but not with RNA from uninduced cells. Furthermore, the size of the RNA detected by IL2.3 is consistent with that estimated for mouse IL2 mRNA by translation of size-fractionated RNA to give biologically active IL2 (35, 36). We conclude that IL2.3 specifically recognizes the mouse IL2 mRNA.

DISCUSSION

The passage of lymphocytes through the thymus involves several developmentally important events in an unknown order. T-cell precursors must home to one or another of the different intrathymic microenvironments (cortex vs. medulla). They proliferate generally and undergo selective clonal amplification. Furthermore, they must differentiate to acquire mature T-cell

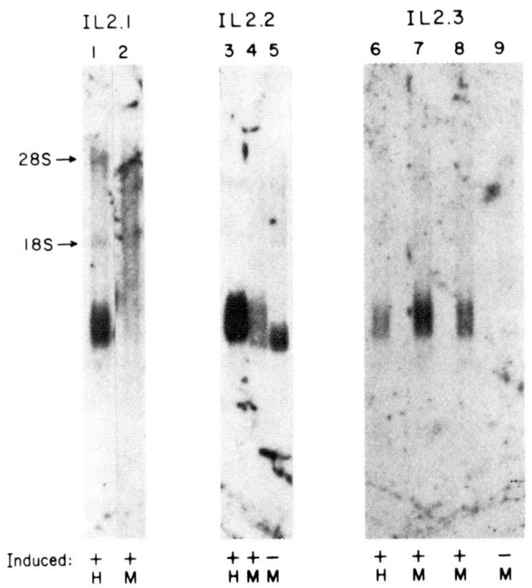

FIGURE 6. Hybridization of mouse and human T-cell RNAs with IL2 probes. Oligonucleotide probes were hybridized to RNA from human (H) cell or mouse (M) T cells as described in the text. (+), after induction. (-), without induction. Equal amounts of polyadenylated RNA from induced and uninduced cells were compared.

characteristics. Our work concerns the relation between the last two events--proliferation and differentiation--with a focus on cells of the helper T lineage.

The most unexpected result is that proliferating as well as postmitotic thymocytes can be induced to secrete IL2, with the proliferating cells the more efficient. All these IL2 producers are apparently within a single lineage as defined by their low PNA binding and by their high expression of the surface glycoprotein Lyt-1 (28). The proliferating cells cannot secrete

IL2 without stimulation and their dose-response optima for Con A and TPA are indistinguishable from those of bona fide mature T cells (unpublished results). At this crude level, at least, their expression of IL2 is inducible as in mature cells. These results argue against a strict linkage between cell division and differentiation. Within the helper lineage, the functionally vital capability to respond to a challenge by making IL2 is well-established in cells that are still being driven to divide in the thymus. This is the first demonstration of function in proliferating thymocytes, but we anticipate that it will not be unique to cells of the helper lineage.

The dividing cells that make IL2 in our system are a distinct minority of the proliferating thymocytes, and they share a characteristic surface phenotype with mature helper T cells. This emphasizes the heterogeneity within the lymphoblast population and suggests that functional properties are established in the helper lineage at least as early as the first expression of typical "helper" surface markers. Depending on the precursor-product relationships among the proliferating cell types, this could mean that the majority of all thymic lymphoblasts, which are PNA^+, are already excluded from giving rise to viable helper-lineage progeny.

We have taken the first steps to study the developmental status of the lymphoblasts that can be induced to make IL2. Work is currently underway to isolate cDNA clones encoding mouse IL2. With these probes, we hope to examine IL2 production in individual cells by in situ hybridization. We can thus look for the earliest appearance of inducible IL2 producers in cell suspensions from the fetal thymus, to determine whether they appear early relative to other classes of thymic lymphoblasts. The Con A and TPA stimulation should enable us to override any immaturity of the nonlymphoid accessory cells in the developing thymus, so that we could unveil any functional reactivity of the pre-T cells themselves as early as possible.

Perhaps the most interesting application of in situ hybridization will be to reexamine the adult thymus for evidence of natural IL2 message expression. With proliferation dissociated from differentiation, intrathymic cell division is more enigmatic than ever. However, as diagrammed in Figure 1, it can only have an "immunological" trigger--i.e., recognition of some antigen--if it is mediated by IL2. At present, it is unknown whether any proliferating thymocytes express IL2 receptors, or whether any cells in the thymus actually secrete IL2 in vivo. By determining whether any IL2-producing cells are present in different lymphoblast domains of the thymus, we can infer when--and in which lineages--developing thymocytes adopt the growth control mechanisms of mature T cells.

ACKNOWLEDGEMENTS

The authors are indebted to Susanna Horvath and Lee Hood for providing the oligonucleotide probes. We also wish to thank Bernard Malissen for helpful advice and stimulating discussions.

REFERENCES

1. Abramson S, Miller RG, Phillips RA (1977). The identification in adult bone marrow of pluripotent and restricted stem cells of the myeloid and lymphoid systems. J Exp Med 145:1567.
2. Smith KA (1980). T-cell growth factor. Immunol Rev 51:337.
3. Howard M, Farrar J, Hilfiker M, Johnson B, Takatsu K, Hamaoka T, Paul WE (1982). Identification of a T-cell derived B-cell growth factor distinct from interleukin-2. J Exp Med 155:914.
4. Burgess AW, Metcalf D (1980). The nature and action of granulocyte-macrophage colony stimulating factors. Blood 56:947.
5. Scollay R (1983). Intrathymic events in the differentiation of T lymphocytes: a continuing enigma. Immunol Today 4:282.
6. Lepault F, Weissman IL (1981). An in vivo assay for thymus-homing bone marrow cells. Nature 293:151.
7. Kindred B (1979). Nude mice in immunology. Prog Allergy 26:137.
8. Scollay R, Chen W.-F., Shortman K (1984). The functional capabilities of cells leaving the thymus. J Immunol 132:25.
9. Gillis S, Scheid M, Watson J (1980). Biochemical and biologic characterization of lymphocyte regulatory molecules. III. The isolation and phenotypic characterization of interleukin-2 producing T-cell lymphomas. J Immunol 125:2570.
10. Glasebrook AL, Fitch FW (1980). Alloreactive cloned T-cell lines. I. Interactions between cloned amplifier and cytolytic T-cell lines. J Exp Med 151:876.

11. Kappler JW, Skidmore B, White J, Marrack P. (1981). Antigen-inducible H-2-restricted, interleukin-2-producing T-cell hybridomas. Lack of independent antigen and H-2 recognition. J Exp Med 153:1198.
12. Robb RJ, Munck A, Smith KA (1981). T-cell growth factor receptors: quantitation, specificity and biological relevance. J Exp Med 154:1455.
13. Cantrell DA, Smith KA (1983). Transient expression of interleukin-2 receptors: consequences for T-cell growth. J Exp Med 158:1895.
14. Schreier MH, Iscove NN, Tees R, Aarden L, von Boehmer H. (1980). Clones of killer and helper T-cells: growth requirements, specificity and retention of function in long-term culture. Immunol Rev 51:315.
15. Gootenberg JE, Ruscetti FW, Mier JW, Gazdar A, Gallo RC (1981). Human cutaneous T-cell lymphoma and leukemia cell lines produce and respond to T-cell growth factor. J Exp Med 154:1403.
16. Gillis S, Smith KA (1977). Long-term culture of tumor-specific cytotoxic T cells. Nature 268:154.
17. Bleackley RC, Havele C, Paetkau V (1982). Cellular and molecular properties of an antigen-specific cytotoxic T-lymphocyte line. J Immunol 128:758.
18. Riendeau D, Harnish DG, Bleackley RC, Paetkau V (1983). Purification of mouse interleukin-2 to apparent homogeneity. J Biol Chem 258:12114.
19. Stern AS, Pan Y-CE, Urdal DL, Mochizuki DY, DeChiara S, Blacker R, Wideman J, Gillis S (1984). Purification to homogeneity and partial characterization of interleukin-2 from a human T-cell leukemia. Proc Natl Acad Sci USA 81:871.
20. Taniguchi T, Matsui H, Fujita T, Takaoka C, Kashima N, Yoshimoto R, Hamuro J (1983). Structure and expression of a cloned cDNA for human interleukin-2. Nature 302:305.
21. Rosenberg SA, Grimm EA, McGrogan M, Doyle M, Kawasaki E, Koths K, Mark DF (1984). Biological activity of recombinant human interleukin-2 produced in Escherichia coli. Science 223:1412.
22. Bryant BJ (1972). Renewal and fate in the mammalian thymus. Mechanisms and inferences of thymocytokinetics. Eur J Immunol 2:38.
23. McPhee D, Pye J, Shortman K (1979). The differentiation of T lymphocytes. Thymus 1:151.

24. Chen W-F, Scollay R, Shortman K (1982). The functional capacity of thymus subpopulations: limit-dilution analysis of all precursors of cytotoxic lymphocytes and of all T cells capable of proliferation in subpopulations separated by the use of peanut agglutinin. J Immunol 129:18.
25. Ceredig R, Glasebrook AL, MacDonald HR (1982). Phenotypic and functional properties of murine thymocytes. I. Precursors of cytolytic T lymphocytes and interleukin-2-producing cells are all contained within a subpopulation of "mature" thymocytes as analyzed by monoclonal antibodies and flow microfluorometry. J Exp Med 155:358.
26. Rothenberg E (1982). A specific biosynthetic marker for immature thymic lymphoblasts. Active synthesis of thymus leukemia antigen restricted to proliferating cells. J Exp Med 155:140.
27. Rothenberg E, Triglia D (1983). Clonal proliferation unlinked to terminal deoxynucleotidyl transferase synthesis in thymocytes of young mice. J Immunol 130:1627-1633.
28. Caplan B, Rothenberg E (1984). High-level secretion of interleukin-2 by a subset of proliferating thymic lymphoblasts. J Immunol., in press.
29. Gillis S, Ferm MM, Ou W, Smith KA (1978). T-cell growth factor: parameters of production and quantitative microassay for activity. J Immunol 120:2027.
30. Chirgwin JM, Przybyla AE, MacDonald RJ, Rutter WJ (1979). Isolation of biologically active ribonucleic acid from sources enriched in ribonuclease. Biochemistry 18:5294.
31. Hansen JA, Martin PJ, Nowinski RC (1980). Monoclonal antibodies identifying a novel T-cell antigen and Ia antigens of human lymphocytes. Immunogenetics 10:247.
32. Farrar JJ, Fuller-Farrar J, Simon PL, Hilfiker ML, Stadler BM, Farrar WL (1980a). Thymoma production of T-cell growth factor (interleukin-2). J Immunol 125:2555.
33. Gonda, TJ, Sheiness DK, Bishop JM (1982). Transcripts from the cellular homologs of retroviral oncogenes: distribution among chicken tissues. Mol Cell Biol 2:617.
34. Reisner Y, Linker-Israeli M, Sharon N (1976). Separation of mouse thymocytes into two subpopulations by the use of peanut aggluntinin. Cell Immunol 25:129.
35. Bleackley RC, Caplan B, Havele C, Ritzel RG, Mosmann TR, Farrar JJ, Paetkau V (1981). Translation of lymphocyte mRNA into biologically-active interleukin-2 in oocytes. J Immunol 127:2432.
36. Efrat S, Pilo S, Kaempfer R (1982). Kinetics of induction and molecular size of mRNAs encoding human interleukin-2 and γ-interferon. Nature 297:236.

Molecular Biology of Development, pages 527–536

ORGANIZATION AND EXPRESSION OF THE MAJOR HISTOCOMPATIBILITY COMPLEX OF THE C57BL/10 MOUSE

James J. Devlin, Claire T. Wake, Hamish Allen, Georg Widera, Andrew L. Mellor, Karen Fahrner, Elizabeth H. Weiss, and Richard A. Flavell

Biogen Research Corp. Cambridge, MA 02142

ABSTRACT We have cloned 900 kilobases of DNA from the major histocompatibility complex of the C57BL/10 mouse (H-2^b), thereby linking the H-2K and I-A regions as well as the H-2D and Qa2,3 regions. Our results suggest that D^b is the allele of L^d and that the H-2K region was generated by the translocation of a pair of genes from the Qa2,3 region. By exchanging exons of the K^b and D^b genes we have demonstrated the importance of interaction between the first and second external domains of class I antigens in cytotoxic T cell recognition. We have also identified 6 class II β-chain sequences as well as provided evidence for gene conversions between class II genes.

INTRODUCTION

Our laboratory has been studying the class I and II genes in the major histocompatibility complex (MHC) of the C57BL/10 (B10) mouse (Fig. 1). The class I genes at the H-2K and H-2D loci encode 45,000 molecular (mw) polymorphic glycoproteins that are expressed as heterodimers, together with β_2-microglobulin, on the surface of virtually all cells. Cytotoxic T cells (CTL) only recognize viral glycoproteins when they are associated with these class I proteins on the cell surface. Several class I genes in the Tla complex encode similar, but less polymorphic molecules, that are only expressed on certain populations of lymphocytes.

The two well characterized class II glycoproteins, A and E, are heterodimers consisting of a 35,000 mw β-chain and a 29,000 mw α-chain. The I-A region of the MHC encodes the Aβ, Aα, and Eβ chains and the I-E region encodes the Eα chain. Expressed primarily on the surface of B lymphocytes and macrophages, these class II molecules, together with foreign antigen, activate the helper T cells required to induce an immune response.

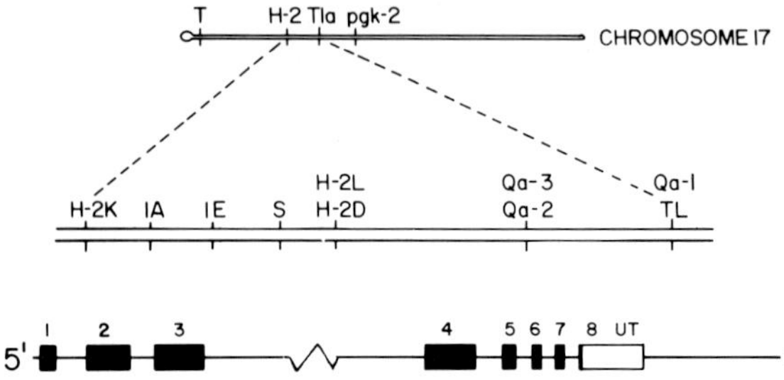

Figure 1. The murine MHC. The H-2 complex is on the left and the Tla complex, containing the Qa2,3 and TL regions, is on the right. Below the MHC map the exon structure of a typical class I gene is shown. The 1st exon encodes the leader peptide; exons 2-4 encode the 1st-3rd external domains, respectively; exon 5 encodes the transmembrane domain; exon 6-8 encode the cyloplasmic domain.

RESULTS AND DISCUSSION

Organization of Class I Genes

By screening genomic libraries with class I gene probes and by subsequent chromosomal walking experiments, we have isolated over 140 cosmids containing 26 different class I genes and organized them into three clusters. The cluster of 13 TL region genes is shown in Figure 2. The genes are defined by virtue of their hybridization to class I gene probes (1); it is therefore not known whether all genes are functional. This cluster was mapped to the TL region, as defined by the B6.K2 mouse, by restriction fragment length polymorphisms.

The second cluster, containing 11 class I genes,

links the H-2D and the Qa2,3 regions (Fig.3). The D^b gene has been expressed in L cells and identified by antibody binding and CTL assays (2,3). It was mapped to the H-2D region using restriction fragment length polymorphisims. The Q1 gene has been mapped by the same method to the Qa2,3 region, as defined by the B6.K1 mouse. Thus, the H-2D and Qa2,3 regions are separated by only about 50 kilobases of DNA.

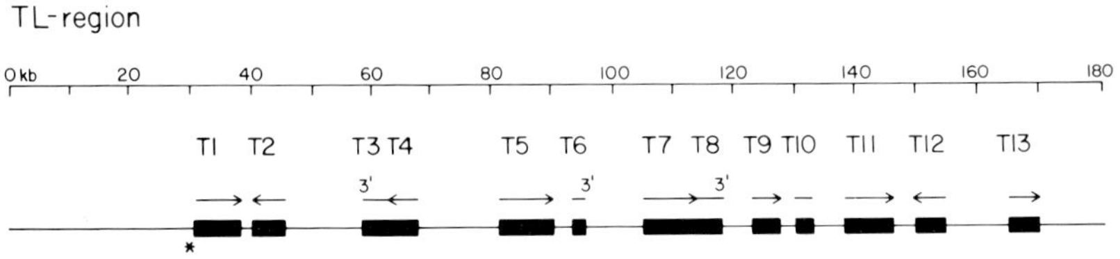

Figure 2. The TL region class I genes. The arrows point to the 3' end of the genes. Genes T3, T6, and T8 do not hybridize to probes for exons 1-3. The * indicates the site of the restriction fragment used to map this cluster.

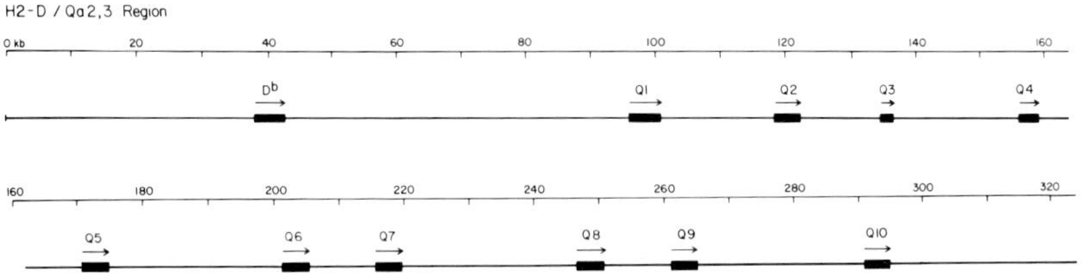

Figure 3. The H-2D and Qa2,3 region genes.

Comparison of the D^b and L^d genes indicates that they, and not the D^b and D^d genes, are alleles. The restriction maps flanking the 3' side of the D^b and L^d genes are quite similar while the 3' map of D^d differs significantly from that of D^b (Fig. 4). In addition, we found that an L^d specific probe (3) hybridizes to an 11 kb KpnI fragment just 5' of the D^b (Fig. 4) gene and to a single band of that size in a KpnI B10 genomic blot. Finally, partial amino acid sequence comparisons indicate that D^b and L^d are 94% homologous while D^b and D^d are only 85% homologous (5).

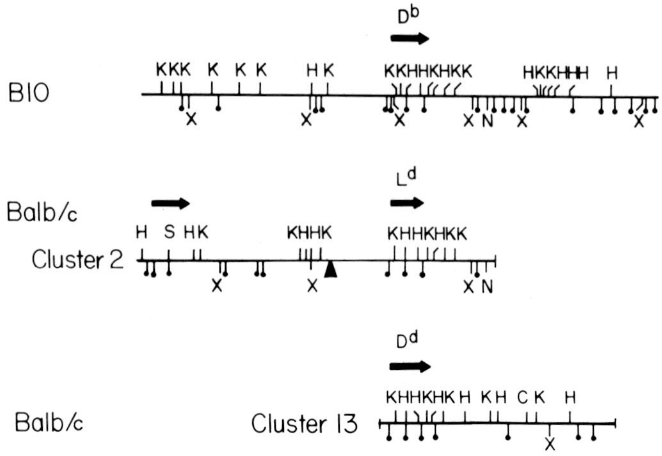

Figure 4. Comparison of restriction maps surrounding the D^b gene (B10 mouse) and the L^d and D^d genes (Balb/c mouse (4)). Genes are marked by arrows with the points on the 3' end. The triangle in cluster 2 indicates the location of the L^d specific probe described in the text. K=Kpnl, H=HpaI, X=XhoI, C=ClaI, S=SacII, N=NarI, lines with dots on them = BamHI.

DNA sequence analysis of the Q10 gene indicates that it may have donated sequence information to create the H-$2K^{bm1}$ mutation by a gene conversion involving 13 to 51 nucleotides (13,14). This gene is also homologous to a liver specific cDNA clone containing a termination codon in the transmembrane exon that was reported by Cosman et al.(6). Thus Q10 may encode a secreted class I protein.

The Q10 end of this cluster contains repeated units of homology suggesting that this region evolved by the duplication of gene pairs (Fig. 5). The brackets over the figure indicate that the unit of duplication may have been either Q8- and Q9-like genes or Q7- and Q8-like genes. The homology of these regions is also demonstrated by the cross hybridization of the DNA flanking these genes on their 3' and 5' sides (Fig. 5). These flanking probes do not hybridize to the corresponding regions of either the Q1 to Q3 genes or the TL region genes.

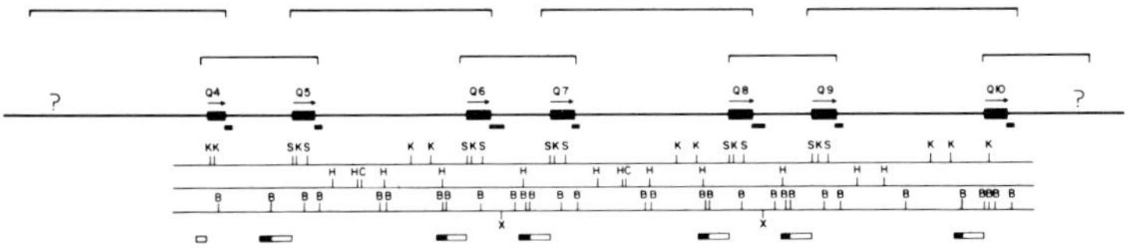

Figure 5. Regions of homology in the Qa2,3 region. Boxes above the restriction map indicate hybridization of a 3' flanking probe. Boxes below the map indicate hybridization of two different 5' flanking probes. B=BamHI, see figure 4 for other enzymes.

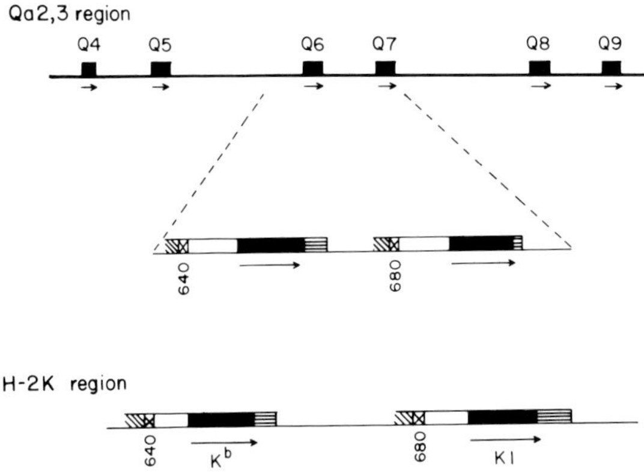

Figure 6. Hybridization of 3' and 5' flanking region probes to Qa2,3 region and H-2K region genes. Genes are represented by black boxes; the other boxes indicate the hybridization of different flanking probes.

Our final pair of genes has been mapped to the H-2K region by restriction fragment length polymorphisms, and one of these genes has been shown to encode the K^b protein (1,7). The flanking regions of this pair of genes show striking similarity to the Q6 and Q7 as well as to the Q8 and Q9 gene pairs (Fig. 6) As an example, one 5' flanking probe hybridizes to a 640-base-pair BamHI fragment flanking the 5' gene of each pair and to a 680-base-pair BamHI frangment flanking the 3' gene of each pair. Because of these homologies, we suggest that the H-2K region was generated by the translocation of a Q8- and Q9-like gene pair from the Qa2,3 region.

Functional Domains of H-2 Class I Proteins

To identify the regions of class I proteins involved in CTL recognition, four hybrid genes were constructed by exchanging exons between the K^b and D^b genes. These class I genes were expressed in L cells and recognition of the hybrid K^b/D^b antigens by CTL and monoclonal antibodies (mAb) specific for either D^b or D^b was investigated. The pattern of CTL and mAb recognition obtained (Fig. 7) showed that the majority of CTL and mAb recognition sites are located on domains 1 and 2. Furthermore, these recognition sites are not influenced by interaction of domains 1 and 2 with polymorphic regions of domain 3. In contrast, interaction betwen domains 1 and 2 alters these recognition sites. The alteration of CTL recognition sites by interaction between domains 1 and 2 suggests that recognition sites may be formed by amino acids from both domains and that the conformation of amino acids at a recognition site may be altered by interactions between domains 1 and 2. These two features may allow the conformation of CTL recognition sites on H-2 class I proteins to be sensitive to alteration by interaction of either domain 1 or 2 with viral antigens.

Organization of Class II Genes

By cosmid cloning we have identified six class II β genes or related sequences and two class II α genes. Maps of the three cosmid clusters containing these class II sequences are shown in Fig. 8. The central cluster contains the well-characterized genes Aβ1, Aα, Eβ1, and Eα, and the homologous sequences Aβ2, and Eβ2. The overall structure of this portion of the $H\text{-}2^b$ I region is quite similar to the published I region maps of the $H\text{-}2^d$ and $H\text{-}2^k$ haplotypes (8). We have also identified two additional β sequences. The Aβ3 sequence was identified by screening a B10 cosmid library with an Aβ1 probe. Aβ3 hybridizes to an Aβ1 probe containing the β2 and transmembrane exons at moderate stringencies, and to the β1 exon of Aβ1 only at very low stringencies. Southern blot hybridizations with an Aβ3 probe indicate that Aβ3 is

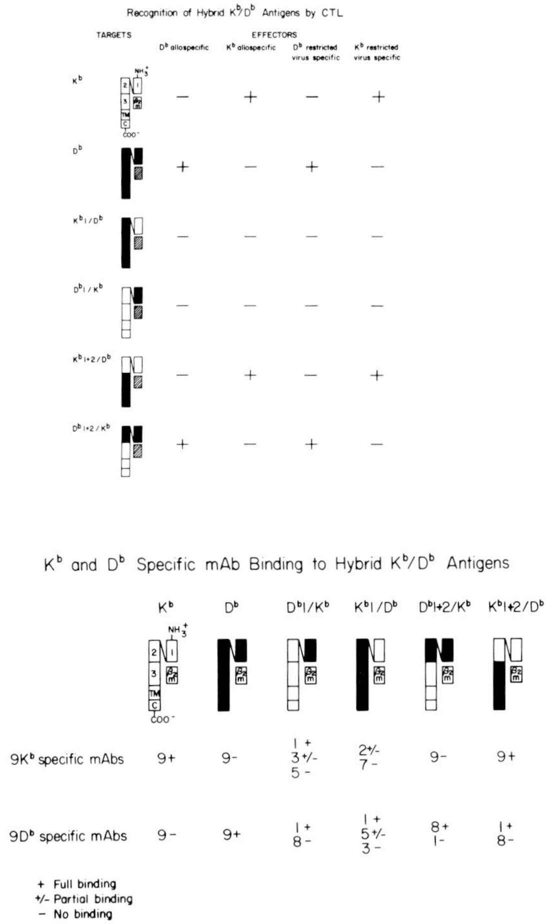

Figure 7. Recognition by CTL (top) and mAb (bottom).

also present in the d and k haplotypes. By chromosomal walking Aβ3 was found to be 70 kilobases from the K^b gene. This is the first time that the DNA linking class I and class II genes has been cloned. The cosmid cluster containing K1, K^b and Aβ3 has not yet been linked to the major I region cluster. However, we have mapped Aβ3 into the I-A subregion, as defined by the AT.L mouse, by restriction fragment length polymorphisms.

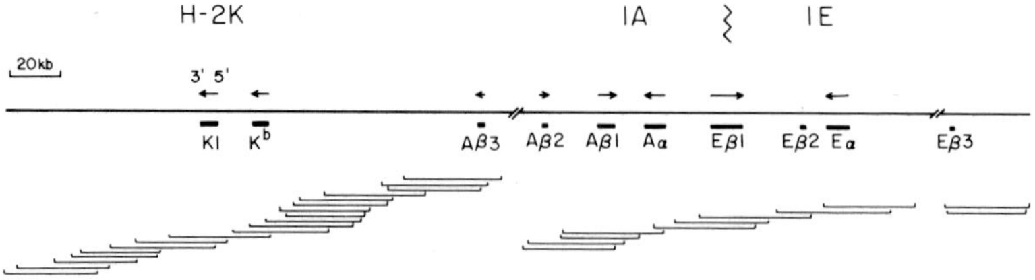

Figure 8. Organization of class II genes.

The Eβ3 cosmids were identified by weak cross-hybridization to a human class II DRβ cDNA. Preliminary experiments map this sequence between I-A and H-2D, that is, into the I-E or the S-region.

Gene Conversion Between Class II Genes

We have previously argued from studies with class I genes that gene conversion may generate polymorphism in this gene family (1,9). Data presented here suggest that gene conversions may also occur between class II genes.

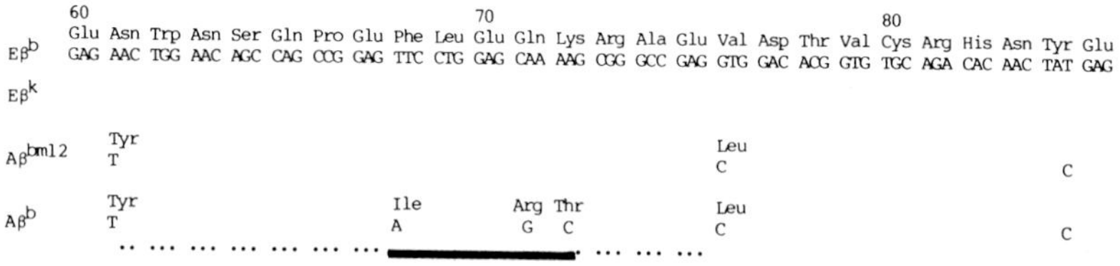

Figure 9. Sequence comparison of part of the β1 domain of the genes $E\beta^b$, $E\beta^k$ (10), $A\beta^{bm12}$ and $A\beta^b$ (11). Only nucleotides and amino acids different from the $E\beta^b$ sequence are shown. The minimum gene conversion is underlined; the maximum is defined by the dotted line.

McIntyre and Seidman (Nature, in press) have shown that the $A\beta^{bm12}$ gene differs from the $A\beta^b$ gene by 3 nucleotides which cause amino acid substitutions at positions 68, 71, and 72 (Fig. 10). We sequenced the $E\beta^b$ gene (12) and found that it is identical to the $A\beta^{bm12}$ gene in the mutated region. These data together suggest that the

$E\beta^b$ gene donated by gene conversion a minimum of 14 nucleotides and a maximum of 44 nucleotides to the $A\beta^b$ gene to generate the $A\beta^{bm12}$ gene. It appears that this transfer of genetic information has also resulted in the transfer of gene function. The immune response to the A-chain loop of sheep insulin is I-E^k restricted in H-2^k mice. C57BL/6 mice (H-2^b) do not express I-E and are nonresponders to this antigen, whereas H-2^{bm12} mice, which also do not express I-E, are responders. Thus, the restriction element for this antigen is the I-A molecule in the H-2^{bm12} mouse (Hochman and Huber, personal communication). The $E\beta^b$ gene is identical to the $E\beta^k$ sequence in the region of the gene conversion event, thus the $E\beta^b$ gene has donated the sequence information for responsiveness even though it is not expressed in the H-2^b mouse.

A Multiplicity of MHC Genes

We have identified 26 class I genes and 8 class II genes in the MHC of the B10 mouse; however, a biological role has been established for only a few of these genes. Our laboratory is currently determining which of these genes are transcriptionally active.

REFERENCES

1. Weiss E, Golden L, Zakut R, Mellor A, Fahrner K, Kvist S, and Flavell RA (1983). The DNA sequence of the H-$2K^b$ gene: evidence for gene conversion as a mechanism for the generation of polymorphism in histocompatibility antigens. The EMBO Journal, 2:453.
2. Oudshoorn-Snoek M, Demant P, Mellor AL, and Flavell RA (1983). Antigenic characterization of products of cloned H-2^b genes transplant. Proc Natl Acad Sci USA, 15:2027.
3. Townsend ARM, Taylor PM, Mellor AL, and Askonas BA (1983). Recognition of D^b and K^b gene products by influenza-specific cytotoxic T cells. Immunogenetics, 17:283.
4. Steinmetz M, Astar W, Minard K, and Hood L (1982). Clusters of Genes Encoding Mouse Transplantation Antigens. Cell, 28:489.

5. Maloy WL, and Coligan J (1983). Primary Structure of the H-2D^b Alloantigen. Immunogenetics, 16:11.
6. Cosman D, Kress M, Khoury G, and Jay G (1982). Tissue specific expression of an unusual H-2 (class I)-related gene. Proc Natl Acad Sci USA, 79:4947.
7. Mellor AL, Golden L, Weiss E, Bullman H, Hurst J, Simpson E, James RFL, Townsend ARM, Taylor PM, Schmidt W, Ferluga J, Leben L, Santamaria M, Atfield G, Fesenstein H, and Flavell RA (1982). Expression of murine H-2^b histocompatibility antigen in cells transformed with cloned H-2 genes. Nature, 298:529.
8. Steinmetz M, Minard K, Horvath S, McNicholas J, Frelinger JG, Wake C, Long E, Mach B, and Hood L (1982). A molecular map of the immune response region from the major histocompatibility complex of the mouse. Nature, 300:35.
9. Weiss EH, Mellor A, Golden L, Fahrner K, Simpson E, Hurst J, and Flavell RA (1982). The structure of a mutant H-2 gene suggests that the generation of polymorphism in H-2 genes may occur by gene conversion-like events. Nature, 301:671.
10. Mengle-Gaw L, and McDevitt HO (1983). Isolation and characterization of a cDNA clone for the murine I-Eβ polypeptide chain. Proc Natl Acad Sci USA, 80:7621.
11. Larhammar D, Hammerling U, Denaro M, Lund T, Flavell RA, Rask L, and Peterson PA (1983). Structure of the murine immune response I-A locus: sequence of the I-Aβ gene and an adjacent β-chain second domain exon. Cell, 34:179.
12. Widera G, and Flavell, RA (1984). The nucleotide sequence of the murine I-Eβ^b immune response gene: evidence for gene conversion events in class II genes of the major histocompatibility complex. The EMBO Journal, In Press.
13. Mellor AL, Weiss E, Ramachandran K, and Flavell RA (1983) A potential donor gene for the bml gene conversion event in the C57BL/10 mouse. Nature, 306:5945.
14. Mellor AL, Weiss E, Kress M, Jay G, and Flavell RA (1984). A non-polymorphic class I gene in the murine major histocompatibility complex. Cell, 36:1.

Molecular Biology of Development, pages 537–547

JOINING OF IMMUNOGLOBULIN HEAVY CHAIN VARIABLE REGION GENE SEGMENTS IN VIVO AND WITHIN A RECOMBINATION SUBSTRATE[1]

T. Keith Blackwell, George D. Yancopoulos, and Frederick W. Alt

Department of Biochemistry
and Institute for Cancer Research,
Columbia University College of Physicians and Surgeons
New York, New York, 10032

ABSTRACT We report here the preferential utilization, in V_HDJ_H rearrangements, of the most D-proximal V_H family within the pre-B cell population in vivo. In addition, we describe the development of an experimental recombination system for the study of the mechanism and regulation of immunoglobulin variable region gene assembly.

INTRODUCTION

The variable region of an immunoglobulin (Ig) heavy chain gene is assembled from three germ-line components--the variable (V_H) segment, the diversity (D) segment, and the joining (J_H) segment--families of which lie in separate clusters in the genome of the mouse (1). During B-cell development, D-to-J_H joining occurs first and on both chromosomes, followed by the addition of a V_H segment to form a complete V_HDJ_H Ig heavy chain variable region gene (2).

Our recent experiments have demonstrated the preferential utilization, in V_HDJ_H rearrangements, of the most D-proximal family of V_H segments in pre-B cell lines derived from BALB/c mice (3). The V_H segment which is most

[1]This work was supported by National Institutes of Health grant AI-20047, American Cancer Society grant NP-393, and an award from the Searle foundation.

frequently utilized in these cells, V_H81X, has never been observed in a V_HDJ_H rearrangement in a mature Ig-secreting cell. These findings suggest that the V_H repertoire in pre-B cells is considerably different from that of mature B-cells. To verify this conclusion, we have examined V_H expression in normal fetal liver, the location of pre-B cells in the developing mouse (4,5). Here we report preferential utilization of this V_H segment family within the pre-B cell population in vivo.

To understand the mechanism and regulation of the process by which Ig heavy and light chain variable region genes are assembled, it is important to develop an experimental recombination system in which both the structure and genomic context of the component segments can be manipulated. Certain pre-B cell lines transformed by Abelson murine leukemia virus (A-MuLV) assemble heavy chain variable region genes during growth in culture (6), thus providing the opportunity to develop such a system through gene transfer technology. We report that molecularly cloned D and J_H segments undergo site-specific recombination, at high frequency, when introduced into the genome of an actively rearranging A-MuLV transformant.

RESULTS

Preferential Utilization of the Most D-Proximal V_H Segments in Fetal Liver

Both productive and non-productive V_HDJ_H rearrangements are transcribed, and germ-line V_H segments generally are not (7). Therefore, within a population of cells, the level of expression of a particular V_H segment is proportional to the number of V_HDJ_H rearrangements utilizing that segment. To assay for specific V_H expression over the course of B-cell development, poly-A^+ RNA was prepared from various stages of murine fetal liver and from four week adult spleen, a rich source of mature cells of the B lineage. Tissues were taken from two strains of mice: BALB/c (B/c), and Columbia (C). Identical RNA slot blots of these samples were probed with several different ^{32}P-labeled DNA fragments specific for the mu heavy chain constant region (Cμ) and for various V_H segments. The V_H-specific probes cross-hybridize to other family members at low stringency (1X SSC, 68°C), but at high stringency (0.15X SSC, 68°C) preferentially hybridize to the V_H segment from which they were prepared (not shown). In a representative experiment (figure 1), RNA samples from fetal liver and spleen were compared for intensity of hybridization, at high stringency, to probes specific for C_u and V_H81X. In both strains of mice, the level of hybridization in fetal liver, relative to spleen, of V_H81X is clearly higher than that of C_μ (figure 1A and 1B). Therefore, V_H81X comprises a greater percentage of $C\mu$-specific mRNA in these stages of fetal liver than in spleen. Such a difference was not seen with the V_HS107 probe (figure 1C).

The developmental expression of several V_H segments was assayed under conditions of low and high stringency as described above, and the signal intensities quantitated by densitometry. Preference (P) values were assigned to each V_H segment according to the following equation:

$$P = \frac{I_{FL}(V_H)/I_{FL}(C\mu)}{I_S(V_H)/I_S(C\mu)}$$

where I_{FL} and I_S denote the intensities of hybridization in fetal liver and spleen respectively. A P value of 1 for a V_H segment signifies equivalent expression, relative to C_μ, in fetal liver and spleen; while a higher value indicates the multiple by which expression is greater in fetal liver. When P values were determined for V_H segments representative of the four most D-proximal V_H families (Brodeur and Riblett, personal communication), only that of V_H81X was significantly greater than 1 (table 1). For the other V_H segments shown, P values were approximately the same at both high and low stringency (not shown); but for V_H81X, P was generally greater at high stringency than at low stringency (table 1). Similar results were obtained for tissue samples prepared from 15.5, 16.5, and 17 day fetal liver (Yancopoulos and Alt, unpublished results). Therefore, the family of V_H segments related to V_H81X is preferentially expressed within the pre-B cell population in fetal liver, compared to the mature B-cell population in spleen, and this preference is even greater for expression of V_H81X alone (table 1). Through comparison to the V_H81X/C_μ signal ratio in a pre-B cell line (18-81A2, see ref.8) that is known to express V_H81X, it was determined that between 20 and 60 percent of the C_μ-specific message in fetal liver contained V_H81X sequences (Yancopoulos and Alt, unpublished results).

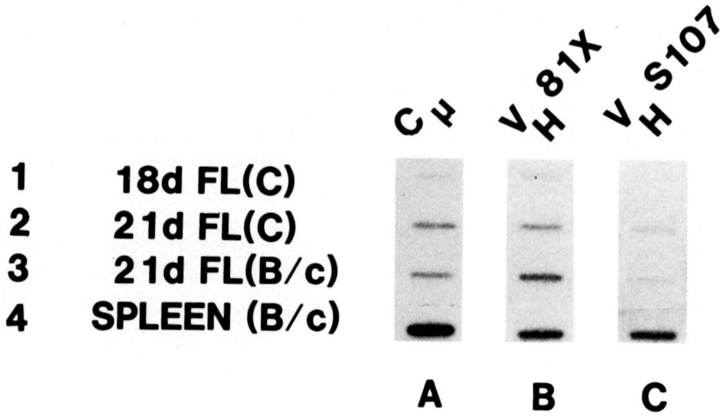

Figure 1. V_H expression in fetal liver and spleen. Poly A^+ RNA was prepared from four week BALB/c (B/C) spleens and from fetal liver (FL) obtained at the indicated days from BALB/c (B/c) and Columbia (C) mice. 2.0 ug of poly A^+ RNA were loaded in slots in rows 1,2,and 3, and 0.01 ug in slots in row 4. Filters were hybridized to the indicated probes as described (3).

Tissue	V_H81X(H)	V_H81X(L)	V_HQ52(H)	V_HS107(H)	V_HNP(H)
18d(C)	3.5-9	5-7	---	---	1-1.3
21d(C)	4.5-7	2.5-3.5	0.77-1.1	1.1-1.7	0.9-1.3
21d(B/c)	7.5-10	2.5-3.3	0.5-0.7	0.75-1	0.9-1.2

Table 1. Preference (P) values for various V_H-specific probes. Values are given for fetal liver taken at the indicated days (d) from BALB/c (B/c) and Columbia (C) mice. Intensities were measured by densitometry. The ranges given indicate minimum and maximum P values calculated from multiple exposures of several experiments. H denotes high stringency; L denotes low stringency.

Site-Specific Recombination between D and J_H Segments That Were Introduced into the Genome of an Actively Rearranging A-MuLV Transformant

A recombinant plasmid, pJT28, was constructed so that the Herpes simplex virus (HSV) thymidine kinase (tk) gene was flanked on one side by the murine DQ52 segment, and on the other by a fragment of DNA containing the murine J_H3 and J_H4 elements (9). This plasmid was introduced, by an unlinked co-transfection with pSV2neo, into a tk- variant of the A-MuLV transformant 38B9, which had been derived from murine fetal liver. The 38B9 line has deleted both copies of DQ52 through D-to-J_H rearrangement on both Ig heavy chain alleles, and is active in further rearrangement of heavy chain V region gene segments (2,6). A cloned line, T2, which was isolated from the transfected population after selection in medium containing G418, was shown to have integrated two copies of pJT28 (9). One copy contained the insert intact (functional form, figure 2), except for the duplication of a short segment of DNA that contained the DQ52 element. The other integration of the pJT28 insert was not intact, having been interrupted within the HSV tk gene.

The structure of the functional form of the integrated pJT28 construct predicts four potential forms of D to J_H rearrangement occurring within the construct (figure 2). These rearrangements appeared spontaneously in the T2 line (figure 2c, lane 2. See reference 9 for detailed analysis). Growth of the T2 line in HAT medium (figure 2a) selected a population of cells (T2H) which had activated the HSV tk gene (figure 2b), and also virtually eliminated those cells which had completed D-to-J_H joining within the construct (figure 2c, lane 3). Passage of T2H into medium containing BUdR selected for inactivation of the transfected tk gene (figure 2b), yielding a population of cells (T2HB) which displayed, with no obvious preference, the four possible forms of D-to-J_H joins (figure 2c, lane 4). Cellular subclones of T2HB each contained a DQ52-to-J_H rearrangement, which had occurred within the functional form of the integrated construct. Each subclone also contained the non-intact integration of the pJT28 insert (representative subclones are shown in figure 2d). Therefore, the only pathway of tk inactivation that was observed within the T2H population was loss of the HSV tk gene through D-to J_H joining.

A total of four D-to-J_H rearrangements were molecularly cloned as Eco RI fragments from the T2HB population or from

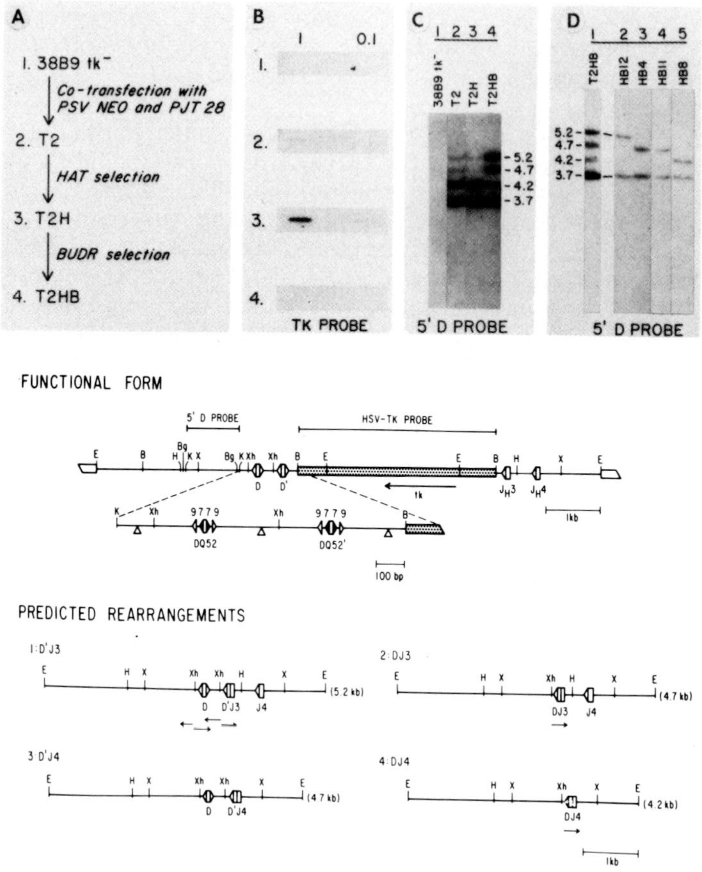

Figure 2. D-to-J_H joining within the integrated construct in the T2 line (9). (A) The isolation of T2 and its derivatives. (B) Expression of the transfected HSV tk gene. Cytoplasmic cell lysates from $2.5X10^5$ cells (lane 1), or from one tenth that amount (lane 0.1), were transferred onto nitrocellulose paper and hybridized to the HSV tk probe. (C) and (D) Genomic DNA from the indicated cell lines was digested with Eco RI, electrophoresed through agarose, blotted onto nitrocellulose, then hybridized to the 5'D probe. Sizes are in kb. Functional form: partial restriction endonuclease map of the functional form of the integrated construct. HSV sequences are indicated with a dotted box. An enlargement of the duplicated region is

shown below the map, with the duplication boundaries denoted by open triangles. The DQ52 segments have been arbitrarily denoted D and D'. The HSV tk and 5'D probes are indicated. Predicted Rearrangements: Partial restriction maps of D-to-J_H rearrangements predicted from the structure of the functional form of the integrated construct. Eco RI sites are indicated as E.

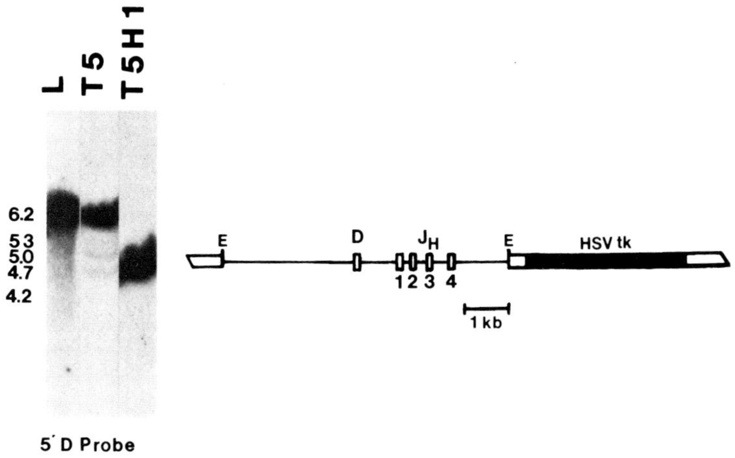

Figure 3: D-to-J_H joining within the integrated construct in the T5 line. Genomic DNA from BALB/c liver (L) and the indicated cell lines was digested with Eco RI and assayed for hybridization to the 5'D probe as described in figure 2. The structure of the integrated construct is diagrammed at the right. HSV tk sequences are indicated by a closed box; pBR322 and uncharacterized flanking sequences by an open box. Eco RI sites are indicated as E. Sizes are in kb.

its cellular subclones, including: one D'J_H3 joint, two DJ_H3 joints, and one DJ_H4 joint (figure 2). Examination of the nucleotide sequences at the point of joining demonstrated that the recombination of these segments had occurred by the mechanism proposed to mediate joining between endogenous V region gene segments (9). In each case, these elements joined at different sites within appropriate regions of the D and J_H coding sequences, indicating clearly that they arose from independent events.

Another plasmid construct, pJT15, contained the HSV tk gene inserted into the Bam H1 site of pBR322, alongside the 6.2 kb Eco RI fragment of murine DNA on which the DQ52 and J_H segments lie. This construct was introduced into the 38B9tk- line as described above. The integrated construct within one transformant, T5, is diagrammed in figure 3. This line contained the 6.2 kb insert of murine DNA intact, with the HSV tk sequences located 3' of the J_H cluster. (figure 3). Genomic DNA from T5 contained submolar 5'D-positive Eco RI fragments of 5.3 kb, 5.0 kb, 4.7 kb, and 4.2 kb, the approximate sizes expected for joining between the DQ52 and the four J_H segments (figure 3, lane T5, the 4.2 kb fragment is apparent with a longer exposure). Both the T5 (figure 3) and T2 (figure 2) genomic DNA samples were prepared approximately 20 cell generations after the appearance of these clones, and both display a similar level of submolar fragments, indicating that these lines undergo rearrangement at comparable frequencies. None of the submolar rearrangements within the T5 line have been isolated and sequenced, but it seems likely, considering the T2 data described above, that these rearrangements represent site-specific D-to-J_H joining. Thus, in two constructs integrated within different cells in presumably different chromosomal locations, D-to-J_H recombination occurs frequently.

When T5 cells were plated in HAT medium more than 90% were killed, a property also observed in the T2 population (9). A line resistant to HAT (T5H1) was derived after three weeks of continuous passage. Cytoplasmic RNA dot blot analysis indicated that selection of T5 in HAT medium was associated with activation of the HSV tk gene (not shown). Genomic DNA from T5H1 contains a 5'D-positive 4.7 kb Eco RI fragment (figure 3). Further mapping of T5H1 DNA indicated that this Eco RI fragment represented a DQ52-to-J_H rearrangement occurring within the integrated construct (not shown). Experiments are presently underway to determine if selection for tk activity in T5 is generally associated with recombination at the contiguous J_H locus. If true, such a correlation would link recombination and the transcriptional activation of a nearby gene, perhaps through a local alteration of chromatin structure.

DISCUSSION

The V_H Repertoire in Pre-B Cells is Different from That of B Cells

We have shown that the most D-proximal V_H segments are preferentially expressed early in normal B cell development in BALB/c mice. Homologous V_H segments are also preferentially expressed in a different mouse strain, Columbia. Therefore, frequent utilization of V_H81X and of related V_H segments, which was first described in pre-B cell lines (3), is apparently a property of normal cells undergoing V_H to DJ_H recombination. A functional role for this phenomenon has not yet been identified, and the mechanism of the subsequent randomization of the B cell V_H repertoire is not yet understood. Various factors could contribute to the frequent utilization of these V_H segments: proximity to the D segments, chromosomal structure around this V_H family at the time of rearrangement, and DNA sequences, specific to these V_H segments, which might be involved in recognition and recombination. For example, it has been shown that the consensus nonamer and heptamer sequences 3' of V_H81X are unique (3). The functional significance of DNA sequences flanking these V_H elements can theoretically be tested in an experimental recombination system such as that described above.

Joining of Ig Variable Region Gene Segments within an Integrated Recombination Substrate

The construct integrated within the T2 line contained only 260 bp of sequence 3' of the DQ52 segment, 138 bp of sequence 5' of the J_H3 segment, and approximately 1 kb of sequence 3' of the J_H4 segment, including the Ig heavy chain enhancer element, but not the Cμ coding region. Therefore, this experiment established the maximum amount of sequence flanking the D and J_H segments required for recombination to occur, and demonstrated that the J_H locus requires neither the contiguous presence of the Cμ coding region nor a specific chromosomal location to be recombinationally active. The introduction, into appropriate cell lines, of other constructs containing further manipulations of Ig variable region gene elements should permit the definition of the mechanistic requirements of the recombinase system.

D-to-J_H joining within the integrated construct occurred frequently in T2 cells (9). Because this joining represented the only observable pathway of inactivation of the HSV tk gene, the plating efficiency of T2 in BUdR medium has been linked to recombinase activity within the line. Therefore, it should be possible to "assay" recombinase activity in such a line under various experimental conditions. Moreover, the linkage of this activity to a selectable marker should permit the isolation of cell lines with genetic alterations in the recombinase apparatus itself.

Experiments underway in the T5 line may provide insight into the importance of chromosomal accessibility in the control of the rearrangement process. Other studies are in progress to determine the tissue specificity of site-specific recombination between transfected V region gene elements. The results described above define a general system for further mechanistic, genetic, and regulatory studies of the recombinase system. An independent study has recently described site-specific joining between Ig light chain V region segments introduced into a more mature A-MuLV transformant (10). Therefore, the general methods we have described should be applicable to the study of other gene rearrangement events, such as Ig class switching, oncogene translocation, and the recombination events associated with the assembly of the T cell antigen receptor.

ACKNOWLEDGEMENTS

T. K. Blackwell and G. D. Yancopoulos were supported by N. I. H. training grant GM-07367. T. K. Blackwell is a student in the Department of Microbiology. F. Alt is an Irma T. Hirschl Career Scientist.

REFERENCES

1. Tonegawa S (1983). Somatic generation of antibody diversity. Nature 302:575.
2. Alt FW, Yancopoulos GD, Blackwell TK, Wood C, Thomas E,

Boss M, Coffman R, Rosenberg N, Tonegawa S, Baltimore D (1984). Ordered rearrangement of immunoglobulin heavy chain variable region segments. EMBO Journal, in press June, 1984.

3. Yancopoulos G, Desiderio S, Paskind M, Kearney JF, Baltimore D, Alt F (1984). Preferential utilization of the most D-proximal V_H segments in pre-B cell lines. Submitted.
4. Levitt D, Cooper MD (1980). Mouse pre-B cells synthesize and secrete u heavy chains but not light chains. Cell 19:617.
5. Siden E, Alt FW, Shinefeld L, Sato V, Baltimore D (1981). Synthesis of immunoglobulin u chain gene products precedes synthesis of light chains during B cell development. Proc Natl Acad Sci USA 78:1823.
6. Alt FW, Rosenberg N, Lewis S, Thomas E, Baltimore D (1981). Organization and reorganization of immunoglobulin genes in A-MuLV transformed cells: rearrangement of heavy but not light chain genes. Cell 27:381.
7. Mather EL, Perry RP (1981). Transcriptional regulation of immunoglobulin V genes. Nuc Acids Res 9:6855.
8. Alt FW, Rosenberg N, Enea V, Siden E, Baltimore D (1982) Multiple immunoglobulin heavy chain transcripts in Abelson murine leukemia virus-transformed cell lines. Mol Cell Biol 2:386.
9. Blackwell TK, Alt FW (1984). Site-specific recombination between immunoglobulin D and J_H segments that were introduced into the genome of a murine pre-B cell line. Cell 37:105.
10. Lewis S, Gifford A, Baltimore D (1984). Joining of V_K to J_K gene segments in a retroviral vector introducod into lymphoid cells. Nature 308:425.

V. MOLECULAR APPROACHES TO NERVE CELL DIFFERENTIATION

Molecular Biology of Development, pages 551–560

TWO INTRONS DEFINE FUNCTIONAL DOMAINS OF A NEUROPEPTIDE PRECURSOR IN *APLYSIA*[1]

Ronald Taussig, Marina R. Picciotto, and Richard H. Scheller

Department of Biological Sciences
Stanford University
Stanford, CA 94305

ABSTRACT Biologically active peptides are synthesized as parts of precursor proteins which are proteolytically processed to generate active molecules. The structure of the gene encoding peptides expressed in *Aplysia* neurons R3-14 suggests that two intervening sequences split the transcript into functional domains. The first exon encodes the 5' untranslated region, the second exon the signal sequence and the bulk of the negative charge of the precursor protein while the third exon encodes the remainder of the precursor and the 3' untranslated region.

INTRODUCTION

The nervous system uses a wide variety of compounds as extracellular chemical messengers. These include molecules which have evolved solely for this purpose, such as acetylcholine or serotonin, and compounds with additional cellular functions, for example, glycine and ATP. The most diverse group of extracellular messengers are the biologically active peptides, chains of amino acids as short as two and as

[1]This work was supported by a grant to Richard H. Scheller from the NIH.

long as few hundred residues (1). Neuropeptides are normally synthesized as components of larger precursor proteins. These precursors are proteolytically cleaved, often liberating more than one active molecule.

Previous structural studies of the endogenous opiates, calcitonin, and ELH gene families suggest that the complex structures of neuropeptide precursors evolve by two general mechanisms. First, duplications within a transcription unit can give rise to genes directing the synthesis of a single precursor protein containing multiple copies of a given neuropeptide. Divergence of these copies from each other could then generate related, yet distinct products. Alternatively, genes encoding multiple peptides could evolve via recombination within the noncoding intervening sequences of unrelated genes. In a single event, these rearrangements would generate new combinations of biologically active molecules encoded by a single precursor.

The Aplysia abdominal ganglion neurons R3-14 modulate cardiac output (2). These cells are thought to use both neuropeptides and the amino acid glycine as extracellular messengers. Using differential screening techniques, we have isolated cDNA and genomic clones encoding the peptides synthesized by these cells (3). The Aplysia haploid genome contains a single copy of the R3-14 neuropeptide gene which is transcribed in neurons of the abdominal ganglion, giving rise to a 1.2 KB poly A+ mRNA. The gene encodes a precursor of 108 amino acid residues which is proteolytically cleaved into the secreted peptides. The precise nature of the peptides cleaved from the precursor remains unclear; however, studies are in progress to resolve this issue. We have used immunocytochemical techniques to localize the cells and processes containing the products of this gene (4). The R3-14 cells each send a single large axon out the branchial nerve which terminates on the efferent vein of the gill at the base of the heart. Neuron R14 is anatomically distinct in that it also sends axons to the vasculature near the ganglion. All of the cells seem to have numerous processes which end in the sheath surrounding the ganglion.

Using electron microscopic techniques, we recently demonstrated that the R3-14 peptide gene contains two large intervening sequences and spans about 7 KB of genomic DNA (3). In this report we present the DNA sequence of the middle exon which allows us to define the precise positions of introns in the mRNA.

RESULTS

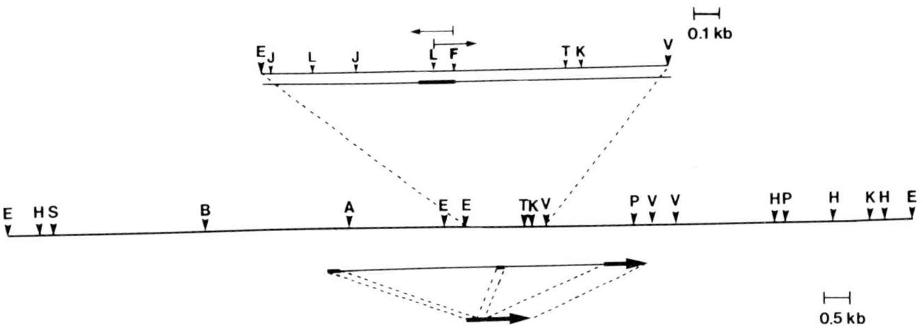

Figure 1 Restriction enzyme map of the R3-14 peptide gene.
The positions of restriction enzyme sites were determined by a combination of partial single and double digests on whole clones or isolated fragments. Locations of the transcript were determined by blotting and electronmicroscopy. The transcribed regions are indicated by the lines below the restriction enzyme maps, the arrows point in the direction of transcription. The Eco RI/ Pvu II fragment containing the middle exon was subcloned in pBR 322 and further mapped. The area encompassed by the arrows above the map was sequenced. E - Eco RI, H - Hind III, S - Sma I, B - Bam HI, A - Sal I, T - Sst I, K - Kpn I, V - Pvu II, P - Pst I, J - Ava II, L - Hae III, F - Hinf I

Figure 1 shows a restriction enzyme map of the region of genomic DNA encoding the the R3-14 peptide gene. The letters represent the positions of various restriction enzyme sites as determined by partial and double digests. To determine the exact location of the two introns we subcloned the Eco RI/Pvu II restriction enzyme fragment into the plasmid cloning vector pBR 322. A detailed restriction map of this subclone was generated. Sites in the middle exon were labelled and the nucleotide sequence was determined extending into both introns.

The sequence of the genomic exon II is compared to the cDNA sequence in figure 2. Several structural features are noted. As expected the intron/exon boarders are flanked by the splice junction consensus sequences, AG at the acceptor sites and GT at the donor sites. Intron II, the 3' intron, contains a dramatic purine/pyrimidine asymmetry. This intervening sequence is composed of 90% purines on the coding strand including a run of the dinucleotide AG repeated seven times. There are a total of 23 AG dinucleotides in the 74 nucleotides sequenced, far beyond random expectation. Conversely, the complementary dinucleotid CT, is overrepresented on the coding strand of intron I. This dinucleotide is present 16 times in the 5' intron, about twice the random expectation. The presence of this asymmetry is further strengthened by the complementary underrepresentation of CT in the 3' intron (2 copies) and the underrepresentation of AG in the 5' intron (3 copies). While the significance of these distributions is not clear, one might expect the presence of large numbers of complementary dinucleotides to favor a base pairing mediated association of these two regions in the primary transcript.

```
          1
GENOME     CTTTTTTTTGTGTCTGTAAATCTCTTGTAACACCTCTAATCTTCGGCACTTATACGACCTTATTGTGTTGCAAGTGTGGTAAAACTTATACTAAAACTTTACTGATTTATTAAATAA

cDNA     CACACGCACTATCTCTCACTAACAGACTCACTCACCCCAAGCAGTGACTTTCAATTCTTTCGCTCGTGCAGAGCTTGGCCGTGAGTCGGTAACAAAGTGAGATCGACTTTCTCATCCAG

         118
GENOME   AATGTACTGGTGTCTAGCCTTGTGTCCAGCC ATG CAA GTC CTC CAC CTG TGT CTA GCG GTG TCC ATC GCT GTG GCC CTC CTG TCC CAG GCT GCG TGG
cDNA                                   CC ATG CAA GTC CTC CAC CTG TGT CTA GCG GTG TCC ATC GCT GTG GCC CTC CTG TCC CAG GCT GCG TGG
         ACATCAAGACATCAACTACCCACAAGCAG

         215
GENOME   TCA GAA GAG GTG TTT GAT GAC ACA GAC GTC GGT GAT GAG CTG ACC AAC GCC TTG GAG TCA GGTGAGAGGGGAGAAACGGGGATAGAGAGAGAGAGAG
cDNA     TCA GAA GAG GTG TTT GAT GAC ACA GAC GTC GGT GAT GAG CTG ACC AAC GCC TTG GAG TCA G
                                                                                          TT CTG ACA GAT TTC AAA GAC AAA CGG GAA

         314
GENOME   AAGGGGGATTGTGGAGAGAAAAAGGGATAGAGATAG

cDNA     GCA GAA GAA CCA TCA GCC TTC ATG ACC
```

Figure 2 The nucleotide sequence of exon II and the corresponding region in the cDNA clone. The nucleotide sequence of untranslated area is contiguous and the sequence of the translated region is separated into codons. The splice junction concensus sequence is underlined. The Eco RI/Pvu II subclone was cleaved with either Hae III or Hinf I and end labeled with 32P-ATP and polynucleotide kinase. The DNA was restricted with a second enzyme to produce asymmetrically labeled fragments and the Maxam and Gilbert chemical modification and cleavage reactions were performed (5). The cleavage products were resolved on 20% and 6% polyacrylamide gels containing 7M urea.

The positions of intervening sequences in the R3-14 gene lend support to the domain hypothesis (6) which states that one of the reasons for having introns is to separate the coding regions for functional units of proteins, thus increasing the probability for recombination between these functional domains. The first intervening sequence occurs two nucleotides before the initiator methionine codon and the second splits codon 43 of the precursor (Figure 3). As a result exon I encodes the complete 5' untranslated region of the mRNA. Exon II encodes the hydrophobic signal sequence characteristic of all membrane bound and secreted proteins, and the bulk of the negative charge of the precursor (Figure 3). We were not able to detect any internal homology resulting from duplications within the transcription unit. Therefore this gene may have evolved by combining previously unrelated exons through recombination within intervening sequences, thus generating a new protein. It is possible that the present day precursor arose from the joining of two domains encoding peptides of essentially opposite charge (Figure 3). This would result in a set of peptides which is suitable for dense packing in secretory vesicles due to ionic interactions between these domains. Ionic associations within a vesicle may dissociate as a result of the pH changes which usually accompany release of the vesicle contents into the extracellular space. The precursor cleavage products would then be free to act independently.

Many genes including the R3-14 peptide precursor have introns in their 5' untranslated region. While there are very few, if any, known functions for the untranslated regions of an mRNA, the site of initiation of transcription would presumably be just distal to the 5' most exon. Regulatory sequences could be located at or near the site of initiation of transcription, therefore this exon could have evolved solely for the purpose of controlling expression of the gene. We have been able to identify only 12 cells in Aplysia which synthesize these peptides, suggesting very specific mechanisms are involved in governing their expression.

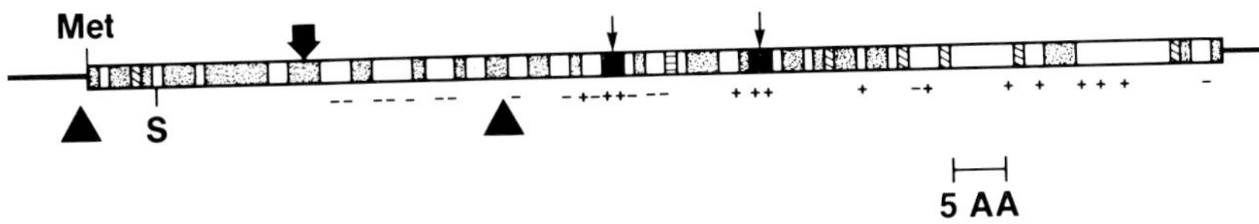

Figure 3 Intron positions in the R3-14 gene. This schematic representation of the R3-14 mRNA shows the untranslated regions, represented by thin lines, and the various classes of amino acids in the precursor. Stippled regions are hyrophobic amino acids (▨), slashed lines are histidine residues (▧), and horizontal lines are proline (▤), solid boxes are dibasic residues (■), and the charge is written below the acidic or basic amino acids (+ or - below). Arrows pointing down are proteolytic cleavage positions (large arrow indicates the approximate position of the signal sequence cleavage and the small arrows are internal cleavage sites). Arrows pointing up are positions of the intervening sequences as determined in this report.

DISCUSSION

The organization of multiple functional products in a single precursor protein, a polyprotein, assures coordinate and stoicheometric synthesis of the molecules. Polyproteins are vital components of the simplest of organisms, viruses (7), and the most complex of all biological systems, the brain. The low complexity of viral genomes makes the polyprotein organization desirable for efficient use of limited genetic information. The functional significance of polyproteins in the brain is less clear. The brain makes use of a vast number of specific interactions to meet its requirements for intercellular communication. Even animal genomes have a limited quantity of genetic information and therefore, polyprotein organization may be useful in specifying the complex network of interactions between neurons.

The endogenous opiate peptide precursors which encode the enkephalins (8) and dynorphins (9) contain multiple copies of related products. Therefore, these genes must have evolved by way of internal duplications within a transcription unit. Similarly, recent cloning and sequencing of two cDNAs encoding the neuropeptide substance P suggests that this gene has evolved by similiar means. These two cDNAs differ by nucleotide sequences encoding a second copy of a tachykinin (10) and it is likely that the two mRNAs arise as differential RNA splicing products. In the case of the opiate peptides, the duplications have arisen within the protein, while in the substance P gene, the duplicated regions are probably separated by intervening sequences. Intron mediated duplications have evolutionary advantages in that they do not require maintenance of the reading frame of the protein.

Several neuropeptide precursors, including R3-14, lack internal homologies. These include the precursor for calcitonin and the calcitonin gene related protein, peptides which arise from a single gene via alternate polyadenylation at sites in different exons (11). This gene, like the R3-14 peptide gene, has probably evolved by the joining of functional units by recombination within introns. In this way, new combinations of peptides can be generated to best cope with the environmental pressures placed on an animal. Unlike the mode of evolution involving internal repetition of active molecules, this mechanism allows for new combinations of peptides and therefore, new interactions between cells in a saltatory event.

The diversity of animal behaviors is at least as great as the diversity of structures. To account for this diversity, mechanisms of evolution of neuronal pathways might involve the reorganization of functional units through events involving duplications and mixing of exons. It is likely that these events not only generate new proteins, but also result in new patterns of gene expression.

ACKNOWLEDGMENTS

This work is supported by grants to Richard H. Scheller from the NIH, The March of Dimes, The McKnight Foundation and The Sloan Foundation.

REFERENCES

1. Bloom, F.E. (1980). Peptides: Integrators of Cell and Tissue Function. New York: Raven Press.
2. Sawada, M., McAdoo, D.J., Blankenship, J.E., and Price, C.H. (1981). Modulation of arterial muscle contraction in Aplysia by glycine and neuron R14. Brain Res. 207:486-490.
3. Nambu, J.R., Taussig, R., Mahon, A.C., and Scheller, R.H. (1983). Gene isolation with cDNA probes from identified Aplysia neurons: Neuropeptide modulators of cardiovascular physiology. Cell 35:47-56.
4. Kreiner, T., Rothbard, J.B., Schoolnik, G.K., and Scheller, R.H. (Submitted) Antibodies to synthetic peptides defined by cDNA cloning reveal a network of peptidergic neurons in Aplysia.
5. Maxam A. and Gilbert, W. (1980). Sequencing end-labeled DNA with base specific chemical cleavages. Meth. Enzymol. 65:499-560.
6. Gilbert, W. (1978). Why genes in pieces? Nature 271:501.
7. Racaniello, V.R. and Baltimore, D. (1981). Molecular cloning of poliovirus cDNA and determination of the complete nucleotide sequence of the viral genome. Proc. Nat. Acad. Sci. USA 8:4887-4891.

8. Noda, M., Furatani, Y., Takahashi, H., Toyosato, M., Hirose, T., Inayama, S., Nakanishi, S., Numa, S. (1982). Cloning and sequence analysis cDNA for Bovine adrenal preproenkephalin. Nature 295:202-204.
9. Kakidani, H., Furatani, Y., Takahashi, H., Noda, M., Mirimoto, Y., Hirose, T., Asai, M., Inayama, S., Nakanashi,S., Numa, S. (1982). Cloning and sequencing analysis for porcine B-neo-endorphin, dynorphin precursor. Nature 298:245-249.
10. Noda, H., Hirose, T., Takashima, H., Inayama, S., Nakanishi, S. (1983). Nucleotide sequences of cloned cDNA's for two types of Bovine brain substance P precursor. Nature 306:32-36.
11. Amara, S.G., Jonas, V., Rosenfeld, M.G., Ong, E.S., and Evans, R.M. (1982). Alternative RNA processing in calcitonin gene expression generates mRNAs encoding different polypeptide products. Nature 298:240-244

Molecular Biology of Development, pages 561–572

REGULATION OF GROWTH HORMONE GENE TRANSCRIPTION BY GROWTH HORMONE RELEASING FACTOR[1]

M. Barinaga[@*2], L.M. Bilezikjian[@], G. Yamomoto[@], C. Rivier[@], W.W. Vale[@], M.G. Rosenfeld[*] and R.M. Evans[@]

[@]The Salk Institute, San Diego, CA 92138, and
[*]University of California, San Diego, La Jolla, CA 92037

ABSTRACT. Growth hormone (GH) is the major product of the somatotrophic cells of the anterior pituitary. The release of GH from the somatotrophs is under the regulation of the hypothalamic peptides, GH releasing factor (GRF) and somatostatin (SS), which respectively stimulate (1-3) and inhibit (4) GH secretion. We have used a nuclear run-off transcription assay to show that GRF rapidly stimulates GH transcription, both in whole animals and in primary cultures of pituitary cells (5). Although GRF stimulated release of GH is Ca^{++} dependent, the transcriptional induction is Ca^{++} independent, and is also independent of a SS block of GH release. GRF raises cyclic AMP (cAMP) levels in pituitary cells, however release of GH can be achieved, without influencing cAMP levels, by treatment of cell cultures with depolarizing concentrations of K^+, or the phorbol ester, phorbol myristate acetate (PMA). Neither K^+ nor PMA influences GH gene transcription.

[1]This work was supported by NIH grant #AM 26741, and conducted in part by the Clayton Foundation for Research, California Division. L.M.B. was supported by NIH postdoctoral fellowship #AM 06864.

[2]Present address: Department of Biological Sciences, Stanford University, Stanford, California.

Forskolin, which stimulates adenylate cyclase and causes a rapid increase in cAMP levels in pituitary cells, stimulates GH gene transcription and GH release. Our data suggest that GRF regulates GH gene transcription by a cAMP mediated mechanism, which can be uncoupled from the GH release response to GRF (6).

INTRODUCTION

The hypothalamic regulatory factors are released by the neurosecretory cells of the hypothalamus into the portal blood system which carries them to the pituitary (7). All but one of the known hypothalamic regulatory factors are peptides. These factors were first identified on the basis of their ability to either stimulate or inhibit secretion of stored hormone from the secretory cells of the anterior pituitary. Several of these regulatory factors have subsequently been shown to affect the synthesis of pituitary hormones (7,8). In three cases, this effect has been demonstrated to occur at the transcriptional level: thyrotropin releasing hormone (TRH) stimulates (9,10), and dopamine inhibits (11) transcription of the prolactin gene, and we have shown that GRF stimulates GH gene transcription (5). The fact that a single polypeptide induces both hormone release and gene transcription raises the important issue of a potential linkage between these two processes. Using primary cultures of rat anterior pituitary cells, which exhibit physiologically normal release and transcriptional responses, we have determined that GRF induction of GH gene transcription occurs independently of GH release (6).

RESULTS

Synthetic Peptides

Vale and coworkers recently purified two polypeptides with GH releasing activity, a 40 amino acid GRF from a human pancreatic islet tumor (1), and a 43 amino acid GRF from rat hypothalamic extracts (3). Synthetic human pancreatic (hp) and rat hypothalamic (rh) GRF were used in the following studies. Rat pituitary cells respond to both peptides, but are 10 times more sensitive to the rhGRF.

GRF Stimulates GH Gene Transcription

We treated 3-day primary cultures of rat anterior pituitary cells with 10 nM hpGRF for one hour. After hormone treatment, nuclei were harvested and the transcription rates of individual genes were directly measured by a run-off transcription assay as previously described (5,6). Specific transcription rates for GH and control genes were determined as parts per million (ppm) of total labelled RNA. GRF treatment caused an approximately two-fold stimulation of GH gene transcription, while transcription of the prolactin gene remained unchanged (figure 1). Transcription rate was also measured for CHOB, a gene which is expressed constitutively in all mammalian tissues and cell lines examined (12). The CHOB transcription rate was about 10 ppm and did not change with GRF treatment (data not shown), indicating that GRF does not have a general stimulatory effect on transcription rates,

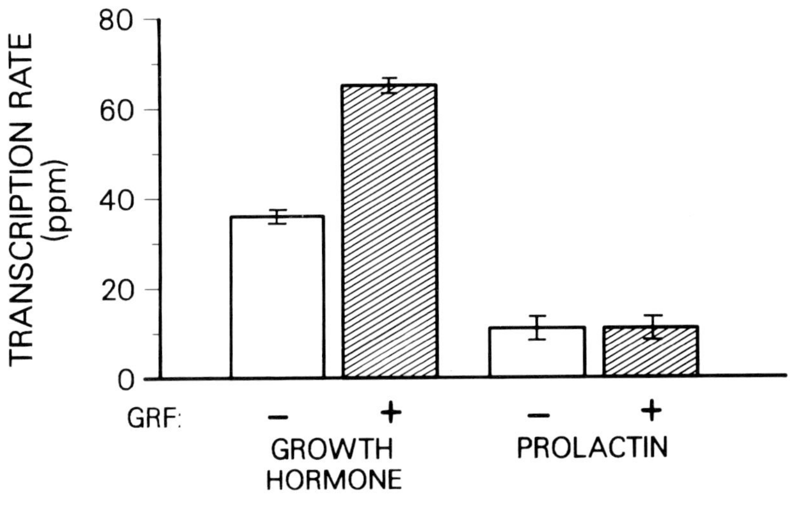

FIGURE 1. Transcriptional response of cultured pituitary cells to hpGRF. Parallel plates of cells were incubated for one hour in the presence (+) or absence (-) of 10 mM hpGRF. Preparation of cells and hormone treatment was as previously described (5). Transcription rates were determined by run-off assays as previously described (5).

but rather specifically stimulates transcription of the GH gene.

To test whether this transcriptional stimulation occurs *in vivo*, rats were anaesthetized with Nembutal, to repress endogenous GRF production (13), and fifteen minutes later, injected with either hpGRF (100 μg/kg, intravenously), or saline. Pituitaries were harvested 10, 15 and 30 minutes after GRF injection, and GH transcription rate was determined. Figure 2 shows that GRF stimulates GH gene transcription *in vivo*, and that the transcriptional stimulation increases over a 10 to 30 minute time period, from 1.8-fold to 2.5-fold.

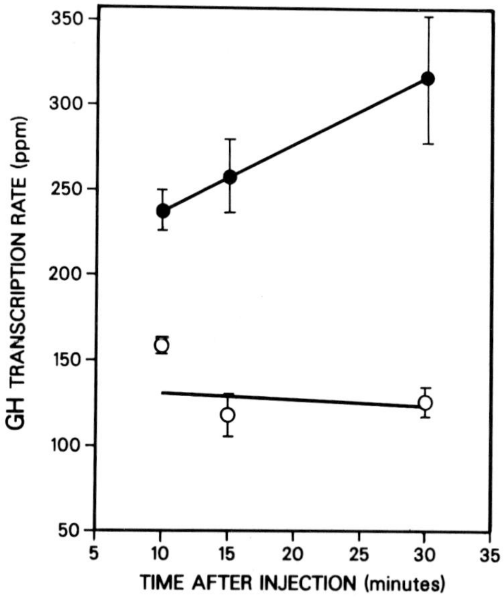

FIGURE 2. *In vivo* response of GH gene transcription rate to hpGRF injection. GH transcription rates in ppm were determined in pituitaries isolated from anaesthetized rats 10, 15 or 30 minutes after intravenous injection of hpGRF (●) or saline (○). Each point represents the mean +/- s.e. of hybridization values for 4 individual rats, except for the 10 minute points, which were determined from 2 individual rats. Treatment of rats and preparation of nuclei was as previously described (5). Transcription rates were determined as in legend to figure 1.

Transcriptional Stimulation Occurs Independently of GH Release

Having determined that GRF stimulates transcription of the GH gene, we designed experiments to determine whether there is a linkage between GH release and this transcriptional stimulation. Newly available synthetic rhGRF (3) was used for the remainder of the experiments reported here, and for each experiment GH release and intracellular cAMP levels were measured, as well as GH gene transcription.

To determine whether GH release is required for GRF stimulation of GH gene transcription, we performed experiments under conditions which block GH release. Secretion of GH is dependent on the presence of free Ca^{++} in the medium (14,15), and can be prevented by incubation of the cells in medium containing EGTA, which chelates Ca^{++}, or verapamil, which blocks voltage dependent Ca^{++} channels. Additionally, GH release can be blocked with SS, the hypothalamic GH release inhibiting factor (4). Figures 3 and 4 show that with each of these treatments, GH release was prevented, however GRF stimulation of GH gene transcription was not diminished, suggesting that the transcriptional stimulation is not dependent on GH release. Neither EGTA, verapamil, nor SS eliminated the GRF stimulated elevation of cAMP levels, although the magnitude of the rise differed with the three treatments. EGTA did not alter the GRF effect on cAMP levels, verapamil seemed to enhance the effect and SS diminished it.

Although GH release seemed not to be required for GRF stimulation of GH gene transcription, the possibility still existed that GH release alone could stimulate transcription of the GH gene. To investigate this possibility, we used three different treatments known to stimulate GH release (figure 5). Incubation of pituitary cells for one hour in depolarizing concentrations of K^+ caused GH release (16,17), but had no effect on cAMP levels or GH gene transcription. The phorbol ester tumor promoter, phorbol myristate acetate (PMA,18,19) caused release of GH (20) without changing cAMP levels or GH gene transcription. Forskolin, a plant diterpene which stimulates adenylate cyclase (21), increased GH release, GH gene transcription and intracellular cAMP levels. Table 1 summarizes the results presented in figures 3, 4 and 5.

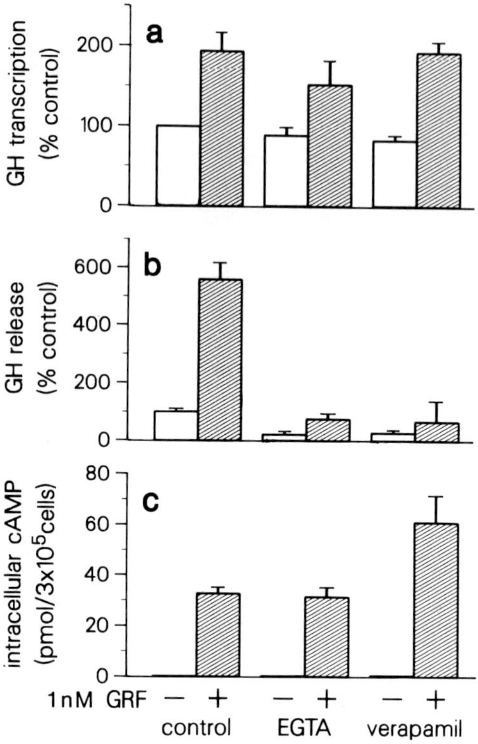

FIGURE 3. Effects of EGTA and verapamil on rhGRF stimulation of GH transcription (a), GH release (b), and intracellular cAMP levels (c). Parallel plates of cells were treated with 1nM rhGRF (+) or GRF dilution buffer (-). Where indicated, the medium contained 2.5 mM EGTA or 100μM verapamil. GH transcription rate and release were determined from the same experiment, and are expressed as % of mock treated control, and cAMP was measured in a separate, parallel experiment, and is expressed as pmol cAMP/3x10^5 cells. Cell preparation, hormone treatment and transcription assays were carried out as in legend to figure 1, with modifications as previously described (6). GH release and cAMP levels were determined by RIA (14).

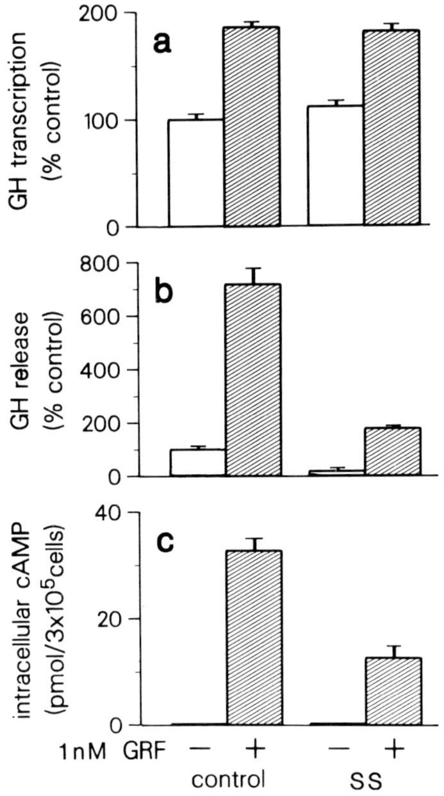

FIGURE 4. Effect of SS on GRF stimulation of GH transcription (a), GH release (b), and intracellular cAMP levels. Parallel plates of cells were treated with rhGRF (+) or GRF dilution buffer (-), in the presence or absence of 10 nM SS in the medium. Methods were the same as in legend to figure 3.

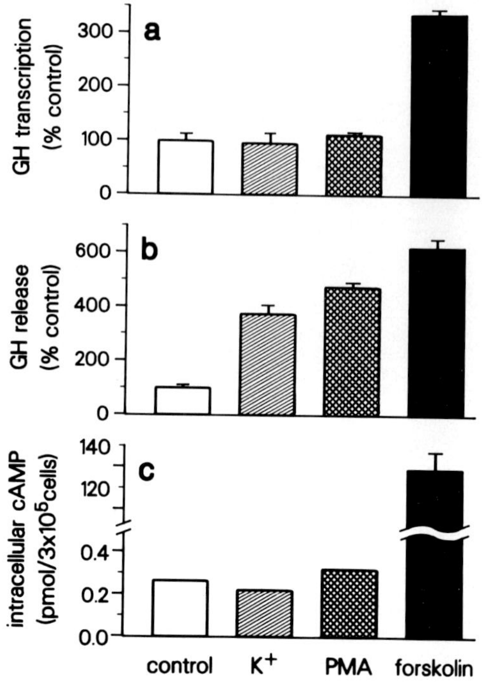

FIGURE 5. Effects of K^+, PMA and forskolin on GH transcription (a), GH release (b), and intracellular cAMP levels (c). Parallel plates of cells were incubated in the presence or absence of 58 mM K^+, 100 nM PMA or 10 μM forskolin. In (a) and (b), values for each treatment are expressed as % of the control value for that treatment. In (c) absolute values are shown. Controls for (c) did not differ significantly. Methods were the same as in legend to figure 3.

DISCUSSION

We have shown that GRF regulates GH synthesis at the level of transcription of the GH gene, and that this transcriptional stimulation occurs both in vivo and in cultures of normal pituitary cells. The rapid onset of the transcriptional response suggests that it is an independent, receptor mediated event.

Table 1

Treatment	GH release	GH gene transcription	cAMP levels
GRF	+	+	+
GRF + EGTA	-	+	+
GRF + verapamil	-	+	+
GRF + SS	-	+	+
K+	+	-	-
PMA	+	-	-
Forskolin	+	+	+

Table 1 Effects of various treatments on GH release, GH gene transcription and intracellular cAMP levels in primary cultures of rat anterior pituitary cells. (-), no effect, (+), stimulation. Concentrations used: GRF, 1nM, EGTA, 2.5mM, verapamil, 100μM, SS, 10nM, K+, 58mM, PMA 100nM and forskolin, 10μM.

We have investigated the possible linkage between the stimulation of GH gene transcription and GH release. Our results, which are summarized in table 1, demonstrate that induction of GH release can readily be uncoupled from stimulation of GH transcription. When cells were treated with GRF under conditions which block GH release, the transcriptional induction was unaffected, and, conversely, treatment with K+ and PMA stimulated GH secretion, but had no effect on transcription of the GH gene. Our conclusion from these results is that GH release is not the stimulus for the increase in GH transcription. Since EGTA and verapamil did not block the transcriptional stimulation, it appears that the events leading to transcriptional stimulation do not include a Ca^{++} dependent step.

The data in table 1 show a correlation between a rise in cAMP levels and stimulation of GH gene transcription. All of the treatments which resulted in elevated cAMP levels increased GH gene trancription, while K^{+} and PMA, which had no effect on cAMP levels, did not influence transcription. This correlation suggests that cAMP might be the intracellular mediator of GRF stimulation of GH gene transcription. Cyclic AMP stimulation of gene transcription might be mediated by phosphorylation of putative nuclear regulatory proteins by cAMP-dependent protein kinase.

Stimulation of prolactin transcription by cAMP analogs has been associated with the phosphorylation of nuclear proteins (22), and we have observed phosphorylation of histone H-1 at serine-37 in normal pituitary cells, in response to treatment with GRF or the cAMP analog, 8-Bromo cAMP (M. Waterman and M.B., unpublished observations). Serine-37 of H-1 histone is the substrate of cAMP dependent protein kinase _in-vitro_ (23), and this result therefore suggests that GRF stimulates cAMP mediated phosphorylation of nuclear proteins.

In figure 6 we propose a model for GRF action on the somatotrophs. Upon binding to its receptor (24), GRF causes a rapid rise in intracellular cAMP levels (14). Cyclic AMP may then mediate the responses of the cell to GRF, via stimulation of cAMP dependent protein kinase, the apparent mediator of all cellular responses to cAMP in eukaryotes (25). Several lines of evidence suggest that a rise in cAMP levels can induce Ca^{++} dependent GH release (14-17), however the possibility also remains that the GRF-receptor interaction may contribute a component to this stimulation, independent of cAMP. We have represented an influx of Ca^{++} as the trigger for GH release, however this has only been indirectly demonstrated (15). Extracellular Ca^{++} is not required for GRF stimulation of GH gene transcription, suggesting that the transcriptional stimulation occurs by a sequence of events that is at least partially independent of

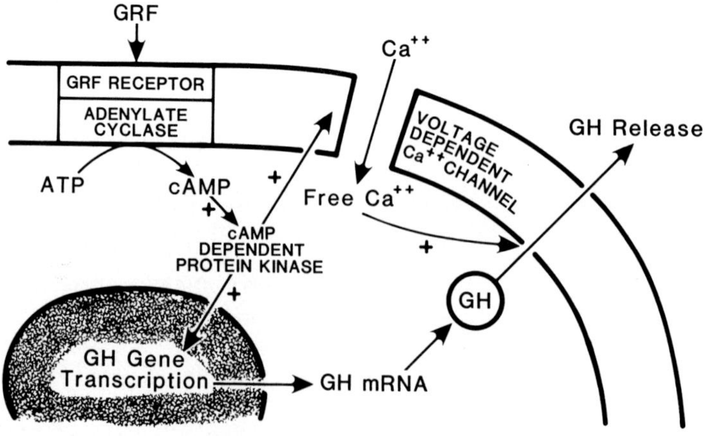

FIGURE 6. A model for GRF action on the somatrophs.

the pathway leading to GH release. We suggest that GRF stimulation of GH gene transcription is mediated by cAMP through the phosphorylation of putative nuclear regulatory proteins.

CONCLUSION

The data presented here demonstrate that GRF stimulates GH gene transcription as well as GH release in primary cultures of normal pituitary cells, and that this reflects the response to GRF in vivo. Moreover, we have shown that the release and transcriptional effects of GRF can be uncoupled from one another; GRF stimulates GH gene transcription in the absence of GH release, and GH release alone does not stimulate GH gene transcription. While GH release is dependent on extracellular Ca^{++}, GRF stimulation of GH gene transcription does not share this requirement, suggesting that these responses occur by at least partially independent mechanisms. We have observed a correlation between elevation of intracellular cAMP levels and stimulation of GH gene transcription, which suggests to us the possibility that cAMP is the intracellular mediator of the transcriptional stimulation by GRF.

REFERENCES

1. Rivier, J., Speiss, J., Thorner, M. and Vale, W. Nature 300, 276-278 (1982).
2. Guillemin, R. et al. Science 218, 585-587 (1982).
3. Speiss, J., Rivier, J. and Vale, W. Nature 303, 532-535 (1983).
4. Brazeau, P., Vale, W., Burgus, N., Ling, N., Butcher, M., Rivier, J. and Guillemin, R. Science 179, 77-78 (1973).
5. Barinaga, M., Yamomoto, G., Rivier, C., Vale, W., Evans, R. and Rosenfeld, M.G. Nature 306, 84-85 (1983).
6. Barinaga, M., Bilezikjian, L.M., Vale, W.W., Rosenfeld, M.G. and Evans, R.M. (manuscript submitted)
7. Vale, W., Rivier, C. and Brown, M. Ann. Rev. Physiol. 39, 473-527 (1977).
8. Vale, W., Vaughan, J., Yamomoto, G., Spiess, J. and Rivier, C. J. Endocrinology 112, 1553-1555 (1983).

9. Potter, E., Nicolaisen, A.K., Ong, E.S., Evans, R.M. and Rosenfeld, M.G. Proc. Nat. Acad. Sci. USA 78, 6662-6666 (1981).
10. Murdoch, G.H., Franco, R., Evans, R.M. and Rosenfeld, M.G. J. Biol. Chem. 258, 15329-15335 (1983).
11. Maurer, R. Nature 294, 94-97 (1981).
12. Harpold, M.M., Evans, R.M., Salditt-Georgieff, M. and Darnell, J.E. Cell 17, 1025-1035 (1979).
13. Arimura, A., Saito, T. and Schally, A.V. Endocrinology 81, 235-245 (1967).
14. Bilezikjian, L.M. and Vale, W.W. Endocrinology 113, 1726-1731 (1983).
15. Spence, J.W., Sheppard, M.S. and Kraicer, J. Endocrinology 106, 764-769 (1980).
16. Kraicer, J. and Spence, J.W. Endocrinology 108, 651-657 (1981).
17. Kraicer, J. and Chow, A.E.H. Endocrinology 111, 1173-1180 (1982).
18. Castagna, M., Takai, Y., Kaibuchi, K., Sano, K., Kikkawa, U. and Nishizuka, Y. J. Biol. Chem. 257, 7847-7851 (1981).
19. Niedel, J.E., Kuhn, L.J. and Vandenbark, G.R. Proc. Natl. Acad. Sci. 80, 36-40 (1983).
20. Smith, M. and Vale, W. Endocrinology 107, 1425-1431 (1980).
21. Seamon, K.B. and Daly, J.W. J. Cyclic Nuc. Res. 7, 201-224 (1981).
22. Murdoch, G.H., Rosenfeld, M.G. and Evans, R.M. Science 218, 1315-1317 (1982).
23. Zeilig, C.E., Langan, T.A. and Glass, D.B. J. Biol. Chem. 256, 994-1001 (1981).
24. Siefert, H. et al, Submitted to Nature (1984).
25. Kuo, J.F. and Greengard, P. Proc. Natl. Acad. Sci. USA 64, 1349-1355 (1969).

Molecular Biology of Development, pages 573–588

ISOLATION OF A cDNA CLONE CODING FOR A KALLIKREIN ENZYME IN MOUSE ANTERIOR PITUITARY TUMOR CELLS[1]

James Douglass, Kathleen Ranney, Michael Uhler,[2] Garrick Little and Edward Herbert

Department of Chemistry, University of Oregon
Eugene, Oregon 97403

ABSTRACT A cDNA clone coding for a member of the kallikrein gene family has been isolated from an AtT-20 (a mouse anterior pituitary tumor cell line) cDNA library. The frequency of appearance of this clone type suggests high levels of a single species of kallikrein mRNA in AtT-20 cells. The deduced amino acid sequence of this kallikrein is different from those previously reported. A synthetic oligonucleotide probe has been designed to distinguish this species of kallikrein mRNA from as many as 30 additional species in the mouse. Northern blot analysis using either cDNA or oligonucleotide as a hybridization probe reveals that while mouse neurointermediate pituitary and pancreas contain kallikrein mRNA, only the neurointermediate pituitary contains the same species found in AtT-20 tumor cells. The possible involvement of this kallikrein in the proteolytic processing of the neuroendocrine peptide precursor, pro-opiomelanocortin is discussed.

INTRODUCTION

In the past 20 years numerous small peptides have been discovered that mediate diverse functions in animals. Some of these peptides may serve both as neurotransmitters

[1]This work was supported by National Institutes of Health Grants AM16879, AM30155, and DA02736.

[2]Present address: Department of Biochemistry, University of Washington, Seattle, Washington 98195.

or neuromodulators and as neurohormones. Almost all of these neuroendocrine peptides are originally synthesized in the form of a large precursor molecule. These precursor proteins are biologically inactive as neuroendocrine peptides and must undergo a series of proteolytic cleavages in order to generate biologically active peptides and to be secreted. The bioactive peptide domains within the majority of neuropeptide precursors characterized to date are flanked by pairs of basic amino acid residues which are thought to be the sites of cleavage by trypsin-like endopeptidases (1).

Pro-opiomelanocortin (POMC) is perhaps the best characterized neuroendocrine peptide precursor. The primary structure of POMC reveals that the neuroendocrine peptides ACTH (adrenocorticotropin), β-endorphin, β- and γ-LPH (lipotropin), and α-MSH (melanocyte stimulating hormones) are contained within the precursor molecule (2). All of the biologically active domains are flanked by pairs of basic amino acids, most commonly lysine and/or arginine residues, suggesting that tryptic-like cleavages are involved in the release of the neuroendocrine peptides from the precursor molecule. Another interesting aspect of POMC processing is that different sets of end products are secreted from the various lobes of the mouse and rat pituitary. In the anterior lobe and AtT-20 cells (a mouse anterior pituitary tumor cell line composed of a pure population of POMC-producing cells) ACTH and β-LPH are the major peptides that are secreted from anterior lobe corticotrophs (3,4) whereas in the neurointermediate lobe these peptides are further cleaved to form α-MSH, corticotropin-like intermediate lobe peptide (CLIP), β-endorphin and derivatives of β-endorphin (3,5). Molecular genetic studies indicate that there is only one POMC gene and one POMC mRNA expressed in mouse (6) and rat (J. Drouin, personal communication) pituitary. Therefore, it appears likely that differences in activity or specificity of processing enzymes or differences in the intracellular environment at the site of enzyme action account for tissue-specific processing of POMC.

Although no definitive identification of a POMC processing enzyme has yet been made, a trypsin-like serine endopeptidase activity has been detected in extracts of the anterior and neurointermediate lobe of rat pituitary (7,8). This protease is capable *in vitro* of converting β-LPH to β-endorphin by specifically cleaving the peptide bond between Arg^{60} - Tyr^{61}. Rat pituitary also has been found to contain a kininogenase enzyme highly concentrated in the neurointermediate lobe (9). This enzyme, which belongs to a large subclass of trypsin-like serine proteases known as kallikreins (10,11), is capable of converting kininogen to peptides (kinins) that regulate blood pressure in the kidney and other tissues (11). The presence of high levels of this enzyme in a gland in which over 30% of the protein is derived from POMC and all of the cells are engaged in synthesizing and processing POMC has suggested to some that a kallikrein-like activity participates in the processing of POMC (9). This suggestion is further supported by the fact that kallikreins are involved in the generation of biologically active growth factors from inactive precursor molecules. The most thoroughly characterized of these kallikreins are epidermal growth factor binding protein (EGF-BP) and nerve growth factor subunits (γ-NGF) (12-14). These proteases are responsible in vivo for the liberation of EGF and NGF from their respective precursor molecules.

Shine and co-workers have demonstrated the existence of as many as 30 different kallikrein genes in the mouse (15). To investigate the possible involvement of the products of one or more of these genes in POMC processing, we have used molecular cloning techniques to determine if kallikrein mRNA is present in AtT-20 cells. This mouse anterior pituitary cell line synthesizes and properly processes large amounts of POMC and has been used as a model system in which to study the proteolytic processing of POMC (16). We report here the identification and analysis of a kallikrein cDNA clone isolated from an AtT-20 cDNA library. An oligonucleotide probe has also been synthesized in order to distinguish this species of mouse kallikrein mRNA from species of kallikrein mRNA which may be transcribed from other mouse kallikrein genes.

MATERIALS AND METHODS

AtT-20 cDNA Library Construction and Screening

Total RNA was prepared from AtT-20 tumor cells according to the guanidine thiocyanate procedure of Chirgwin et al. (17). Following the isolation of poly A+ mRNA from total AtT-20 RNA, cDNA was synthesized and cloned into the plasmid vector pBR322 via the use of EcoR1 and HindIII linkers (18). The AtT-20 cDNA library consists of approximately 50,000 recombinant cDNA clones. Bacterial colonies were transferred to 0.4 μm nitrocellulose filters and hybridized to ^{32}P-labelled pMK1 cDNA (19). Following purification of hybridization-positive clones, plasmid DNA was isolated and subjected to restriction analysis. cDNA inserts were subsequently subcloned into EcoR1/HindIII restricted M13 RF DNA (20) and the nucleotide sequence was determined according to the dideoxy chain termination method of Sanger (21).

Oligodeoxynucleotide Synthesis

The oligonucleotide 5' CATCTGGGTATTCATATT 3' was synthesized by the phosphotriester method (22) on a Biosearch Sam I automated DNA synthesizer. The oligonucleotide was purified by polyacrylamide gel-electrophoresis followed by gel-filtration on a Sephadex G-50 superfine column (Pharmacia).

DNA Dot Blot and Northern Blot Analysis

For DNA dot blot analysis, 1 μg of pBR322, pMAK3 or pMK1 was denatured in .5M NaOH followed by neutralization in 1M ammonium acetate. The samples were applied to .4 μm nitrocellulose and baked onto the filter. Hybridization with ^{32}P-labelled oligonucleotide was carried out overnight at the temperatures indicated in Figure 4. The filters were then washed at room temperature with 3 changes of 6 x SSC before autoradiography.

For Northern blot analysis, the RNA samples were prepared and electrophoresed on 1.7% agarose gels, transferred to nitrocellulose and hybridized with ^{32}P-labelled pMK1 insert according to Thomas (23). Hybridization with radiolabelled oligomer was carried out at the temperatures indicated in 4 x SSC, 1x Denhardt's, 10mM EDTA, 200 μg/ml single stranded herring sperm DNA, and 0.1% sarcosyl. Blots probed with the oligomer were washed in 4 x SSC, 1mM EDTA, 0.1% pyrophosphate and 0.1% sarcosyl for 3x 10 minutes at a temperature 10°C below the hybridization temperature. Blots were exposed to x-ray film for 16 hours at -70°C with intensifying screens.

RESULTS

AtT-20 tumor cells have been used as a model system in which to study the expression and proteolytic processing of POMC (16). If a kallikrein is involved in any of these proteolytic processing reactions, then kallikrein mRNA should be present in this cell line. Accordingly, an AtT-20 cDNA library was constructed using poly A+ mRNA isolated from AtT-20 cells. Following the generation of cDNA, Eco R1 and HindIII linkers were used to insert the cDNA into the plasmid vector, pBR322. As a result of this procedure, any cDNA containing endogenous EcoR1 and/or HindIII sites would be restricted at internal sites, thus precluding the isolation of a full-length cDNA clone. This cDNA library was screened with the HindIII insert from a kallikrein cDNA clone (pMK1) previously isolated from a mouse submaxillary gland cDNA library (27) (kindly provided by Drs. R. Richards and J. Shine). Approximately 0.1% of the cDNA clones hybridized with this cDNA probe, suggesting that kallikrein mRNA is present at high levels in AtT-20 cells. Northern blot analysis (not shown) confirmed the presence of a species of mRNA in AtT-20 cells which hybridized to the same kallikrein cDNA probe. From these data, we estimated that approximately 0.5% of the total AtT-20 poly A+ mRNA population codes for kallikrein. These values are approximately 6-fold lower than the level of POMC mRNA present in the same cell type. Approximately 3% of the total poly A+ mRNA population in AtT-20 cells codes for POMC (21) while 0.6% of the cDNA clones from the identical AtT-20 cDNA library contain POMC inserts. Thus, both POMC and kallikrein mRNA are present in relatively high levels in AtT-20 cells.

In order to determine how many species of kallikrein mRNA are synthesized in AtT-20 cells, 42 individual kallikrein-positive cDNA clones were isolated. The plasmid DNA was purified and subjected to restriction analysis while the cDNA inserts were subcloned into M13 RF DNA for full nucleotide sequence analysis. Figure 1 shows that 5 different classes of cDNA clones were observed as judged by restriction analysis. However, DNA sequencing revealed that each clone type contained a cDNA insert that was generated from the same species of kallikrein mRNA (nucleotide sequence analysis to be published elsewhere). Thus, AtT-20 cells contain relatively high levels of a unique species of kallikrein mRNA, the sequence of which is different from those of previously published mouse

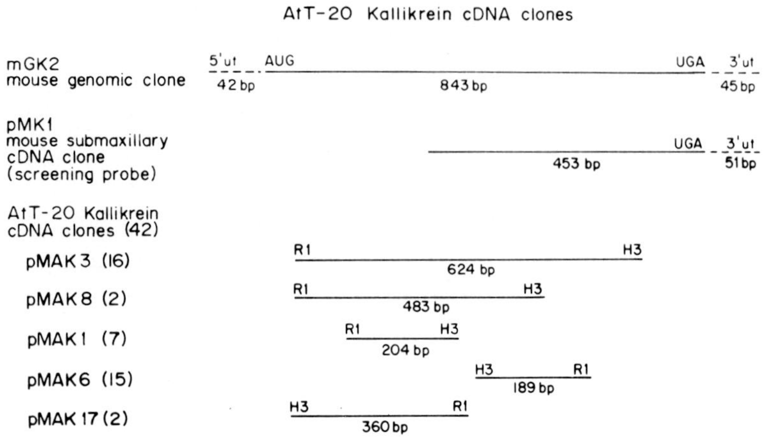

FIGURE 1. Schematic diagram of AtT-20 kallikrein cDNA clones. The structure of a full-length kallikrein mRNA is represented by mGK2 (15). Solid line represents coding region while dotted lines represent untranslated regions (ut). The partial kallikrein cDNA clone used to screen an AtT-20 cDNA library is designated as pMK1 (27). The classes of AtT-20 kallikrein cDNA clones and the numbers of each isolated class are shown. The restriction sites present at the termini of these clones are designated as R1 (EcoR1) or H3 (HindIII).

kallikrein cDNAs. This conclusion is more clearly illustrated in Figure 2. Here, the predicted amino acid sequence of the AtT-20 kallikrein (as determined from the nucleotide sequence of pMAK3) is compared to the primary amino acid sequence of other mouse kallikreins. Some of these amino acid sequences were deduced from the nucleotide sequence of mouse kallikrein genomic DNA clones (mGK-1 and mGK-2 from ref. 15), while others were deduced from the nucleotide sequence of kallikrein cDNA clones isolated from mouse submaxillary gland cDNA libraries (EGF-BP, ref. 25; γ-NGF, ref. 26; pMK-1, ref. 27; pMK-2, ref. 28). Although the AtT-20 kallikrein exhibits a great deal of amino acid

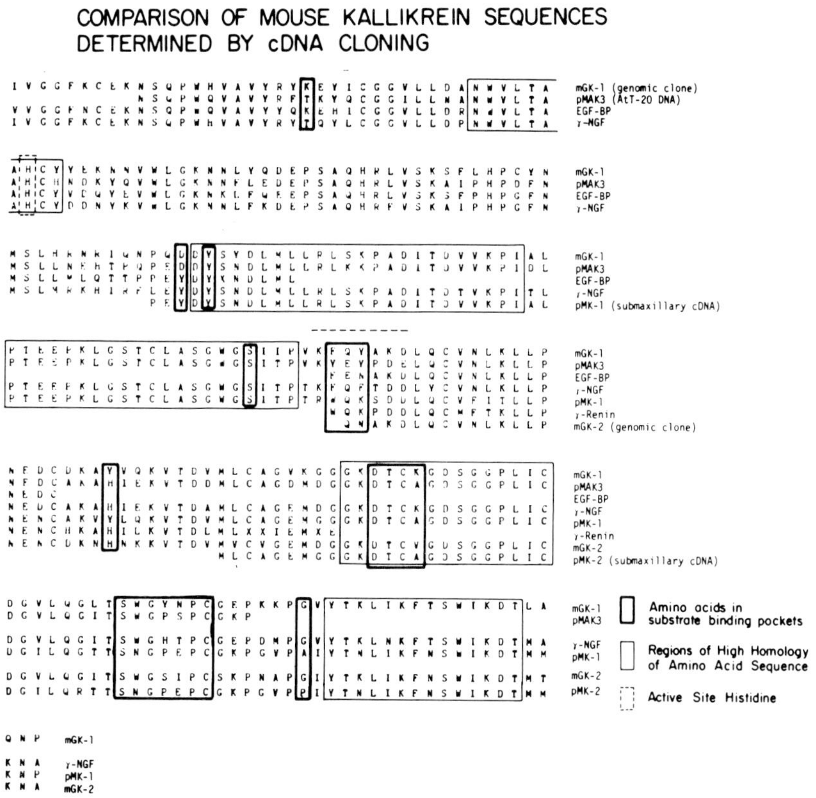

FIGURE 2. Comparison of mouse kallikrein amino acid sequences as determined by cDNA cloning. The cDNA sequences from which these amino acid sequences were deduced are noted and referenced in the text. Heavily boxed regions represent amino acids believed to line the substrate binding pocket of serine proteases (15). Lightly boxed regions represent portions of the kallikrein exhibiting a high degree of amino acid sequence homology. The dashed box denotes the position of a highly conserved histidine residue present in the active site of the molecules (29). The dotted line overlines a region of pronounced amino acid sequence heterogeneity.

sequence homology (approximately 80%) when compared with other mouse kallikreins,there are localized regions of sequence heterogeneity. Some of these regions comprise amino acid residues which are believed to constitute the substrate-binding pocket of serine proteases (15). Indeed, it may be that these regions of the protein are involved in determining the substrate specificity of the kallikreins because there is clearly less homology in this region than in the remainder of the sequence. The regions which are highly homologous between AtT-20 kallikrein and those listed may be necessary for the formation of a proper 3-dimensional structure which may in turn confirm the role of these proteins as serine proteases. The major point to be made from this figure, however, is that AtT-20 cells contain high levels of a species of mRNA which codes for a member of the kallikrein family. This kallikrein is different from those previously identified and amino acid sequence comparison suggests that it may also have a unique substrate specificity.

Because of the high degree of homology seen at the nucleotide level, a single kallikrein cDNA probe can hybridize to numerous species of kallikrein mRNA. Therefore, cDNA probes cannot be used to study the distribution and expression of this particular species of kallikrein mRNA (to be referred to as AtT-20 kallikrein mRNA). For this reason, a synthetic oligonucleotide was designed which could be used to distinguish AtT-20 kallikrein mRNA from other species of kallikrein mRNA (Figure 3). A segment of seven amino acid residues showing a high degree of heterogeneity (overlined in Figure 2) was chosen and the nucleotide sequence coding for these residues was analyzed. Over an 18 base region, there is an average of 40% mismatch of bases between AtT-20 kallikrein mRNA and 5 other species of mouse kallikrein mRNA. Thus, an oligonucleotide 18 bases long with the sequence 5' CATCTGGGTATTCATATT 3' was synthesized. This oligomer should be capable of distinguishing AtT-20 kallikrein mRNA from other species of kallikrein mRNA when hybridization is carried out under the proper conditions. The plasmid dot blot in Figure 4 shows that when hybridization is carried out at 40° or 45°C, the oligomer hybridizes with complementary sequences present in pMAK3 (AtT-20 kallikrein cDNA)but gives no hybridization signal with pMK1 (mouse submaxillary kallikrein cDNA) or the plasmid vector, pBR322. At 50°C the oligomer can no longer hybridize with pMAK3 cDNA as this temperature is above the Tm for

Strategy for Selecting AtT-20 Kallikrein mRNA...
Oligonucleotide Probe Design

pMAK3/AtT-20 Kallikrein	5'	AA TAT GAA TAC CCA GAT G	3'
pMK3/pancreatic Kallikrein		GA TTG GAA TTT AGT GAT G	
mGK1/genomic clone		AG TCA CAA TAT GCA AAA G	
pMK1/submaxillary Kallikrein		GA TGG CAA AAG TCA GAT G	
pGK2/genomic clone		AA TTC CAA AAT GCA AAA G	
pmγN/submaxillary Kallikrein (γ subunit of nerve growth factor)		AA TTC CAA TTC ACA GAT G	
pMAK3 oliogonucleotide (AtT-20 Kallikrein mRNA specific probe)	3'	TT ATA CTT ATG GGT CTA C	5'

Mismatches are underlined
Average mismatch is 40%

FIGURE 3. Strategy for selectively monitoring AtT-20 kallikrein mRNA. . . oligonucleotide probe design. An 18 base nucleotide sequence (5' to 3')coding for the amino acid residues overlined in Figure 2 is shown for four species of mouse kallikrein cDNA and two different mouse kallikrein genes. The cDNA or genomic DNA fragments from which these sequences were obtained are noted in Figure 2 and in the text. The synthetic oligonucleotide, 18 bases long, which is capable of distinguishing AtT-20 kallikrein mRNA from other species of kallikrein mRNA is designated as pMAK3 oligonucleotide (3' to 5'). Mismatches to this complementary probe are underlined and average out to 40% in this particular 18 base sequence.

base pairing (based on calculated values). Thus, the oligomer is capable of distinguishing AtT-20 kallikrein mRNA from other species of kallikrein mRNA when hybridization is carried out in the 40° to 45°C temperature range.

If the kallikrein coded for by AtT-20 kallikrein mRNA is involved in the proteolytic processing of POMC, then tissues containing POMC may also be expected to contain AtT-20 kallikrein. In the mouse, the neurointermediate lobe of the pituitary contains high levels of POMC mRNA while no POMC mRNA is observed in the pancreas (21). The Northern blot shown in Figure 5 reveals that while kallikrein mRNA is present in both mouse neurointermediate pituitary and pancreas, only the neurointermediate lobe appears to contain AtT-20 kallikrein mRNA. When the cDNA insert from pMK1 was used as a hybridization probe, a uniquely-sized hybridization signal representing kallikrein mRNA was observed in the neurointermediate lobe (NIL). A relatively small amount of total NIL RNA was present on this blot, suggesting that the NIL contains high levels of kallikrein mRNA, perhaps higher than the levels seen in AtT-20 cells. We cannot say at this time how many species of kallikrein mRNA are present in the NIL due to the high degree of homology at the nucleotide sequence level among kallikrein mRNAs and the fact that many species of kallikrein mRNA

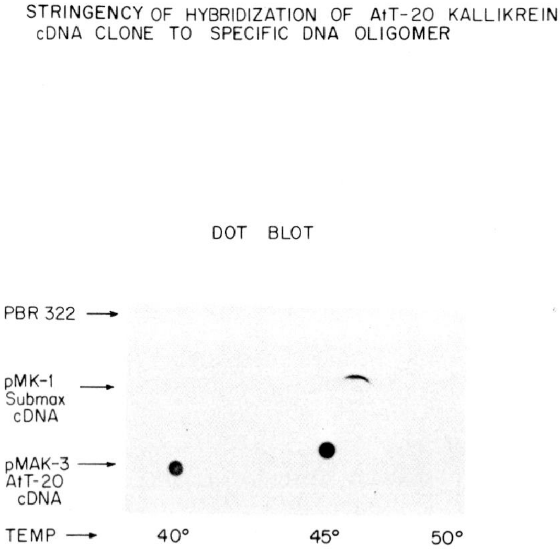

FIGURE 4. Hybridization of ^{32}P-labelled oligomer to pBR322, pMK1 and pMAK3 plasmid DNA. Dot blot analysis was performed under the conditions described in Materials and Methods. The temperatures at which hybridization was carried out are noted.

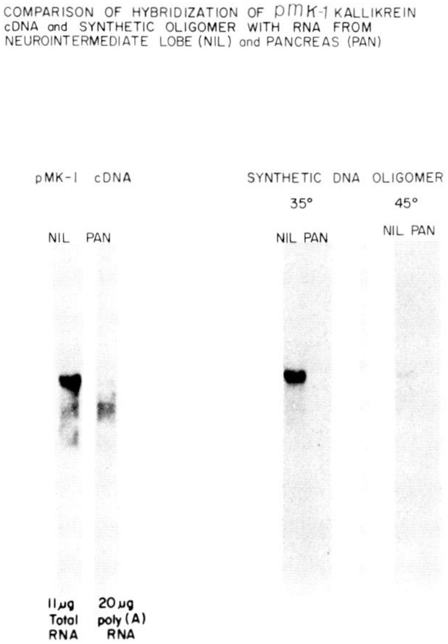

FIGURE 5. Northern blot analysis of mouse neurointermediate lobe and pancreas RNA. The RNA samples (total or poly A+) were prepared, electrophoresed on a 1.5% agarose gel and transferred to nitrocellulose paper as described in Materials and Methods. The filter was hybridized with 10^7 cpm of either nick-translated pMK1 insert (cDNA) or ^{32}P-labelled oligomer. Hybridization was performed overnight at 65°C with the cDNA probe or 35°/45°C with the oligomer.

are of similar size. The Northern blot also reveals the presence of kallikrein mRNA in the mouse pancreas. A similar size kallikrein mRNA is observed in this tissue as well as one or more smaller species of kallikrein mRNA. These smaller species may, however, represent degradation products of the larger form. The intensity of the hybridization signal for pancreas is much lower than that seen for the NIL, even though more poly A+ mRNA is present in the pancreas sample. This data again suggests that the NIL contains high levels of kallikrein mRNA.

When the oligomer, specific for AtT-20 kallikrein mRNA, was used as a hybridization probe on the identical Northern blot a strong hybridization signal was seen in the NIL RNA sample. The intensity of the signal was reduced as the hybridization temperature was raised from 35° to 45°C. This was not observed when the oligomer was hybridized to a complementary pMAK3 cDNA sequence (Figure 4), indicating that the oligomer·DNA hybrid is slightly more stable than an oligomer·RNA hybrid. An alternative explanation is that there are multiple species of kallikrein mRNA in the NIL, one of which does not form a perfect hybrid with the oligomer. This hybrid would then become destabilized upon increasing the temperature of hybridization. The size of the NIL mRNA that hybridizes to the oligomer is identical to that which hybridizes to the cDNA probe. These data suggest that AtT-20 kallikrein mRNA is present in fairly high levels in the mouse NIL. In contrast, no detectable AtT-20 kallikrein mRNA appears to be present in the mouse pancreas. At both hybridization temperatures with the oligomer no signal was apparent, even when the autoradiograph was overexposed. Thus, although pancreas contains kallikrein mRNA, it does not contain detectable levels of AtT-20 kallikrein mRNA.

DISCUSSION

Bioactive peptide domains in polyproteins are flanked by basic amino acid residues suggesting that trypsin-like endopeptidases are involved in cleaving these small peptides out of their respective precursors. In mammals, many enzymes have been implicated in precursor processing but to date only a handful of processing enzymes have actually been identified (EGF-binding protein and the γ subunit of the 7S NGF complex). Several candidate processing enzymes have been localized in mature secretory granules and shown to be active under conditions believed to exist in these granules. However, the processing by endopeptidases may occur at other sites in the cell where the conditions are different from those in mature secretory granules. POMC, for example, undergoes a series of cleavages and modifications which occur in a specific order as the polyprotein moves through the secretory pathway. Some of the cleavage reactions may occur in the endoplasmic reticulum, others in the Golgi or immature secretory granules. Since we do not know the conditions that exist in all of these sites in the cell, it is difficult to develop strict criteria for identifying processing enzymes on this basis.

The two identified precursor processing enzymes mentioned above belong to the subgroup of mammalian serine proteases known as kallikreins (30). These enzymes, which are glycoproteins, have a molecular weight range of 25-40,000, exhibit strict substrate specificity in vivo and are characterized by their ability to release kinins from kininogens in vitro. The diversity of action of these enzymes is consistent with their presence in a wide variety of tissues.

As mentioned before, Shine and co-workers have recently shown that there are at least 30 distinct kallikrein genes in the mouse and rat genome that exhibit a high degree of homology (15). Sequence analysis of mouse genomic DNA and cDNA clones has revealed that the limited variability in the coding region of the kallikrein genes lies in the amino acid sequences involving the substrate binding sites of the enzymes. This implies that kallikrein enzymes differ from one another in the type of substrate sequences that serve as recognition and cleavage sites (15).

The occurrence of a kallikrein-like activity in highly concentrated form in the neurointermediate lobe of rat pituitary (a major site of synthesis of POMC), suggests that this enzyme is involved in POMC processing (9). The isolation of a kallikrein cDNA clone from an AtT-20 cDNA library is consistent with this hypothesis. The level of kallikrein mRNA in AtT-20 cells is quite high. Up to 0.5% of the total poly A+ mRNA population may code for kallikrein. This value is about 6-fold lower than the level of POMC mRNA in the same cell line. Assuming that the translational efficiencies of these mRNAs are similar, for every six molecules of POMC there would be one kallikrein molecule. This observation is not surprising in view of the finding that the kallikrein involved in NGF processing is present at the same level as the precursor. Even though no large POMC-containing complex has been isolated to date, there is no evidence to suggest that the processing enzyme is present in limited quantity in cells containing POMC.

Although 42 out of 42 randomly picked kallikrein-positive cDNA clones appear to have been derived from the same species of kallikrein mRNA, we cannot rule out the possibility that another kallikrein mRNA is present in AtT-20 cells at low levels. By analyzing more kallikrein-positive cDNA clones with both cDNA and oligomer probes we should be able to determine if another kallikrein mRNA is present in these cells. This result will be critical for evaluating the role of kallikrein in POMC processing. Also, more nucleotide sequence data on the twenty or so uncharacterized mouse kallikrein genes is needed to determine unambiguously if our oligonucleotide probe is recognizing only one species of kallikrein mRNA. However, since the average mismatch between our oligonucleotide probe and the complementary region of six other mouse kallikrein genes is nearly 40%, we feel reasonably confident that our probe is specific for AtT-20 kallikrein mRNA.

Northern blot results with the oligomer probe reveal that the same species of kallikrein mRNA present in AtT-20 cells is also present in the mouse neurointermediate pituitary at high levels. Using the same technique, we also find this species of kallikrein mRNA in mouse anterior pituitary (data not shown). Both of these tissues synthesize and proteolytically process POMC, thus strengthening the speculation that this particular kallikrein is involved in POMC processing. The anterior pituitary in rodents is highly vascularized and since glandular kallikreins play an important role in regulating local blood flow, it could be argued that this particular kallikrein is performing this function. The neurointermediate lobe, however, is only slightly vascularized, if at all, so it seems unlikely that the highly abundant kallikrein in this particular tissue is playing a role in blood pressure regulation. Also, the number of species of kallikrein mRNA in these tissues has not been investigated and it may well be that the pituitary, the anterior lobe in particular, contain several species of kallikrein mRNA since this tissue is composed of a variety of cell types.

Many neuroendocrine peptide precursor molecules are synthesized in the brain. In the future, it will be extremely interesting to study the variety of kallikreins which may be present in this highly complex tissue. Already, a monoclonal antibody against purified rat urinary kallikrein has been coupled to agarose and used to isolate kallikrein from total rat brain (31). We have also isolated several cDNA clones from a rat preoptic cDNA library (courtesy P. Seeburg) that hybridize strongly with the AtT-20 kallikrein cDNA clone. Thus, it appears that one or more kallikrein genes are expressed in the rat brain.

ACKNOWLEDGEMENTS

JD was supported by an American Cancer Society postdoctoral fellowship PF-2051 and NIH Research Grant F32 GM09807-01. MU was supported by Molecular Biology Training Grant GM07759. EH(principal investigator) research is supported by National Institutes of Health grants NIDA DA02736, NIADDK AM16879 and NIADDK AM30155. The authors would also like to thank Drs. R. Richards and J. Shine for their generous gift of pMK1, and S. Engbretson for expert manuscript preparation.

REFERENCES

1. Docherty K, Steiner D (1982). Ann Rev Physiol 44:625.
2. Nakanishi S, Inoue A, Kita T, Nakamura M, Chang ACY, Cohen S, Numa S (1979). Nature 278:423.
3. Roberts JL, Phillips M, Rosa P, Herbert E (1978). Biochemistry 17:3609.
4. Hinman M, Herbert E (1980). Biochemistry 19:5395.
5. Mains RE, Eipper BA (1979). J Biol Chem 254:7885.
6. Uhler M, Herbert E, D'Eustachio P, Ruddle F (1983). J Biol Chem 258:9444.
7. Kenessey A, Graf L, Palkovits M (1977). Brain Res Bull 2:247.
8. Graf L, Kenessey M (1976). FEBS Lett 69:255.
9. Powers CA, Nasjletti A (1983). Endocrinology112:1194.
10. Fiedler F (1979).In Erdos E (ed): "Handbook of Experimental Pharmacology," Berlin: Springer-Verlag, p 103.
11. Eisen V, Vogt W (1970). In Erdos E (ed): "Handbook of Experimental Pharmacology," Berlin: Springer-Verlag, p 82.
12. Bothwell MA, Wilson WH, Shooter EM (1979). J Biol Chem 254:7287.
13. Frey P, Forand R, Maciag T, Shooter EM (1979). Proc Natl Acad Sci USA 76:6294.
14. Ullrich A, Gray A, Berman C, Dull TJ (1983). Nature 303:821.
15. Mason AJ, Evans BA, Cox DR, Shine J, Richards RI (1983). Nature 303:300.
16. Eipper BA, Mains RE (1980). Endocrine Reviews 1:1.
17. Chirgwin JM, Przybyla AE, McDonald RJ, Rutter WJ (1979). Biochemistry 24:5294.

18. Kurtz DT, Nicodemus CF (1981). Gene 13:145.
19. Grunstein M, Hogness DD (1975). Proc Natl Acad Sci USA 72:3961.
20. Messing J, Crea R, Seeburg PH (1981). Nucleic Acids Research 9:309.
21. Sanger F, Nicklen S, Coulson AR (1977). Proc Natl Acad Sci USA 74:5463.
22. Efimov VA, Reverdatto SV, Chakhmakhcheva OG (1982). Nucleic Acids Research 10:6675.
23. Thomas PS (1980). Proc. Natl Acad Sci USA 77:5201.
24. Uhler M (1982). PhD Thesis, University of Oregon, Eugene, Oregon.
25. Ronne H, Lundgren S, Severinsson L, Rask L, Peterson PA (1983). EMBO Jour 2:1561.
26. Ullrich A, Gray A, Hayflick J, Seeburg PH (1984). Proc Natl Acad Sci USA (in press).
27. Richard RE, Catanzara DF, Mason AJ, Morris BJ, Baxter JD, Shine J (1982). J Biol Chem 257:2758.
28. Nordeen SK, Mason AJ, Richards RE, Baxter JD, Shine J (1982). DNA 1:309.
29. DeHaen C, Neurath H, Teller DC (1975). J Mol Biol 92:225.
30. Schacter M (1980) Pharmacol Rev 31:1.
31. Chao J, Woodley C, Chao L, Margolius HS (1983). J Biol Chem 258:15173.

VI. MOLECULAR ASPECTS OF PLANT DEVELOPMENT

Molecular Biology of Development, pages 591–599

GENE SETS ACTIVE IN COTTONSEED EMBRYOGENESIS[1]

Leon Dure III, Caryl A. Chlan, Jean C. Baker, and Glenn A. Galau[2]

Department of Biochemistry, University of Georgia
Athens, Georgia 30602

ABSTRACT Several sets of genes active in the embryogenesis/early germination of the cottonseed have been defined. Two of these sets, one coding for developmentally regulated genes (seed storage proteins) and the other coding for genes that are up modulated in expression in late embryogenesis, were chosen to serve as tools for searching for sequences involved in regulating their expression. cDNA representing members of these two sets have been produced and used to isolate genomic DNA containing the genes and flanking and intron regions. Nearly full length cDNA clones representing each of the three storage protein gene families have been sequenced and the amino acid sequences deduced. The relatedness of these families and the processing steps involved in the conversion of these proteins to their final form can be deduced from these sequences.

INTRODUCTION

For the past several years we have studied some of the sets of genes that are expressed during the embryogenesis/early germination of the cottonseed. We have been interested in defining several sets of genes, each of which is regulated independently; that is, the individual members of each gene

[1]This work was supported by NSF, NIH and Agrigenetics Research Corporation.

[2]Present address; Department of Botany, University of Georgia, Athens, Georgia 30602.

set are expressed coordinately, whereas the sets themselves are each expressed for different periods during the developmental scheme.

We have utilized the 2D electrophoretic separation of extant protein, in vivo labelled protein, and in vitro labelled protein (from isolated mRNA) to define these gene sets. This was done by comparing the 2D protein patterns derived from seed tissue at a number of points during embryogenesis/germination. Proteins that changed in abundancy during this period were grouped into sets based on the coordination of their appearance/disappearance or rise/fall. Utilizing 2D electrophoresis limited this study to those genes whose protein products constitute the most abundant 300-400 proteins of the seed tissue. Furthermore, the bulk of this work has been confined to the cotyledon tissue of the seed.

Seven sets of genes have been recognized by these means (1,2,3). The temporal patterns of expression of these sets are not surprising from a developmental point of view; i.e., sets appear to rise/fall in expression in line with major developmental changes during this period of ontogeny. Two sets of genes appear to be expressed in unchanging levels throughout embryogenesis. One of these sets continues to be expressed in relatively high levels in early germination. The actin genes are members of this group (Figure 1). The proteins representing the other gene set are no longer synthesized in detectable amounts in germination, indicating that the expression of this gene set is embryo specific.

The high level of expression of a third gene set appears restricted to the period in which cotyledon cells are dividing, a fourth set is not expressed until germination commences, while the expression of two other sets is not detectable until late in embryogenesis. One of these sets possibly is regulated by the plant growth regulator abscisic acid (ABA) (1,2,3). The final set is comprised of the genes for the seed storage proteins whose synthesis far surpasses all other protein synthesis during most of embryonic cotyledon development (4,5).

The existence of these seven developmentally distinct gene sets has been confirmed by the analysis of the kinetics of cDNA/mRNA reassociation (6).

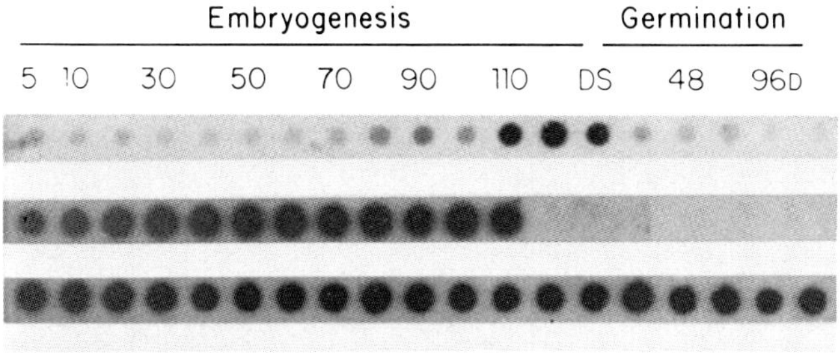

FIGURE 1. Dot blot of RNA obtained from developmental time points given at the top of the figure. Numbers referring to embryogenesis are the wet weight of the embryo, whereas those referring to germination development are hours of germination. The dots were hybridized with cloned probes representing gene sets as described in the text.

STORAGE PROTEIN VS ABA-REGULATED GENES

Our motive for describing these gene sets has been to use individual genes from several of the sets as tools for searching for DNA sequences that are involved in regulating their expression. To date, two sets have been chosen for this work. These are the storage protein gene set and the set that may be regulated by ABA. The rationale behind these choices is based on the conceptual dicotomy between intrinsic developmentally regulated genes and genes that are constitutively "open" but whose expression is up/down modulated. This distinction is illustrated in Figure 1, which is a "dot" blot of RNA, extracted from points during embryogenesis/germination, hybridized with cDNA probes representing a member of the storage protein gene set, a member of the putative ABA set, and an actin cDNA probe which shows an example of a constitutive highly expressed gene. Clearly the levels of the mRNA for the putative ABA gene are up/down modulated from a basal level during the developmental period, whereas the expression of the storage protein genes appears to be shut off in late embryogenesis. This relationship between the two types of genes was confirmed by cDNA/mRNA solution hybridization kinetic studies; however, it was found that traces of the storage protein mRNA could be detected by this technique in cotyledons after 24 hr of germination.

cDNAs representing members of these two gene sets were used to isolate genomic DNA sequences from several types of phage lambda/cotton DNA libraries. Preliminary experiments have shown that the cottonseed storage proteins emanate from three families of genes that are distinguishable on both a nucleotide and amino acid sequence level, although they share enough sequence homology to have arisen from a common gene ancestor. Several copies of each family exist in the genome although the precise number and genomic arrangement have not been determined to date. Southern blots suggest the existence of introns in the genes of each family as well. The mRNAs emanating from two of the families constitute about 15% each of the total mRNA in mid-embryogenic cotyledons, whereas the other family gives rise to mRNAs that constitute about 5% of the total (5).

In contrast, the putative ABA genes appear to exist as single-copy sequences/haploid genome. Since the species of cotton used in this work is a natural allotetraploid, each cDNA probe representing members of this set has been found in "hybrid arrest" experiments to arrest the synthesis of two proteins that are separable on 2D gels. When the various cDNAs representing members of this gene set were used in solution hybridization experiments to determine the degree of coordination in expression among these genes, the set clearly became divisible into two subsets, one set being up modulated in expression before the other in embryogenesis. The subset that is most likely regulated by the cellular ABA concentration has not been determined to date. The genomic DNA fragments representing members of each of the storage protein families and members of the two subsets up modulated in late embryogenesis are currently being sequenced in the search for sequence elements involved in the regulation of their synthesis.

SOME CHARACTERISTICS OF THE STORAGE PROTEINS

cDNAs representing the three storage protein families were obtained that are approximately equal in size to the mRNAs for these proteins. Assuming that nearly full length amino acid sequences could be deduced from the nucleotide sequences of these cDNAs, they were sequenced in order to further characterize these proteins. Three of the cloned cDNAs, each representing a storage protein family, that were sequenced are shown in Figure 2 and compared in length to their putative mRNAs (4). Two of the cDNAs (C 72, C 94)

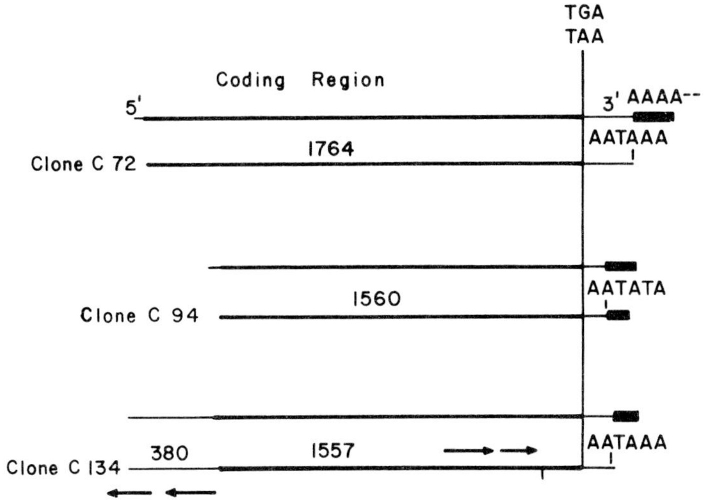

FIGURE 2. Characteristics of three cDNAs, each representing one of the seed storage protein gene families. Above the cDNAs (labelled C 72, C 94 and C 134) are the putative dimensions of their mRNAs of origin.

do not contain the 5' untranslated region, but come within 2-3 amino acids of the N-terminal methionine. The third cDNA (C 134) contains an extensive 5' untranslated region that contains in the inverse compliment orientation two perfectly repeated sequences found in the translated region. These inverted repeats, indicated in Figure 2, are 202 and 83 nucleotides in length and suggest the occurrence of genomic arrangements in the evolution of these genes. All three cDNAs contain the 3' untranslated region as far as the AATAAA poly(A) signal sequences; one cDNA (C 94) contains 66 bases of the 3' poly(A) tail. The lengths of the putative coding regions predict proteins that correspond in size to the known initial translation products of the storage protein mRNAs (7).

A number of aspects of storage protein processing could be deduced from these sequences including the sites of cleavage that result in the conversion of the preproproteins to the mature proteins of the dry cottonseed. The amino acid compositions of the proteins compared well with those arrived at by protein hydrolysis (8) and established the concentration of Gln and Asn in these proteins, which cannot be done by protein hydrolysis. Almost 50% of the nitrogen

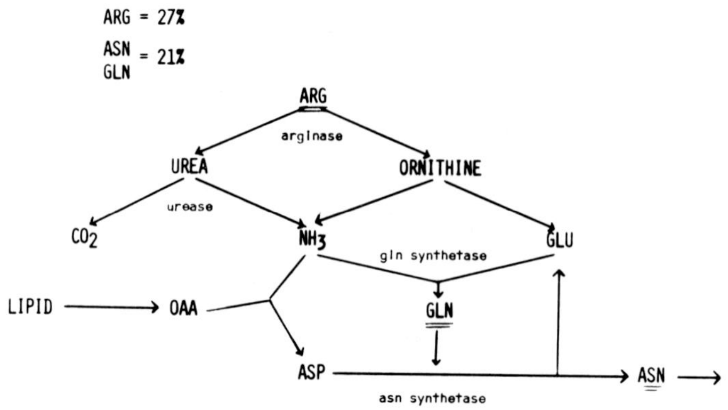

FIGURE 3. Likely pathway for the movement of storage protein nitrogen to Asn from Arg and Gln in the cotyledons of germinating cotyledons.

carried by these proteins is in only three amino acids (27% in Arg and 21% in Gln + Asn). Since nitrogen is transported from cotyledons to the rapidly growing shoot-root axis during germination as Asn, the movement of nitrogen from Arg and Gln to Asn must be a major metabolic pathway in early germination. Some of the enzymes of this pathway (Figure 3) were investigated several years ago (9). Arginase, urease and Gln synthetase activities were all found at substantial levels in mature seed cotyledons and found to increase in germination. This increase could only be partially inhibited by actinomycin D at that time. Asn synthetase, however, could not be detected in mature seed cotyledons and its appearance in germination was totally inhibited by actinomycin D. The catalytic rate (turnover number) of Asn synthetase per cotyledon pair was found to be only about 1/100th that of the other enzymes studied suggesting that this enzyme is the "pacemaker" in this pathway in germinating cotyledons. This suggestion was strengthened by measurements of the free amino acid content of germinating cotyledons that were carried out some years ago (10). Asn was found to accumulate far in excess of the other amino acids in early germination; however, in seeds incubated in actinomycin D, Asn disappeared and Gln, which is the nitrogen donor for the cotton Asn synthetase, accumulated in great excess. These results are presented in cartoon form in Figure 4. Since the likely source of nitrogen for

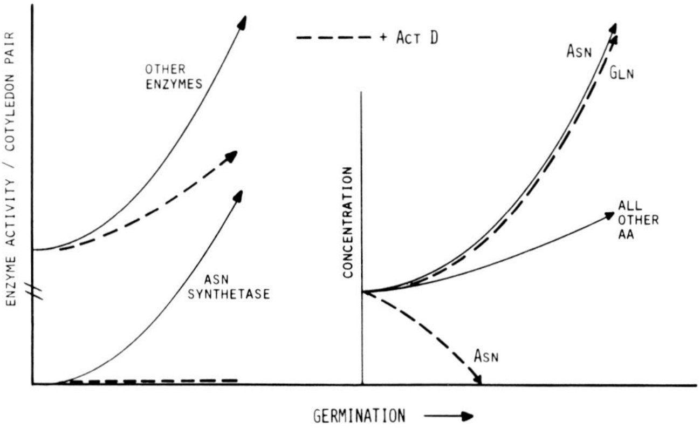

FIGURE 4. Schematic representation of enzyme activites (left) and free amino acid levels (right) in germinating cotton cotyledons. Text gives explanation of experiments.

the transamination of OAA (Figure 3) is Glu, the movement of nitrogen from storage protein to Asn is seen to be a tight, metabolically economic process involving alpha ketoglutarate, Glu and Gln.

The amino acid sequences show that two of the storage protein families share a great deal of amino acid homology whereas the third family has evolved to be quite distinct. However, the most striking aspect of these amino acid sequences is the occurrence of clusters of poly Glu in all three families. Examples of these clusters are given below.

```
               +   +   -   +   -   -   -       -   -   -   -       -   -       -       -
C 72 (69 kD)   ARG ARG GLU ARG GLU GLU GLU ALA GLU GLU GLU GLU THR GLU GLU GLY GLU GLN GLU

               +   -   -   -   -               -       -       -   -   -   +   +
               LYS GLU GLU GLU GLU GLN GLN GLN GLU GLN GLU GLN GLU GLU GLU ARG ARG


               +       -   -       -   -   -   -       -       -   -   -   -   -   -   -
C 94 (60 kD)   ARG GLY GLU GLU SER GLU GLU GLU GLU GLY GLU GLY GLU GLU GLU GLU GLU GLU ASP

               -   -   -   -
               GLU GLU GLU GLU

               -   -   -   -   +   -   -   +
               GLU GLU GLU GLU ARG GLU GLU ARG


               +   -   -   -       -   -   -   +
C 135 (60 kD)  ARG GLU GLU GLU GLY GLU GLU GLU ARG

               +       -   +   -   -   -   -
               ARG GLY GLU ARG GLU GLU GLU GLU

               -   -   -   -   +   -   -   +
               GLU GLU GLU GLU ARG GLU GLU ARG
```

The isoelectric points of these proteins are near neutrality because the high concentration of Arg residues offset the Glu clusters. These proteins obviously are salt bridged to form insoluble complexes in the seed storage protein vesicles. It is not known if the Glu clusters serve other functions in the seed, such as carriers of Ca^{++} or Mg^{++}.

The sequencing of the genomic fragments has not progressed to the point where an analysis of possible consensus homologous regions, conceivably involved in the regulation of expression of the two gene sets, can be conducted.

REFERENCES

1. Dure III L, Greenway SC, Galau GA (1981). Developmental biochemistry of cottonseed embryogenesis and germination XIV Changing messenger ribonucleic acid populations as shown by in vitro and in vivo protein synthesis. Biochemistry 20: 4162-4268.
2. Dure III L, Chlan CA, Galau GA (1983). Cottonseed storage proteins as a tool for developmental biology. In Ciferri O, Dure III L (eds): "The Structure and Function of Plant Genomes" New York: Plenum Press, p 113.
3. Dure III L, Galau GA, Chlan CA, Pyle J (1983). Developmentally regulated gene sets in cotton embryogenesis. In Goldberg RJ (ed): "Plant Molecular Biology" New York: Alan R. Liss, Inc, p 331.
4. Galau GA, Chlan CA, Dure III L (1983). Ibid XVI Analysis of the principal cotton storage protein gene family with cloned cDNA probes. Plant Mol Biol 2: 189-198.
5. Dure III L, Pyle JB, Chlan CA, Baker JC, Galau GA (1983). Ibid XVII Developmental expression of genes for the principal storage proteins. Plant Mol Biol 2: 199-206.
6. Galau GA, Dure III L (1981). Ibid XV Changing messenger ribonucleic acid populations as shown by reciprocal heterologous complementary deoxyribonucleic acid hybridization. Biochemistry 20: 4169-4178.
7. Dure III L, Galau GA (1981). Ibid XIII Regulation of the biosynthesis of the principal storage proteins. Plant Physiol 68: 187-194.
8. Dure III L, Chlan CA (1981). Ibid XII Purification and properties of the principal storage proteins. Plant Physiol 68: 187-194.

9. Dilworth MF, Dure III L (1978). Ibid X Nitrogen flow from arginine to asparagine in germination. Plant Physiol 61: 698-702.
10. Capdevilla AM, Dure III L (1977). Ibid VIII Free amino acid pool composition during cotyledon development. Plant Physiol 59: 268-273.

Molecular Biology of Development, pages 601–610

THE ROLE OF CHLOROPLAST DEVELOPMENT IN NUCLEAR GENE EXPRESSION[1]

William C. Taylor, Stephen P. Mayfield[2] and Belinda Martineau

Department of Genetics, University of California
Berkeley, CA 94720

ABSTRACT Treatment of maize seedlings with an herbicide which completely inhibits carotenoid biosynthesis also blocks the accumulation of mRNA encoding the light-harvesting chlorophyll a/b protein (LHCP), a nuclear-encoded constituent of the chloroplast photosystem II complex. Plastid development is arrested at a very early stage when herbicide-treated plants are grown in high light intensity. Plastid development can proceed when treated plants are grown in very low light and LHCP mRNA accumulates to normal levels. Events in the early stages of chloroplast development influence the accumulation of a nuclear-encoded mRNA which codes for a chloroplast destined polypeptide.

INTRODUCTION

The structural and functional development of chloroplasts are major components of leaf cell development. Chloroplast development involves the

----

1. This work was supported by the Competitive Research Grants Office of USDA

2. Supported by a McKnight Foundation Grant

assembly of extensive internal membrane structures in which the pigment - protein complexes responsible for the harvesting of light energy and its conversion into chemical energy are found. This chemical energy is used to drive the fixation and reduction of atmospheric CO_2 to form carbohydrates. The enzymes catalyzing CO_2 fixation and carbon reduction are located in the soluble fraction of the chloroplast.

Maize has two photosynthetically-active cell types which are structurally and functionally differentiated from one another. All of the enzymes involved in CO_2 fixation and carbon reduction are compartmentalized in one or the other cell type. At the heart of this specialized form of carbon reduction, called C4 metabolism, is a photosynthetic phosphoenolpyruvate carboxylase (PEPCase) which is responsible for the initial fixation of atmospheric CO_2. PEPCase is found only in leaf mesophyll cells (1), while the second CO_2 fixing enzyme, ribulose 1,5-bisphosphate carboxylase (RuBPCase) is restricted to bundle sheath cells (2).

Small, undifferentiated meristematic cells are located primarily at the base of a maize leaf. As these cells successively cease division, they expand and elongate, pushing older, more differentiated cells ahead. A positional gradient is thus formed, with undifferentiated cells located at the base and older cells toward the tip. Morphological evidence of bundle sheath or mesophyll cell differentiation is first found about one-fourth of the distance from base to tip (3). We have recently shown that enzymes of the C4 pathway are first detectable by immunological procedures at this same position (4, Nelson and Taylor, unpublished). Accumulation of C4 enzymes follow very similar patterns, regardless of cell type, or chloroplastic or cytosolic localization. RuBPCase consists of nuclear-encoded small subunits (SSu) and chloroplast-encoded large subunits (LSu). Both subunits show similar patterns of accumulation (4).

Bundle sheath and mesophyll chloroplasts show differences in their light harvesting and electron transport activities. Bundle sheath chloroplasts lack photosystem II activities and the protein-chlorophyll complex associated with these activities (5, 6). The light-harvesting chlorophyll a/b protein (LHCP) is an integral component of the antenna-chlorophyll complex of photosystem II. LHCP is nuclear encoded and cytosolically translated as a soluble 32 kd precursor. The mature 28 kd form of LHCP is the most prevalent membrane protein within the leaf. In the fully developed portion of maize leaves, LHCP is compartmentalized within mesophyll chloroplasts (Ohad, personal communication) and its mRNA is detectable only in mesophyll cells (Martineau and Taylor, unpublished). Because the C4 enzymes are also compartmentalized, one might expect to find LHCP accumulation delayed until the two cell types become morphologically differentiated. However, LHCP is first detectable at the base of the leaf and accumulation continues as cell development proceeds (4). As will be discussed later, this pattern of accumulation may tell something about the processes of bundle sheath and mesophyll cell differentiation.

For enzymes of the C4 pathway, light plays a major role in determining the rate of protein accumulation but does not influence the timing of accumulation. Significant levels of both carboxylases accumulate in the complete absence of light, but only after cellular differentiation has occured (7). The primary effect of light is to increase the rate of mRNA accumulation.

The accumulation of LHCP, however, appears to be regulated by light. Seedlings grown in the dark have no detectable LHCP. When shifted into the light, LHCP accumulates rapidly after a lag of 12 - 24 hours (8-11). The increase in LHCP has been shown to correlate with the increase in chlorophylls a and b. The final steps in chlorophyll biosynthesis require light. It has been proposed that chlorophylls a and b are necessary for the stable integration of LHCP into the thylakoid membrane and that in the absence of chlorophyll, LHCP is degraded within the chloroplast (12-15). LHCP accumulation is thus controlled by the accumulation of chlorophyll, which is in turn controlled by the quantity and quality of light. Light also induces the accumulation fo LHCP mRNA from very low levels to very high, representing 0.5-1.0% of leaf poly(A) RNA (7).

Although the accumulation of LHCP polypeptides is dependent upon the accumulation of chlorophylls a and b, the accumulation of LHCP mRNA is not. Mutants which are blocked in the biosynthesis of chlorophyll produce seedlings with dark yellow leaves, indicating the presence of high levels of carotenoids (16). While these leaves contain no detectable LHCP and no detectable chlorophyll, they contain normal levels of LHCP mRNA (17). Mutants which are blocked in the accumulation of carotenoids produce seedlings with white leaves. Many of these mutant seedlings are capable of chlorophyll synthesis, but chlorophyll is rapidly photooxidzed in the absence of carotenoids (18). White carotenoid-deficient leaves contain very low levels of LHCP mRNA, similar to the amount found in dark-grown leaves (15,17).

RESULTS

White seedlings can also be produced by treating normal seedlings with an herbicide, such as norflurazon, which blocks the conversion of phytoene to lycopene in the carotenoid biosynthesis pathway (19). Seeds of a genotypically normal inbred line (B73, Pioneer Seed, Johnston, Iowa) were germinated in a 10^{-5} M solution of norflurazon (Sandoz 9789) and watered with this same solution for 7 days. This treatment produces white seedlings with no spectrophotometrically detectable chlorophyll or carotenoids (17). Soluble and membrane protein samples and total RNA samples were prepared from herbicide-treated and untreated, phenotypically normal seedlings using methods described by Nelson et al. (7). A blot of gel-fractionated protein samples was reacted with maize LHCP antiserum. Significant levels of membrane-associated LHCP are detectable in green leaves of the untreated seedlings (Figure 1A). LHCP polypeptides are not detectable in white leaves of herbicide-treated plants (Figure 1B). Hybridization of a blot of gel-fractionated total leaf RNA with a maize LHCP cDNA clone shows that the level of LHCP mRNA is high in green untreated leaves but very low in white leaves (Figure 1C,D). The amount of LHCP mRNA in herbicide-treated white leaves is several-fold higher than the amount in dark-grown leaves of the same genotype (15).

Carotenoid deficiencies, whether caused by a mutational lesion or an herbicide, have a number of pleiotropic effects. Plastid development is arrested at a very early stage such that the plastids contain no internal membrane structures (20). White leaves also are deficient in plastid ribosomes (21). For these reasons it is apparent that the failure of white, herbicide-treated leaves to accumulate normal levels of LHCP mRNA is not necessarily related to the absence of carotenoids in these leaves. When herbicide-treated leaves are grown in very low light intensities the photooxidation of chlorophyll is minimized and leaves become pale green. Plastid development proceeds in carotenoid-deficient leaves grown in very low light (20). Thylakoid membranes are

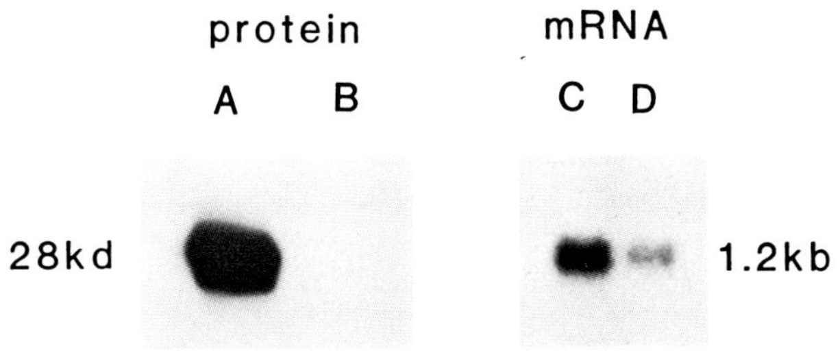

Figure 1. LHCP polypeptides are absent in herbicide-treated white leaves and LHCP mRNA is significantly reduced. Seeds were germinated and grown in 10^{-5} M norflurazon under normal daylight conditions. Leaves of treated and untreated seedlings were harvested after 7 days. RNA and protein isolation and blotting as described (7). Blot of proteins from (A) untreated seedlings or (B) white, herbicide-treated seedlings reacted with LHCP antiserum. Blot of leaf RNA from (C) untreated seedlings or (D) white, herbicide-treated seedlings hybridized with LHCP cDNA clone pM7 (17).

assembled and chloroplast protein synthesis occurs, but carotenoids do not accumulate.

The following experiment was performed to determine whether the absence of carotenoids or the block in plastid development was responsible for the low level of LHCP mRNA in white leaves. Seeds of the same inbred line were germinated and grown with or without norflurazon under very low light intensities. After 7 days half of the seedlings in each group were shifted to high light intensity for 24 hours. The amount of LHCP mRNA and the amount of chlorophyll was determined in each sample. Untreated seedlings shifted from very low to high light had the highest levels of chlorophylls a and b. By comparison, untreated leaves grown in very low light contained 23 % of chlorophyll a and 18 % of chlorophyll b, while herbicide-treated seedlings grown in very low light had 30 % of a and 21 % of b. Herbicide-treated plants shifted from very low to high light contained 2 % of a and 13 % of b.

There were high levels of LHCP mRNA in both leaf samples from untreated plants, with a several-fold difference between plants grown in very low light and those shifted from very low to high light (Figure 2C,D). The significant result was that herbicide-treated plants grown in very low light contained significant levels of LHCP mRNA (Figure 2A). However, when these plants were shifted from very low to high light, almost all

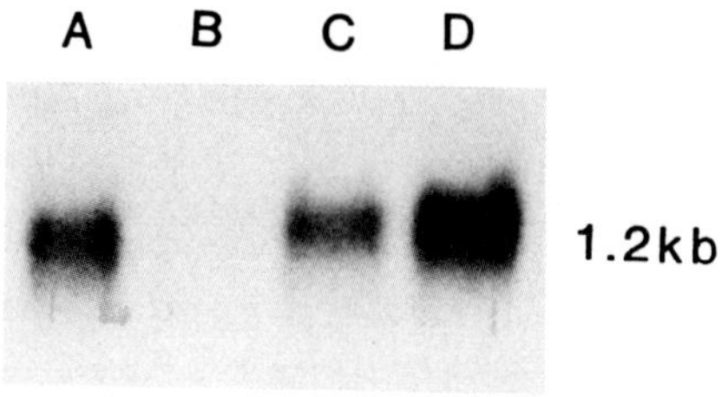

Figure 2. LHCP mRNA accumulates in herbicide-treated plants grown in very low light. Blot of total RNA hybridized with LHCP cDNA clone pM7. RNA from herbicide-treated plants (A) grown in very low light, then (B) shifted to high light, or untreated plants (C) grown in very low light, then (D) shifted to high light.

LHCP mRNA disappeared within 24 hours (Figure 2B). Both treated and untreated plants contained remarkably similar amounts of LHCP mRNA when grown in very low light (Figure 2A,C).

DISCUSSION

C4 photosynthesis, the pathway of CO_2 fixation found in maize and other tropical and desert plants, is carried out by a set of specialized enzymes which are compartmentalized in either of two cell types having specialized anatomies. It is the sequestering of C4 enzymes in the two cell types which is believed to be responsible for the high photosynthetic efficiency exhibited by C4 plants, an adaptation which allows these plants to exploit what might otherwise be marginal environments (22,23). Just as structurally differentiated mesophyll and bundle sheath cells are an important functional component of the C4 pathway, cellular differentiation is required for the synthesis and assembly of the pathway during leaf cell development (4).

There are other proteins, in addition to the C4 enzymes, which are compartmentalized in one of the two cell types of a maize leaf. Photosystem II, with its light-harvesting antenna complex, is found largely, if not exclusively, in mesophyll chloroplasts (5). LHCP, the protein component of the photosystem II antenna, is also sequestered in mesophyll chloroplasts. In spite of its compartmentalization, LHCP accumulates at very early stages of leaf cell development, well before mesophyll and bundle sheath cells show any evidence of differentiation (4,7). For this and other reasons, it is clear that LHCP accumulation is regulated by an entirely different set of signals than those regulating the accumulation of C4 enzymes.

The data presented here show that blocking plastid development at a very early stage also blocks the accumulation of LHCP mRNA, cytosolically localized transcripts of a nuclear gene family (Mayfield and Taylor, unpublished). When herbicide-treated plants were grown in very low light, they accumulated chlorophylls a and b. Robertson and his collaborators (20) have shown that carotenoid-deficient plants grown in very low light develop thylakoid membranes, indicating that the block in plastid development seen at high light has been bypassed. Once plastid development proceeds past this block, LHCP mRNA accumulates to high levels. However, when these carotenoid-deficient plants are shifted to high light, LHCP mRNA accumulation ceases and existing mRNA is degraded within a 24 hour period. Whatever information passes from the developing chloroplast to induce the accumulation of LHCP mRNA, it appears to be sensitive to photooxidation. Continual chloroplast development is necessary for continual accumulation of LHCP mRNA.

In light-grown seedlings, LHCP accumulation begins at very early stages of leaf cell development. There are two important signals at this stage which regulate accumulation. One is the initiation of the early stages of chloroplast differentiation. The other is light, which is responsible for increasing the quantity of LHCP mRNA and for inducing the synthesis of chlorophylls a and b. The quantities of both mRNA and polypeptides are regulated by these signals, chlorophyll regulating only the accumulation of polypeptides.

The mechanism by which LHCP and its mRNA ultimately will be sequestered in mesophyll cells is unknown. Cytological observations may provide some clues. Cells which ultimately develop into bundle sheath cells appear in several aspects to develop initially as mesophyll cells. Their chloroplasts develop thylakoid membranes with stacked regions, called grana, in which photosystem II complexes are located. These structures suggest that these pre-bundle sheath cells have photosystem II and contain LHCP. As these cells associate with developing intermediate vascular bundles, they disassemble their grana and develop into bundle sheath cells with their characteristic agranal chloroplasts. Whatever signal initiates the bundle sheath cell program of differentiation also shuts off expression of the LHCP gene family. Existing LHCP mRNA and polypeptides are degraded. The accumulation of LHCP is not dependent upon the differentiation of the two cell types, but rather is influenced by it.

REFERENCES

1. Perrot-Rechenmann C, Vidal J, Brulfert J, Burlet A, Gadal P (1982). Planta 155:24.
2. Huber SC, Hall TC, Edwards GE (1976). Plant Physiol 57:730.
3. Leech RM, Rumsby MG, Thomson WW (1973). Plant Physiol 52:240.
4. Mayfield SP, Taylor WC (1984). Planta. In press.
5. Anderson JM, Woo KC, Boardman NK (1971). Biochim Biophys Acta 245:398.
6. Kirchanski SJ, Park RB (1976). Plant Physiol 58:45.
7. Nelson T, Harpster MH, Mayfield SP, Taylor WC (1984). J Cell Biol 98:558.
8. Schmidt GW, Bartlett SG, Grossman AR, Cashmore AR, Chua N-H (1980). In Leaver CJ (ed): "Genome Organization and Expression in Plants," New York: Plenum Press, p 337.
9. Apel K, Kloppstech K (1978). Eur J Biochem 85:581.
10. Cuming AC, Bennett J (1981). Eur J Biochem 118:71.
11. Tobin E (1981). Plant Physiol 67:1078.
12. Apel K, Kloppstech K (1980). Planta 150:426.
13. Viro M, Kloppstech K (1982). Planta 154:18.
14. Bennett J (1981). Eur J Biochem 118:61.
15. Harpster MH, Mayfield SP, Taylor WC (1984) Plant Molec Biol. In press.
16. Mascia PN, Robertson DS (1980). J Heredity 71:19.
17. Mayfield SP, Taylor WC. Submitted.
18. Anderson IC, Robertson DS (1960). Plant Physiol 35:531.
19. Bartels PG, McCullough C (1972). Biochem Biophys Res Comm 48:16.
20. Bachmann MD, Robertson DS, Bowen CC, Anderson IC (1973). J Ultrastruct Res 45:384.
21. Troxler RF, Lester R, Craft FD, Albright JT (1969). Plant Physiol 44:1609.

22. Laetsch WM (1974). Ann Rev Plant Physiol 25:27.
23. Downton WJS (1971). In Hatch MD, Osmond CB, Slayter RO (eds): "Photosynthesis and Photorespiration," New York: Wiley-Interscience, p 3.

Molecular Biology of Development, pages 611–617

A RHIZOBIUM MELILOTI SYMBIOTIC GENE THAT REGULATES OTHER NITROGEN FIXING GENES[1]

Wynne W. Szeto, J. Lynn Zimmerman[2], Venkatesan Sundaresan[3] and Frederick M. Ausubel

Department of Molecular Biology,
Massachusetts General Hospital, Boston, MA 02114

ABSTRACT A *Rhizobium meliloti* regulatory gene required for the expression of two closely linked symbiotic operons, the nitrogenase (*nif*HDK) and P2 operons, has been identified and characterized. This regulatory locus codes for an activator and may be related to the genes that control the nitrogen fixation pathway of *Klebsiella pneumoniae*.

INTRODUCTION

Only certain procaryotes are capable of nitrogen (N_2) fixation (the reduction of atmospheric N_2 to ammonia by the enzyme nitrogenase). Some organisms that fix N_2 do so during symbiosis with specific host plants (e.g. *Rhizobium meliloti* and alfalfa), while others carry out the N_2 reduction reaction only under certain physiological conditions (e.g. *Klebsiella pneumoniae*).

The establishment of the *Rhizobium*-alfalfa symbiosis is an interactive process requiring the ordered expression of both bacterial and plant genes. The symbiotic association begins with the invasion of alfalfa root hairs by *R. meliloti* and ends with the formation of root nodules --

[1]This work was supported by funds provided by Hoechst A.G.

[2]Present address: Department of Biological Sciences, University of Maryland Baltimore County, Catonsville, MD 21228

[3]Present address: Department of Genetics, University of California at Berkeley, Berkeley, CA 94720

specialized organs designed to maximize the efficiency of N_2 reduction. Within the nodules, Rhizobia differentiate into morphologically distinct forms called bacteroids, which synthesize large quantities of the enzyme nitrogenase.

In contrast to R. meliloti, the enteric bacteria Klebsiella pneumoniae fixes N_2 in the free-living state under conditions of low oxygen tension and ammonia starvation. Although Rhizobium and Klebsiella are quite different organisms, certain aspects of the N_2 fixation apparatus in the two species are remarkably similar. The genes encoding the nitrogenase polypeptides nifH, nifD and nifK are highly conserved (1). Furthermore, in both species, the nifHDK genes comprise an operon transcribed in the order nifHDK (2,3). In K. pneumoniae, the nifHDK operon is located within a larger cluster of at least 17 contiguous genes, all of which are essential for N_2 fixation (the nif genes), organized into 7 or 8 operons (4,5).

In R. meliloti, the nifHDK operon is also closely linked to at least 2 transcription units containing genes essential for N_2 reduction (2,6). Unlike Klebsiella, the Rhizobium nif genes are not located on the bacterial chromosome, but rather reside on a large indigenous plasmid (7,8).

The K. pneumoniae nif operons are positively regulated by the product of the nifA gene (9,10,11). In addition, the Klebsiella nif genes are indirectly regulated by a centralized regulatory system that activates several nitrogen assimilation pathways (12,13). Under conditions of ammonia starvation, the ntrC(glnG) and ntrA(glnF) products activate transcription of the nifA gene; the nifA product together with the glnF product then activates all of the other nif operons (14,15). Because the nifA protein can activate the same promoters as the glnG protein, it is likely that these genes are evolutionarily related (14,15).

The mechanisms that regulate the expression of the Rhizobium nif genes are unknown. However, DNA sequence analyses have shown that the R. meliloti nifH promoter and several K. pneumoniae nif promoters have a similar structure, which is characterized by two consensus sequences, 5'-TTGCA-3' at -10 to -15 and 5'-CTGG-3' at -23 to -26 (16,17,18). Thus, it is possible that these promoters may be regulated by similar proteins. Moreover, a R. meliloti nifH::lacZ fusion plasmid is activated in $glnF^+$ E. coli strains that produce the glnG or nifA proteins constitutively (17,19). This suggests that Rhizobium may synthesize a product related to the glnG and nifA proteins during symbiosis for the regulation of other N_2 fixing genes. Some studies that we have conducted that enabled us to identify and characterize a R. meliloti symbiotic regulatory gene are described below.

RESULTS AND DISCUSSION

The Rhizobium Symbiotic Regulatory Locus

To search for potential Rhizobium regulatory genes, the assumption was made that such genes would be closely linked to the nifHDK operon, as is the case in Klebsiella. Thus, a series of R. meliloti strains containing Tn5 insertions outside of but within 20 kb of the nifHDK operon [generated by site-directed mutagenesis (20)] were screened for their failure to synthesize nifHDK polypeptides during symbiosis (3). Two strains, 1352 and 1354, carrying Tn5 insertions at a region approximately 5 to 5.5 kb upstream of the nifHDK operon (see Figure 1), were found to differentiate into bacteroids that did not accumulate detectable amounts of nifHDK polypeptides or mRNA. In contrast, insertion No. 30

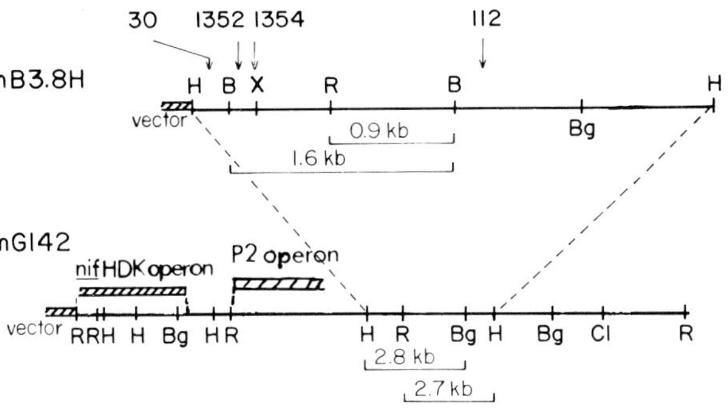

FIGURE 1. Restriction map of a region of the R. meliloti megaplasmid DNA that includes the nitrogenase operon and the regulatory locus. The arrows indicate the positions of Tn5 insertions in the corresponding strains. The sizes of the restriction fragments that hybridized to glnG/nifA DNA are indicated. pRmB3.8H and pRmG142 are plasmids containing the indicated Rhizobium DNA regions. R=EcoRI, H=HindIII, B=BamHI, Bg=BglII and Cl=ClaI.

had no apparent effect on nodule formation nor on the expression of nif genes (including nifHDK). Insertion No. 112 formed ineffective nodules (Fix^-) that contained wild-type levels of nitrogenase polypeptides. Genetic and physical analyses of these Rhizobium strains indicate that insertion No. 30 delimits the "left" boundary of the presumptive regulatory locus and insertion No. 112 delimits its "right" boundary. Furthermore, the regulatory gene has a maximum size of 1.8 kb (21).

The R. meliloti Regulatory Gene Codes for an Activator of the nifHDK Operon

A R. meliloti nifH::E. coli lacZ fusion plasmid was actively transcribed in bacteroids formed by Rhizobium strains KS1 (Tn5::lacZ) and 1491 (Tn5::nifH). In contrast, the same plasmid was not expressed in bacteroids formed by strain 1354 (the regulatory mutant) (21). These results are consistent with the interpretation that the regulatory gene codes for a positive transcriptional activator of the nifHDK operon rather than a protein that specifically prevents the degradation of nifHDK mRNA.

The Regulatory Locus is Pleiotropic

RNA hybridization studies (21) indicate that, in addition to nifHDK mRNA, 1354 bacteroids do not accumulate transcripts corresponding to the EcoRI-HindIII fragment of plasmid pRmG142 (see Figure 1), which contains another symbiotic operon - the P2 operon (22). Thus, the product of the R. meliloti regulatory gene appears to control the expression of both the nifHDK and P2 operons during symbiosis. These results imply that the nifH and P2 promoters should have a similar structure. Recent DNA sequencing data confirms this hypothesis (22).

The Nature of the Regulatory Locus

Several lines of evidence suggest that the R. meliloti regulatory gene is related to the K. pneumoniae nifA and E. coli glnG genes. (nifA and glnG are the principal regulatory loci of the Klebsiella nif genes.) First, DNA of the Rhizobium regulatory locus hybridizes to glnG and, to a lesser extent, nifA DNA (21). Second, the nifA and glnG proteins can substitute for the function of the product of the Rhizobium regulatory locus in that these proteins can activate the transcription of the nifH and P2 promoters in free-living Rhizobium cells (21). Third, the R. meliloti nifH (17,19) and P2 (22) promoters (both under the control of the regulatory gene) share the consensus sequences 5'-TTGCA-3' (-10 to -15) and 5'-CTGG-3' (-23 to -26), which are common to promoters regulated by the nifA and glnG proteins (16,18). Fourth, DNA sequencing studies show that the Rhizobium regulatory locus has the capacity to code for a protein whose size is similar to that of the nifA and glnG proteins (approximately 55 kd) (W. Buikema, unpublished).

It is likely that the R. meliloti regulatory gene (nifR) is a nif-specific regulatory gene like nifA rather than a general nitrogen regulatory gene like glnG. This is because the growth of strains 1352 and 1354 (the regulatory mutants) on media containing proline, histidine, arginine and glutamine as the sole N_2 source (a glnG controlled function) is not impaired (21). Furthermore, when glnG DNA is hybridized to EcoRI- digested total Rhizobium DNA, hybridization to a 3.7 kb DNA band rather than to nifR is detected. This suggests that the 3.7 kb EcoRI band contains the true Rhizobium glnG gene.

The mechanism by which the R. meliloti regulatory gene is activated is unknown. It is possible that the Rhizobium nif regulatory gene, unlike the Klebsiella nifA gene, is activated independently of a centralized ammonia control system. One plausible mechanism is that the R. meliloti regulatory gene is activated by a centralized symbiotic regulatory system that is responsible for activating many symbiotic genes.

REFERENCES

1. Ausubel FM (1982). Molecular genetics of symbiotic nitrogen fixation. Cell 29:1.
2. Ruvkun GB, Sundaresan V, Ausubel FM (1982). Directed transposon Tn5 mutagenesis and complementation analysis of Rhizobium meliloti symbiotic nitrogen fixation genes. Cell 29:551.
3. Zimmerman JL, Szeto WW, Ausubel FM (1983). Molecular characterization of Tn5 induced, symbiotic (Fix^-) mutants of Rhizobium meliloti. J Bacteriol 156:1025.

4. Roberts GP, Brill WJ (1981) Genetics and regulation of nitrogen fixation. Ann Rev Microbiol 35:207.
5. Ausubel FM, Cannon FC (1981). Molecular genetic analysis of Klebsiella pneumoniae nitrogen fixation (nif) genes. Cold Spring Harbor Symp Quant Biol 45:487.
6. Corbin D, Barren L, Ditta G (1983). Organization and expression of Rhizobium meliloti nitrogen fixing genes. Proc Nat Acad Sci USA 80:3005.
7. Buikema WJ, Long SR, Brown SE, van den Bos RC, Earl CD, Ausubel FM (1983). Physical and genetic characterization of Rhizobium meliloti symbiotic mutants. J Molec Appl Genet 2:249.
8. Rosenberg C, Boistard P, Denarie J, Casse-Delbart F (1981). Genes controlling early and late functions in symbiosis are located on a megaplasmid in Rhizobium meliloti. Mol Gen Genet 184:326.
9. Dixon R, Eady RR, Espin G, Hill S, Iaccarino M, Kahn D, Merrick M (1980). Analysis of regulation of Klebsiella pneumoniae nitrogen fixation (nif) gene cluster with gene fusions. Nature 286:128.
10. Buchanan-Wollaston V, Cannon MC, Beynon JC, Cannon FC (1981). Role of the nifA gene product in the regulation of nif expression in Klebsiella pneumoniae. Nature 294:776.
11. Buchanan-Wollaston V, Cannon MC, Cannon FC (1981). The use of cloned nif (nitrogen fixation) DNA to investigate transcriptional regulation of nif expression in Klebsiella pneumoniae. Mol Gen Genet 184:102.
12. de Bruijn FJ, Ausubel, FM (1981). The cloning and transposon Tn5 mutagenesis of the glnA region of Klebsiella pneumoniae: identification of glnR, a gene involved in the regulation of the nif and hut operons. Mol Gen Genet 183:289.
13. Magasanik B (1982). Nitrogen assimilation in bacteria. Ann Rev Genet 16:135.
14. Ow DW, Ausubel FM (1983). The nifA gene which regulates the Klebsiella pneumoniae nif gene cluster can substitute for the nitrogen regulatory gene glnG(ntrC). Nature 301:307.
15. Merrick M (1983). Nitrogen control of the nif regulon in Klebsiella pneumoniae: involvement of the ntrA gene and analogies between ntrC and nifA. EMBO Journal 2:31.
16. Ow DW, Sundaresan V, Rothstein DM, Brown SE, Ausubel FM (1983). Promoters generated by the glnG(ntrC) and nifA gene products share a heptameric consensus sequence in the -15 region. Proc Nat Acad Sci USA 80:2524.

17. Sundaresan V, Jones JDG, Ow DW, Ausubel FM (1983). Conservation of nitrogenase promoters from Rhizobium meliloti and Klebsiella pneumoniae. Nature 301:728.
18. Beynon JC, Cannon MC, Buchanan-Wollaston V, Cannon FC (1983). The nif promoters of Klebsiella pneumoniae have a characteristic primary structure. Cell 34:665.
19. Sundaresan V, Ow DW, Ausubel FM (1983). Activation of Klebsiella pneumoniae and Rhizobium meliloti nitrogenase promoters by gln(ntr) regulatory proteins. Proc Nat Acad Sci USA 80:4030.
20. Ruvkun GB, Ausubel FM (1981). A general method for site-directed mutagenesis in prokaryotes. Nature 289:85.
21. Szeto WW, Zimmerman JL, Sundaresan V, Ausubel FM (1984) A Rhizobium meliloti symbiotic regulatory gene. Cell 36:535.
22. Better M, Lewis B, Corbin D, Ditta G, Helinski D (1983). Structural relationships among Rhizobium meliloti symbiotic promoters. Cell 35:479.

VII. TRANSFORMATION IN WHOLE ORGANISMS AND CELLS

Molecular Biology of Development, pages 621-632

GENE TRANSFER IN THE SEA URCHIN[1]

Constantin N. Flytzanis, Andrew P. McMahon[2]
Barbara R. Hough-Evans, Karen S. Katula, Roy J. Britten
and Eric H. Davidson

Division of Biology, California Institute of Technology
Pasadena, California 91125 USA

INTRODUCTION

The isolation of many genes that function during the early stages of development of the sea urchin opens to experimental investigation several fundamental problems. Among these are the identification of the genomic sequences required for lineage-specific gene activation (1), the functional significance of structural features of maternal transcripts (2,3), the molecular nature of the process of commitment in the early embryo, and the role of given gene products in early morphogenesis. In order to approach such issues, we have developed a sea urchin gene transfer system, which includes methods of introducing cloned DNA into sea urchin eggs and of monitoring its subsequent physical fate from the developing embryos and larvae through metamorphosis to adulthood. Our goal is to study the transcriptional activity of transferred cloned genes, both in the embryos deriving from the injected eggs, and ultimately in embryos deriving from sexually mature animals that bear germ

[1]This research was supported by the National Institute of Child Health and Human Development (HD 05753). C.N.F. is a Lievre Fellow of the American Cancer Society, California Division, A.P.M. was supported by the same Society (Fellowship S-11-83 and J-33-82, respectively) and K.S.K. by an NIH postdoctoral training grant (HD 07257).
[2]Present address: National Institute for Medical Research, The Ridgeway, Mill Hill, London NW7 1AA, England.

line integrations of the cloned sequences. The power of this approach has been well illustrated in recent studies on the function of stably integrated genes transferred into the genome of Drosophila and mice (4-12). In both organisms transferred genes have been shown to function in the correct terminally differentiated tissues.

We describe here a simple and rapid microinjection technique for introduction of DNA (or other molecules) into the cytoplasm of the sea urchin egg. We show that the injected sequences are retained throughout embryogenesis, and that they undergo a dramatic net replication if introduced in the proper physical conformation. We show, furthermore, that the replicated DNA persists stably in as many as 80% of the advanced larvae and 4-16% of the juvenile urchins descending from the injected eggs. We also describe the sequence organization of an integrated concatenate and the host genome locus into which it had inserted. The data summarized are from studies by McMahon et al. (13) and Flytzanis et al. (14). Space does not permit a complete exposition of these experiments here, and for this, as well as a more detailed interpretation, the reader is referred to these sources.

Microinjection of DNA into Unfertilized Eggs

The sea urchin egg is the first marine egg to be utilized for studies of this nature, and the procedure that has emerged takes advantage of several of the biological properties of this material. The egg jelly coat is first removed by treatment with acid sea water, and the eggs, which are about 80 μm in diameter, are then fixed in rows on a dish suitable for mounting on the stage of an inverted dissection microscope. The surface of the egg is negatively charged at about -60 mV, and the eggs can be firmly immobilized by electrostatic attraction to a protamine sulfate coat applied to the surface of the injection dish. A holding micropipette is thus not required for the micromanipulation. The fixed lines of eggs are rapidly injected into the cytoplasm with ~2 pl (1.5% of the egg volume) of a DNA solution using a continuously flowing micropipette of about 0.5-1.0 μm tip diameter. The eggs are then fertilized in situ by addition of a drop of sperm. With some practice 200-300 eggs can be injected per hr, with 90-95% undergoing normal fertilization. The injected eggs proceed to cleave while remaining attached by the elevated fertilization membrane to the protamine sulfate coated surface. At the mesenchyme blastula stage, 18-20 hr post-fertilization, the embryos produce a hatching enzyme which dissolves the fertilization membrane, and the ciliated blastulae swim free from the surface of the dish. They can then be collected and cultured by ordinary methods. Survival of embryos which developed from injected eggs was acceptably high. Over 20,000 injected, fertilized eggs were studied in the course of this work and 40 to 50% of these completed normal embryogenesis to the pluteus stage, 72-96 hr post-fertilization. It is clear that there is some variability between batches of eggs in their response to microinjection. Frequently, the early cleavages of injected eggs are retarded the equivalent of one to four cleavage divisions, although this apparently has no adverse effect on subsequent embryonic development. In addition, embryonic development is sensitive to the mass of DNA injected. Survival figures indicated above are for eggs injected with DNA solutions at concentrations up to 32 μg/ml (~12,000 molecules of a 5 kb sequence in 2 pl). Injection of DNA solutions at concentrations equal to or greater than 320 μg/ml is completely toxic to normal post-fertilization development.

Concatenate Formation and Amplification of the Injected DNA During Early Embryogenesis

To determine the physical fate during embryogenesis of DNA molecules injected into the egg cytoplasm, we extracted the total DNA from embryos that had been allowed to develop to various stages and examined it for the presence of the exogenous sequences. For most of the work described here we utilized a 5.1 kb plasmid called pISA. This construct, which is diagrammed in Figure 1a, includes a 5' flanking sequence from the S. purpuratus CyI actin gene (15), ligated to the Tn5 aminoglycoside 3' phosphotransferase (neomycin resistance) gene (16), which is flanked in turn by a sequence including the 3' poly(A) addition site from the Herpes virus thymidine kinase gene (17). pISA contains a single BamHI site, and for the experiment shown in Figure 1b, approximately 9,000 molecules of the plasmid linearized at this site were injected per egg. Late gastrulae were collected 48 hr post-fertilization, the DNA was extracted, digested where indicated and run on a 0.8% agarose gel. After transfer of the DNA to nitrocellulose, the blot was hybridized with a probe that contains all of the aminoglycoside 3' phosphotransferase sequences present in pISA, as well as some additional Tn5 sequences. Lane 2 of Figure 1b contains DNA isolated from 64 uninjected gastrulae and lane 3 contains DNA isolated from 67 gastrulae injected with BamHI linearized pISA.

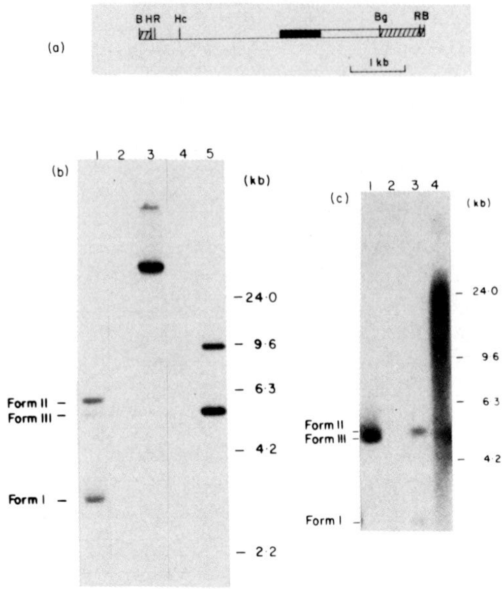

Fig. 1. Fate of injected DNA during early embryogenesis. (a) Linear map of the 5.1 kb pISA. B: BamH1, Bg: BglII, R: EcoRI, H: HindIII, Hc: HincII. Single line, pBR322 sequences; hatched bar, sea urchin actin sequences from gene CyI; open bar, aminoglycoside 3' phosphotransferase gene from the Tn5 transposable element; solid bar, poly(A) addition sequences from Herpes simplex virus thymidine kinase gene. (b) Lane 1, 3×10^6 molecules of pISA in the three indicated conformational states (form I: supercoiled; form II: relaxed circular; form III: linear). Lane 2, undigested DNA from 64 uninjected gastrulae. Lane 3, undigested DNA from 67 gastrulae injected with linearlized pISA that had been digested with BamHI. Lane 4, BglII digestion of DNA from 55 uninjected gastrulae. Lane 5, BglII digestion of DNA from 55 gastrulae injected with linearized pISA. The positions and lengths of molecular weight standards are given in kilobases. (c) Lane 1, 1.4×10^6 molecules of pISA that had been linearized by digestion with BamHI (form III). The positions of migration of forms I and II are indicated. Lane 2, undigested DNA from 35 uninjected blastulae. Lane 3, undigested DNA from 35 blastulae that had been injected with supercoiled pISA. Lane 4, undigested DNA from 35 blastulae that had been injected with BamHI linearized pISA. Data are collated from McMahon et al. (13).

The position of the labeled fragmented is the same as that of the total high molecular weight embryo DNA observed by ethidium bromide staining (not shown). While no hybridization of the probe can be detected in lane 2, a clear reaction occurs with the DNA from the injected gastrulae in lane 3. Moreover, the sequences homologous to the probe are seen to integrate as high molecular weight DNA rather than in the position of the linear molecules that had been injected (lane 1, form III). As shown in Figure 1b the hybridization of the aminoglycoside 3' phosphotransferase gene probe is observed in the high molecular weight DNA fraction of the undigested sample (lane 3), but on digestion with BglII the probe displays instead two prominent lower molecular weight bands of 8.8 and 5.1 kb (lane 5). These fragments correspond to those expected, had a random, end-to-end concatenation of the linear pISA molecules occurred within the injected eggs. As the probe used in the experiment described in Figure 1b is homologous only to the Tn5 sequence, it can react exclusively with the two of the three fragments (the 8.8 kb and the 5.0 kb) generated after a BglII digestion of a concatenate produced by a random end-to-end ligation of the linearized with BamHI injected molecules. The amount of hybridization obtained with the probe sequences in the experiment presented in Figure 1 indicates that a large increase in the quantity of the exogenous DNA had occurred by the late gastrula stage. This can be seen by comparison of the embryo DNA lanes with the standards present in lane 1, which contains about 6X more DNA than what was injected in the experiments shown in lanes 3 and 5. We estimated that in these embryos at least a 10-fold net amplification of the injected DNA had occurred. Estimates from other experiments (not shown) are based either on scintillation counting of the hybridizing bands of the gel blots, or densitometry of the autoradiograms, in comparison with known serial dilutions of DNA standards run in parallel. It is clear that in all stages of embryonic development the exogenous DNA is present in far greater amounts (up to 100-fold) than was originally injected. Moreover, replication occurs irrespective of whether the injected linear plasmids contain sea urchin or other eukaryotic sequences or no eukaryotic sequences at all. Most of the replication occurs within the first 24 hr of development, which coincides with the period when most of the cell divisions carried out by the S. purpuratus embryo take place. Cell number increases logarithmically between 5 hr (16 cells) and 18 hr (~450 cells) in this species, and after this many cells withdraw from the mitotic cycle (there are ~1500 cells in the 72 hr pluteus). The extensive early replication of injected DNA apparently requires specifically the high molecular weight concatenate formed after injection of linear DNA molecules, since when supercoiled plasmid DNA is injected it fails to amplify detectably. This is shown in Figure 1c. Here we injected approximately 9,000 molecules of pISA per egg, either in linear or supercoiled form and examined the blastulae 24 hrs post-fertilization for the Tn5 sequence as above. Weakly hybridizing bands in the position of both supercoiled and nicked circular DNA were detected in extracts from the blastulae injected with supercoiled molecules (lane 3). Comparison with the DNA standards run in parallel (lane 1), which contain 5- to 10-fold more DNA than was injected into the pooled blastula sample, indicates that DNA injected as supercoils persists in the embryos, mainly in supercoiled or relaxed circular form, but there is no net increase in mass. In contrast, the undigested DNA sample from the blastulae injected with linear molecules in the same experiment displays a greatly augmented hybridization, as before.

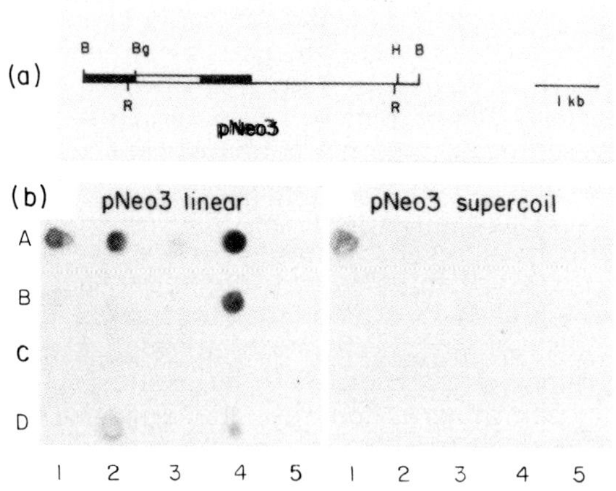

Fig. 2. Persistence of exogenous DNA in 5 weeks larvae. (a) Linear map of the 5.2 kb plasmid pNeo3. Abbreviations are as noted in the legend to Fig. 1a. Single line, pBR322 sequences; open bar, aminoglycoside 3' phosphotransferase gene from the Tn5; solid bar, the promoter and the poly(A) addition sequences from the Herpes simplex virus thymidine kinase gene. (b) Column 1 of each panel contains standard quantities of the injected pNeo3 plasmid, which had been linearized and spotted on the nitrocellulose filters: positions A, 1.25×10^6 molecules; B, 2.5×10^5 molecules; C, 5×10^4 molecules; D, 1×10^4 molecules. Columns 2, 3 and 4 contain the DNA of larvae grown from injected eggs, and column 5 the DNA of larvae grown from uninjected eggs. Data are collated from Flytzanis et al. (14).

The replicated molecules are present in this experiment as a smear of varying length fragments (Fig. 1c, lane 4). Quantitation by densitometry indicates that the linear DNA was amplified by replication about 28-fold.

Persistence of Exogenous DNA in 5 Week Larvae

Although DNA injected into the cytoplasm of unfertilized eggs amplifies and then persists throughout embryogenesis, there is no net increase in embryo mass during this period, and our next object was to determine whether the exogenous DNA would be maintained and further replicated after feeding and growth begin. Larvae were raised from eggs injected with linearized pNeo3 plasmid (Fig. 2a). pNeo3 contains the Tn5 aminoglycoside 3' phosphotransferase gene linked to the promoter and the poly(A) addition site from the Herpes simplex virus thymidine kinase gene (tk), cloned into pBR322 (Dr. B. Wold, personal communication). Five weeks after the initiation of feeding, the DNA from individual larvae was extracted and spotted onto nitrocellulose filters. These were then hybridized with radioactive probes representing the injected sequences. Figure 2b displays a typical series of results. In each panel column 1 contains DNA standards, consisting of the injected plasmid in decreasing concentrations, and column 5 contains DNA extracted from uninjected embryos. The remaining 12 positions contain DNAs extracted from the individual experimental larvae. In this experiment (panel A), seven out of the 12 larvae tested (~58%) were positive for the injected pNeo3 sequence. From the results of these and many other experiments (data not shown) we estimate that on the average 55% of the larvae raised from eggs injected with linear DNAs display the exogenous sequences and in some experiments this value may rise as high as 80%. We have shown in Figure 1 that injected linear but not supercoiled DNA will undergo concatenation and amplification early in development. Figure 2b (panel B) demonstrates that when supercoiled molecules of the plasmid pNeo3 are injected, the 5 week larvae retain no detectable exogenous DNA sequences whatsoever. The same results has been obtained with several different supercoiled plasmids. The limit of sensitivity in these experiments is about equal to the content of the lowest DNA standard on the filters (~10^4 molecules/larva), or about equal to the amount injected. Thus, we cannot say whether the supercoiled molecules are retained in the larvae, but it is clear that no net amplification of these molecules has occurred by 5 weeks of larval growth. We conclude that the presence of detectable amounts of the linear injected DNA in most 5 week larvae is correlated with its amplification in the embryo. Furthermore, except for the all-important conformational requirement for linear molecules, we have observed no reproducible differences among the injected plasmids in either efficiency of exogenous DNA amplification during embryogenesis or the fraction of exogenous DNA of individual 5 week larvae retaining detectable quantities of this DNA. To obtain an approximate comparison between the amounts of exogenous DNA persisting in 5 week larvae and the amounts initially generated during embryonic DNA amplification, we quantitated autoradiographs such as those shown in Figure 2b by densitometry (data not shown). We estimated that for about half of the larvae there have been relatively little additional DNA amplification after embryogenesis. For the other half of the larvae the quantities of exogenous DNA present indicate that many additional rounds of replication must have occurred. We believe the major factor underlying this difference is the distribution of the initial concatenates to cell lineages that either do or do not undergo extensive expansion during larval growth.

Integration of Injected Sequences in the Genomes of Post-Metamorphosis Juveniles

Two months after metamorphosis S. purpuratus juveniles contain a sufficient quantity of DNA to perform individual genome blots. Using this method we examined the DNA of a total of 58 individual juveniles (data not shown). The frequency of positive animals for the injected sequences ranged in three different experimental groups, from 4 to 16%. This is to be compared to the average value of 55% of 5 week larvae, and in some cases up to 80%, that retain the injected sequences. Apparently, only a minor fraction of these larvae contain exogenous DNA that will be passed on and continue to replicate in the juvenile sea urchins emerging from metamorphosis. An example of a genome blot prepared from the DNA of a two month old juvenile descended from an egg injected with the plasmid pNeo3 is shown in Figure 3a. After digestion of the genomic DNA with BglII, for which there is a single site in the pNeo3 sequence (Fig. 2a), hybridization with the pNeo3 probe displays the bands (heavy arrows) expected if the DNA consists of a random end-to-end concatenate of the linear molecules originally injected. From the intensity of the major bands in Figure 3a we calculate that there are about 50 copies of pNeo3 per average cell in this juvenile sea urchin. A few additional minor bands (fine arrows) can also be detected in this genome blot, which do not correspond

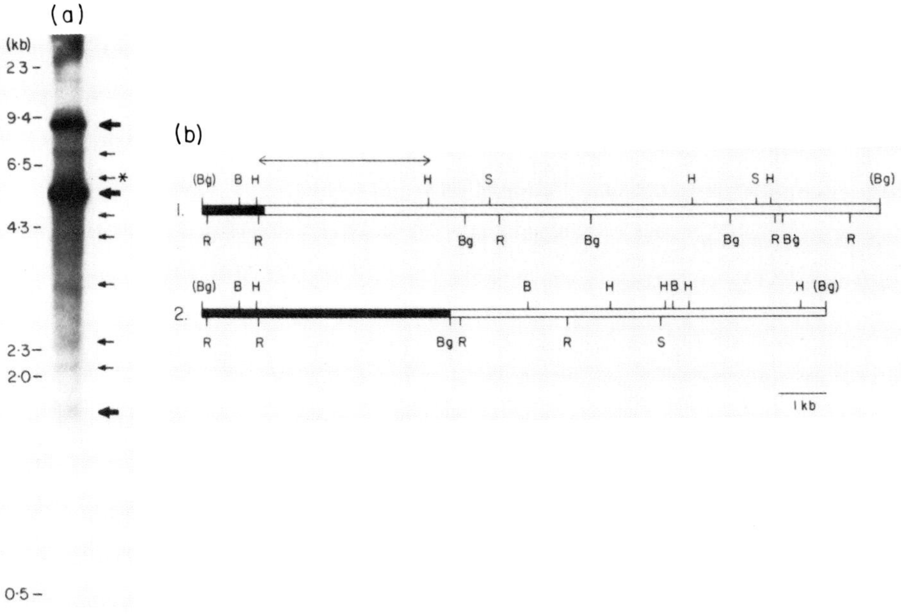

Fig. 3. Integration of injected DNA in the genome of a juvenile sea urchin. (a) Genome blot obtained from BglII digested DNA extracted from a 2 month old juvenile, grown from an egg injected with 9,000 molecules of BamHI linearized pNeo3. Heavy arrows indicate the bands expected to be generated on BglII digestion of a random end to end concatenate of the plasmid pNeo3. Other light bands are marked by fine arrows. (b) Restriction enzyme maps of the inserts of the genomic clones 1 and 2. Abbreviations are as noted in the legend to Fig. 1a. The open bar represents sea urchin genomic DNA and the solid bar pNeo3 sequence. The double headed arrow shows a HindIII fragment that was used to determine the sequence of the junction. Data are collated from Flytzanis et al. (14).

in length to any of those that would be generated by BglII digestion of a concatenate of complete pNeo3 sequences. These light bands could represent the terminal fragments of pNeo3 concatenates that are integrated into the genome. The small number of these bands suggests that if there are indeed junction fragments, the integrations occurred early and there are not many different sites of integration. To determine the actual organization of the pNeo3 sequences detected in the genome of this animal, we constructed a recombinant DNA library from this DNA, and examined the arrangement of pNeo3 sequences in the recombinants that we recovered. This genomic DNA library was constructed by digesting partially the DNA of the juvenile urchin presented in Figure 3a with the restriction enzyme BglII and ligating the obtained fragments to the BamHI cloning site of the λ cloning vector EMBL3 (18). Figure 3b shows the restriction enzyme maps of two recombinant genomic clones isolated from the genomic library, after screening with the pNeo3 plasmid as a probe. Further characterization of these clones was obtained by hybridization of pNeo3 and total sea urchin DNA probes, to blots containing different fragments generated by restriction enzyme digests. Since the sea urchin genome contains repetitive sequences interspersed at 1-2 kb intervals, the sea urchin DNA probe will react with any genomic sequence more than a few kb in length. Figure 3b displays evidence that the recombinant named clone 1 contains a junction fragment, which reacts with the pNeo3 probe and with the sea urchin DNA as well (double headed arrow). In recombinant clone 2 a BglII site separates the pNeo3 positive sequences from the sea urchin DNA. This clone was probably created by an adventitious ligation that occurred during the cloning process. The recombinant clones 1 and 2 contain fragments of the concatenate where the linear injected molecules were ligated into a head to tail arrangement. Thus, clone 2 contains the 5.2 kb fragment as shown in the genome blot of Figure 3a. The BglII junction fragment of clone 1 is of the same size as one of the minor bands (5.4 kb) observed in Figure 3a (marked by asterisk). Thus, at least one, and possibly the other minor bands in the genome blot may also represent bona fide junction fragments. The exact locus of integration in clone 1 was further analyzed by sequencing (data not shown). The breakpoint in the plasmid sequence is 55 nt from the EcoRI site in the pBR322 element of pNeo3. The integration event occurred within a highly repetitive region of the sea urchin genome, as evidenced by genomic blots carried out with the sea urchin DNA fragment of clone 1 adjacent to pNeo3 sequence, as a probe.

In another set of animals, which were raised from eggs injected with the plasmids pTFN and pπ25.1 (14), we investigated the possibility that integrated exogenous sequences would be retained in the germ line. Sea urchins raised in the laboratory usually reach sexual maturity about 10-12 months after fertilization, although some animals take considerably longer. Gametes were obtained from 12 animals (11 males and 1 female) between 12 and 20 months after fertilization. DNA was isolated from the sperm of the male animals and from embryos developed from the eggs of the female sea urchin. This DNA was used for genomic blots which were hybridized to radioactive probes prepared from the plasmid DNAs that were used for micro-injection. The sperm DNA from one male animal contained the injected sequences, whereas the gametes of the other 11 animals were negative. Our preliminary analysis strongly suggests that the exogeneous sequences are integrated in the sperm DNA of this animal and are present at approximately one copy per haploid genome. It remains to be seen if this organization is the same in the other tissues of this animal, if all the sperm contain the same sequences, and if they are heritably transferred to offspring.

The observations reported here suffice to reconstruct the probable fate of exogenous DNA molecules introduced into the cytoplasm of the unfertilized sea urchin egg by microinjection. We demonstrated that injected linear DNA molecules undergo a rapid end-to-end ligation, and that during embryogenesis the resulting concateners are on the average amplified about 25-fold. We have shown that a majority of the individual mature larvae raised from such injected embryos contain amplified plasmid DNA. Finally, we have shown that at least a fraction of these larvae give rise to postmetamorphosis juveniles in which the exogenous DNA is integrated into the sea urchin genome. Recent studies (19; and unpublished data) show that exogenous amplified DNA sequences introduced by our procedures are transcriptionally active in embryos and in feeding larvae. Thus the work briefly summarized here, and described in more detail elsewhere by McMahon et al. (13) and Flytzanis et al. (14) may provide the basis for a useable gene transfer system for the analysis of embryonic gene activation in the sea urchin.

ACKNOWLEDGEMENTS

We thank Frances Teng and Edward Nolan for technical assistance.

REFERENCES

1. Angerer, RC, Davidson, EH (1984). Science, in press.
2. Costantini, FD, Britten, RJ, Davidson, EH (1980). Nature 287:111-117.
3. Anderson, DM, Richter, JD, Chamberlin, ME, Price, DH, Britten, RJ, Smith, LD, Davidson, EH (1982). J molec Biol 155:281-309.
4. Spradling, AC, Rubin, GM (1982). Science 218:341-347.
5. Rubin, GM, Spradling, AC (1982). Science 218:348-353.
6. Goldberg, DA, Posakony, JW, Maniatis, T (1983). Cell 34:59-73.
7. Richards, G, Cassab, A, Bourouis, M, Jarry, B, Dissous, C (1983). EMBO J 2:2137-2142.
8. Scholnick, SB, Morgan, BA, Hirsh, J (1983). Cell 34:37-45.
9. Spradling, AC, Rubin, GM (1983). Cell 34:47-57.
10. Hazelrigg, T, Levis, R, Rubin, GM (1984). Cell 36:469-481.
11. McKnight, GS, Hammer, RE, Kuenzel, EA, Brinster, RL (1983). Cell 34:335-341.
12. Brinster, RL, Ritchie, KA, Hammer, RE, O'Brien, RL, Arp, B, Storb, U (1983). Nature 306:332-336.
13. McMahon, AP, Flytzanis, CN, Hough-Evans, BR, Katula, KS, Britten, RJ, Davidson, EH (1984). Submitted for publication.
14. Flytzanis, CN, McMahon, AP, Hough-Evans, BR, Katula, KS, Britten, RJ, Davidson, EH (1984). Submitted for publication.
15. Lee, JJ, Shott, RJ, Rose, SJ, Thomas, TL, Britten, RJ, Davidson, EH (1984). J molec Biol 172: 149-176.
16. Davies, J, Smith, DI (1978). Ann Rev Microbiol 32:469-518.
17. McKnight, SL (1980). Nucleic Acids Res 8:5949-5964.
18. Frischauf, A-M, Lehrach, H, Poustkd, A, Murray, N (1983). J molec Biol 170:827-842.
19. McMahon, AP, Novak, T, Britten, RJ, Davidson, E (1984). Proc Natl Acad Sci, in press.

Molecular Biology of Development, pages 633–645

DNA-MEDIATED TRANSFORMATION IN DICTYOSTELIUM

Wolfgang Nellen, Colleen Silan, and Richard A. Firtel

Department of Biology
University of California, San Diego
La Jolla, California 92093

ABSTRACT We have constructed a new vector and modified an existing transformation protocol for *Dictyostelium discoideum*. The vector consists of a fusion of the Actin 6-5' untranslated and flanking regions and the first 24 nucleotides of the N-terminal coding region of the Tn5 neomycin resistance gene. The vector which gives the highest frequency of transformation has the initiating AUGs from the actin and neomycin resistance genes in frame. Using this vector and the new protocol, we obtained transformation frequencies of $\sim 10^{-4}$ of the input cells. We have isolated and examined the DNA from populations of transformants. All contain vector DNA, at ~5 copies per haploid genome. The DNA is stably replicated for at least 30 generations after the selection has been removed. We have also examined transcription of the neomycin resistance gene fusion in transformed cells. The gene is expressed in vegetative cells grown axenically. Upon starvation, the relative level of RNA complementary to the neomycin resistance gene decreases as expected if the *Dictyostelium* actin 6 promoter on the gene fusion is being regulated in a manner similar to the endogenous Actin 6 promoter. We expect that this new system can be used for examining the control of gene activity in *Dictyostelium*.

INTRODUCTION

Dictyostelium discoideum represents an excellent system to examine differential gene regulation during development. When the food source is removed from vegetative amoebae, the cells initiate a defined developmental program involving the formation of a multicellular aggregate which differentiates and forms two discrete cell populations, prespore and prestalk cells, in a defined spatial arrangement (1). Under normal developmental conditions, this differentiation requires the formation of a multicellular aggregate; however, it is now also possible to induce in vitro the expression of genes normally preferentially expressed in either prespore or prestalk cells under defined conditions in single cells (2-4). During culmination, a fruiting body is formed in which the prespore cells and prestalk cells differentiate into spores and stalk cells respectively. The simplicity of the developmental program, with the presence of only two major cell types, allows one to ask questions concerning the molecular control mechanism involved in cell type differentiation. In order to be able to examine these mechanisms, it is necessary to have a functional DNA-mediated transformation system.

Previously, we reported a DNA-mediated transformation system for Dictyostelium vegetative cells (5,6). The construction of vector CERF.DRp14 consisted of fusing the neomycin resistance gene from Tn5 to the 5′ flanking region of the Dictyostelium Actin 8 gene, which is known to be expressed at a relatively high level in vegetative cells (7). The fusion was immediately 3′ to the mapped cap site of the Actin 8 gene and contained ~70 nucleotides upstream from the proposed TATA box. The potential 5′ untranslated region and the AUG initiation codon of the fused gene came from the bacterial Tn5 gene. In addition, the vector contained a 2.7 kb fragment of Dictyostelium DNA (DRp14) which could function as an origin of replication in yeast. Previously, this sequence was shown to be necessary for transformation. Methods of introducing the DNA into Dictyostelium cells, subsequent selection, and identification of clones derived from single transformed cells were established (5).

Using the established protocol, the vector could transform Dictyostelium cells, albeit with variable reproducibility. Vector DNA was shown to be transiently extrachromosomal and then integrated in more stable transformants (5). Approximately 6-12 months subsequent to publishing these results, we were no longer able to transform Dictyostelium consistently. Attempts to improve the transformation protocol, as well as the use of new derivatives of the vector pCERF.DRp14 and of vector constructions published by Barkley and Meller (8) did not lead to more reliable results. Since new comparative data on the structure of Dictyostelium genes were available (9,10, Romans and Firtel, manuscript in preparation), we decided to construct a new vector based on this information. Analysis of >20 Dictyostelium genes has indicated that the 5′ untranslated regions are extremely AT rich and that the AUG initiation codon is preceded by one or more A residues. In pCERF.DRp14 this 5′ untranslated region was provided by the bacterial Neo^R gene which is more GC rich and the AUG initiation codon is preceded by a C residue. Furthermore, analysis of 15 of the 17-20 *Dictyostelium* actin genes shows that many have conserved GC rich sequences with dyad symmetry in the 5′ flanking region (Romans and Firtel, manuscript in preparation). Such sequences are lacking in the short 5′ flanking region of Actin 8 used in pCERF.DRp14 (11).

In the new vectors, the neo^R gene from Tn5 is fused to the Actin 6 gene at a position 24 nucleotides 3′ to the AUG initiation codon. Such a construction conserves the proposed translation initiation signals from the *Dictyostelium* gene. Like the Actin 8 gene, the Actin 6 mRNA represents a relatively high level of vegetative actin mRNA (20-25%) (7,12). Moreover, the actin 6 fragment contains approximately 600 nucleotides upstream from the TATA box, including GC rich sequences with dyad symmetry, similar to sequences found upstream from some *Dictyostelium* actin genes, but not other *Dictyostelium* genes.

RESULTS

In *Dictyostelium*, actin is encoded by 17-20 genes (10,13). It is known that actin protein synthesis is differentially regulated during development and that various actin genes are also differentially transcribed both qualitatively and quantitatively through *Dictyostelium* development (see 10). Fifteen of the 17-20 members of the actin gene family have been cloned, the 5′ ends of several of these have been mapped, and their relative expression through *Dictyostelium* development has been determined (7,10,12, Romans and Firtel, manuscript in preparation). Actin 6 was chosen since it is expressed at a relatively high level in vegetative cells (7). Messenger RNA from this gene comprises 20-25% of the actin messenger RNA in vegetative cells or an estimated 0.2-0.5% of total $poly(A)^+$ RNA. Later in development, RNA complementary to Actin 6 decreases appreciably, and by 10 hours represents ~5% of total actin mRNA. We have also determined that there is a series of GC-rich regions having diad symmetry lying -100 to -250 nucleotides upstream from the AUG in many of the actin genes that have been analyzed, including Actin 6 (Romans and Firtel, manuscript in preparation).

The construction of the gene fusion is shown in Figure 1. An Eco RI/Hind III fragment containing 24 nucleotides

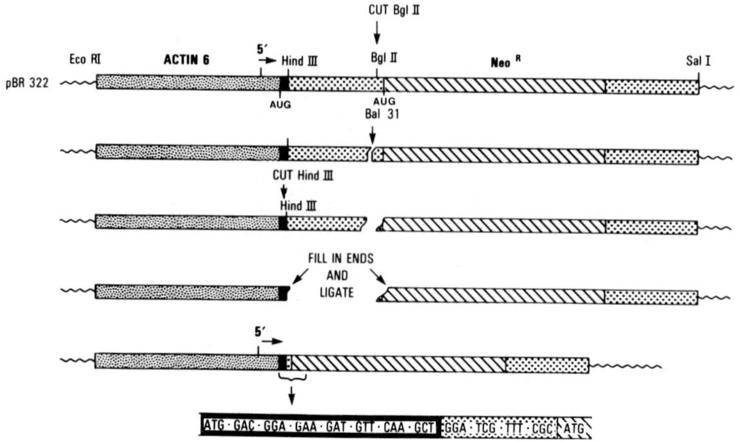

Figure 1. Construction of vector B10. Details of construction of vector B10 are given in the text. The sequence between the AUG of Actin 6 and the AUG of the Tn5 neomycin resistance gene are shown. The sequence surrounded by a thick black line is the first 24 nts of Actin 6 coding region; the sequence in the stippled box is the 12 nts 5′ to the neo^R gene AUG (region with stripes). ∿ pBR322; [stippled box] Actin 6 5′ untranslated and flanking regions; [black box] actin coding sequences; [dotted box] noncoding regions from Tn5; [striped box] Tn5 neomycin resistance gene coding region.

coding for the N-terminus of Actin 6, 5′ untranslated region, and approximately 600 nucleotides of the 5′ flanking region was ligated to a Hind III/Sal I fragment containing the neomycin resistance gene from Tn5 (14). In this construct, there are ~350 nucleotides between the Hind III site of the fusion and the AUG of the neomycin resistance gene. The DNA was digested with Bgl II and resected with Bal 31 in order to delete stop codons between the Bgl II site and the AUG of the Neo^R gene and to bring the gene fusion into the correct reading frame. The DNA was then digested with Hind III, the ends repaired with the large fragment of DNA polymerase I, and ligated together. After transformation into E. coli, the DNA from a series of clones was analyzed by restriction enzyme digest followed by gel electrophoresis. Clones in which the plasmid DNA contained ~10-25 nucleotide Bal 31 deletion were chosen for further analysis.

In order to determine which gene fusions might functionally express the neomycin resistance gene in Dictyostelium, DNA from these clones was used to transform

Dictyostelium using a modification of previous methods (5) to select for cells resistant to the aminoglycoside G418. While the starting vector prior to Bal 31 resection did not give an increase in G418 resistant colonies over the background, two of the plasmids (B10 and B12) carrying 10-15 nucleotide deletions gave a significant number of G418-resistant colonies. A series of transformations were done utilizing B10 and B12, with pBR322 as a control. All transformations were with circular DNA purified by banding once or twice in cesium chloride with ethidium bromide followed by extraction with phenol and chloroform and precipitation with ethanol. In a series of experiments using 12 μg of DNA, B10 gave transformation frequencies ranging between $4x10^2$-$2x10^3$ G418-resistant colonies per 10^7 input cells. Within three experiments, the transformation frequency varied by only a factor of 2 between duplicates or triplicates. pBR322, used as a control, gave between 0-15 colonies after selection on G418, approximately 100-fold lower than B10. The efficiency for B12 was more variable and normally ranged 10-50-fold below that for B10.

To determine the position of the gene fusion in B10 and B12, the nucleotide sequence of the region containing the 5' untranslated region of Actin 6 and N-terminal region of Tn5 neo^R gene was determined. The results showed that B10 contains an in-frame gene fusion between the Dictyostelium Actin 6 AUG and the Neo^R gene AUG (see Figure 1). The region between the two AUGs contained 24 nucleotides from the N-terminal coding region of Actin 6 and 12 nucleotides from the region between the Bgl II site and the AUG of the Tn5 neomycin resistance gene. This suggests that translation initiates at the actin AUG and reads through the neomycin resistance gene, forming a fusion protein, although initiation could begin at either AUG. Sequence analysis of B12 showed that it contained the 24 nucleotides of the N-terminal region of Actin 6 and 18 nucleotides 5' to the AUG of the neomycin resistance gene, thus putting the AUG of the neomycin resistance gene out of frame with the actin AUG. Because of the higher frequency of transformation with B10, further experiments were pursued with this vector.

Previously we showed that a 2.7 kb Dictyostelium Eco RI fragment denoted DRp14 was necessary for transformation with our original vector pCERF (5). This Dictyostelium fragment was selected because it can act as an ARS in yeast. Two new vectors, called B10.DRp14P and B10.DRp14D, were made in which the DRp14 Eco RI fragment was cloned into the B10 vector in both orientations (see Figure 2). The orientation was determined using the asymmetric Bgl II site and denoted proximal (P) when the Bgl II site was closer to the Actin 6

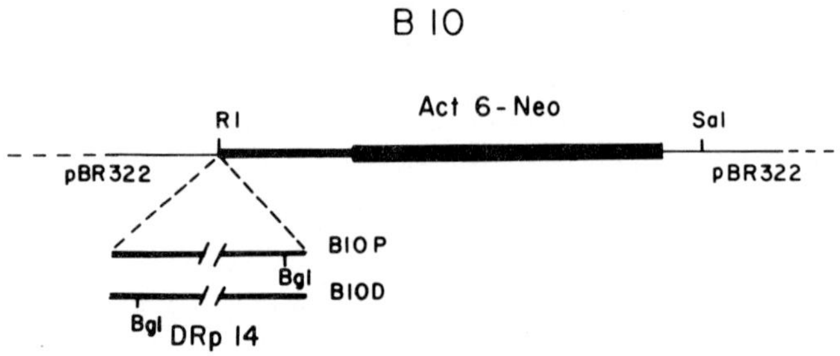

Figure 2. Map of B10.DRp14P and B10.DRp14B. The two vectors are described in text. Both consist of inserting the 2.7 kb DRp14 fragment into the Eco RI site of B10. The intermediate thick black line represents the 5′ untranslated and flanking regions from Actin 6. The thickest black line represents the Actin 6-neomycin resistance gene fusion. The Bgl II site in the DRp14 sequence in the B10.DRp14 vectors is shown.

region or distal (D) when the restriction site was further away. These two vectors were also used in the transformation. The transformation frequency for B10.DRp14P was between 50-100% of that of B10, while B10.DRp14D gave slightly lower frequencies. Therefore, in contrast to previous results with pCERF.DRp14, the DRp14 fragment does not appear to aid the transformation frequency with B10.

We examined whether cultures transformed with B10 or B10.DRp14 contained vector DNA. Populations containing 50-500 transformants were grown in the absence of selection in axenic medium or in the presence of G418 with autoclaved *E. coli*. DNA was isolated, digested with Eco RI and Sal I, which excises the Actin 6-neomycin resistance gene fusion and the DRp14 sequence from pBR322 (see Figure 2), size fractionated on agarose gels, and blotted. The filter was then probed with nick translated DNA containing sequences complementary to both the Tn5 neo^R gene and pBR322. As is shown in Figure 3A, DNA complementary to the gene fusion and pBR322 was found in all the transformed cultures, and averaged ~5 copies of vector DNA per cell (unpublished observations).

Figure 3. Genomic Southern blots of DNA isolated from transformants. Cells were transformed with the appropriate vector and plated on black Millipore filters in association of K. aerogenes to select for transformants (5). After transformants replicated sufficiently to form a visible colony, a group of colonies were isolated from the filters and put directly into either axenic medium or autoclaved E. coli in phosphate buffer with 20 µg/ml G418. Cells were grown for 4-6 generations until there were a sufficient number of cells for the isolation of DNA. Bb and Ba: DNA isolated from cells transformed with B10 grown axenically (Bb) or in association with autoclaved E. coli and G418 (Ba). BPb and BPa: DNA isolated from cells transformed with B10.DRp14P grown axenically (BPb) or in association with E. coli and G418 (BPa). DNA was also isolated from the Ba and Bb cultures grown for an additional 30 generations (Ba-30 and Bb-30). Lanes labeled Dd contain 2 µg of Dictyostelium DNA isolated from untransformed cells grown axenically.

The DNA was digested with restriction enzymes, size fractionated on agarose gels, blotted to nitrocellulose, and probed with nick translated plasmid which carries the coding region from Tn5 neomycin resistance gene cloned into a derivative of pBR322. The vector used for the probe did not contain sequences complementary to the actin portion of B10.

Some of the differences in the relative intensity of the bands hybridizing to the probe in part A and part B are due to unequal amounts of DNA loaded on each lane. The amount of miniprep DNA from transformants loaded on each lane, estimated by staining, ranged from 1/4 to 3/4 µg. This variation may account for some of the differences in intensity of the hybridizing fragments; however, these intensity differences could also be due to different average copy numbers of the vector in different populations of transformants.

A. DNA was digested with Eco RI plus Sal I. V = pBR322 sequences (3.7 kb). F = sequences from gene fusion (~2 kb). Note that Eco RI excises DRp14 from the gene fusion and pBR322 sequences.

B. DNA was digested with Sal 1. The size of the bands in kilobases is shown. M-2X represents marker DNA equivalent to 2 copies of the vector per genome per 2 µg of Dictyostelium DNA.

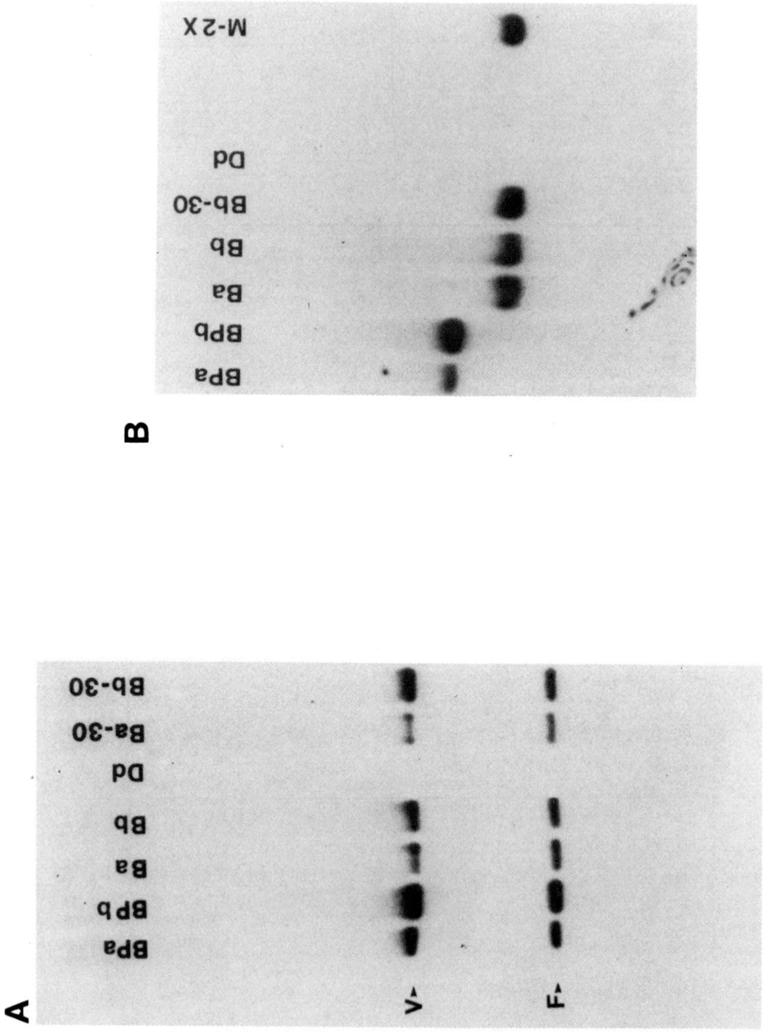

Figure 3.

Both the gene fusion and the pBR322 sequences are present at approximately equal concentrations, suggesting that the DNA may either be integrated as a tandem array in some of the cells, or found extrachromosomally as a tandem array or circular molecule. To test this further, DNA was digested with Sal I, which cleaves once within B10 or B10.DRp14 (see Figure 2), size fractionated, blotted, and analyzed in a fashion to that in Figure 3A. As can be seen from Figure 3B, a single band of molecular weight equal to the size of the input plasmid is found. The relative intensity of this band is approximately the same as the vector band observed in the Eco RI-Sal I double digests, indicating that one of the above models on the physical state of the DNA in the cells is correct, and that there do not appear to be multiple, individually integrated copies. DNA was also isolated from cells grown for an additional 30 generations and analyzed as described above. As can be seen in Figures 3A and 3B, vector DNA sequences are found at approximately the same level in DNA from transformants grown for an addition 30 generations as is seen in the DNA isolated from cells for a few generations after transformants appear (5, see legend to Figure 3).

We have also examined the expression of the gene fusion in transformants. For these the Dde I-Pst I fragment containing the 5' end of the Tn5 Neo^R, the short coding region of Actin 6, and the 5' untranslated and some of the 5' flanking regions of the Actin 6 gene was subcloned into M13mp8 and SP64 (Melton and Green, personal communication) (see Figure 4). Single stranded DNA and RNA complementary to the proposed fusion mRNA were made using the Klenow fragment of pol I or SP6 RNA polymerase respectively (15) and hybridized to RNA isolated from vegetative cultures of transformed populations. These were then treated with either S_1 nuclease or ribonuclease A respectively, and the protected fragments were then sized by gel electrophoresis. The results of probing with RNA complementary to the sequence shown in Figure 4 are shown in Figure 5. RNAs used for these experiments were isolated from vegetatively growing transformed cells (0 hours), control (non-transformed) cells, and cells starved for either 5 or 6 hours. These results show that in all transformed populations the Actin 6-neo^R gene fusion is being transcribed, as is expected from our data on the resistance of these clones to G418. However, the relative level of protection of the probe is less than expected if the Actin 6 promoter of the gene fusion were being transcribed at the same rate as the endogenous promoter and if the fusion mRNA were equally stable as actin 6 mRNA (unpublished observation). Results shown in this

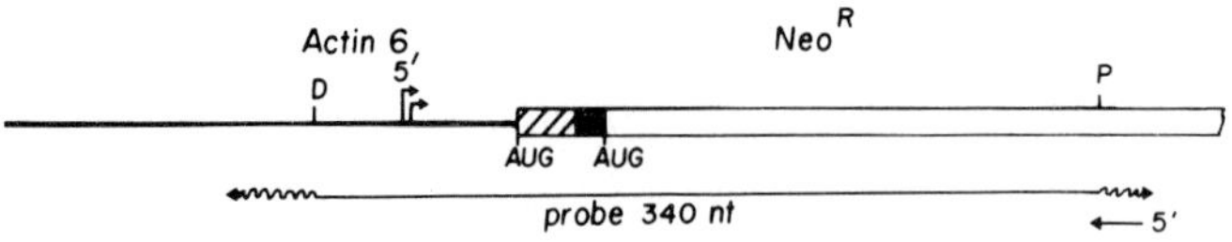

Figure 4. Map of the probe used for transcription studies. The region from the Dde I restriction site (D) located upstream from the previously mapped cap sites for Actin 6 (7) and the first Pst I restriction site in the Tn5 neomycin resistance gene was subcloned into the Pst I-Hinc II restriction sites of M13mp8. The Hind III-Eco RI fragment from this was then recloned into the vector SP64, which carries the promoter for SP6 polymerase (Melton and Green, personal communication). Approximate locations of the mapped cap sites of the Actin 6 mRNA are shown. represents the N-terminal coding region for Actin 6. represents the 12 nucleotides from the 5' untranslated region of the neomycin resistance gene. The long open rectangle represents the N-terminal region from the neomycin resistance gene. Probes made from this vector are ~340 nucleotides in length. The represents sequences from the linker of M13mp8. The 5' end of the probe (complementary to gene fusion RNA) is indicated.

figure also suggest that the relative levels of RNA complementary to the gene fusion in vegetative and 5 or 6 hour starving cells parallels the normal 4-5 fold decrease in Actin 6 mRNA levels during this part of development (7,12).

DISCUSSION

We have constructed new vectors for transforming Dictyostelium consisting of a fusion between the Actin 6 N-terminal and 5' untranslated and flanking regions with the Neo^R gene from Tn5. These vectors confer resistance to G418 in Dictyostelium and can be effectively used to select transformants. The frequency is $\sim 10^{-4}$ of the input cells. DNA sequencing indicates that the AUG of the actin and the neomycin resistance gene are in frame in one vector, B10, which gives the best frequency of transformation of the vectors tested. At the present time, we have not directly determined whether translation for the proposed protein inactivating G418 initiates at the actin AUG or the AUG of

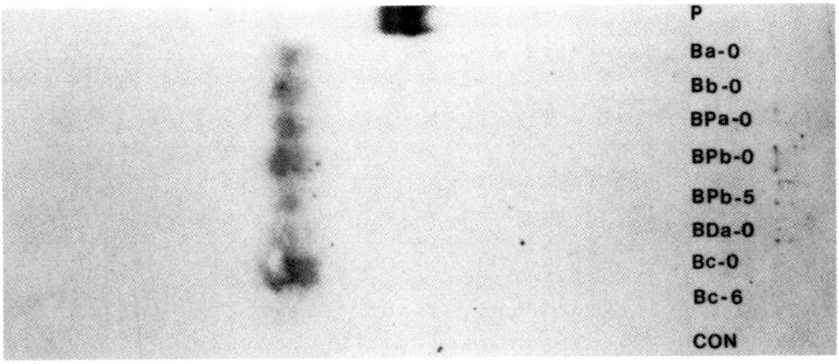

Figure 5. Expression of B10 vectors in transformants. An RNA probe as described in Figure 4 was synthesized using SP6 polymerase (15, Melton and Green, personal communication). The probe was hybridized in excess to RNA isolated from several populations of Dictyostelium cells transformed with B10 (B series), B10.DRp14P (BP series), or B10.DRp14D (BDa). RNA isolated from vegetative cells is indicated by a 0. RNA isolated from cells starved for either 5 or 6 hours is indicated by a 5 or a 6 respectively. After hybridization, the reaction mixture was treated with ribonuclease A followed by proteinase-K, extracted with phenol and chloroform and ethanol precipitated. The precipitated material was then sized on a 1.6% agarose gel containing formaldehyde. The gel was then dried and autoradiographed. The first lane (P) shows the molecular weight of the probe prior to hybridization. The majority of the material represents a full-length transcript of the insert sequences cloned in SP64 and is ~340 nt. The next 8 lanes show hybridization with RNA isolated from transformants. The molecular weight of the band is ~260 nt. The last lane (CON) contains probe hybridized to RNA isolated from untransformed vegetative Dictyostelium cells. The probe was constructed such that endogenous Actin 6 mRNA would also hybridize to the 5' untranslated and first 24 nts of the coding region of Actin 6. This material would have run off this gel.

the Neo^R gene. However, the higher frequency of transformation with B10 over B12 suggests that translation initiation predominantly begins at the Actin 6 AUG. Since B12 does appear to have some level of transformation, we would also suggest that at least some initiation of translation from B12 RNA initiates at the AUG of the neomycin resistace gene.

DNA blot data indicate that, on the average, there are

~5 copies of vector DNA per transformed cell (quantitation not shown). Additional blotting experiments using restriction enzymes that cleave once within the vector DNA show a single band (see Figure 3B, unpublished observation). At the present time, these experiments do not distinguish between extrachromosomal copies, circular or in tandem arrays, or between tandemly repeated copies which are integrated. We have also examined DNA isolated from cells after nonselective growth for an additional 30 generations. The relative level of DNA per cell is approximately the same as that in the original population, suggesting that the DNA is stable. If the DNA is replicating extrachromosomally, then the Actin 6 sequences may fortuitously contain an origin of replication which also allows proper segregation during cell division.

Preliminary expression studies indicate that there is RNA complementary to the gene fusion, but at a level below that of the endogeneous Actin 6 mRNA (Firtel and Silan, unpublished observation). It should be pointed out that the vector as constructed does not contain a Dictyostelium intron or a 3' untranslated region and polyadenylation signal. We feel, however, that an intron may not be necessary since no intervening sequences have been identified in the Dictyostelium genes. It is very possible that the low level of complementary RNA may be due to a shorter than normal half-life and that adding a Dictyostelium 3' untranslated region may allow the RNA to accumulate to a higher steady state level.

With the new constructions and transformation protocol, we can obtain reproducible transformation at a reasonable frequency. Such a system will be essential in working out the molecular mechanisms involved in differential gene expression in Dictyostelium. It is hoped that this system will open up the avenues which are necessary to pursue such studies.

ACKNOWLEDGEMENTS

WN was supported by a post-doctoral fellowship from the Dewtsche Forschungsgemeinschaft. RAF was a recipient of a ACS Faculty Research Award. This work was supported by grants from USPH-NIGMS to RAF.

REFERENCES

1. Loomis, WF (1975). "Dictyostelium discoideum: A Developmental System." New York: Academic Press.
2. Mehdy MC, Ratner D, Firtel RA (1983). Cell 32:762.
3. Mehdy MC, Saxe CL III, Firtel, RA (1984). In Davidson, EH and Firtel, RA (ed): "Molecular Biology of Development," New York: Alan R. Liss, Inc. (in press)
4. Kay RR (1982). Proc Nat Acad Sci 79:3228.
5. Hirth K-P, Edwards CA, Firtel RA (1982). Proc Nat Acad Sci 79:7356.
6. McKeown M, Hirth D-P, Edwards C, Firtel RA (1982). In "Embryonic Development: Gene Structure and Function."
7. McKeown M, Firtel RA (1980). Cell 24:799.
8. Barclay SL, and Meller E. (1983) Mol. Cell. Biol. 3:2117.
9. Kimmel AR, Firtel RA (1983). Nucl. Acids Res. 11:541.
10. Kimmel AR, Firtel RA (1982). In Loomis WF (ed): "The Development of Dictyostelium discoideum," New York: Academic Press, pp 233.
11. Firtel RA, Timm R, Kimmel AR, McKeown M (1979). Proc Nat Acad Sci 76:6206.
12. McKeown M, Kimmel AR, Firtel RA (1981). In Brown D, Fox CF (ed): "Developmental Biology Using Purified Genes," pp107.
13. Kindle KL, Firtel RA (1978). Cell 15:763.
14. Southern P, Berg P (1982). J Mol Appl Genet 1:327.
15. Green MR, Maniatis T, Melton DA (1983) Cell 32:681.

Molecular Biology of Development, pages 647–656

DEVELOPMENTALLY REGULATED EXPRESSION OF CHIMERIC MUSCLE GENES TRANSFERRED INTO MYOGENIC CELLS

Uri Nudel, Danielle Melloul, Batya Aloni, David Greenberg and David Yaffe

Department of Cell Biology
The Weizmann Institute of Science
Rehovot 76100, Israel

ABSTRACT We have constructed two chimeric genes, one containing two-thirds of the rat skeletal muscle actin gene plus 730 bp 5' flanking DNA spliced to the 3' end region of human ε-globin gene; and a second one containing only the 5' region plus 730 bp flanking DNA of the skeletal muscle actin gene spliced to the bacterial chloramphenicol acetyltransferase structural gene. The two genes were introduced into myogenic L8 cells. In many of the clones containing these genes we observed a large increase in the amounts of the transferred gene products following differentiation of the cultures (ranging from severalfold to more than 50-fold). In contrast, in clones transfected with plasmids containing nonmuscle genes, only small changes in the amounts of the transferred gene products were observed following differentiation of the cultures. The results demonstrate that DNA sequences in or near the promoter region of the rat skeletal muscle actin gene are involved in the tissue- and stage-specific expression of the gene.

INTRODUCTION

Muscle cell cultures provide a very convenient model system for the study of the molecular mechanism of control of gene expression during differentiation. During terminal differentiation, precursor muscle cells fuse to form multinucleated fibers. This process is accompanied by a rapid accumulation of muscle-specific mRNAs (1) and muscle-specific contractile proteins and enzymes which are involved in the energy supply machinery for the contracting muscle (for review see Ref. 2). It has been shown that the DNAase I sensitivity of genes coding for the muscle-specific isoforms of contractile proteins increases during cell fusion, thus strongly suggesting transcriptional activation of these genes (3).

We have recently isolated and sequenced several the genes coding for rat skeletal muscle actin and cytoplasmic β-actin (4, 5). In this communication we describe experiments in which we reintroduced modified actin genes into myogenic cells and studied their expression during differentiation. We found that sequences in or near the promoter region of the skeletal muscle actin gene are involved in the developmentally regulated expression of the transferred muscle gene.

MATERIALS AND METHODS

The construction of the chimeric plasmids used in this study was described previously (6). Myogenic L8 cells were grown in cultures and were induced to differentiate as described (7). Transfection of cells with the chimeric plasmids was done by the calcium phosphate precipitation technique (8) with some modifications (6). RNA was prepared from cultures by the lithium chloride/urea extraction technique (9). S1 endonuclease analysis was done as described by Berk and Sharp (10) and modified by Weaver and Weissmann (11). Preparation of cell extracts and assay of CAT activity were done according to Gorman et al. (12) and Melloul et al. (6). Creatine kinase (CK) activity in cell extracts was assayed as described by Shainberg et al. (13).

RESULTS

We first studied the question of whether the information for tissue- and stage-specific expression of a muscle gene is an intrinsic property of the gene, determined by DNA sequences in the structural gene region. In order to be able to identify the RNA product of the transferred muscle gene we constructed a recombinant plasmid, pCV,

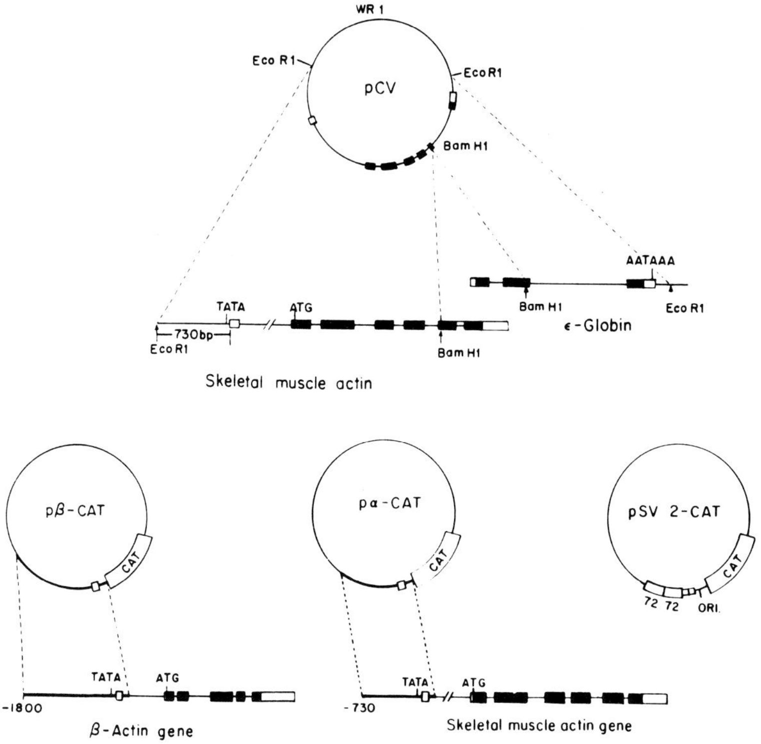

FIGURE 1. Structure of the plasmids containing the chimeric genes. For details on the construction see Ref. 6. In plasmids pCV, pα-CAT, and pβ-CAT open and black bars represent untranslated and translated regions, respectively. WR1 = sequences of the plasmid pWR1.

containing the 5' 2/3 of the rat skeletal muscle actin gene plus 730 bp flanking DNA spliced to the 3' 1/3 of the human ε-globin gene (Fig. 1 and Ref. 6). L8 myogenic cells were co-transfected with this plasmid and the plasmid pIPB1 containing the neomycin resistant marker from Tn5. Clones resistant to the neomycin derivative G418 were isolated. We found that most of the clones could be induced to differentiate. The copy number of the actin/globin chimeric gene varied between 1 and 50 (one clone contained more than 500 copies of the transferred gene).

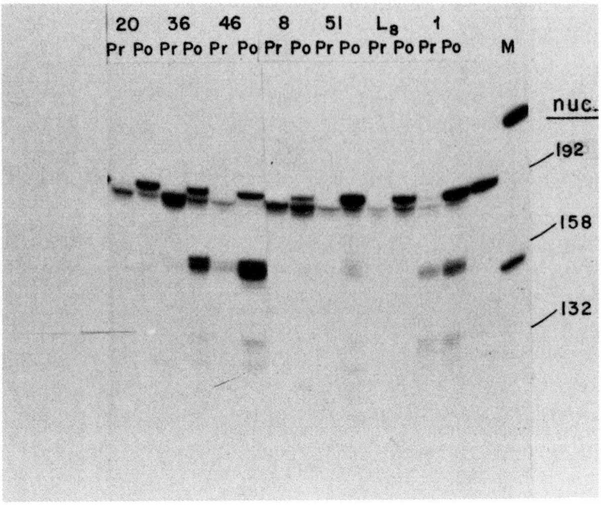

FIGURE 2. S1 endonuclease analysis of RNA from clones containing the actin/globin chimeric gene. The probe used in the S1 endonuclease analysis was an end-labeled DNA fragment isolated from a rat skeletal muscle actin cDNA clone (6). The entire probe (192 nucleotides) is protected by the endogenous rat skeletal muscle actin mRNA. A 158 nucleotide DNA fragment is protected by the actin/globin chimeric mRNA (6). Samples containing 40 ug of total RNA extracted from undifferentiated (Pr) or differentiated (Po) cultures were hybridized with approximately 20 ng end-labeled probe (18 h, 53°C). After hybridization, the mixtures were treated with S1 endonuclease (500 units, 30 min at 37°C), precipitated and electrophoresed on a polyacrylamide/urea sequencing gel. The gel was then fluorographed (from Melloul et al., 6).

We analyzed the expression of the actin/globin chimeric gene in undifferentiated and differentiated cultures of 9 clones by the S1 endonuclease mapping technique, using both actin and globin end-labeled DNA probes. Figure 2 shows an S1 endonuclease analysis of several representative clones, using an end-labeled DNA probe derived from a rat

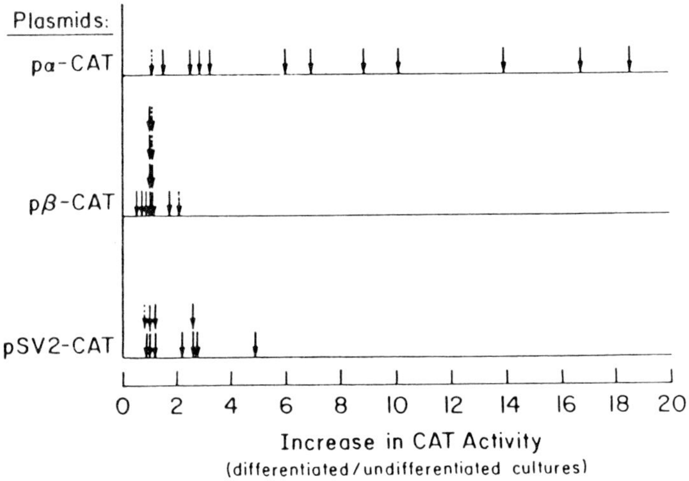

FIGURE 3. Increase in CAT activity following differentiation of the clones containing the plasmids pα-CAT, pβ-CAT and PSV2-CAT. Conversion of chloramphenicol to its acetylated forms was measured as previously described (6). The numbers on the abscissa indicate the ratio between CAT activity in differentiated cultures and CAT activity in undifferentiated cultures of the same clone. Each arrow represents the ratio in a single clone. Broken arrows represent clones in which no cell fusion occurred after growth for 72 h in the differentiation stimulating 2HI medium (from Melloul et al., 6).

actin cDNA clone. In 5 clones, a marked increase in the amount of the actin/globin RNA product during cell fusion was observed. In two of these (L8-36, L8-46), the increase was more than 50-fold. Moreover, in these two clones the amount of the actin/globin gene product in differentiated cultures was greater than the amount of the native actin mRNA. Using probes derived from the 5' end of the actin gene and from the 3'-end of the globin gene, we found that the chimeric mRNA was initiated at the authentic actin mRNA cap site and terminated at the authentic globin mRNA poly(A)-addition site. The kinetics of accumulation during differentiation of the actin/globin chimeric mRNA (in clone

L8-46) was very similar to that of the native actin mRNA in the same cultures (6). These results strongly suggest that both the actin/globin chimeric gene and the native actin gene are regulated by a common mechanism.

As a control experiment we measured the level of expression of a mouse/human β-globin chimeric gene stably integrated into L8 myogenic cells. It has previously been reported that the expression of this gene is developmentally regulated when introduced into mouse erythroleukemia cells (14). The mouse/human chimeric β-globin gene was expressed in 6 out of 8 clones tested. However, no increase in the amount of chimeric β-globin mRNA after cell fusion was observed. In fact, in all of these clones a 2-3-fold decrease in this mRNA occurred after cell fusion (not shown).

To determine whether the DNA sequences involved in the regulated expression are located in or near the promoter region of the skeletal muscle actin gene, we constructed the plasmids pα-CAT and pβ-CAT containing the promoter region, and the exon at the 5' untranslated region and 5' flanking DNA of rat skeletal muscle actin gene and β-actin gene, respectively, spliced 5' to bacterial chloramphenicol acetyl transferase (CAT) gene, as described in Fig. 1 and Ref. 6. We established myogenic cell lines containing plasmids pα-CAT, pβ-CAT or pSV2-CAT (a plasmid containing the promoter and enhancer regions of SV40; Ref. 12). Southern blot analysis of DNA from these clones showed a great variability in the transferred gene copy number, which varied from only a few to several hundreds of copies per haploid genome (not shown).

Cell extracts from undifferentiated and differentiated cultures of these cell lines were prepared and CAT activity was assayed. In 7 out of 11 clones transformed with the plasmid p -CAT, the CAT activity in extracts from differentiated cultures was 6-19 times higher than in extracts from undifferentiated cultures (Fig. 2). In contrast, in all 10 clones transfected with plasmids p -CAT, and in 9 clones transfected with the plasmid pSV2-CAT, the CAT activity in extracts from differentiated cultures was almost the same as in extracts from undifferentiated cultures (Fig. 4). (A 5-fold increase in CAT activity was observed in one of the pSV2-CAT containing clones.)
S1 endonuclease analysis of RNA from one of the -CAT containing clones revealed that the differences in CAT activity observed between undifferentiated and

differentiated cultures was also expressed at the mRNA level (not shown). The kinetics of accumulation of CAT activity in a clone containing the plasmid pα-CAT is very similar to that of accumulation of creatine kinase activity (a marker for muscle cell differentiation) in the same cultures (Fig. 4).

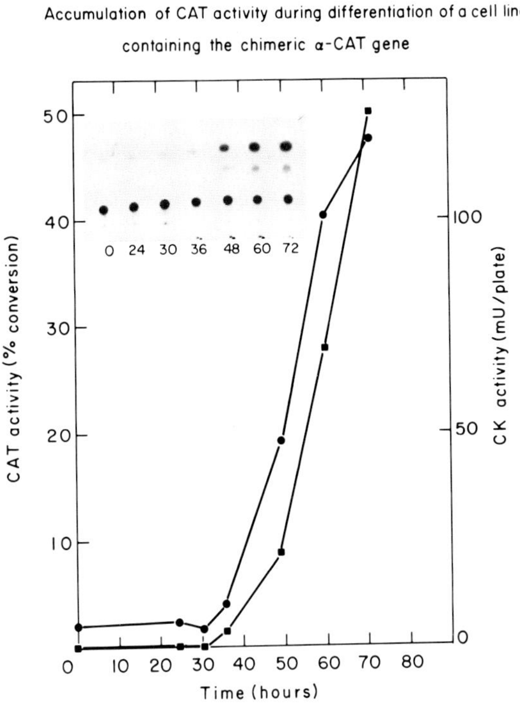

FIGURE 4. Kinetics of accumulation of CAT (●) and CK (■) activity during differentiation of a clone containing the plasmid p α-CAT. At zero time, cultures were stimulated to differentiate, cell extracts were prepared at the indicated times, and CAT and creatine kinase activity was measured. The insert shows the autoradiogram of the thin layer chromatographic plate used in the CAT assay.

The experiments described above were done with established cell lines in which the transferred genes were integrated into the genome. In order to exclude the effects of the site of integration on the expression of the transferred genes, we performed a series of transient transfection experiments in which several types of cells were transfected with the plasmids p -CAT, p -CAT, and pSV2-CAT, and the CAT activity was measured 20 h later. We found only low levels of CAT activity in cultures of the nonmuscle L cells and 3T3 cells and in cultures of undifferentiated L8 cells transfected with the plasmid p -CAT, and a much higher CAT activity in differentiated myogenic L8 cultures transfected with the same plasmid. In contrast, in cultures transfected with the plasmids p -CAT and pSV2-CAT, the CAT activity detected in nonmuscle cells was about the same as that detected in differentiated L8 cells.

DISCUSSION

In this communication we described experiments showing that DNA sequences or structures at or near the promoter region of the skeletal muscle actin gene contain information for tissue- and stage-specific expression of the gene. Comparison of the nucleotide sequence of the 5' flanking DNA of the rat and chick skeletal muscle actin genes reveals several conserved regions, especially that around the CAAT box (16, 17). Such a high degree of conservation in two genes that separated 250-300 million years ago strongly suggest a biological role for the sequence. It is therefore quite likely that these and several other conserved regions upstream from the CAAT box, between positions -100 and -220 (U. Nudel and C. Ordahl, unpublished) are involved in the tissue-specific expression of the genes and/or in other regulatory functions.

We found that both the level of expression and the degree of induction of the transferred genes during differentiation varied considerably between different clones. Similar findings were reported also for the developmentally regulated expression of globin genes introduced into murine erythroleukemia cells (14, 15). In addition, in most of the clones containing the chimeric actin genes we observed a low level of expression of the transferred genes in undifferentiated cultures in which the native skeletal muscle actin gene was not expressed. It seems, therefore, that in addition to the regulatory sequences located near the promoter region of the skeletal muscle actin gene, other factors may affect their expression, such as the chromatin domain into which the gene becomes integrated. It should also be pointed out that the introduction of genes into somatic cells may circumvent processes related to the control of gene expression occurring during early development (such as DNA methylation).

ACKNOWLEDGEMENTS

We wish to thank Ms. Zehava Levi, Sara Neuman and Ora Saxel for their excellent technical assistance; Dr. J. Calvo and X. Young for constructing the plasmid pCV; Drs. C.M. Gorman and B.H. Howard for the plasmids pSV0-CAT and pSV2-CAT; Dr. Silverstein for plasmid pIPB1 and the Scherring Corporation for a gift of G418. This work was supported by the Muscular Dystrophy Association, Inc., USA, the National Institutes of Health, USA, the U.S. Israel Binational Science Foundation, Jerusalem and the Leo and Julia Forchheimer Center for Molecular Genetics of the Weizmann Institute of Science. U. Nudel is the incumbent of the A. and E. Bloom Career Development Chair for Cancer Research.

REFERENCES

1. Shani M, Zevin-Sonkin D, Saxel O, Carmon Y, Katcoff D, Nudel U, Yaffe D (1981). The correlation between the synthesis of skeletal muscle actin, myosin heavy chain and myosin light chain and the accumulation of the corresponding mRNA sequences during myogenesis. Dev Biol 86:483.
2. Buckingham ME, Minty AJ (1983). Contractile protein genes. In Maclean N, Gregory S, Flavell RA (eds) "Eucaryotic genes: Their Structure, Activity and Regulation," London: Butterworth.
3. Carmon Y, Czosnek H, Nudel U, Shani M, Yaffe D (1982). DNAase I sensitivity of genes expressed during myogenesis. Nucl Acids Res 10:3085.
4. Nudel U, Zakut R, Shani M, Neuman S, Levi Z, Yaffe D (1983). The nucleotide sequence of the rat cytoplasmic β-actin gene. Nucl Acids Res 11:1759.
5. Zakut R, Shani M, Givol D, Neuman S, Yaffe D, Nudel U (1982). The nucleotide sequence of the rat skeletal muscle actin gene. Nature 298:857.

6. Melloul D, Aloni B, Calvo J, Yaffe D, Nudel U (1984). Developmentally regulated expression of chimeric genes containing muscle actin DNA sequences in transfected myogenic cells. EMBO J, in press.
7. Yaffe D, Saxel O (1977). A myogenic cell line with altered serum requirements for differentiation. Differentiation 7:159.
8. Graham FL, Van der Eb AJ (1973). A new technique for the assay of human adenovirus 5 DNA. Virology 52:456.
9. Auffray C, Nageotte R, Chambraud B, Rougeon F (1980). Mouse immunoglobulin genes: A bacterial plasmid containing the entire coding sequence for a pre- 2a heavy chain. Nucl Acids Res 8:1231.
10. Berk AJ, Sharp PA (1977). Sizing and mapping of early adenovirus mRNA by gel electrophoresis of S1 endonuclease-digested hybrids. Cell 12:721.
11. Weaver RF, Weissmann C (1979). Mapping of RNA by a modification of the Berk and Sharp procedure: The 5' termini of the 15S β-globin mRNA precursor and mature 10S β-globin mRNA have identical map coordinates. Nucl Acids Res 7:1175.
12. Gorman CM, Moffat LF, Howard BH (1982). Recombinant genomes which express chloramphenicol acetyltransferase in mammalian cells. Mol Cell Biol 2:1044.
13. Shainberg A, Yagil G, Yaffe D (1971). Alteration of enzymatic activities during muscle differentiation in vitro. Dev Biol. 25:1.
14. Chao MV, Mellon P, Charnay F, Maniatis T, Axel R (1983). The regulated expression of β-globin genes introduced into mouse erythroleukemia cells. Cell 32:483.
15. Wright S, de Boer E, Grosveld FG, Flavell RA (1983). Regulated expression of the human β-globin gene family in murine erythroleukemia cells. Nature 305:333.
16. Nudel U, Mayer Y, Zakut R, Shani M, Czosnek H, Aloni B, Yaffe D (1984). The structure and expression of rat actin genes. In Eppenberger HM, Perriard JC (eds) "Experimental Biology and Medicine, Vol 9," Basel: Karger Press, p 219.
17. Ordahl CP, Cooper TA (1983). Strong homology in promoter and 3' untranslated regions of chick and rat α-actin genes. Nature 303:348.

Molecular Biology of Development, pages 657–665

THE LTR OF FELINE LEUKEMIA VIRUS ENHANCES THE EXPRESSION OF THE BACTERIAL NEO^R GENE IN HUMAN CELLS

Nevis Fregien and Norman Davidson

California Institute of Technology,
Department of Chemistry
Pasadena, California 91125

ABSTRACT

A variety of recombinant DNAs were constructed to place the bacterial neomycin resistance under the control of the feline leukemia virus long terminal repeat (FeLV-LTR). These plasmids were tested for their ability to transform mouse, Ltk$^-$, cells and human, RD (Rhabdomyosarcoma)tk$^-$ cells, and compared to plasmids using the HSV tk and SV40 promoters. The mouse cells transformed equally well with all promoters. The human cells showed increased transformation with FeLV-LTR containing constructions. RNA blots showed the transcripts to be of the proper length. RNA processing was observed for one plasmid which contained the FeLV splice. Enzyme assays indicated the amount of neo^r protein was proportional to the amount of mRNA.

INTRODUCTION

In the many cases that have been studied, the long terminal repeats (LTRs) of retroviral genomes contain both promoter elements and enhancer elements (1,2,3). The latter DNA segments can function as transcriptional activation elements for a nearby promoter. They can act in either orientation with respect to the direction of transcription, at positions either upstream or downstream of the transcription unit and, at a

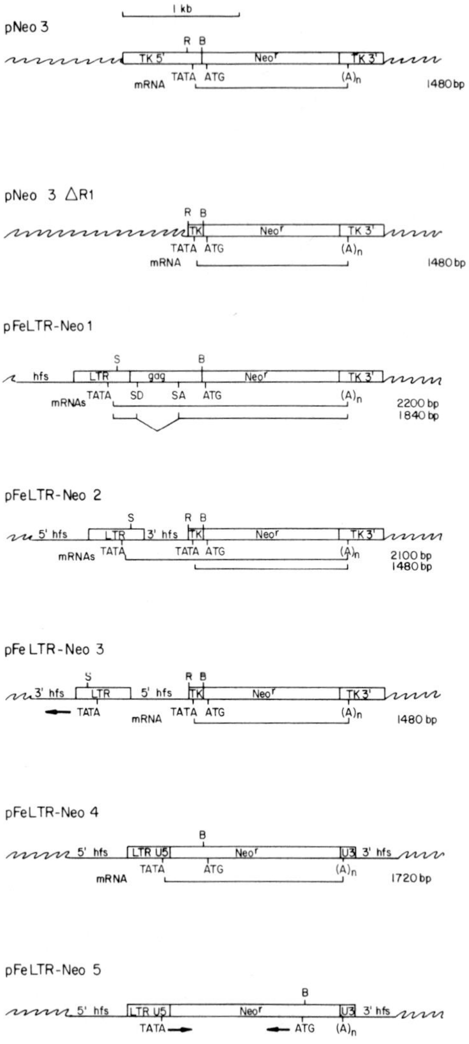
1 kb
pNeo 3
R B
TK 5'
Neo^r
TK 3'
mRNA TATA ATG (A)n 1480 bp
pNeo 3 ΔR1
TK
mRNA
pFeLTR-Neo 1
S B
hfs LTR gag
mRNAs TATA SD SA ATG (A)n
2200 bp
1840 bp
pFeLTR-Neo 2
5' hfs LTR 3' hfs
2100 bp
1480 bp
pFe LTR-Neo 3
3' hfs 5' hfs
pFeLTR-Neo 4
LTR U5 U3 3' hfs
1720 bp
pFe LTR-Neo 5

Figure 1.

considerable distance from the start site of transcription.

In the present communication, we have studied the effectiveness of the LTR from feline leukemia virus (FeLV) to provide both promoter function and enhancer activity for the expression of a gene introduced into mammalian cells by gene transfer methods. We have compared the effectiveness of the FeLV LTR to other enhancers and promoters in both murine and human cells.

RESULTS

The plasmid constructions we have studied are shown in Figure 1. In all cases the gene being expressed is the bacterial neo^r gene coding for the enzyme aminoglycoside 3'-phosphotransferase type II (APH(3')II) derived from the transposon Tn5. This gene, when expressed, protects mammalian cells from the toxic effects of the antibiotic G418 (4). Many of our constructions were derived from the plasmid pNeo 3 (constructed and kindly provided by B. Wold). In this plasmid, expression of the neo^r gene is driven by the herpes simplex virus thymidine kinase (HSV tk) promoter, and the transcript terminates at the HSV tk poly(A) site. By deleting the 5' BamHI to EcoRI fragment of the complete HSV tk promoter region (BamHI to BglII), a promoter with greatly reduced activity is produced (5). This deletion of pNeo3 is denoted pNeo3ΔR1. Expression by our FeLV

Figure 1. Structure of neo^r constructions.

The origins of the DNAs in the constructions are indicated. The pBR322 vector DNA (∿∿) is not completely shown. Restriction enzyme sites are shown above as R=EcoR1, B=BglII, and S=SmaI. Other abbreviations are SD, splice donor; SA, splice acceptor; $(A)_n$, polyadenylation signal. Predicted mRNAs sizes assume a 200 nucleotide poly(A) tail.

constructions has also been compared to that from the plasmid pSV2-Neo, in which the neo^r gene is driven by the SV40 early promoter (6).

The construction of a complete FeLV LTR flanked by human genomic sequences from a proviral clone has been described (7). In pFeLTR-Neo2 and pFeLTR-Neo3, this LTR has been placed close to the truncated HSVtk promoter in both possible orientations. The LTR is separated from the tk promoter by 500 bases and 375 bases of human genomic sequences, respectively.

In pFeLTR-Neo1, the entire tk promoter fragment has been replaced by the viral LTR plus 620 bases of FeLV provirus sequence (up to the BglII site, which is within the gag sequence).

The constructions pFeLTR-Neo4 and pFeLTR-Neo5 are the most unusual ones studied here. By digestion with SmaI, the transcription initiation (promoter) and termination (polyadenylation) functions of the LTR can be split. In pFeLTR-Neo4 and 5 the neo^r gene, sans the tk promoter and tk poly(A) site, has been inserted at this SmaI site within the LTR in both possible orientations.

The effectiveness of the several constructions for expression of the neo^r gene has been compared by three different methods: a) transformation efficiency; b) RNA accumulation; and c) enzyme accumulation. Since FeLV replicates efficiently in human cells, we have compared the expression properties in both murine and human cells.

a) Transformation efficiency.

The several DNAs were tested for the frequency with which they conferred resistance to the cytotoxic neomycin analog, G-418. The results in Table 1 show that in mouse Ltk$^-$ cells the tk promoter, the FeLV promoter and the SV40 promoter all give approximately the same efficiency when the protein coding sequences are oriented correctly relative to the promoter (pNeo3, pFeLTR-Neo's 1, 2 and 4, pSV2-Neo). With the incorrect orientation and no tk promoter (pFeLTR-Neo5) transformation efficiencies are low.

The truncated tk promoter alone (pNeo3ΔR1) shows reduced transformation efficiencies. However, by inserting the FeLV LTR in either orientation in the proximity of the promoter (pFeLTR-Neo 2 and 3) an enhancement of transformation efficiency is seen.

Table 1. Transformation efficiencies of neo^r constructions.

Plasmid	Ltk^- Cells[a]	$RDtk^-$ Cells[b]
pBR Neo	0.1	<0.2
pNeo3	2.9	4.0
pNeo3 ΔR1	0.1	<0.2
pFeLTR-Neo1	1.9	5.0
pFeLTR-Neo2	1.5	6.5
pFeLTR-Neo3	0.725	9.0
pFeLTR-Neo4	3.9	57.8
pFeLTR-Neo5	0.2	<0.2
pFeLTR-Neo6	4.0	70.0
pFeLTR-Neo7	0.1	<0.2
pSV2-Neo	3.6	26.3

[a]Colonies/10^6 cells/ng plasmid.

[b]Colonies/10^6 cells/μg plasmid.

In human cells the effects of the FeLV LTR are much greater. The LTR in the promixity of the truncated promoter gives even higher transformation efficiencies than the parental tk promoter. Even greater enhancement of transformation is obtained when the neo^r gene is inserted within the LTR. This construction is even more efficient than pSV2-Neo.

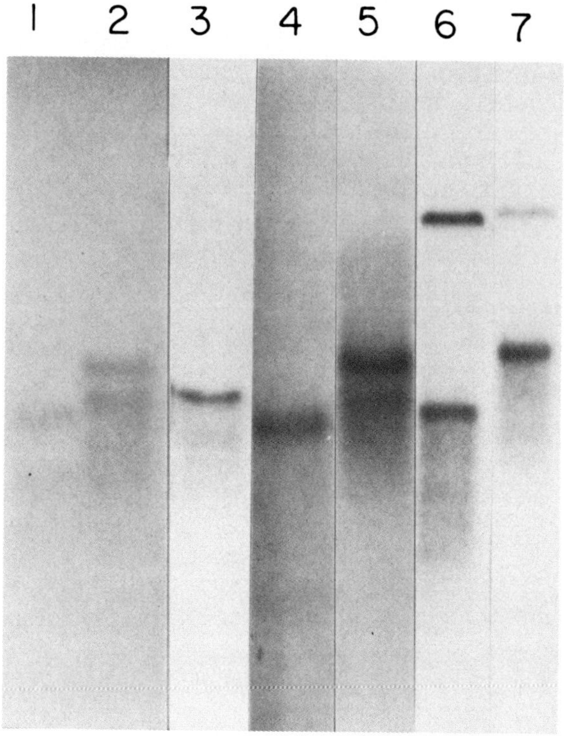

Figure 2. Gel blot hybridizations to RNA from cloned transformants

Poly(A)+ RNA was separated on 1% agarose gels containing 10 mM methylmercuric hydroxide and then transferred to nitrocellulose. The filters were hybridized with a 900 bp DNA fragment from within the neo^r coding region. Lane 1, Ltk^- pNeo3 (1.6 kb); lane 2, Ltk^- pFeLTR-Neo1 (2.3 kb, 2.0 kb); lane 3, Ltk^- pFeLTR-Neo2 (2.0 kb); lane 4, $RDtk^-$ pNeo3 (1.6 kb); lane 5, $RDtk^-$ pFeLTR-Neo1 (2.3 kb, 2.0 kb); lane 6, $RDtk^-$ pFeLTR-Neo4 (5.0 kb, 1.7 kb); lane 7, $RDtk^-$ pSV2-Neo (5.2 kb, 2.5 kb).

b) RNA and Protein Accumulation

Individual clones were isolated and assayed for steady state levels of RNA and enzyme. Total poly(A) RNA was examined by RNA gel blot hybridizations. Most of the transformants showed bands of the appropriate size and in a few cases an additional band was detected (Figure 2). Transformants produced by pFeLTR-Neo1 showed two sizes of corresponding to the precursor and product by the utilization of the FeLV gag splice included in this plasmid. In the single pFeLTR-Neo2 transformant examined, the only detectable RNA was the size expected from the FeLV LTR promoter and not the tk promoter.

Quantitatively, the amount of RNA in the different clones varied greatly. While, generally, the amount of RNA from the FeLV LTR promoter was greater in human cells, there is no direct correlation between transformation efficiency and the amount of RNA found in stable clones.

The amount of enzyme produced in these clones was measured using an in situ assay on total proteins separated by electrophoresis in non-denaturing acryamide gels (to be described elsewhere). These measurements show that the amount of enzyme in the clones corresponds reasonably well with the amount of mRNA in the cell (Figure 3).

DISCUSSION

The FeLV LTR contains an enhancer that can augment expression from the nearby truncated tk promoter irrespective of the orientation of the LTR (pFeLTR-Neo2 and 3). It can also provide promoter function itself (pFeLTR-Neo1 and 5). The FeLV LTR promoter elements are more effective than the HSVtk promoter in human cells, while murine cells show smaller differences among promoters.

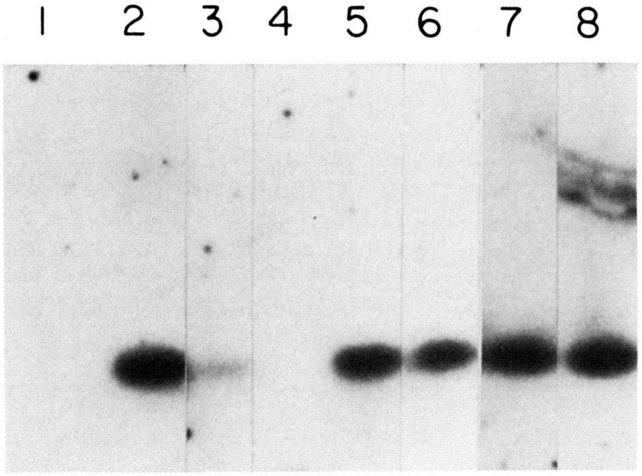

Figure 3. In situ assay of APH(3')II activity from transformed clones.

Total cytoplasmic proteins from cloned transformants were separated in non-denaturing acrylamide gels and incubated with kanamycin and $\gamma^{32}P$-ATP. The autoradiograph shows the position of labeled kanamycin. Lane 1, Ltk^- untransformed; lane 2, Ltk^-, pNeo3; lane 3, Ltk^-, pFeLTR-Neo1; lane 4, $RDtk^-$ untransformed; lane 5, $RDtk^-$ pNeo3; lane 6, $RDtk^-$ pFeLTR-Neo1; lane 7, $RDtk^-$ pFeLTR-Neo4; lane 8, $RDtk^-$ pSV2-Neo. The clones are the same used for the RNA in Figure 2.

ACKNOWLEDGEMENTS

We would like to thank Marie Krempin for expert technical assistance and advice.

REFERENCES

1. Luciw, P. A., Bishop, J. M., Varmus, H. E. and Capecchi, M. R. (1983) Location and function of retroviral and SV40 sequences that enhance biochemical transformation after microinjection of DNA. Cell 33, 705-716.
2. Chandler, V. L., Maler, B. A., and Yamamoto, K. R. (1983) DNA sequences bound specifically to glucocorticoid receptor in vitro render a heterologous promoter hormone response in in vivo. Cell 33, 489-499.
3. Lalmins, L. A., Khoury, G., German, C., Howard, B. and Gruns, P. (1982) Host-specific activation of transcription by tandem repeats from simian virus 40 and Moloney murine sarcoma virus. Proc. Natl. Acad. Sci. USA 79, 6453-6457.
4. Colbere-Garapin, F., Horodniceanu, F., Kourilsky, P. and Garapin, A. (1981) A new dominant hybrid selective marker for higher eukaryotic cells. J. Mol. Biol. 150, 1-14.
5. McKnight, S. L., Gavis, E. R., Kinasbury, R. and Axel, R. (1981) Analysis of transcriptional regulatory signals of the HSV thymidine kinase gene: identification of an upstream control region. Cell 25, 385-398.
6. Southern, P. J. and Berg, P. (1982) Transformation of mammalian cells to antibiotic resistance with a bacterial gene under control of the SV40 early region promoter. J. Mol. Appl. Genet. 1, 327-341.
7. Casey, J. W., Roach, A., Mullins, J. I., Burck, K. B., Nicolson, M. O., Gardner, M. B. and Davidson, N. (1981) The U3 portion of feline leukemia virus DNA identifies horizontally acquired proviruses in leukemic cats. Proc. Natl. Acad. Sci. USA 78:12, 7778-7782.

Index